Monographs in Theoretical Computer Science
An EATCS Series

More information about this series at http://www.springer.com/series/776

Hartmut Ehrig • Claudia Ermel • Ulrike Golas
Frank Hermann

Graph and Model Transformation

General Framework and Applications

 Springer

Hartmut Ehrig
Institut für Softwaretechnik
  und Theoretische Informatik
Technische Universität Berlin
Berlin, Germany

Claudia Ermel
Institut für Softwaretechnik
  und Theoretische Informatik
Technische Universität Berlin
Berlin, Germany

Ulrike Golas
Institut für Informatik
Humboldt-Universität zu Berlin
Berlin, Germany
and
Konrad-Zuse-Zentrum für
  Informationstechnik
Berlin, Germany

Frank Hermann
Interdisciplinary Centre for Security
  Reliability and Trust
University of Luxembourg
Luxembourg
and
Carmeq GmbH
Berlin, Germany

Series Editors
Monika Henzinger
Faculty of Science
Universität Wien
Wien, Austria

Juraj Hromkovič
ETH Zentrum
Department of Computer Science
Swiss Federal Institute of Technology
Zürich, Switzerland

Mogens Nielsen
Department of Computer Science
Aarhus Universitet
Aarhus, Denmark

Grzegorz Rozenberg
Leiden Centre of Advanced
  Computer Science
Leiden University
Leiden, The Netherlands

Arto Salomaa
Turku Centre of Computer Science
Turku, Finland

ISSN 1431-2654   ISSN 2193-2069   (electronic)
Monographs in Theoretical Computer Science. An EATCS Series
ISBN 978-3-662-56910-8   ISBN 978-3-662-47980-3   (eBook)
DOI 10.1007/978-3-662-47980-3

Springer Heidelberg New York Dordrecht London
© Springer-Verlag Berlin Heidelberg 2015
Softcover re-print of the Hardcover 1st edition 2015

Printed on acid-free paper

Springer-Verlag GmbH Berlin Heidelberg is part of Springer Science+Business Media (www.springer.com)

Preface

Graphs are important structures in mathematics, computer science and several other research and application areas. The concepts of graph transformation and graph grammars started in the late 1960s and early 1970s to become of interest in picture processing and computer science. The main idea was to generalise well-known rewriting techniques from strings and trees to graphs. Today, graph transformation techniques are playing a central role in theoretical computer science, as well as in several application areas, such as software engineering, concurrent and distributed systems, and especially visual modelling techniques and model transformations.

The state of the art of graph transformation techniques was presented in the "Handbook of Graph Grammars and Computing by Graph Transformation" in 1997, and later, especially for algebraic graph transformation, in the EATCS monograph "Fundamentals of Algebraic Graph Transformation" in 2006. In that monograph, called the FAGT-book, the important application area of model transformations was presented as a detailed example only. Since then, the algebraic approach of triple graph grammars has been developed and is presented in this book, which allows not only model transformation, but also model integration and synchronisation as well as analysis techniques, including correctness, completeness, functional behaviour and conflict resolution. Moreover, the theory of algebraic graph transformation presented in the FAGT-book is extended in this book with regard to the abstract framework based on M-adhesive categories, and by multi-amalgamated transformations, including the powerful concept of (nested) application conditions. The theory is applied in this book to self-adaptive systems and enterprise modelling, and it is supported by various tools, extending the tool AGG, well-known from the FAGT-book. Altogether this new book can be considered as a continuation of the FAGT-book, leading to a new state of the art of graph and model transformation in 2014.

The material of this book was developed by the groups in Berlin and Luxembourg in close cooperation with several international partners, including Gabriele Taentzer, Karsten Ehrig, Fernando Orejas, Reiko Heckel, Andy Schürr, Annegret Habel, Barbara König, Leen Lambers, and Christoph Brandt. Many thanks to all of them. Chap. 10 on self-adaptive systems is co-authored by Antonio Bucchiarone, Patrizio Pelliccione, and Olga Runge.

Finally, we thank Grzegorz Rozenberg and all other editors of the EATCS Monographs series, and those of Springer, especially Ronan Nugent, for smooth publication.

Berlin, Spring 2015   Hartmut Ehrig
                                                                       Claudia Ermel
                                                                       Ulrike Golas
                                                                       Frank Hermann

Contents

Part I
Introduction to Graph and Model Transformation

This first part of the book provides a general introduction to graph transformation and model transformations. After a general introduction in Chap. 1, we present in Chap. 2 graphs, typed graphs and attributed graphs in the sense of [EEPT06] and graph transformation with application conditions. In contrast to basic application conditions in [EEPT06] we introduce the more powerful nested application conditions in the sense of [HP05] and present the following main results in this more general framework: Local Church–Rosser and Parallelism Theorem, Concurrency, Amalgamation, Embedding and Extension Theorem as well as Critical Pair Analysis and Local Confluence Theorem. All these results have been shown without application conditions in [EEPT06], except amalgamation, which is an important extension in this book. These theorems are carefully motivated by running examples, but they are stated without proofs in Chap. 2, because they are special cases of corresponding results in the general framework of M-adhesive transformation systems presented in Part II. In Chap. 3, we introduce model transformations in general and model transformation based on graph transformation as motivated in Sect. 1.1.4. Especially, we introduce triple graph grammars and show how they can be used to define model transformation, model integration and model synchronisation. Moreover, the main results concerning analysis of model transformations are illustrated by running examples, while the full theory, including proofs, is given in Part III.

Chapter 1
General Introduction

1.1 General Overview of Graph and Model Transformation

In this general introduction we give a general overview of graph and model transformation and a short overview of the parts and chapters of this book. The main questions are the following:

- What is graph transformation?
- What is the algebraic approach to graph transformation?
- What is model transformation?
- How can algebraic graph transformation support model transformation?

1.1.1 What Is Graph Transformation?

Graphs are important structures in mathematics, computer science and several other research and application areas. A graph consists of nodes, also called vertices; edges; and two functions assigning source and target nodes to each edge. In fact, there are several variants of graphs, like labelled, typed, and attributed graphs, which will be considered in this book, because they are important for different kinds of applications. Properties of graphs, like shortest paths, are studied within graph theory, where in general the structure of the graph is not changed. Graph transformation, in contrast, is a formal approach for structural modifications of graphs via the application of transformation rules. A graph rule, also called production $p = (L, R)$, consists of a left-hand side graph L, a right-hand side graph R, and a mechanism specifying how to replace L by R as shown schematically in Fig. 1.1.

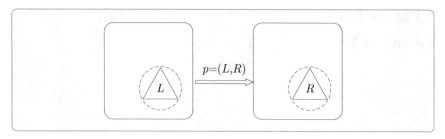

Fig. 1.1 Rule-based modification of graphs

This graph replacement mechanism is different in each of the following main graph transformation approaches presented in Volume 1 of the Handbook of Graph Grammars and Computing by Graph Transformation [Roz97]:

- Node Label Replacement Approach
- Hyperedge Replacement Approach
- Algebraic Approach
- Logical Approach
- Theory of 2-Structures
- Programmed Graph Replacement Approach

In all approaches, a graph transformation system consists of a set of rules; moreover, a graph transformation system together with a distinct start graph forms a graph grammar.

1.1.2 What Is the Algebraic Approach to Graph Transformation?

In this book, we present the algebraic approach of graph transformation, where a (basic) graph $G = (V, E, s, t)$ is an algebra with base sets V (vertices), E (edges), and operations $s: E \rightarrow V$ (source) and $t: E \rightarrow V$ (target). Graph morphisms are special cases of algebra homomorphisms $f = (f_V: V_1 \rightarrow V_2, f_E: E_1 \rightarrow E_2)$. This means that a graph morphism is required to be compatible with the operations source and target. It is important to note that graphs and graph morphisms define a category **Graphs**, such that categorical constructions and results are applicable in the algebraic approach of graph transformation. In fact, an important concept is the gluing construction of graphs, which corresponds to the pushout construction in the category **Graphs**. Pushouts are unique up to isomorphism and have useful composition and decomposition properties. The main conceptual idea of gluing is the following: Given graphs G_1 and G_2 with common intersection G_0, the gluing G_3 of G_1 and G_2 along G_0, written $G_3 = G_1 +_{G_0} G_2$, is given by the union G_3 of G_1 and G_2 and shown in the gluing diagram in Fig. 1.2, which is a pushout diagram in the category **Graphs**.

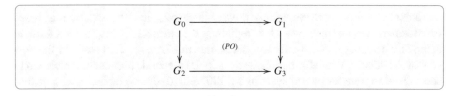

Fig. 1.2 Gluing (pushout) diagram of graphs

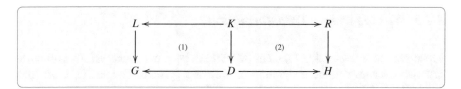

Fig. 1.3 Direct graph transformation

A production $p = (L \leftarrow K \rightarrow R)$ in the algebraic approach is given not only by left- and right-hand side graphs L and R, but, in addition, by a gluing graph K and (injective) graph morphisms from K to L and R. Given a context graph D with morphism $K \rightarrow D$, a direct graph transformation from a graph G to a graph H via a production p, written $G \Rightarrow H$ via p, is given by two gluing (pushout) diagrams as shown in Fig. 1.3. This means that G is the gluing of L and D along K, and H the gluing of R and D along K. In other words, L is replaced by R, while the context D remains unchanged. This definition of direct graph transformation is elegant, be-cause it is well defined (up to isomorphism) and symmetric. However, it leaves open how to apply a production p to a given host graph G and how to calculate the host graph H. In order to apply a production $p = (L \leftarrow K \rightarrow R)$ to a graph G, we first have to find an occurrence of L in G, given by a graph morphism $m: L \rightarrow G$, called match morphism. Then, we have to construct D and H in such a way that (1) and (2) become gluing (pushout) diagrams in Fig. 1.3. Roughly spoken, D is constructed by deleting from G all parts of L which are not in K, written $G \backslash (L \backslash K)$. In order to avoid that D becomes a partial graph, where some edges have no source or target, a certain gluing condition (see Chap. 2) has to be satisfied, which makes sure that D becomes a well-defined graph, and diagram (1) in Fig. 1.3 is a pushout diagram. This means, given a production $p = (L \leftarrow K \rightarrow R)$ and a match $m: L \rightarrow G$ satisfying the gluing condition, we obtain in a first step the context graph D and gluing (pushout) diagram (1) and in a second step diagram (2) by gluing (pushout) construction. The first step corresponds to the deletion of $L \setminus K$ from G and the second step to the addition of $R \setminus K$ leading to H, written $G \Rightarrow H$ via p and m. This algebraic approach is called double pushout (DPO) approach, because a direct transformation consists of two pushouts in the category **Graphs** (see Fig. 1.3). An important variant of the alge-braic approach is the single pushout (SPO) approach, where a direct transformation is defined by a single pushout in the category **PGraphs** of graphs and partial graph morphisms. In this book, we mainly present the algebraic DPO approach of graph transformation. Moreover, we allow replacing the category of graphs by a suitable axiomatic category (see $\mathcal{M}$-adhesive categories in Chap. 4). This leads to the concept of $\mathcal{M}$-adhesive transformation systems in the algebraic approach, which can be specialised to transformation systems for different kinds of graphs, Petri nets, and other kinds of high-level replacement systems.

1.1.3 What Is Model Transformation?

Model-driven software development (MDD) has been used successfully within the last two decades for the generation of software systems. Especially, UML diagrams [UML15] are useful for modelling different views of systems on an abstract level independently of specific implementations. In this case, models are UML diagrams, but in general models can be any kind of visual or textual artefacts. This culminates in the well-known slogan "Everything is a Model" stated in [Béz05]. Model transformation means defining transformations between (different) models. It plays a central role in MDD and several other applications. Model transformations in MDD are especially used to refactor models, to translate them to intermediate models, and to generate code. According to [CH06], we distinguish between endogenous and exogenous transformations. Endogenous transformations take place within one modelling language and exogenous ones are translations between different model languages. Moreover, model-to-model transformations are usually distinguished from model-to-text transformations. Typical examples of model-to-model transformations are the transformation S2P from statecharts to Petri nets in [EEPT06] and CD2RDBM from class diagrams to relational database models in this book. Important properties for most kinds of model transformations are type consistency, termination, syntactical and semantic correctness, completeness, functional behaviour and information preservation. We will discuss this topic in the next subsection and in Part III of this book.

1.1.4 How Can Algebraic Graph Transformation Support Model Transformation?

In [CH06], an overview of various model transformation approaches is given following object-oriented, rule-based, constraint-based and imperative concepts. In the following, we show how algebraic graph transformation can support the definition and analysis of rule-based model transformations [Tae10]. Especially for visual models, graph transformation is a natural choice for manipulating their underlying graph structures. The double pushout (DPO) approach introduced above can be interpreted as a kind of in-place transformation, where the source graph is transformed step by step into the target graph. Using the DPO approach for typed graphs—with different type graphs for source and target domain—allows us to ensure type consistency by construction [EEPT06]. The rich theory of the DPO approach provides support for the verification of other properties of model transformations discussed above [EE08]. Even better support for the verification of these properties is given by the triple graph grammar (TGG) approach [KS06, EEE+07] presented in Chap. 3 and Part III of this book. A triple graph consists of a source graph, a target graph, and a correspondence graph. The last one is mapped to the source and the target graph in order to establish a correspondence between elements of these graphs. The TGG approach is closely related to the DPO approach, in the way that graphs are replaced by triple graphs and TGG rules are usually nondeleting. The main additional idea is the following: From each TGG rule, a forward and a backward rule can be derived automatically, which allows us to construct type-consistent and syntactically correct forward and backward transformations between the source model and target model domains.

1.1.5 Historical Notes

Historically, graph grammars and transformations were first studied as "web grammars" by Pfalz and Rosenfeld [PR69] in order to define rule-based image recognition. Pratt [Pra71] used pair graph grammars for string-to-graph translations, similar to the concept of the triple graph grammar approach. The historical roots of the algebraic approach were presented by Ehrig, Pfender, and Schneider [EPS73]. The first introduction to the DPO approach—including the well-known Local Church–Rosser Theorem—was presented by Ehrig and Rosen in [ER76, Ehr79]. The first book on graph grammars was published by Nagl [Nag79] with its main focus on the Chomsky hierarchy, implementation and applications. The concept of graph transformation has at least three different historical roots:

1. from Chomsky grammars on strings to graph grammars,
2. from term rewriting to graph rewriting,
3. from textual description to visual modelling.

Motivated by these roots, the concept of "Computing by Graph Transformation" was developed as a basic paradigm in the ESPRIT Basic Research Actions COMPUGRAPH and APPLIGRAPH, and continued in the TMR Networks GETGRATS and SEGRAVIS in the period 1990–2006. The state of the art of graph transformation and their applications of 15 years ago is documented in three volumes of the "Handbook of Graph Grammars and Computing by Graph Transformation" [Roz97, EEKR99, EKMR99], where [Roz97] includes an introduction to the algebraic SPO and DPO approaches. A first detailed part of the theory of the DPO approach was published in the EATCS Monographs in TCS [EEPT06], while the newer developments are presented in this book. We present its main concepts, based on the extended theory of $\mathcal{M}$-adhesive transformation systems [EGH10], including results for parallelism, concurrency and amalgamation [EGH+14]; results for systems with nested application conditions concerning embedding, critical pairs and local confluence [EGH$^+$12]; characterisations of constructions based on the notion of finitary $\mathcal{M}$-adhesive categories [GBEG14]; multi-amalgamation [GHE14]; concurrency based on permutation equivalence [HCE14]; and model transformation and model synchronisation based on triple graph grammars [HEGO14, HEO$^+$13].

1.2 The Chapters of This Book and the Main Results

1.2.1 Part I–Introduction to Graph and Model Transformation

Part I of this book is an introduction to graph and model transformation based on the algebraic approach to graph grammars in the classical sense of [Ehr79] and triple graph grammars introduced in [Sch94], respectively. After a general introduction in Chap. 1 we present in Chap. 2 graphs, typed graphs and attributed graphs in the sense of [EEPT06] and graph transformation with application conditions. In contrast to basic application conditions in [EEPT06] we introduce the more powerful nested application conditions in the sense of [HP05] and present the following main results in this more general framework: Local Church–Rosser and Parallelism Theorem, Concurrency, Amalgamation, Embedding and Extension Theorems as well as Critical Pair Analysis and Local Confluence Theorem. All these results have been shown without application conditions in [EEPT06], except amalgamation, which is an important extension in this book. All these results are carefully motivated by running examples, but they are stated without proofs in Chap. 2, because they are special cases of corresponding results in the general framework of $\mathcal{M}$-adhesive transformation systems presented in Part II. In Chap. 3, we introduce model transformations in general and model transformation based on graph transformation as motivated in Sect. 1.1.4. In particular, we introduce triple graph grammars and show how they can be used to define model transformation, model integration and model synchronisation. The main results concerning analysis of model transformations are illustrated by running examples, while the full theory including proofs is given in Part III.

1.2.2 Part II–$\mathcal{M}$-Adhesive Transformation Systems

The algebraic approach to graph transformation is not restricted to graphs of the form $G = (V, E, s, t)$, as considered in Sect. 1.1.1, but has been generalised to a large variety of different types of graphs and other kinds of high-level structures, such as labelled graphs, typed graphs, hypergraphs and different kinds of low and high-level Petri nets. The extension from graphs to high-level structures was introduced in [EHKP91a, EHKP91b], leading to the theory of high-level replacement (HLR) systems. In [EHPP04] the concept of HLR systems was joined to that of adhesive categories of Lack and Sobocinsky in [LS04], leading to the concepts of adhesive HLR categories used in [EEPT06] and M-adhesive categories in this book, where all these concepts are introduced in Chap. 4. Moreover, this chapter includes an overview of different adhesive and HLR notions and several results concerning HLR properties, which are used in the general theories of Chapters 5 and 6 and for the construction of M-adhesive categories. In fact, M-adhesive categories and transformation systems constitute a suitable general framework for an abstract theory of graph and model transformations, which can be instantiated to various kinds of high-level structures, especially to those mentioned above. All the concepts and results—introduced for graph transformation in Chap. 2—are carefully presented and proven in Chap. 5 for M-adhesive transformation systems and in Chap. 6 for multi-amalgamated transformations. Finally it is shown in Chap. 6 how multi-amalgamation can be used to define the semantics of elementary Petri nets.

1.2.3 Part III–Model Transformation Based on Triple Graph Grammars

Following the informal introduction to model transformation in Chap. 3 of Part I, we present the formal theory of graph transformation based on triple graph grammars in Part III. In Chap. 7, we give the foundations of triple graph grammars leading to model transformation and model integration. It is important to note that transformation and integration are based on operational rules, which can be generated automatically from the triple graph grammar rules. A flattening construction allows us to show the equivalence of model transformations based on triple graph grammars and plain graph grammars. In Chap. 8, we present several analysis techniques for model transformations, which are supported by tools discussed in Part II. Important properties, which can be guaranteed or analysed in Chap. 8, include correctness and completeness, functional behaviour and information preservation, as well as conflict resolution and optimisation. In Chap. 9 model transformation techniques are applied to model synchronisation, which is an important technique to keep or gain consistency of source and target models after changing one or both of them. This leads to unidirectional and concurrent model synchronisation, respectively.

1.2.4 Part IV–Application Domains, Case Studies and Tool Support

In Part IV we present different application domains and case studies according to different parts of the theory given in Parts II and III, respectively. Moreover we give an overview of different tools, which support modelling and analysis of systems using graph transformation techniques presented in this book. In Chap. 10, we introduce self-adaptive systems and show how they can be modelled and analysed using graph transformation systems in Chap. 2, including a case study concerning business processes. The application domain of enterprise modelling is considered in Chap. 11, based on Chapters 3, 7 and 8, together with a case study on model transformation between business and IT service models. Chap. 12 includes a discussion of the following tools :

1. The Attributed Graph Grammar System AGG 2.0
2. ActiGra: Checking Consistency between Control Flow and Functional Behaviour
3. Controlled EMF Model Transformation with *EMF Henshin*
4. Bidirectional EMF Model Transformation with *HenshinTGG*

1.2.5 Appendices A and B

Appendix A presents basic notions of category theory and provides a short summary of the categorical terms used throughout this book. We introduce categories, show how to construct them, and present some basic constructions such as pushouts and pullbacks. In addition, we give some specific categorical results which are needed for the main part of the book. For a more detailed introduction to category theory see [EM85, EM90, AHS90, EMC+01]. Appendix B provides different properties as well as some more technical proofs and additional properties for Parts II and III.

1.2.6 Hints for Reading This Book

For a gentle introduction to graph transformation from an application point of view we propose starting with Chap. 2 and continuing with Chapters 10 and 11. The general framework in Part II requires some knowledge in category theory as given in Appendix A. For readers interested mainly in model transformation, we propose starting with Chap. 3 and continuing with Chapters 7, 8 and 11, where some parts require basic knowledge of graph transformation presented in Chap. 2. Finally model synchronisation in Chap. 9 should be studied after Chapters 3, 7 and 8.

The main parts of the theory for graph transformation systems without application conditions were presented already in our first book [EEPT06]. This first book includes also a discussion and case study for model transformations, but the theory based on triple graph grammars in Part III is not included in [EEPT06].

Chapter 2
Graph Transformation

In this chapter, we introduce graphs and graph transformation. In Sect. 2.1, we define graphs, typed graphs, and typed attributed graphs with their corresponding morphisms. Transformations of these graphs are introduced in Sect. 2.2, together with application conditions and two shift properties. In Sect. 2.3, important results for graph transformations are motivated and explained.

2.1 Graphs, Typed Graphs, and Attributed Graphs

Graphs and graph-like structures are the main basis for (visual) models. Basically, a graph consists of nodes, also called vertices, and edges, which link two nodes. Here, we consider graphs which may have parallel edges as well as loops. A graph morphism then maps the nodes and edges of the domain graph to those of the codomain graph such that the source and target nodes of each edge are preserved by the mapping.

Definition 2.1 (Graph and graph morphism). A *graph* $G = (V_G, E_G, s_G, t_G)$ consists of a set V_G of nodes, a set E_G of edges, and two functions $s_G, t_G : E_G \to V_G$ mapping to each edge its source and target node.

Given graphs G_1 and G_2, a *graph morphism* $f : G_1 \to G_2$, $f = (f_V, f_E)$, consists of two functions $f_V : V_{G_1} \to V_{G_2}$, $f_E : E_{G_1} \to E_{G_2}$ such that $s_{G_2} \circ f_E = f_V \circ s_{G_1}$ and $t_{G_2} \circ f_E = f_V \circ t_{G_1}$.

Graphs and graph morphisms form the category **Graphs**, together with the componentwise compositions and identities. △

An important extension of plain graphs is the introduction of types. A type graph defines a node type alphabet as well as an edge type alphabet, which can be used to assign a type to each element of a graph. This typing is done by a graph morphism into the type graph. Type graph morphisms then have to preserve the typing.

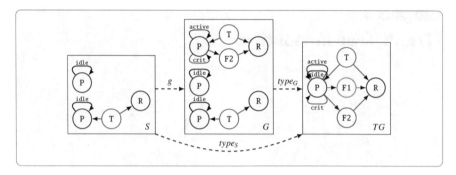

Fig. 2.1 Example typed graph and typed graph morphism

Definition 2.2 (Typed graph and typed graph morphism). A *type graph* is a distinguished graph TG. Given a type graph TG, a tuple $G^T = (G, type_G)$ of a graph G and a graph morphism $type_G : G \to TG$ is called a *typed graph*.

Given typed graphs G_1^T and G_2^T, a *typed graph morphism* $f : G_1^T \to G_2^T$ is a graph morphism $f : G_1 \to G_2$ such that $type_{G_2} \circ f = type_{G_1}$. Given a type graph TG, typed graphs and typed graph morphisms form the category **Graphs$_{TG}$**, together with the componentwise compositions and identities.

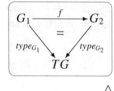

△

If the typing is clear in the context, we may not explicitly mention it and consider only the typed graph G with implicit typing $type_G$.

Example 2.3. To illustrate our definitions and results in the following sections, we introduce an example describing a mutual exclusion algorithm closely following Dijkstra's work [Dij65] and extending our example in [EGH+14]. In our system, we have an arbitrary number of processes P and resources R. To each resource, a turn variable T may be connected assigning it to a process. Each process may be `idle` or `active` and has a flag with possible values $0, 1, 2$, initially set to 0, which is graphically described by no flag at all at this process. Moreover, a label `crit` marks a process which has entered its critical section actually using the resource. Thus, the type graph used for our example is $TG = (V_{TG}, E_{TG}, s_{TG}, t_{TG})$ with $V_{TG} = \{P, T, R, F1, F2\}$ and $E_{TG} = \{active, idle, crit\}$, as shown in the right of Fig. 2.1. In the left of this figure, a system S is modelled containing a resource and two `idle` processes, where one of them is connected via a turn variable to the resource. There is an injective graph morphism $g : S \to G$ extending S by another `active` process with a flag and a turn to an additional resource. Both S and G are typed over TG.

In drawings of graphs, nodes are drawn by circles and edges by arrows pointing from the source to the target node. The actual mapping of the elements can be concluded by positions or is conveyed by indices, if necessary.   △

Attributed graphs are graphs extended by an underlying data structure given by an algebra (see [EEPT06]), such that nodes and edges of a graph may carry attribute values. For the formal definition, these attributes are represented by edges into the corresponding data domain, which is given by a node set. An attributed graph is based on an E-graph that has, in addition to the standard graph nodes and edges, a set of data nodes as well as node and edge attribute edges.

Definition 2.4 (Attributed graph and attributed graph morphism). An *E-graph*
$G^E = (V_G^G, V_D^G, E_G^G, E_{NA}^G, E_{EA}^G, (s_i^G, t_i^G)_{i \in \{G, NA, EA\}})$ consists of graph nodes V_G^G, data nodes V_D^G, graph edges E_G^G, node attribute edges E_{NA}^G, and edge attribute edges E_{EA}^G, according to the following signature.

For E-graphs G_1^E and G_2^E, an *E-graph morphism* $f : G_1^E \to G_2^E$ is a tuple $f = ((f_{V_i} : V_i^{G_1} \to V_i^{G_2})_{i \in \{G, D\}}, (f_{E_j} : E_j^{G_1} \to E_j^{G_2})_{j \in \{G, NA, EA\}})$ such that f commutes with all source and target functions.

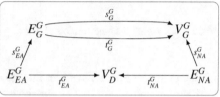

An *attributed graph* G over a data signature $DSIG = (S_D, OP_D)$ with attribute value sorts $S_D' \subseteq S_D$ is given by $G = (G^E, D_G)$, where G^E is an E-graph and D_G is a *DSIG*-algebra such that $\cup_{s \in S_D'} D_{G,s} = V_D^G$.

For attributed graphs $G_1 = (G_1^E, D_{G_1})$ and $G_2 = (G_2^E, D_{G_2})$, an *attributed graph morphism* $f : G_1 \to G_2$ is a pair $f = (f_G, f_D)$ with an E-graph morphism $f_G : G_1^E \to G_2^E$ and an algebra homomorphism $f_D : D_{G_1} \to D_{G_2}$ such that $f_{G,V_D}(x) = f_{D,s}(x)$ for all $x \in D_{G_1,s}, s \in S_D'$.

Attributed graphs and attributed graph morphisms form the category **AGraphs**, together with the componentwise compositions and identities.   △

As for standard typed graphs, an attributed type graph defines a set of types which can be used to assign types to the nodes and edges of an attributed graph. The typing itself is done by an attributed graph morphism between the attributed graph and the attributed type graph.

Definition 2.5 (Typed attributed graph and morphism). An *attributed type graph* is a distinguished attributed graph $ATG = (TG, Z)$, where Z is the final *DSIG*-algebra.

A tuple $G^T = (G, type_G)$ of an attributed graph G together with an attributed graph morphism $type_G : G \to ATG$ is then called a *typed attributed graph*.

Given typed attributed graphs $G_1^T = (G_1, type_{G_1})$ and $G_2^T = (G_2, type_{G_2})$, a *typed attributed graph morphism* $f : G_1^T \to G_2^T$ is an attributed graph morphism $f : G_1 \to G_2$ such that $type_{G_2} \circ f = type_{G_1}$.

For a given attributed type graph ATG, typed attributed graphs and typed attributed graph morphisms form the category **AGraphs$_{ATG}$**, together with the componentwise compositions and identities.   △

Example 2.6. Considering the model from Ex. 2.3, we may also use attributes to model the state of a process instead of the connected loop. In addition, a Boolean attribute of the resource can describe if it is currently in use. Moreover, only one type

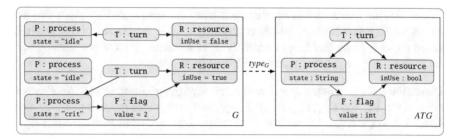

Fig. 2.2 Example with typed attributed graphs

of flag connects P and T with an integer value of 1 or 2. The corresponding typed attributed graphs G and ATG are shown in Fig. 2.2 in a UML class diagram-like style.   △

2.2 Graph Transformation with Application Conditions

In [EEPT06], transformation systems based on a categorical foundation were introduced which can be instantiated to various graphs and graph-like structures. In this section, we present the implementation of this theory for transformations of typed graphs using rules extended with application conditions. Those have been introduced in [EEPT06], but no full theory was developed there.

Basically, a condition describes whether a graph contains a certain structure as a subgraph.

Definition 2.7 (Graph condition). A *(nested) graph condition ac* over a graph P is of the form $ac = \text{true}$, $ac = \neg ac'$, $ac = \exists (a, ac'')$, $ac = \wedge_{i \in I} ac_i$, or $ac = \vee_{i \in I} ac_i$, where ac' is a graph condition over P, $a : P \rightarrow C$ is a morphism, ac'' is a graph condition over C, and $(ac_i)_{i \in I}$ with an index set I are graph conditions over P.   △

For simplicity, false abbreviates $\neg\text{true}$, $\exists a$ abbreviates $\exists (a, \text{true})$, and $\forall (a, ac)$ abbreviates $\neg \exists (a, \neg ac)$.

A graph condition is satisfied by a morphism into a graph if the required structure exists, which can be verified by the existence of suitable morphisms.

Definition 2.8 (Satisfaction of graph conditions). Given a graph condition ac over P, a morphism $p : P \rightarrow G$ *satisfies ac*, written $p \models ac$, if

$$ac \triangleright P \xrightarrow{a} C \triangleleft ac'$$
$$p \searrow \quad \swarrow q$$
$$G$$

- $ac = \text{true}$,
- $ac = \neg ac'$ and $p \not\models ac'$,
- $ac = \exists (a, ac')$ and there exists an injective morphism q with $q \circ a = p$ and $q \models ac'$,

- $ac = \wedge_{i \in I} ac_i$ and $\forall\, i \in I : p \models ac_i$, or
- $ac = \vee_{i \in I} ac_i$ and $\exists\, i \in I : p \models ac_i$. △

A rule is a general description of local changes that may occur in graphs. Mainly, it consists of a deletion and a construction part, defined by the rule morphisms l and r, respectively. In addition, an application condition restricts the application of this rule to certain graphs.

Definition 2.9 (Rule). A *rule* $p = (L \xleftarrow{l} K \xrightarrow{r} R, ac)$ consists of graphs L, K, and R, called left-hand side, gluing, and right-hand side, respectively, two injective morphisms l and r, and a graph condition ac over L, called *application condition*.

△

A transformation describes the application of a rule to a graph via a match. It can only be applied if the match satisfies the application condition.

Definition 2.10 (Transformation). Given a rule $p = (L \xleftarrow{l} K \xrightarrow{r} R, ac)$, a graph G, and a morphism $m : L \to G$, called match, such that $m \models ac$, a *direct transformation* $G \xRightarrow{p,m} H$ from G to a graph H is given by the pushouts (1) and (2).

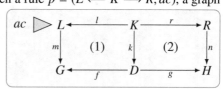

A sequence of direct transformations is called a *transformation*.

△

Remark 2.11. Note that for the construction of pushout (1) we have to construct the pushout complement of $m \circ l$, which is only possible if the so-called gluing condition is satisfied (see [EEPT06]). Intuitively, gluing points are all elements in L that are preserved in K. A dangling point is a node x in L such that $m(x)$ in G is the source or target of an edge with no preimage in L. In addition, identification points are elements in L that are mapped noninjectively by m. The gluing condition is fulfilled if all dangling and identification points are also gluing points. △

Example 2.12. Now we introduce the rules for the mutual exclusion algorithm. Its main aim is to ensure that at any time at most one process is using each resource. A different variant of this algorithm implemented by graph transformation can be found in [EEPT06], where the lack of application conditions induces a much more complex model including more types and additional rules for handling a single resource. Using application conditions we can simplify the models, forgo additional edges representing the next executable step of the system, and extend the context to an arbitrary number of resources. This example is based on and extends the one presented in [EGH+14].

Initially, each process is `idle` and for each resource the turn variable is connected to an arbitrary process, enabling it to use that resource. If a process P wants to use some resource R it becomes active and points the flag F1 to R. If, in addition, it has the turn for R, it may proceed to use it, which is described by an F2-flag to the resource and a `crit` loop at the process. Otherwise, if the turn for R belongs to

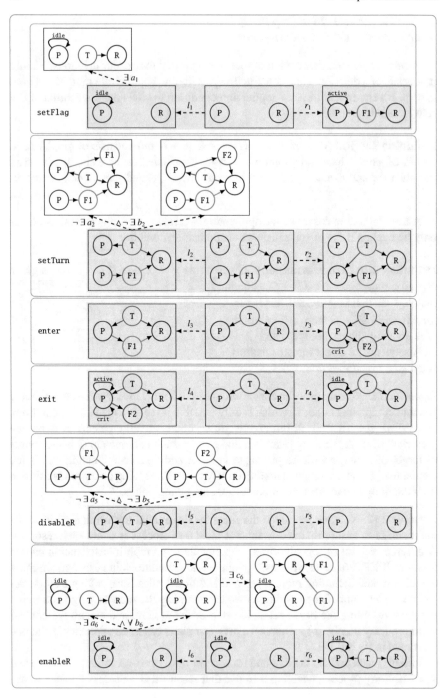

Fig. 2.3 The rules for the mutual exclusion algorithm

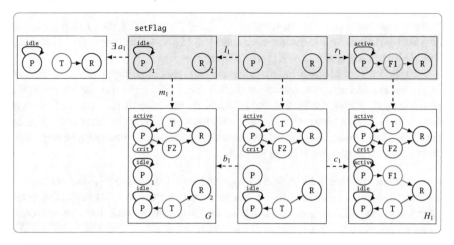

Fig. 2.4 A rule application

another process P', P must wait until P' is not flagging R. At this point the process may get the turn for R and start using it. When P has finished using R, the flag and crit are removed, and the process is idle again. As an extension of this normal behaviour, a resource may be disabled, denoted by eliminating its turn variable, if there is no flag present for it. Moreover, a resource may be enabled again if all other resources have at least two requests waiting.

The rules setFlag, setTurn, enter, and exit in Fig. 2.3 describe the standard behaviour of the system. With setFlag, a process becomes active and sets its F1-flag to a resource. Note that this rule has a positive application condition requiring that the resource has a turn variable noting it as enabled. If a process has set an F1-flag to a resource whose turn variable points to another process with no flag to the resource, the turn variable can be assigned to the first process via setTurn. Here, the application condition forbids the process having the turn of the resource from flagging it. The rule enter describes that, if a process has the turn of and points to a resource R with an F1-flag, then this flag is replaced by an F2-flag, and a loop crit is added to the process. When the process is finished, the rule exit is executed, deleting the loop and the flag, making the process idle again. Moreover, with the rules disableR and enableR, a resource can be disabled or enabled if the corresponding application conditions are fulfilled.

In the figures, the application condition true is not drawn. Application conditions $Q(a, ac)$, with $Q \in \{ \exists, \neg \exists, \forall \}$, are drawn by the morphism a marked by Qa and combined with a drawing of ac, and conjunctions of application conditions are marked by $\wedge$ between the morphisms.

Consider the rule setFlag with the match m_1 depicted in the left of Fig. 2.4. Note that m_1 matches the process and resource of the rule setFlag to the middle process and lower resource in G, respectively, as indicated by the small numbers, such that m_1 satisfies the gluing condition as well as the application condition $\exists a_1$. This leads to the direct transformation $G \xrightarrow{\text{setFlag},m_1} H_1$ inserting an F1-flag from the now active process to the resource, as shown in Fig. 2.4. The graph H_1 is obtained from G by removing $m_1(L_1 - K_1)$ and adding $R_1 - K_1$.

Note that we could easily have a rule `setFlag` without any application condition. In particular it is enough to include in the left-hand side of the rule the turn variable pointing to R. In contrast to that, the application condition $\forall (b_6, \exists c_6)$ of the rule `enableR` cannot be removed, although it is also a positive application condition. In particular, this condition is nested twice, which is needed to specify that every other enabled resource has two waiting processes.   △

Graph conditions can be shifted over graph morphisms into equivalent conditions over the codomain [HP09, EHL10]. For this shift construction, all surjective over-lappings of the codomains of the shift and condition morphisms have to be collected. Here, we only explain this construction and give an example; for the full definition see Def. 5.3. The shift construction is recursively defined. For a graph condition $ac = \exists (a, ac')$ and a shift morphism b we construct the set $\mathcal{F} = \{(a', b') \mid (a', b')$ jointly surjective, b' injective, $b' \circ a = a' \circ b\}$ and define $\text{Shift}(b, ac) = \bigvee_{(a',b') \in \mathcal{F}} \exists (a', \text{Shift}(b', ac'))$.

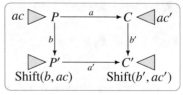

Example 2.13. Consider the application condition $\forall (b_6, \exists c_6)$ of the rule `enableR`, which is an application condition over the left-hand side of this rule. We want to shift this condition over the morphism v shown at the top of Fig. 2.5. The first step of the construction is shown in the upper part of Fig. 2.5, it results in the in-termediate application condition $\text{Shift}(v, \forall (b_6, \exists c_6)) = \forall (d_1, \text{Shift}(v_1, \exists c_6)) \land \forall (d_2, \text{Shift}(v_2, \exists c_6))$. Since v_i has to be injective and the resulting graph has to be an overlapping of the codomains of v and b_6 such that the diagram commutes, only these two solutions are possible. In a second step, the second part of the application condition has to be shifted over the two new morphisms v_1 and v_2. The result is shown in the lower part of Fig. 2.5, leading to the resulting application condition $\text{Shift}(v, \forall (b_6, \exists c_6)) = \forall (d_1, \exists e_1 \lor \exists e_2) \land \forall (d_2, \exists e_3)$.   △

Similarly to the shift construction, we can also merge a graph condition over a graph morphism. The difference lies in different injective morphisms to be required, with a' being injective instead of b'. Additionally, b' has to be a match morphism for the merge construction, which is no restriction at all if the class of match morphisms contains all morphisms. Again, here we only explain this construction and give an example; for the full definition see Def. 5.5. The merge construction is recursively defined. For a graph condition $ac = \exists (a, ac')$ and a merge morphism b we construct the set $\mathcal{F}' = \{(a', b') \mid (a', b')$ jointly surjective, a' injective, b' match, $b' \circ a = a' \circ b\}$ and define $\text{Merge}(b, ac) = \bigvee_{(a',b') \in \mathcal{F}'} \exists (a', \text{Merge}(b', ac'))$.

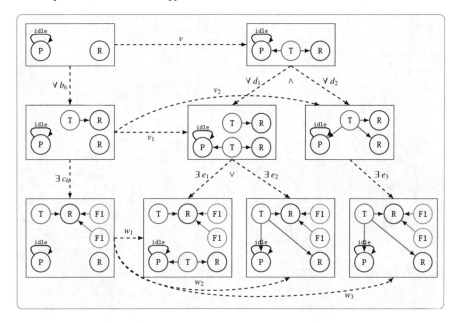

Fig. 2.5 Shift of the application condition $\forall\,(b_6,\,\exists\,c_6)$ over a morphism

Intuitively, the constructions "merge" and "shift" differ in the directions of identifications. If the merge construction yields the depicted diagram, it means that identifications along the given graph morphism $b : P \rightarrow P'$ must subsume the identification performed via $a : P \rightarrow C$. Further identifications along b' may occur on elements in $C\backslash a(P)$, if the class of matches permits those. In contrast to that, the shift construction requires that the identifications along b be subsumed by those via a, and it generally permits identifications along a' on elements in $P'\backslash b(P)$.

Example 2.14. On the left of Fig. 2.6, the shift of the graph condition $ac = (\,\exists\,a : P \rightarrow C, \text{true})$ along the graph morphism $b : P \rightarrow P'$ is depicted. The shift construction yields a graph condition over P' with $\text{Shift}(b, ac) = \vee_{i=1,\dots,4}(\,\exists\,a'_i : P' \rightarrow C'_i, \text{true})$. The four graphs C'_i, with corresponding graph morphisms $a'_i : P' \rightarrow C'_i$ and $b'_i : C \rightarrow C'_i$, are obtained by all jointly surjective pairs ensuring that the diagram commutes with injective b'_i. In particular, C'_3 is the pushout object of a and b, while for the graphs C'_1 and C'_2 the node 4 is glued together with both nodes 1 and 2 without or with additional gluing of both edges. For the graph C'_4, node 4 is glued together with node 3.

Consider the class of match morphisms given by all morphisms. On the right of Fig. 2.6, the merge of the graph condition ac along the graph morphism b is depicted. The merge construction yields a graph condition over P' with $\text{Merge}(b, ac) = \vee_{i=1,\dots,3}(\,\exists\,a''_i : P' \rightarrow C''_i, \text{true})$. The identification of the nodes 1 and 2 along the graph morphism b is transferred to the graph morphisms $b''_i : C \rightarrow C''_i$. The three graphs C''_i, with corresponding graph morphisms $a'_i : P' \rightarrow C'_i$ and $b'_i : C \rightarrow C'_i$, are

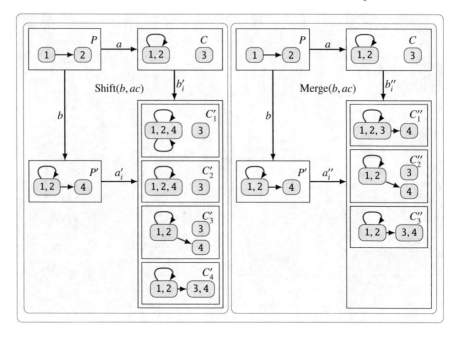

Fig. 2.6 Comparison of shift and merge construction

obtained by all jointly surjective pairs making the diagram commute with injective a_i''. In this case, C_2'' is the pushout object of (a, b), while for the graphs C_1'' and C_3'' the node 3 is glued together with nodes $(1, 2)$ or 4, respectively.   △

Remark 2.15. In the context of model transformations based on typed attributed graphs with inheritance in Part III, we may consider match morphisms that are injective on the graph part, but may refine types and may identify data values. In this case, the merge construction yields conditions where node types can be refined via b', but only on nodes that do not occur in P due to the requirement that a' has to be injective. This choice of match morphisms allows us to identify only graph nodes via b that are also identified via a. But note that identifications on data values would still be possible via b.   △

The satisfaction of a graph condition $ac = (\exists a : P \rightarrow C, ac')$ is defined for arbitrary matches $p : P \rightarrow G$. This general definition is important, because a restriction to injective matches would be problematic for several application domains. For example, if objects are attributed graphs and a condition contains variables, it should be possible to evaluate some of these variables to the same value, which is forbidden by injectivity.

More specifically, in the category **AGraphs**$_{ATG}$, graph conditions are often attributed via a term algebra with variables $T_{OP}(X)$ and instance graphs are attributed via a concrete data algebra A. In most cases, $T_{OP}(X)$ is not isomorphic to A and non-injective matches p may refine types along a type inheritance relation. This means

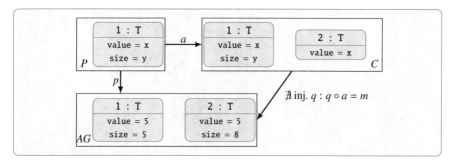

Fig. 2.7 Nonsatisfaction of a condition for a noninjective match

that the required injective graph morphism $q : C \rightarrow G$ according to Def. 2.8 does not exist and a graph condition of the form $ac = (\exists a : P \rightarrow C, ac')$ is never satisfied if $T_{OP}(X)$ is not isomorphic to A.

Example 2.16. Consider the graph condition $ac = (\exists a : P \rightarrow C, \text{true})$ shown in the top row of Fig. 2.7. It specifies that for any node of type T (graph P) there has to be a second node of the same type with the same value for the attribute `value` (graph C). The graphs of the condition are attributed via the term algebra $T_{OP}(X)$ with variables x and y. In contrast to that, the instance graph $AG = (G, D)$ is attributed via a data algebra D using integers for the values of the attributes `value` and `size`. For the node 1 of type T in AG, both attributes are assigned the same value, 5. Therefore, the variables x and y in C have to be evaluated to 5 for any morphism $q : C \rightarrow AG$. This means that q is not injective, and therefore the graph condition ac is not satisfied by AG.   △

In order to overcome this general problem for the satisfiability of graph conditions for noninjective matches, we need to derive sub-conditions that handle each of the specific cases of possible noninjective matches $p : P \rightarrow G$. Instead of specifying these sub-conditions explicitly, we will provide the general concept of AC schemata, where we specify a base condition and provide a general construction from which the induced concrete conditions can be derived. Such an AC schema consists of the disjunction of all merges of an application condition along surjective morphisms starting from its domain.

Definition 2.17 (AC schema). Given a condition ac over P and the set $\mathcal{E}_P = \{e \mid e \text{ surjective}, \text{dom}(e) = P\}$ of all surjective morphisms with domain P, the *AC schema* $\overline{ac}$ over P is given by $\overline{ac} = \bigvee_{f \in \mathcal{E}_P} \exists (f, \text{Merge}(f, ac))$.   △

The satisfaction of an AC schema by a graph morphism p only depends on the satisfaction of one component of the corresponding epi–mono factorisation $m \circ e = p$, i.e., $p \models \overline{ac}$ if and only if 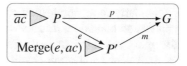 if $m \models \text{Merge}(e, ac)$ (see Fact 5.8). Although Def. 2.17 specifies that an AC schema

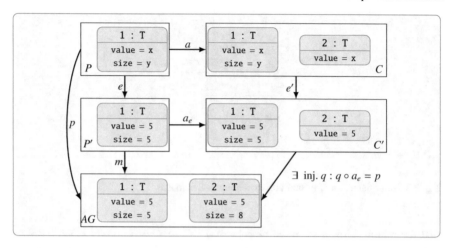

Fig. 2.8 Satisfaction of an AC schema for a noninjective match

induces a possibly infinite disjunction, this means that only one of these elements has to be constructed for checking satisfaction of the condition for a concrete match. The epi–mono factorisation of this match yields the epimorphism that is used for the merge construction. Intuitively, AC schemata are a way to specify a graph condition that allows for further identifications by the match, because these identifications are transferred recursively to the subcomponents of the graph condition. Moreover, if p is injective, then the satisfaction of an AC schema coincides with classical satisfaction, because the factorisation is trivially $p = p \circ id$.

Example 2.18. From the graph condition $ac = (\exists a : P \to C, \text{true})$ in Fig. 2.7 we construct the AC schema $\overline{ac} = \bigvee_{f \in \mathcal{E}_P} \exists (f, \exists (a_f, \text{true}))$, where all $f \in \mathcal{E}_P$ are surjective and represent different instantiations of the variables x and y. Moreover, the graph condition $\exists (a_f, \text{true})$ represents the corresponding instantiation for the additional node, 2.

To check if the graph morphism $p : P \to AG$ satisfies $\overline{ac}$, we construct the epi–mono factorisation $p = m \circ e$ in Fig. 2.8, where the morphism $e \in \mathcal{E}_P$ is used for the merge construction. The graphs P' and C' share the same algebra with AG. The graph constraint $ac' = \exists (a_e, \text{true})$ represents an instance of the AC schema $\overline{ac}$ that is used for checking satisfiability. The identification of the variables x and y is transferred to this instance, and we find an injective graph morphism $q : C' \to AG$ such that $q \circ a_e = m$. This means that the AC schema $\overline{ac}$ is satisfied by p, while the underlying graph condition ac is not satisfied by p, as shown in Ex. 2.16.   △

Similarly to an application condition over the left-hand side L, which is a pre-application condition, it is also possible to define post-application conditions over the right-hand side R of a rule. These application conditions over R can be translated to equivalent application conditions over L (and vice versa) [HP09] using a shift construction. Therefore, we can restrict our rules to application conditions over L.

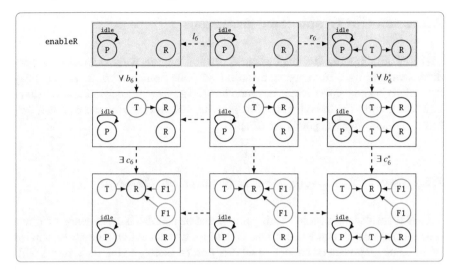

Fig. 2.9 Shift of the application condition from the left- to the right-hand side

As for the shift over morphisms, here we only motivate the construction and give an example; for the complete definition see Def. 5.15.

The shift of a graph condition $ac_R = \exists\,(a, ac'_R)$ over a rule p is recursively defined by $L(p, ac_R) = \exists\,(b, L(p^*, ac'_R))$

$$L(p, ac_R) \triangleright L \xleftarrow{\;l\;} K \xrightarrow{\;r\;} R \triangleleft ac_R$$
$$\quad\;\; b\downarrow \quad (2) \quad c\downarrow \quad (1) \quad a\downarrow$$
$$L(p^*, ac'_R) \triangleright Y \xleftarrow{\;l^*\;} Z \xrightarrow{\;r^*\;} X \triangleleft ac'$$

if $a \circ r$ has a pushout complement (1) and $p^* = (Y \xleftarrow{l^*} Z \xrightarrow{r^*} X)$ is the derived rule by constructing pushout (2), or $L(p, ac_R) = $ false otherwise,

Example 2.19. Suppose we want to translate the application condition $\forall\,(b_6, \exists\,c_6)$ of the rule enableR to the right-hand side. Basically, this means applying the rule to the first graph of the application condition, leading to a span which is applied as a rule to the second graph. The result is shown in Fig. 2.9, i.e., the translated application condition is $\forall\,(b_6^*, \exists\,c_6^*)$.   △

A set of rules constitutes a graph transformation system, and, combined with a start graph, a graph grammar. The language of such a grammar contains all graphs derivable from the start graph.

Definition 2.20 (Graph transformation system and grammar). A *graph transformation system GS* $= (P)$ consists of a set of rules P.

A *graph grammar GG* $= (GS, S)$ consists of a graph transformation system GS and a start graph S.

The *language L* of a graph grammar GG is defined by

$$L = \{G \mid \exists \text{ transformation } S \overset{*}{\Rightarrow} G \text{ via } P\}.$$ △

2.3 Results for Graph Transformations

In [EEPT06], important results for transformation systems without application conditions were proven. Here, we motivate and state the results and as far as necessary the underlying concepts for the corresponding theorems with application conditions, based on graphs. For the full definitions, results, and proofs in the more general setting of $\mathcal{M}$-adhesive categories see Chap. 5.

2.3.1 Local Church–Rosser and Parallelism Theorem

This first result is concerned with parallel and sequential independence of direct transformations. We study under what conditions two direct transformations applied to the same graph can be applied in arbitrary order leading to the same result. This leads to the Local Church–Rosser Theorem. Moreover, the corresponding rules can be applied in parallel in this case, leading to the Parallelism Theorem.

First, we define the notion of parallel and sequential independence. Two direct transformations $G \xrightarrow{p_1,m_1} H_1$ and $G \xrightarrow{p_2,m_2} H_2$ are parallel independent if p_1 does not delete anything p_2 uses and does not create or delete anything to invalidate ac_2, and vice versa.

Definition 2.21 (Parallel independence). Two direct transformations $G \xrightarrow{p_1,m_1} H_1$ and $G \xrightarrow{p_2,m_2} H_2$ are *parallel independent* if there are morphisms $d_{12} : L_1 \to D_2$ and $d_{21} : L_2 \to D_1$ such that $f_2 \circ d_{12} = m_1$, $f_1 \circ d_{21} = m_2$, $g_2 \circ d_{12} \models ac_1$, and $g_1 \circ d_{21} \models ac_2$.

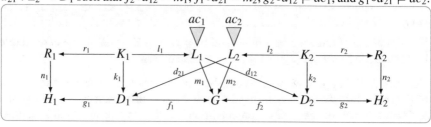

$\triangle$

Analogously, two direct transformations $G \xrightarrow{p_1,m_1} H_1 \xrightarrow{p_2,m_2} G'$ are sequentially independent if p_1 does not create something p_2 uses, p_2 does not delete something p_1 uses or creates, p_1 does not delete or create anything, thereby initially validating ac_2, and p_2 does not delete or create something invalidating ac_1.

Definition 2.22 (Sequential independence). Two direct transformations $G \xrightarrow{p_1,m_1} H_1 \xrightarrow{p_2,m_2} G'$ are *sequentially independent* if there are morphisms $d_{12} : R_1 \to D_2$ and $d_{21} : L_2 \to D_1$ such that $f_2 \circ d_{12} = n_1$, $g_1 \circ d_{21} = m_2$, $f_1 \circ d_{21} \models ac_2$, and $g_2 \circ d_{12} \models L((R_1 \xleftarrow{r_1} K_1 \xrightarrow{l_1} L_1), ac_1)$.

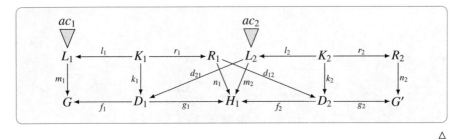

$\triangle$

Example 2.23. The pair $H_1 \xleftarrow{\text{setFlag},m_1} G \xrightarrow{\text{exit},m'} G'$ of direct transformations in Fig. 2.10 is parallel independent. The left rule application is the one already considered in Fig. 2.4, while m' matches the process of the rule exit to the uppermost process in G. The morphisms d_{12} and d_{21} exist such that $b_1 \circ d_{21} = m'$, $b_2 \circ d_{12} = m_1$, and $c_2 \circ d_{12} \models \exists a_1$.

The sequence $H_1 \xrightarrow{\text{setTurn},m_2} H_2 \xrightarrow{\text{setFlag},m_3} H_3$ of direct transformations in Fig. 2.11 is sequentially dependent. Note that m_2 matches the processes of the rule setTurn to the lower processes in H_1, but in reverse order, while m_3 maps the process of the rule setFlag to the lowest process in H_2. The morphisms d_{12} and d_{21} exist such that $c_1 \circ d_{21} = m_3$, $b_2 \circ d_{12} = m_2^*$, and $b_1 \circ d_{21} \models \exists a_1$, but $c_2 \circ d_{12} \not\models R(\text{setTurn}, \neg \exists a_2 \wedge \neg \exists b_2)$. The transformations are sequentially dependent, since the rule setFlag adds a second flag, which is forbidden by the application condition $\neg \exists a_2$ of the rule setTurn. Note that the transformations without application conditions would be sequentially independent. $\triangle$

The idea of a parallel rule is, in case of parallel independence, to apply both rules in parallel. For two rules p_1 and p_2, the parallel rule $p_1 + p_2$ is constructed as the

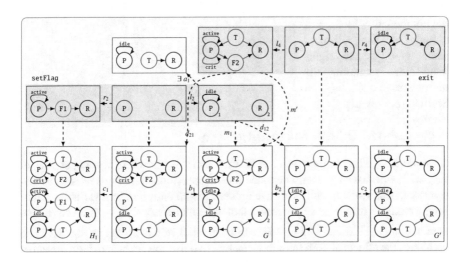

Fig. 2.10 Parallel independent transformations

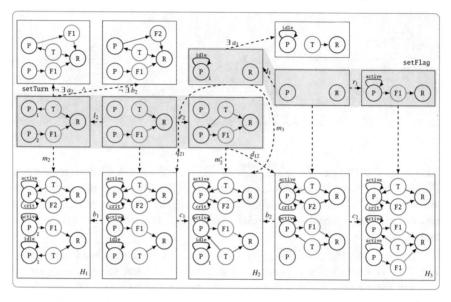

Fig. 2.11 Sequentially dependent transformations

disjoint union of all three components of the rules, denoted by +. For the application
conditions we have to make sure that both single rules can be applied in any order.

Definition 2.24 (Parallel rule). Given rules $p_1 = (L_1 \xleftarrow{l_1} K_1 \xrightarrow{r_1} R_1, ac_1)$ and
$p_2 = (L_2 \xleftarrow{l_2} K_2 \xrightarrow{r_2} R_2, ac_2)$, the *parallel rule* $p_1 + p_2 = (L_1 + L_2 \xleftarrow{l_1+l_2} K_1 + K_2 \xrightarrow{r_1+r_2} R_1 + R_2, ac)$ is defined by
the componentwise disjoint
unions of the left-hand
sides, gluings, and right-
hand sides including the
morphisms, and $ac = $
$\mathrm{Shift}(i_{L_1}, ac_1) \wedge \mathrm{L}(p_1 + p_2,$
$\mathrm{Shift}(i_{R_1}, \mathrm{R}(p_1, ac_1))) \wedge$
$\mathrm{Shift}(i_{L_2}, ac_2) \wedge \mathrm{L}(p_1 + p_2, \mathrm{Shift}(i_{R_2}, \mathrm{R}(p_2, ac_2))).$ △

Example 2.25. The parallel rule `setFlag + exit` is shown in the upper row of
Fig. 2.12, where we have only depicted those application conditions which are reasonable in our system, while we have ignored illegal ones like turn variables pointing to multiple resources or processes that are simultaneously idle and active. The
application $G \xRightarrow{\text{setFlag}+\text{exit},m_1+m'} H'$ of this parallel rule is shown in Fig. 2.12 and
combines the effects of both rules to G leading to the graph H', where both the upper
process became idle and the middle process became active.   △

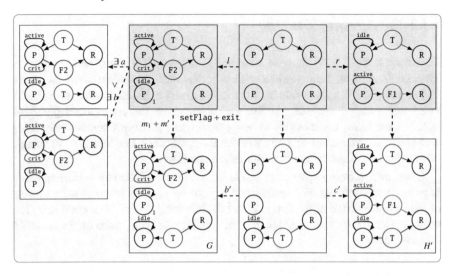

Fig. 2.12 Parallel rule and transformation

With these notions of independence and parallel rule, we are able to formulate the Local Church–Rosser and Parallelism Theorem. Note that this theorem is the instantiation of Theorem 5.26 to graphs.

Theorem 2.26 (Local Church–Rosser and Parallelism Theorem). *Given two parallel independent direct transformations $G \xRightarrow{p_1,m_1} H_1$ and $G \xRightarrow{p_2,m_2} H_2$, there is a graph G' together with direct transformations $H_1 \xRightarrow{p_2,m_2'} G'$ and $H_2 \xRightarrow{p_1,m_1'} G'$ such that $G \xRightarrow{p_1,m_1} H_1 \xRightarrow{p_2,m_2'} G'$ and $G \xRightarrow{p_2,m_2} H_2 \xRightarrow{p_1,m_1'} G'$ are sequentially independent.*

Given two sequentially independent direct transformations $G \xRightarrow{p_1,m_1} H_1 \xRightarrow{p_2,m_2'} G'$, there is a graph H_2 together with direct transformations $G \xRightarrow{p_2,m_2} H_2 \xRightarrow{p_1,m_1'} G'$ such that $G \xRightarrow{p_1,m_1} H_1$ and $G \xRightarrow{p_2,m_2} H_2$ are parallel independent.

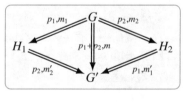

In any case of independence, there is a parallel transformation $G \xRightarrow{p_1+p_2,m} G'$ and, vice versa, a direct transformation $G \xRightarrow{p_1+p_2,m} G'$ via the parallel rule $p_1 + p_2$ can be sequentialised both ways. △

2.3.2 Concurrency Theorem

In contrast to the Local Church–Rosser and Parallelism Theorem, the Concurrency Theorem is concerned with the execution of sequentially dependent transformations. In this case, we cannot commute subsequent direct transformations, as done for independent transformations, nor are we able to apply the corresponding parallel rule. Nevertheless, it is possible to apply both transformations concurrently using a so-called E-concurrent rule and shifting the application conditions of the single rules to an equivalent concurrent application condition.

Given an arbitrary sequence $G \xstackrel{p_1,m_1}{\Longrightarrow} H \xstackrel{p_2,m_2}{\Longrightarrow} G'$ of direct transformations it is possible to construct an E-concurrent rule $p_1 *_E p_2$. The graph E is an overlap of the right-hand side of the first rule and the left-hand side of the second rule. The construction of the concurrent application condition is again based on the two shift constructions.

Definition 2.27 (Concurrent rule). Given rules $p_1 = (L_1 \xstackrel{l_1}{\longleftarrow} K_1 \xstackrel{r_1}{\longrightarrow} R_1, ac_1)$ and $p_2 = (L_2 \xstackrel{l_2}{\longleftarrow} K_2 \xstackrel{r_2}{\longrightarrow} R_2, ac_2)$, a graph E with jointly surjective morphisms $e_1 : R_1 \to E$ and $e_2 : L_2 \to E$ is an E-*dependency relation* of p_1 and p_2 if the pushout complements (1) of $e_1 \circ r_1$ and (2) of $e_2 \circ l_2$ exist.

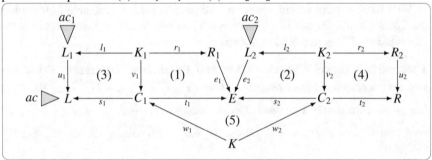

Given an E-dependency relation (E, e_1, e_2) of p_1 and p_2, the E-*concurrent rule* $p_1 *_E p_2 = (L \xstackrel{s_1 \circ w_1}{\longleftarrow} K \xstackrel{t_2 \circ w_2}{\longrightarrow} R, ac)$ is constructed by pushouts (1), (2), (3), (4), and pullback (5), with $ac = \text{Shift}(u_1, ac_1) \wedge L(p^*, \text{Shift}(e_2, ac_2))$ for $p^* = (L \xstackrel{s_1}{\longleftarrow} C_1 \xstackrel{t_1}{\longrightarrow} E)$.

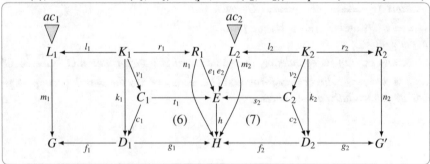

A sequence $G \xrightarrow{p_1,m_1} H \xrightarrow{p_2,m_2} G'$ is called *E-related* if there exist $h : E \to H$, $c_1 : C_1 \to D_1$, and $c_2 : C_2 \to D_2$ such that h is injective, $h \circ e_1 = n_1$, $h \circ e_2 = m_2$, $c_1 \circ v_1 = k_1$, $c_2 \circ v_2 = k_2$, and (6) and (7) are pushouts   △

Example 2.28. In Fig. 2.13, the *E*-concurrent rule construction is depicted, leading to the *E*-related sequence $H_1 \xrightarrow{\text{setTurn},m_2} H_2 \xrightarrow{\text{setFlag},m_3} H_3$ of the direct transformations already considered in Fig. 2.11. Note that e_1 matches the processes of setTurn to the two processes in the same order and e_2 matches the process of setFlag to the upper process. Moreover, $ac_L = \text{Shift}(u_1, ac_2) \land \text{L}(p^*, \text{Shift}(e_2, ac_1))$ is not depicted explicitly. Leaving out invalid models like idle processes with flags, it evaluates to true.   △

For a sequence $G \xrightarrow{p_1,m_1} H \xrightarrow{p_2,m_2} G'$ of direct transformations we can construct an *E*-dependency relation such that the sequence is *E*-related. Then the *E*-concurrent rule $p_1 *_E p_2$ allows us to construct a direct transformation $G \xrightarrow{p_1 *_E p_2} G'$ via $p_1 *_E p_2$. Vice versa, each direct transformation $G \xrightarrow{p_1 *_E p_2} G'$ via the *E*-concurrent rule $p_1 *_E p_2$ can be sequentialised, leading to an *E*-related transformation sequence $G \xrightarrow{p_1,m_1} H \xrightarrow{p_2,m_2} G'$ of direct transformations via p_1 and p_2. Note that this theorem is the instantiation of Theorem 5.30 to graphs.

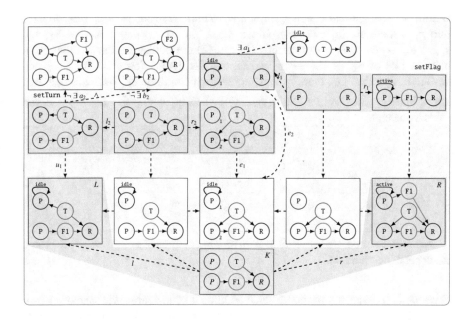

Fig. 2.13 *E-concurrent rule construction*

Theorem 2.29 (Concurrency Theorem). *For rules p_1 and p_2 and an E-concurrent rule $p_1 *_E p_2$ we have:*

- *Given an E-related transformation sequence* $G \xrightarrow{p_1,m_1} H \xrightarrow{p_2,m_2} G'$, *there is a* synthesis construction *leading to a direct transformation $G \xrightarrow{p_1 *_E p_2,m} G'$ via the E-concurrent rule $p_1 *_E p_2$.*

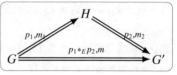

- *Given a direct transformation $G \xrightarrow{p_1 *_E p_2,m} G'$, there is an* analysis construction *leading to an E-related transformation sequence $G \xrightarrow{p_1,m_1} H \xrightarrow{p_2,m_2} G'$.*
- *The synthesis and analysis constructions are inverse to each other up to isomorphism.*   △

2.3.3 Amalgamation

With amalgamation, we synchronise a number of rule applications. The idea is to model a certain number of actions which are similar for each step with a subrule, while corresponding complement rules describe the effects of each rule application outside this subrule.

Definition 2.30 (Subrule and complement rule). A rule p_0 is a *subrule* of a rule p_1 if there are injective morphisms $s_{1,L} : L_0 \to L_1$, $s_{1,K} : K_0 \to K_1$, and $s_{1,R} : R_0 \to R_1$ such that diagrams (1) and (2) are pullbacks, the pushout complement (1') of $K_0 \to L_0 \to L_1$ exists, and the application conditions ac_0 and ac_1 are *compatible*, i.e., there is some application condition ac_1' over L_{10} such that $ac_1 \equiv \mathrm{Shift}(s_{1,L}, ac_0) \wedge \mathrm{L}(p_1^*, \mathrm{Shift}(v_1, ac_1'))$, where $p_1^* = (L_1 \xleftarrow{u_1} L_{10} \xrightarrow{v_1} E_1)$ and (3) is a pushout.

$$
\begin{array}{ccccc}
ac_0 \triangleright L_0 & \xleftarrow{l_0} K_0 & \xrightarrow{r_0} R_0 & & \\
s_{1,L}\downarrow \quad (1) \quad s_{1,K}\downarrow \quad (2) \quad \downarrow s_{1,R} & & & \\
ac_1 \triangleright L_1 & \xleftarrow{l_1} K_1 & \xrightarrow{r_1} R_1 & &
\end{array}
\qquad
\begin{array}{ccc}
L_0 & \xleftarrow{l_0} K_0 & \xrightarrow{r_0} R_0 \\
s_{1,L}\downarrow \quad (1') \quad \downarrow w_1 \quad (3) \quad \downarrow e_{11} \\
L_1 & \xleftarrow{u_1} L_{10} & \xrightarrow{v_1} E_1 \\
& ac_1' \triangle &
\end{array}
$$

A rule $\bar{p}_1$ is a *complement rule* of p_1 with respect to p_0 if $p_1 = p_0 *_{E_1} \bar{p}_1$ for some E_1-dependency relation.   △

Example 2.31. We want to model some additional behaviour of the system. A process with an F1-flag to this resource can be redirected to a different resource (rule p_7), and a resource may be disabled and marked for update if the process having its turn is not active (rule p_8). Note that we also have to adapt the type graph, adding the update-loop for a resource.

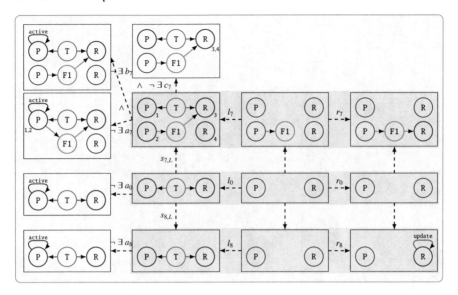

Fig. 2.14 The subrule p_0 of the rules p_7 and p_8

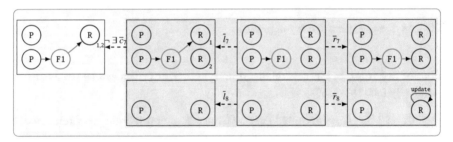

Fig. 2.15 The complement rules $\overline{p}_7$ and $\overline{p}_8$

Rule p_7 is depicted at the top of Fig. 2.14 and shows the redirection of the F1-flag of the process. Rule p_8 is shown at the bottom of this figure and adds the update-marker. In the middle row of Fig. 2.14, the subrule p_0 is shown which disables the resource. This rule is actually a valid subrule of p_7 and p_8, because the given squares are pullbacks, in both cases the pushout complements exist, and for the application conditions we have that $ac_i \equiv \mathrm{Shift}(s_{i,L}, ac_0) \wedge \mathrm{L}(p_i^*, \mathrm{Shift}(v_i, ac_i'))$ for $i = 7, 8$.

The complement rules $\overline{p}_7$ and $\overline{p}_8$ are given in Fig. 2.15. Note, that the application condition $\neg \exists \, c_7$ is translated into an application condition $\neg \exists \, \overline{c}_7$, while we do not need an application condition for $\overline{p}_8$.   △

The construction of an amalgamated rule generalises the one of a parallel rule, where all rules are glued together along the subrule. Here, we only give the construc-

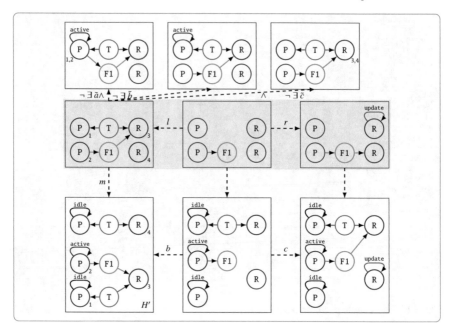

Fig. 2.16 The amalgamated rule $\tilde{p}$ of p_7 and p_8

tion for two rules, but in general an arbitrary number of rules can be amalgamated iteratively (see Def. 6.9).

Definition 2.32 (Amalgamated rule). Given a common subrule p_0 of rules p_1 and p_2, the *amalgamated rule* $\tilde{p} = p_1 \oplus_{p_0} p_2$ is given by $\tilde{p} = (\tilde{L} \xleftarrow{\tilde{l}} \tilde{K} \xrightarrow{\tilde{r}} \tilde{R}, \tilde{ac})$, where $\tilde{L}$, $\tilde{K}$, and $\tilde{R}$ are the pushouts of the left-hand sides, gluings, and right-hand sides, respectively, $\tilde{l}$ and $\tilde{r}$ are the uniquely existing morphisms, and $\tilde{ac} = \mathrm{Shift}(t_{1,L}, ac_1) \wedge \mathrm{Shift}(t_{2,L}, ac_2)$.

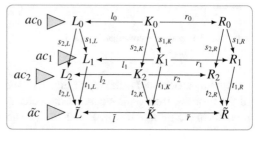

$\triangle$

Example 2.33. The amalgamated rule $\tilde{p} = p_7 \oplus_{p_0} p_8$ is shown in the upper rows of Fig. 2.16. It combines the effects of p_7 and p_8, where both rules disable the same resource. The application of this amalgamated rule to the graph H' from Fig. 2.12 is shown in Fig. 2.16. Note that m maps the left-hand side of the rule to the lower part of H', but in reverse order. Simultaneously, the resource is disabled, its update-flag is set, and the F1-flag of the process is redirected. $\triangle$

Two direct transformations of the same graph are amalgamable if both matches agree on the subrule and are independent outside. In this case, the amalgamation theorem states that we can apply the amalgamated rule to realise the effects of both rules in one step. Note that this theorem is the instantiation of Theorem 6.17 to graphs for two amalgamable rules.

Theorem 2.34 (Amalgamation Theorem). *For rules p_1 and p_2 with amalgamated rule $\tilde{p} = p_1 \oplus_{p_0} p_2$, consider the complement rules q_i of $\tilde{p}$ w. r. t. p_i, i.e., $\tilde{p} = p_1 *_{E'_1} q_1 = p_2 *_{E'_2} q_2$. Then we have:*

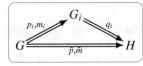

- *Given amalgamable direct transformations $G \xRightarrow{p_1,m_1} G_1$ and $G \xRightarrow{p_2,m_2} G_2$, there is an amalgamated transformation $G \xRightarrow{\tilde{p},\tilde{m}} H$ and direct transformations $G_1 \xRightarrow{q_1} H$ and $G_2 \xRightarrow{q_2} H$ such that $G \xRightarrow{p_1,m_1} G_1 \xRightarrow{q_1} H$ and $G \xRightarrow{p_2,m_2} G_2 \xRightarrow{q_2} H$ are decompositions of $G \xRightarrow{\tilde{p},\tilde{m}} H$.*
- *Given an amalgamated direct transformation $G \xRightarrow{\tilde{p},\tilde{m}} H$, there are transformations $G \xRightarrow{p_1,m_1} G_1 \xRightarrow{q_1} H$ and $G \xRightarrow{p_2,m_2} G_2 \xRightarrow{q_2} H$ such that the direct transformations $G \xRightarrow{p_1,m_1} G_1$ and $G \xRightarrow{p_2,m_2} G_2$ are amalgamable.*
- *The synthesis and analysis constructions are inverse to each other up to isomorphism.* △

2.3.4 Embedding and Extension Theorem

For the Embedding and Extension Theorem, we analyse under what conditions a transformation $t : G_0 \xRightarrow{*} G_n$ can be extended to a transformation $t' : G'_0 \xRightarrow{*} G'_n$ via an extension morphism $k_0 : G_0 \to G'_0$ (see Fig. 2.17). The idea is to obtain an *extension diagram* (1), which is defined by pushouts (2_i)–(5_i) for all $i = 1, \ldots, n$, where the same rules $p_1, \ldots, p_n$ are applied in the same order in t and t'.

It is important to note that this is not always possible, because there may be some elements in G'_0 invalidating an application condition or forbidding the deletion of something which can still be deleted in G_0. But we are able to give a necessary and sufficient consistency condition to allow such an extension. This result is important for all kinds of applications where we have a large graph G'_0, but only small subparts of G'_0 have to be changed by the rules $p_1, \ldots, p_n$. In this case, we choose a suitable small subgraph G_0 of G'_0 and construct a transformation $t : G_0 \xRightarrow{*} G_n$ via $p_1, \ldots, p_n$ first. Then we compute the *derived span* of this transformation, which we extend in a second step via the inclusion $k_0 : G_0 \to G'_0$ to a transformation $t' : G'_0 \xRightarrow{*} G'_n$ via the same rules $p_1, \ldots, p_n$. Since we only have to compute the small transformation from G_0 to G_n and the extension of G_n to G'_n, this makes the computation of $G'_0 \Rightarrow G'_n$ more efficient.

The derived span connects the first and the last graph of a transformation and describes in one step, similarly to a rule, the changes between them. Over the derived

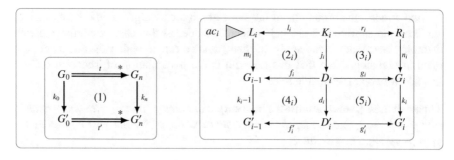

Fig. 2.17 Embedding and extension: sequence (left) and intermediate step (right)

span we can also define a derived application condition which becomes useful later for the Local Confluence Theorem.

Definition 2.35 (Derived span and application condition). Given a transformation $t : G_0 \stackrel{*}{\Rightarrow} G_n$ via rules $p_1, \ldots, p_n$, the *derived span der(t)* is inductively defined by

$$
der(t) = \begin{cases}
G_0 \stackrel{f_1}{\leftarrow} D_1 \stackrel{g_1}{\rightarrow} G_1 & \text{for } t : G_0 \stackrel{p_1,m_1}{\Longrightarrow} G_1 \\[2mm]
G_0 \stackrel{d_0' \circ d}{\leftarrow} D \stackrel{g_n \circ d_n}{\longrightarrow} G_n & \text{for } t : G_0 \stackrel{*}{\Rightarrow} G_{n-1} \stackrel{p_n,m_n}{\Longrightarrow} G_n \text{ with} \\[2mm]
& der(G_0 \stackrel{*}{\Rightarrow} G_{n-1}) = (G_0 \stackrel{d_0'}{\leftarrow} D' \stackrel{d_{n-1}'}{\longrightarrow} G_{n-1}) \\
& \text{and pullback } (PB)
\end{cases}
$$

$$
G_0 \xleftarrow{\quad d_0' \quad} D' \xrightarrow{\quad d_{n-1}' \quad} G_{n-1} \xleftarrow{\quad f_n \quad} D_n \xrightarrow{\quad g_n \quad} G_n
$$
$$
(PB)
$$
$$
D
$$

Moreover, the *derived application condition ac(t)* is defined by

$$
ac(t) = \begin{cases}
\text{Shift}(m_1, ac_1) & \text{for } t : G_0 \stackrel{p_1,m_1}{\Longrightarrow} G_1 \\[2mm]
ac(G_0 \stackrel{*}{\Rightarrow} G_{n-1}) & \text{for } t : G_0 \stackrel{*}{\Rightarrow} G_{n-1} \stackrel{p_n,m_n}{\Longrightarrow} G_n \\
\wedge L(p_n^*, \text{Shift}(m_n, ac_n)) & \text{with } p_n^* = der(G_0 \stackrel{*}{\Rightarrow} G_{n-1})
\end{cases}
$$

$\triangle$

Example 2.36. Consider the transformation $G \stackrel{setFlag}{\Longrightarrow} H_1 \stackrel{setTurn}{\Longrightarrow} H_2 \stackrel{setFlag}{\Longrightarrow} H_3$ from Figs. 2.4 and 2.11. The derived span of this transformation and its derived application condition are shown in Fig. 2.18 and combine all changes applied in the single transformation steps. Note that the derived application condition actually forbids matching both resources or idle processes noninjectively. $\triangle$

For the consistency condition, we need the concept of initial pushouts. This is a categorical formalisation of boundary and context leading to the smallest pushout over a morphism. Intuitively, the boundary contains all elements of the domain new

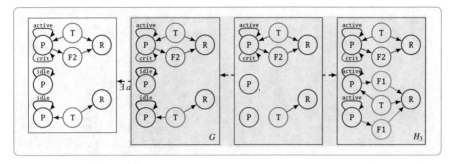

Fig. 2.18 The derived span of $G \stackrel{*}{\Rightarrow} H_3$

elements in the codomain are connected to. All these new elements and their connections are then collected in the context.

For k_0 to be *boundary-consistent*, we have to find a morphism from the boundary to the derived span, which means that no element in the boundary is deleted by the transformation. Moreover, we need *AC consistency*; therefore k_0 has to fulfill a summarised set of application conditions collected from all rules and shifted to G_0. We say that k_0 is *consistent* with respect to t if it is both boundary-consistent and AC-consistent. Consistency of k_0 is both necessary and sufficient for embedding a transformation $t : G_0 \stackrel{*}{\Rightarrow} G_n$ via k_0. Note that this theorem is the instantiation of Theorem 5.34 to graphs.

Theorem 2.37 (Embedding and Extension Theorem). *Given a transformation $t :$ $G_0 \stackrel{*}{\Rightarrow} G_n$ and a morphism $k_0 : G_0 \to G'_0$ which is consistent with respect to t, there*

exists an extension diagram (1) over t and k_0.

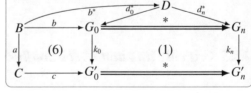

Vice versa, given a transformation $t : G_0 \stackrel{*}{\Rightarrow} G_n$ with an extension diagram (1) and an initial pushout (6) over $k_0 : G_0 \to G'_0$, as motivated above, we have that:

1. k_0 is consistent with respect to $t : G_0 \stackrel{*}{\Rightarrow} G_n$.

2. There is a rule $p^* = (der(t), ac(t))$ leading to a direct transformation $G'_0 \stackrel{p^*}{\Longrightarrow} G'_n$.

3. G'_n is the pushout of the context C and G_n along the boundary B, i.e., $G'_n = G_n +_B C$. △

Example 2.38. We embed the graph $G_0 = G$ from Fig. 2.4 into a larger context graph G'_0, where an additional process has an F1-flag pointing to the lower resource. The boundary B and context graph C are shown in the left of Fig. 2.19, where the boundary only contains the lower resource to which the new process is connected. Since this resource is not deleted, the extension morphism k_0 is boundary-consistent. Moreover, it is AC-consistent, because the derived application condition is fulfilled.

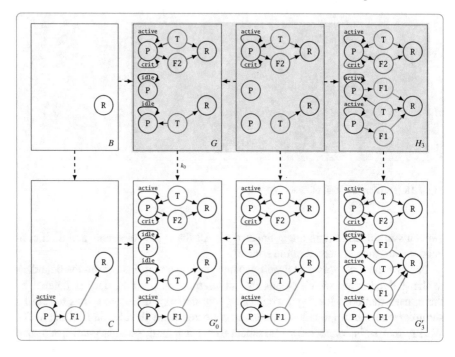

Fig. 2.19 The embedding and extension of G into G'_0

Therefore, we have consistency and can construct the transformation $G'_0 \Longrightarrow^* G'_3$ as shown in Fig. 2.19, where G'_3 is the pushout of H_3 and C along B. △

2.3.5 Critical Pairs and Local Confluence Theorem

A transformation system is called *confluent* if, for all transformations $G \overset{*}{\Rightarrow} H_1$ and $G \overset{*}{\Rightarrow} H_2$, there is an object X together with transformations $H_1 \overset{*}{\Rightarrow} X$ and $H_2 \overset{*}{\Rightarrow} X$. *Local confluence* means that this property holds for all pairs of direct transformations $G \xrightarrow{p_1,m_1} H_1$ and $G \xrightarrow{p_2,m_2} H_2$.

Confluence is an important property of a transformation system, because, in spite of local nondeterminism concerning the application of a rule, we have

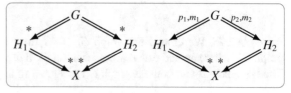

global determinism for confluent transformation systems. *Global determinism* means that, for each pair of terminating transformations $G \overset{*}{\Rightarrow} H$ and $G \overset{*}{\Rightarrow} H'$

with the same source object, the target objects H and H' are equal or isomorphic. A transformation $G \overset{*}{\Rightarrow} H$ is called *terminating* if no rule is applicable to H anymore. This means that each transformation sequence terminates after a finite number of steps.

The Local Church–Rosser Theorem shows that, for two parallel independent direct transformations $G \overset{p_1,m_1}{\Longrightarrow} H_1$ and $G \overset{p_2,m_2}{\Longrightarrow} H_2$, there is a graph G' together with direct transformations $H_1 \overset{p_2,m_2'}{\Longrightarrow} G'$ and $H_2 \overset{p_1,m_1'}{\Longrightarrow} G'$. This means that we can apply the rules p_1 and p_2 with given matches in an arbitrary order. If each pair of rules is parallel independent for all possible matches, then it can be shown that the corresponding transformation system is confluent.

In the following, we discuss local confluence for the general case in which $G \overset{p_1,m_1}{\Longrightarrow} H_1$ and $G \overset{p_2,m_2}{\Longrightarrow} H_2$ are not necessarily parallel independent. According to a general result for rewriting systems, it is sufficient to consider local confluence, provided that the transformation system is terminating [Plu95].

The main idea is to study critical pairs. For a pair $P_1 \overset{p_1,o_1}{\Longleftarrow} K \overset{p_2,o_2}{\Longrightarrow} P_2$ of direct transformations to constitute a critical pair, the matches o_1 and o_2 are allowed to violate the application conditions, while they induce new ones that have to be respected by a parallel dependent extension of the critical pair. These induced application conditions make sure that the extension respects the application conditions of the given rules and that there is indeed a conflict.

Definition 2.39 (Critical pair). Given rules $p_1 = (L_1 \overset{l_1}{\leftarrow} K_1 \overset{r_1}{\rightarrow} R_1, ac_1)$ and $p_2 = (L_2 \overset{l_2}{\leftarrow} K_2 \overset{r_2}{\rightarrow} R_2, ac_2)$, a pair $P_1 \overset{\overline{p}_1,o_1}{\Longleftarrow} K \overset{\overline{p}_2,o_2}{\Longrightarrow} P_2$ of direct transformations without application conditions is a *critical pair* (for transformations with application conditions) if (o_1, o_2) are jointly surjective and there exists an extension of the pair via an injective morphism $m : K \rightarrow G$ such that $m \models ac_K = ac_K^E \wedge ac_K^C$, with

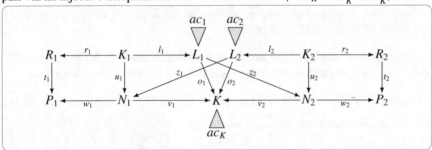

- *extension application condition:* $ac_K^E = \text{Shift}(o_1, ac_1) \wedge \text{Shift}(o_2, ac_2)$ and
- *conflict inducing application condition:* $ac_K^C = \neg(ac_{z_1} \wedge ac_{z_2})$, with
  if $\exists z_1 : v_1 \circ z_1 = o_2$ then $ac_{z_1} = \text{L}(p_1^*, \text{Shift}(w_1 \circ z_1, ac_2))$ else $ac_{z_1} = \text{false}$,
$$\text{with } p_1^* = (K \overset{v_1}{\leftarrow} N_1 \overset{w_1}{\rightarrow} P_1)$$
  if $\exists z_2 : v_2 \circ z_2 = o_1$ then $ac_{z_2} = \text{L}(p_2^*, \text{Shift}(w_2 \circ z_2, ac_1))$ else $ac_{z_2} = \text{false}$,
$$\text{with } p_2^* = (K \overset{v_2}{\leftarrow} N_2 \overset{w_2}{\rightarrow} P_2) \qquad \triangle$$

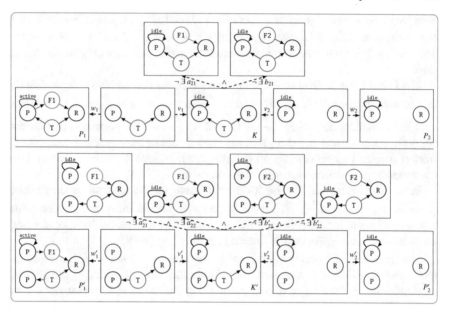

Fig. 2.20 The critical pairs of `setFlag` and `disableR`

Example 2.40. In Fig. 2.20, the two critical pairs of the rules `setFlag` and `disableR` are shown, together with the application conditions ac_K and $ac_{K'}$. Both critical pairs overlap the resources of the two rules, leading to a dependency, since activating the idle process by `setflag` forbids the disabling of the resource by `disableR`. Note, that ac_K^C and $ac_{K'}^C$ evaluate to true, meaning that whenever we find a match from K or K' respecting the application condition ac_K^E or $ac_{K'}^E$, respectively, we definitely have a conflict of this sort.   △

Every pair of parallel dependent direct transformations is an extension of a critical pair. Note that this theorem is the instantiation of Theorem 5.41 to graphs.

Theorem 2.41 (Completeness Theorem). *For each pair of parallel dependent direct transformations $H_1 \overset{p_1,m_1}{\Longleftarrow} G \overset{p_2,m_2}{\Longrightarrow} H_2$ there is a critical pair $P_1 \overset{p_1,o_1}{\Longleftarrow} K \overset{p_2,o_2}{\Longrightarrow} P_2$ with induced application condition ac_K and an injective morphism $m : K \to G$ with $m \models ac_K$ leading to extension diagrams (1) and (2).*   △

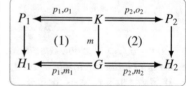

In order to show local confluence it is sufficient to show strict AC confluence of all its critical pairs. As discussed above, confluence of a critical pair $P_1 \Leftarrow K \Rightarrow P_2$ means the existence of an object K' together with transformations $P_1 \overset{*}{\Rightarrow} K'$ and $P_2 \overset{*}{\Rightarrow} K'$. Strictness is a technical condition which means, intuitively, that the parts which are preserved by both transformations of the critical pair are also preserved in the common object K'. For strict AC confluence of a critical pair, the transformations of the strict solution of the critical pair must be extendable to G, which means that each application condition of both transformations must be satisfied in the bigger context.

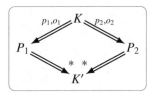

Definition 2.42 (Strict AC confluence). A critical pair $P_1 \xLeftarrow{p_1,o_1} K \xRightarrow{p_2,o_2} P_2$ with induced application conditions ac_K is strictly AC confluent if it is

1. confluent without application conditions, i.e., there are transformations $P_1 \xRightarrow{*} K'$ and $P_2 \xRightarrow{*} K'$ eventually disregarding the application conditions,
2. strict, i.e., given derived spans $der(K_i \xRightarrow{p_i,o_i} P_i) = (K \xleftarrow{v_i} N_i \xrightarrow{w_i} P_i)$ and $der(P_i \xRightarrow{*} K') = (P_i \xleftarrow{v_{i+2}} N_{i+2} \xrightarrow{w_{i+2}} K')$ for $i = 1, 2$ and pullback (1), there exist morphisms z_3, z_4 such that diagrams (2), (3), and (4) commute, and

3. for $\bar{t}_i : K \xRightarrow{p_i,o_i} P_i \xRightarrow{*} K'$ it holds that $ac_K \Rightarrow ac(\bar{t}_i)$ for $i = 1, 2$, where $ac(\bar{t}_i)$ is the derived application condition of $\bar{t}_i$.

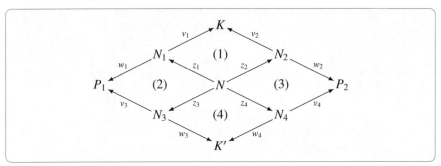

Using this notion of strict AC confluence we can state a sufficient condition for a transformation system to be locally confluent. Note that this theorem is the instantiation of Theorem 5.44 to graphs.

Theorem 2.43 (Local Confluence Theorem). *A transformation system is locally confluent if all its critical pairs are strictly AC confluent.* △

Example 2.44. The critical pairs in Ex. 2.40 as well as all other critical pairs of our mutual exclusion example are strictly confluent. Therefore, the transformation system is locally confluent. Although it is not terminating, it is also confluent. △

Chapter 3
Model Transformation

This chapter is an introduction to model transformation, which is a key component of model-driven development. Sect. 3.1 describes the relevance and concepts of model transformations in general. Using the notions of typed attributed graphs in Chap. 2, Sect. 3.2 presents the main aspects of model transformations based on graph transformation on a general level. As a specific instantiation of these concepts, Sect. 3.3 introduces triple graph grammars (TGGs) as a powerful technique for bidirectional model transformations. Sect. 3.3 provides an overview of how these concepts are used as a foundation for Part III. This chapter is based on previous work [Erm09, EE10, HHK10, EEE+07, HEGO14].

3.1 Introduction to Model Transformation

Model transformations are a key concept for modular and distributed model-driven development (MDD). They are used thoroughly for model optimisation and other forms of model evolution. Moreover, model transformations are used to map models between different domains in order to perform code generation or to apply analysis techniques.

In this multi-domain context, *modelling languages* are the primary way in which system developers communicate and design systems. Many domain-specific modelling languages (DSMLs) are *visual modelling languages* (that contain also textual elements) providing a highly specialised set of domain concepts [CSW08]. A visual syntax presents a model in a diagrammatical (two-dimensional) way and is often suitable for presenting models in an intuitive and understandable way.

In the MDD context, DSMLs are proposed whose type systems formalise the application structure, behaviour, and requirements within particular domains. DSMLs are described using *meta models*, which define the relationships among concepts in a domain by class diagrams and specify language properties by constraints using OCL [Obj14b] associated with these domain concepts. Graphical modelling features enable the integration of concepts that are commonly used by domain experts. This

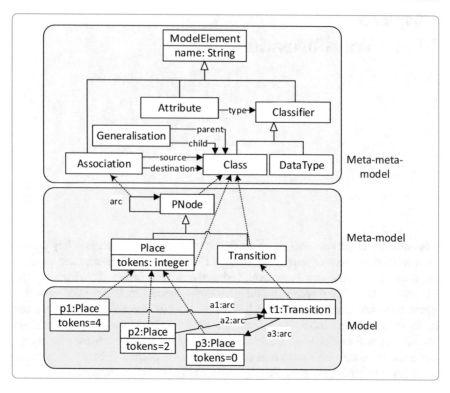

Fig. 3.1 Relations between meta model levels

helps flatten learning curves, eases the accessibility to a broader range of domain
experts, such as system engineers and experienced software architects, and helps
them to ensure that software systems meet user needs. A meta model describes the
various kinds of model elements of a DSML, and the way they are arranged, related,
and constrained. Notably, meta models are notated as class diagrams and hence are
visual models. Meta model elements provide a typing scheme for model elements.
This typing is expressed by the meta relation between a model element and its meta
element (from the meta model). A model *conforms* to a meta model if each model element has its meta element defined within the meta model, and the given constraints
are satisfied.

The growing number of meta models has emphasised the need for an integration
framework for all available meta models by providing a new item, the meta meta
model, dedicated to the definition of meta models. In the same way models are
defined in conformance with their meta model, meta models are defined by means
of the meta meta model language. An overview of different levels of meta modelling
is given in Fig. 3.1, where the sample model is a Petri net, the meta model is a
class diagram defining the structural knowledge of Petri net concepts (*Place, tokens,*

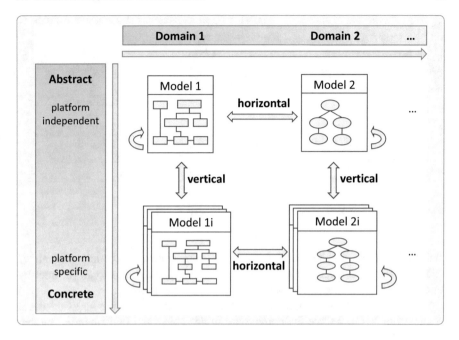

Fig. 3.2 Model transformations in model-driven development

Transition and *arc*), and the meta meta model shows the key model elements for modelling class diagrams (the meta model language).

Model transformations are the links between the artefacts of MDD. Different kinds of model transformations are proposed that cover different application domains and different steps of a sound software production process, including business modelling, requirements engineering, conceptual modelling and model-based code generation techniques. According to the taxonomy of model transformations by Mens et. al. [MG06], they can be categorised in two dimensions, as depicted in Fig. 3.2.

First of all, exogeneous model transformations take as input a model of one language and produce as output a model of another language, while input and output models of an endogeneous model transformations belong to the same language. The second dimension separates horizontal model transformations that do not change the level of abstraction from vertical model transformations, which explicitly do change the level of abstraction. Examples of model transformations in these dimensions as listed in [MG06] are: model refactoring (endogeneous, horizontal), formal model refinement (exogeneous, horizontal), language migration (endogeneous, vertical) and code generation (exogeneous, vertical).

In MDD, model transformations are (partially) automated. This automation reduces the required amount of manual work for software development, such that the software engineering process is supposed to become less error-prone and more ef-

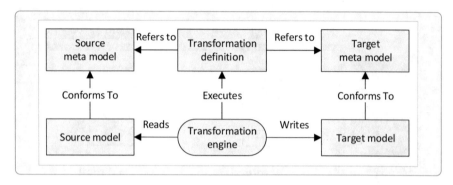

Fig. 3.3 Basic principle of model transformations (from [CH06])

ficient. The issue is "the model is the code" rather than "the code is the model" [PM07]. The automated transformation process is also referred to as "correct-by-construction", as opposed to "construct-by-correction" [Sch06].

The vision of MDD proposes the following principles for the software development process to reduce the complexity in designing and implementing modern software [Sel08]:

- **Modelling at different abstraction levels**
  MDD promotes the extensive and systematic use of models from a very early phase of the design cycle. System requirements and design are captured by high-level (often visual) engineering models (using popular and standardised modelling languages like UML [UML15], XML Schemes [WWW04], SysML [Sys14] or BPMN [OMG14]) or by *Domain-Specific Languages* (DSLs) whose concepts closely reflect the concepts of the problem domain at a *suitable abstraction* level abstracting away technological implementation detail. For the design of DSLs in the context of MDD, *meta modelling* is used [CSW08]. A meta model is a model of the concepts expressed by a modelling language.
- **Automating model transformations**
  In MDD, the key idea is to automate model transformations. This includes (but is not limited to) the ability to *generate code*, i.e., to transform models expressing high-level domain-specific concepts into equivalent programs (e.g., Java code, XML documents, XML schemes, etc.). Fig. 3.3 shows the basic principle underlying model transformations [CH06].
  Source as well as target models have to conform to their respective meta models. Therefore, the definition of the transformation has to be compatible with the meta models involved. The transformation definition is applied to a concrete model using a transformation engine. Model transformations between different modelling languages are called *exogenous*. For certain model transformations (e.g., model refactoring), the source and target meta models may be the same. Such model transformations are called *endogenous*.

A model-to-model transformation may also have the purpose of utilising *analysis* techniques available for the target domain. The formal analysis of design models can be carried out by generating appropriate mathematical models using automated model transformations. Problems detected by automated analysis can be fixed by model refinement prior to investing in manual coding for implementation.

For executable modelling languages, automated model transformations can also be used to *simulate* high-level models in order to *validate* the suitability of the modelled system behaviour in an early development phase.

• **Using tools that adhere to open industry standards**

  Model-Driven Architecture (MDA) [Sel08] is OMG's initiative to support MDD with a set of open industry standards, including the ability to exchange models between different tools. Standards allow tool builders to focus on their specific area of expertise. The basic OMG standards refer to modelling languages (UML [UML15], MOF [MOF15]) and model transform definitions (QVT [QVT15]).

The promises of MDD are manifold [VSB$^+$06]: development time will be reduced; the model becomes "timeless" and is never outdated since it changes with the domain and not with the technology; the generated code is correct if the model is; documentation may be generated from the model and is thus consistent with the code (which is also generated); systems are easier adaptable (platform-independent); tasks can be easier divided in complex projects; analysis and validation on model basis lead to earlier error detection before code is generated.

Model transformations appear in several contexts, e.g., in the various facets of model-driven architecture [MDA15] encompassing model refinement and interoperability of system components. The involved languages can be closely related or they can be more heterogeneous, e.g., in the special case of model refactoring, the source language and the target language are the same. From a general point of view, a model transformation $MT : \mathcal{L}_S \Rightarrow \mathcal{L}_T$ transforms models from the (domain-specific) source language $\mathcal{L}_S$ to models of the target language $\mathcal{L}_T$.

In the following, we list main challenges for model transformations as presented in [HHK10]. They concern *functional* as well as *nonfunctional aspects*. Some of these challenges were initially presented by Schürr et. al. [SK08] for the specific scope of model transformation approaches based on triple graph grammars (TGGs).

At first, we consider the dimension of functional aspects, which concern the reliability of the produced results. Depending on the concrete application of a model transformation $MT : \mathcal{L}_S \Rightarrow \mathcal{L}_T$, some of the following properties may have to be ensured.

1. *Syntactical Correctness:* For each model $M_S \in \mathcal{L}_S$ that is transformed by MT the resulting model M_T has to be syntactically correct, i.e., $M_T \in \mathcal{L}_T$.
2. *Semantical Correctness:* The semantics of each model $M_S \in \mathcal{L}_S$ that is transformed by MT has to be preserved or reflected, respectively.
3. *Completeness:* The model transformation MT can be performed on each model $M_S \in \mathcal{L}_S$. Additionally, it can be required that MT reaches all models $M_T \in \mathcal{L}_T$.

4. *Functional Behaviour:* For each source model $M_S \in \mathcal{L}_S$, the model transformation *MT* will always terminate and lead to the same resulting target model M_T.

In the second dimension, we treat nonfunctional aspects. They concern usability and applicability properties of model transformations. Depending on the application domain, some of the following challenges may be required in addition to the functional ones listed above.

1. *Efficiency:* Model transformations should have polynomial space and time complexity. Furthermore, there may be further time constraints that need to be respected, depending on the application domain and the intended way of use.
2. *Intuitive specification:* The specification of model transformations can be performed based on patterns that describe how model fragments in a source model correspond to model fragments in a target model. If the source (or target) language is a visual language, then the components of the model transformation can be visualised using the concrete syntax of the visual language.
3. *Maintainability:* Extensions and modifications of a model transformation should only require little effort. Side effects of local changes should be handled and analysed automatically.
4. *Expressiveness:* Special control conditions and constructs have to be available in order to handle more complex models, which e.g., contain substructures with a partial ordering or hierarchies.
5. *Bidirectional model transformations:* The specification of a model transformation should provide the basis for both a model transformation from the source to the target language and a model transformation in the reverse direction.

The power of *bidirectional model transformations* is based on the simultaneous support of transformations in both forward and backward direction. This way, models can be maintained in two repositories—one for diagrams in the source domain and one for diagrams in the target domain. The modellers can work in separate teams, and the specified model transformations are used to support the interchange between these groups and their models. In particular, a modeller can generate models in one domain from models in another domain using the concepts for model transformation in Chap. 7 and he can validate and ensure syntactical correctness and completeness using the results and analysis techniques in Chap. 8.

In Sects. 3.2 and 3.3, we introduce suitable techniques for the specification of model transformations based on graph transformation. Part III presents the formal techniques for model transformations based on triple graph grammars (TGGs), which provide validated and verified capabilities for a wide range of the challenges listed above.

3.2 Model Transformation by Graph Transformation

While meta modelling provides a declarative approach to DSML definition, grammars are more constructive, i.e., closer to the implementation. Due to their appeal-

Table 3.1 Mapping meta modelling notions to graph terminology

Meta modelling notion	Graph terminology
Model	Type graph TG
Inheritance	Node type inheritance in TG
Class	Node in type graph TG
Association	Edge in type graph TG
Multiplicities	Node and edge type multiplicities in TG
Class attributes	Attribute types belonging to node types
Model instance	TG-typed, attributed graph G with typing morphism $G \rightarrow TG$
Object	Node in TG-typed graph G
Reference	Edge in TG-typed graph G that must not violate certain multiplicity constraints.

ing visual form, graph grammars can directly be used as high-level visual specification mechanism for DSMLs [BEdLT04]. Defining the abstract syntax of visual models as graphs, a graph grammar defines directly the DSML grammar. The induced graph language determines the corresponding DSML. Visual language parsers can be immediately deduced from such a graph grammar. Furthermore, abstract syntax graphs are also the starting point for visual modelling of model behaviour [Erm06, dLVA04, Var02, HKT02] and model transformations [MVVK05, MVDJ05, SAL$^+$03, Sch94].

Meta modelling is closely related to graph typing, where a type graph takes the role of the meta model, and an instance graph, typed over the type graph, corresponds to a model conforming to a meta model. In order to better map meta modelling concepts to typed, attributed graphs, the graph transformation theory has been enhanced in [BEdLT04] with node type inheritance facilities, and it has been shown how typed graph transformation with inheritance can be flattened to simple typed graph transformation. Meta modelling and graph transformation can be integrated by identifying symbol classes with node types and associations with edge types.

Table 3.1 shows a comparison of main meta modelling notions to their counterparts in the terminology of typed graphs. Classes in a meta model correspond to nodes in a type graph. Associations between classes can be seen as edges in a type graph. Class inheritance and multiplicity constraints of association ends can be defined in the type graph by node type inheritance and graph constraints for specifying edge multiplicities. Objects as instantiations of classes of a meta model are comparable to nodes in a graph which is typed over a type graph. Objects can be linked to each other by setting reference values. Such references correspond to edges in a typed attributed graph. The notion of being a model conforming to a meta model can be adequately formalised for typed graphs by the existence of a typing morphism from an instance graph to the type graph. OCL constraints can be translated to graph constraints or to graph transformation rules (syntax rules), as has been shown in [WTEK08, BKPPT00, AHRT14]. Thus, declarative as well as constructive

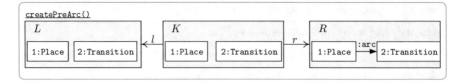

Fig. 3.4 Syntax rule adding a Petri net arc

elements may be used for DSML definition based on typed graph transformation. An example for a rule from the syntax grammar for Petri nets is shown in Fig. 3.4, where an arc is inserted between a place node and a transition node. Note that the graphs in this rule conform to the meta model (are typed over the type graph) shown in the center of Fig. 3.1.

The classical approach to model transformation based on typed, attributed graph transformation [EE05, Erm09] does not require any specific structuring techniques or restrictions on transformation rules. The type graph is given by an integrated

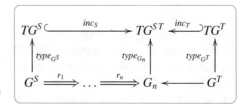

type graph TG^{ST} consisting of the type graphs for the source and target language, and, additionally, reference nodes with arcs mapping source elements to target elements. It constitutes a single graph, such that the division in source and target components is not explicit. We express model transformations directly by TG^{ST}-typed graph transformation rules $L \leftarrow K \rightarrow R$ where basically L contains relevant source model elements and some context, while $R \setminus K$ basically represents target model elements that correspond to the relevant source model elements. The model transformation starts with graph G^S typed over TG^S. As TG^S is a subgraph of TG^{ST}, G^S is also typed over TG^{ST}. During the model transformation process, the intermediate graphs $G^S = G_1; \ldots; G_n$ are all typed over TG^{ST}. To delete all items in G_n which are not typed over TG^T, we can construct a restriction (a pullback in the category **Graphs$_{ATG}$**), which deletes all those items in one step. In addition to normal graph transformation rules, we also use rule schemes to express parallel transformation steps of arbitrary many similar model element patterns.

Remark 3.1 (Notation for domain components). In the context of model transformations, we always differentiate between source and target domains. For this reason, we denote the specific domain of graphs and graph morphisms in superscript (e.g., source graph G^S). Thereby, that we can place all further information as index (e.g., the first graph of a sequence G_1).   △

The following general concept of model transformations integrates the described constructions and defines special properties that are relevant for model transformations. By $t^{S>}(G)$, we denote the retyping of G that is initially typed over TG^S to a model that is typed over the integrated type graph TG^{ST}. Furthermore, $t^{T<}(G)$ specifies a restriction of G typed over TG^{ST} to a model G' typed over TG^T only, which can be constructed as pullback of $G \to TG^{ST} \leftarrow TG^T$. The execution of the graph transformation systems may be controlled via a control condition restricting the possible transformations. Such conditions will be explained exemplarily based on triple graph grammars thereafter.

Definition 3.2 (General concept of model transformations based on graph transformation). Let **GRAPHS** be the category of plain graphs **Graphs** or triple graphs **TrGraphs**.

1. Given a source language $\mathcal{L}_S \subseteq \mathbf{GRAPHS}_{TG^S}$ and a target language $\mathcal{L}_T \subseteq \mathbf{GRAPHS}_{TG^T}$, a *model transformation* $MT : \mathcal{L}(TG^S) \Rightarrow \mathcal{L}(TG^T)$ from $\mathcal{L}(TG^S)$ to $\mathcal{L}(TG^T)$ is given by $MT = (\mathcal{L}(TG^S), \mathcal{L}(TG^T), TG^{ST}, t_S, t_T, \mathrm{GTS})$ where TG^{ST} is an integrated type graph with injective type graph morphisms $(TG^S \xrightarrow{t_S} TG^{ST} \xleftarrow{t_T} TG^T)$, and GTS a graph transformation system with nondeleting rules R typed over TG^{ST} and a control condition for GTS-transformations. Moreover, a consistency relation $MT_C \subseteq \mathcal{L}(TG^S) \times \mathcal{L}(TG^T)$ for MT defines all consistent pairs (G^S, G^T) of source and target models.
2. A *model transformation sequence* via MT, in short, MT-sequence, is given by $(G^S, G_1 \Rightarrow^* G_n, G^T)$, where $G^S \in \mathcal{L}(TG^S), G^T \in \mathcal{L}(TG^T)$ and $G_1 \Rightarrow^* G_n$ is a GTS-transformation satisfying the control condition of GTS with $G_1 = t^{S>}(G^S)$ and $G^T = t^{T<}(G_n)$, as defined above.
3. The *model transformation relation* $MT_R \subseteq \mathcal{L}(TG^S) \times \mathcal{L}(TG^T)$ defined by MT is given by: $(G^S, G^T) \in MT_R \Leftrightarrow \exists MT - \text{sequence } (G^S, G_1 \Rightarrow^* G_n, G^T)$.
4. $MT : \mathcal{L}(TG^S) \Rightarrow \mathcal{L}(TG^T)$ is called

   a. syntactically correct, if for all $(G^S, G^T) \in MT_R$ we have $G^S \in \mathcal{L}_S, G^T \in \mathcal{L}_T$ and $(G^S, G^T) \in MT_C$,
   b. total, if for each $G^S \in \mathcal{L}_S$ there is a pair $(G^S, G^T) \in MT_R$,
   c. surjective, if for each $G^T \in \mathcal{L}_T$ there is a pair $(G^S, G^T) \in MT_R$,
   d. complete, if MT_R is total and surjective,
   e. functional, if MT_R is right unique.   △

Most examples of model transformations based on plain graph transformation considered in the literature adhere to this general concept. A typical example is the model transformation *SC2PN* from state charts to Petri nets (see Chapter 14 of [EEPT06] with restriction construction instead of deleting rules): The control condition is given by layers, where the rules with negative application conditions are applied as long as possible in one layer, and suitable termination criteria have to be satisfied before switching to the next layer. But also model transformations based on triple rules adhere to this concept.

We implemented this approach in our tool AGG [AGG14, BEL+10, RET12] (see also Chap. 12), supporting the definition of type graphs, typed attributed graph rules and constraints. Further graph transformation systems for domain-specific model transformations are VIATRA2 [BNS+05] and the Graph Rewriting and Transformation Language (GReAT) [SAL+03]. In VIATRA2, developers define graph patterns and graph transformation rules as components using a textual domain-specific programming language. The components are assembled into complex model transformations by abstract state machine rules. In GReAT, meta models of the source and target models are used to establish the vocabulary of L and R and to ensure that the transformation produces a well-formed target model.

The classical approach on graph transformation-based model transformation has been extended to support the transformation of EMF models in Eclipse, thus bridging the gap between MDD tools and those for graph transformation. The Eclipse Modeling Framework (EMF) [EMF14] has evolved to one of the standard technologies to define modelling languages. EMF is based on MOF [MOF15] and provides a (meta) modelling and code generation framework for Eclipse applications based on structured data models. Containment relations, i.e., aggregations, define an ownership relation between objects. Thereby, they induce a tree structure in instance models, implying some constraints that must be ensured at runtime. As semantical constraints for containment edges, the MOF specification states that *"an object may have at most one container"*, and that *"cyclic containment is invalid"*.

A transformation framework for EMF models is presented in [ABJ$^+$10, BEK$^+$06], where containment edges are modelled as graph edges of a special *containment* type. The problem is guaranteeing that EMF model transformations defined by graph transformation always satisfy the EMF containment constraints. In [BET12], these constraints are translated to special kinds of EMF model transformation rules such that their application leads to consistent transformation results only. EMF model transformation is supported by our tool Henshin (formerly called EMF Tiger) [BEL$^+$10, ABJ$^+$10] and its extension HenshinTGG for handling model transformations based on triple graph grammars, which are introduced in the next section. The tools AGG, Henshin and HenshinTGG are described in more detail in Chap. 12.

3.3 Introduction to Triple Graph Grammars

Triple graph grammars (TGGs) are a well established concept for the specification and execution of bidirectional model transformations within model-driven software engineering. They form a specific case of model transformation based on graph transformation described in Sect. 3.2 before. Since their introduction by Schürr [Sch94], TGGs have been applied in several case studies and they have shown a convenient combination of formal and intuitive specification abilities. In addition to having the general advantages of bidirectional model transformations, TGGs simplify the design of model transformations. A single set of triple rules is sufficient to generate the operational rules for the forward and backward model transformations. Thereby, TGGs enhance usability as well as consistency in MDD. Furthermore, model transformations based on TGGs preserve a given source model by creating a separate target model together with a correspondence structure. This way, the given models are not modified, which is especially important for database-driven model repositories. Moreover, TGGs specify model transformations based on the abstract syntax of DSLs and are therefore not restricted to specific types of modelling languages.

The key idea for the execution of model transformations via TGGs is to preserve the given source model and to add the missing target and correspondence elements in separate but connected components. For this reason, the transformation rules add new structures and do not necessarily need to delete existing elements. The resulting target model is obtained by type restriction. Indeed, nondeleting triple rules are sufficient for many case studies. However, in general it may be very difficult, if not impossible, to specify a model transformation whose validity depends on some global properties of the given models. An example may be automata minimisation, where we transform a finite automaton into an automaton with the same behaviour, but with the smallest possible set of states. In this case, the transformation should translate any two states with the same behaviour into a single state. However, knowing if two states have the same behaviour is a global property of the given automaton. Nevertheless, a possible way of simplifying the model transformation is performing some additional preprocessing of the source model or postprocessing of the target model. For this reason and as it is common praxis for TGGs, we consider transformation rules that are nondeleting.

Triple graph grammars [Sch94] are a well-known approach for bidirectional model transformations. In [KS06], the basic concepts of triple graph grammars are formalised in a set-theoretical way, which is generalised and extended in [EEE$^+$07] to typed attributed graphs. In this section, we present triple graph transformation systems with application conditions. They form an instantiation of $\mathcal{M}$-adhesive transformation systems defined in Chap. 5, as we show in Chap. 7. We can use any kind of graph inside triple graphs, as long as they form an $\mathcal{M}$-adhesive category (see Chap. 4). This means that we can have triple graphs (or, better, triple structures) consisting of many kinds of graphical structures. In this book, we use typed attributed triple graphs.

Definition 3.3 (Triple graph). A *triple graph* $G = (G^S \xleftarrow{s_G} G_C \xrightarrow{t_G} G^T)$ consists of graphs G^S, G^C, and G^T, called source, correspondence, and target graphs, and two graph morphisms s_G and t_G mapping the correspondence to the source and target graphs. A *triple graph morphism* $m = (m_S, m_C, m_T) : G \to H$ matches the single graphs and preserves the correspondence part. Formally, a triple graph morphism m consists of graph morphisms $m_S : G_S \to H_S$, $m_C : G_C \to H_C$ and $m_T : G_T \to H_T$ such that $m_S \circ s_G = s_H \circ m_C$ and $m_T \circ t_G = t_H \circ m_C$.

$$
\begin{array}{ccccc}
G = (& G^S & \xleftarrow{s_G} & G^C & \xrightarrow{t_G} & G^T) \\
m \downarrow & m^S \downarrow & & m^C \downarrow & & m^T \downarrow \\
R = (& H^S & \xleftarrow{s_H} & H^C & \xrightarrow{t_H} & H^T)
\end{array}
$$

A *typed triple graph* $(G, type_G)$ is given by a typing morphism $type_G : G \to TG$ from the triple graph G into a given *triple type graph* TG. A typed triple graph morphism $f : (G_1, type_{G_1}) \to (G_2, type_{G_2})$ is a triple graph morphism f such that $type_{G_2} \circ f = type_{G_1}$. △

Triple graphs and typed triple graphs, together with the componentwise compositions and identities, form the categories **TrGraphs** and **TrGraphs**$_{TG}$, where

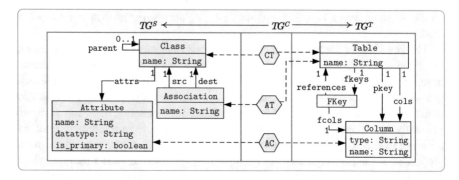

Fig. 3.5 Triple type graph for *CD2RDBM*

Graphs is the category of graphs (see Def. 2.1). Using the category of attributed graphs **AGraphs** (see Def. 2.4) for the triple components, we derive the categories **ATrGraphs** and **ATrGraphs**$_{ATG}$ of (typed) attributed triple graphs.

Definition 3.4 (Category of typed attributed triple graphs). Typed attributed triple graphs and typed attributed triple graph morphisms form the category **ATrGraphs**$_{ATG}$. △

Definition 3.5 (Base categories of triple graphs). Triple graphs and triple graph morphisms form the category **TrGraphs**. Analogously, category **ATrGraphs** of attributed triple graphs is given by attributed triple graphs and attributed triple graph morphims. Moreover, given a triple graph *TG* in **TrGraphs**, we obtain category **TrGraphs**$_{TG}$ consisting of typed triple graphs typed over *TG*. △

Example 3.6 (Triple type graph). Fig. 3.5 shows the triple type graph *TG* of the triple graph grammar *TGG* for the running example of a model transformation *CD2RDBM* from class diagrams to database models, which is based on the example presented in [EEE+07, HEGO14]. The source component TG^S defines the structure of class diagrams while the target component TG^T specifies the structure of relational database models. Classes in the source component (node type Class) correspond to tables in the target component, attributes correspond to columns, and associations to foreign keys. Throughout the example, elements are arranged left, center, and right according to the component types source, correspondence and target. Attributes of structural nodes and edges are depicted within their containing structural nodes and edges. Note that the correspondence component is important for the relation of the source elements to their aligned target elements. For this reason, it is used in practical scenarios to navigate via the traceability links from source structures to target structures and vice versa. The morphisms between the three components are illustrated by dashed arrows. The depicted multiplicity constraints are ensured by the triple rules of the grammar shown in Fig. 3.8. Moreover, the source language contains only those class diagrams in which each class hierarchy has at most one primary attribute. △

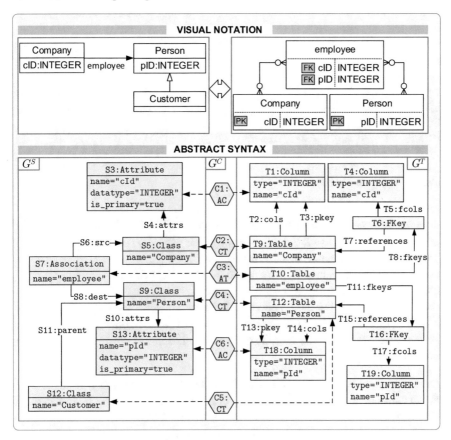

Fig. 3.6 Triple graph instance for *CD2RDBM*

Example 3.7 (Triple graph). Fig. 3.6 shows an instance triple graph $G = (G^S \leftarrow G^C \rightarrow G^T)$ that is typed over *TG* from Ex. 3.6. The lower part of the figure shows the triple graph G that specifies the abstract syntax of the class diagram, the database model and the correspondence links. The corresponding visualisation is provided at the top of Fig. 3.6, where foreign keys are marked with FK and primary keys are marked with PK. The triple graph specifies a company (class Company and table Company) and its employees (class Person and table Person) as well as its customers (class Customer and table Person). Customers have a dedicated ID (attribute cust_id and column cust_id).   △

Triple rules construct the source, correspondence, and target parts in one step. Each triple rule specifies a pattern of consistent corresponding fragments in its required context. In Sect. 7.3, we derive the operational rules for model transformation from these triple rules.

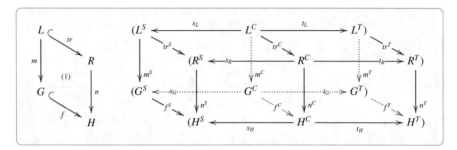

Fig. 3.7 Triple transformation step

A *triple rule tr* is given by an $\mathcal{M}$-morphism $tr = (tr^S, tr^C, tr^T) : L \to R$ (injective on the graph component and an isomorphism on the data part), and without loss of generality we assume tr to be an inclusion. Since triple rules are nondeleting, there is no need for the more general concept of

$$L = (L^S \xleftarrow{s_L} L^C \xrightarrow{t_L} L^T)$$
$$tr\downarrow \quad tr^S\downarrow \qquad tr^C\downarrow \qquad tr^T\downarrow$$
$$R = (R^S \xleftarrow{s_R} R^C \xrightarrow{t_R} R^T)$$

a span of morphisms for a rule $p = (L \xleftarrow{l} K \xrightarrow{r} R)$ in an $\mathcal{M}$-adhesive transformation system (see Chap. 5). The morphism l can be chosen as $l = id$ and we omit the intermediate object K.

Definition 3.8 (Triple rule and transformation). A *triple rule* $tr = (tr : L \to R, ac)$ consists of triple graphs L and R, an $\mathcal{M}$-morphism $tr : L \to R$, and an application condition ac over L. Given a triple graph G, a triple rule $tr = (tr, ac)$ and a match $m: L \to G$ with $m \models ac$, a *direct triple transformation* $G \xRightarrow{tr,m} H$ of G via tr and m is given by pushout (1) in Fig. 3.7, which is constructed as the componentwise pushouts in the S-, C-, and T-components, where the morphisms s_H and t_H are induced by the pushout of the correspondence component. In addition to H, we obtain co-match $n : R \to H$ and transformation inclusion $f : G \hookrightarrow H$. A direct transformation is also called triple graph transformation (TGT) step.   △

Application conditions of triple rules are used to restrict the application of the rules to certain contexts as defined in Chap. 2. A triple rule $tr: L \to R$ with an application condition ac over L is applicable if the match $m: L \to G$ satisfies the application condition ac, i.e., $m \models ac$. Roughly spoken, an application condition is a Boolean formula over some additional context structure for the left-hand side L (see Defs. 2.7 and 2.8). In our running example, we will exlusively use negative application conditions (NACs) [HHT96]. A NAC has the form $ac = \neg(\exists a: L \to C)$ and is used to avoid the application of a rule if the match for L can be extended to the forbidden context C. A match $m: L \to G$ satisfies a NAC $ac = \neg(\exists a: L \to C)$ if there is no embedding $q: C \to G \in \mathcal{M}$ that is compatible with m, i.e., such that $q \circ a = m$. From now on, a triple rule denotes a rule with application conditions, while the absence of application conditions is explicitly mentioned.

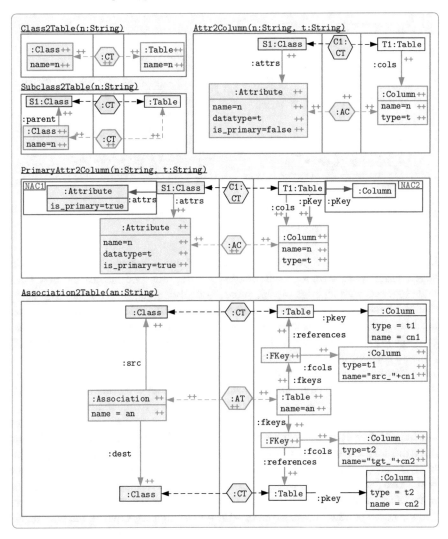

Fig. 3.8 Rules for the model transformation *CD2RDBM*

Note that, due to the structure of the triple rules, both double [EEPT06] and single pushout approach [EHK⁺96] are equivalent in this case.

Example 3.9 (Triple rules and triple transformation step). The triple rules in Fig. 3.8 are the rules of the grammar *TGG* for the model transformation *CD2RDBM*. They are presented in short notation, i.e., left- and right-hand sides of a rule are depicted in one triple graph. Elements which are created by the rule are labelled with "++" and additionally marked by green line colouring. Rule Class2Table synchronously creates a class with name n together with the corresponding table in

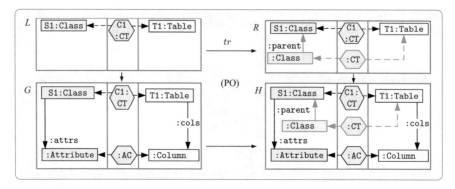

Fig. 3.9 Triple graph transformation step via rule Subclass2Table (without data values)

the relational database. Accordingly, subclasses are connected to the tables of its superclasses by rule Subclass2Table. Note that this rule creates the new class node together with an edge of type parent, implying that our compact case study does not handle the case of multiple inheritance. Finally, rule Attr2Column creates attributes with type t together with their corresponding columns in the database component. Rule PrimaryAttr2Column extends Attr2Column by creating additionally a link of type pkey for the column and by setting the attribute value is_primary = true. This rule contains NACs, which are specified in short notation. The NAC-only elements are specified by red line colouring and additionally with a surrounding frame with label NAC. A complete NAC is obtained by composing the left-hand side of a rule with the marked NAC-only elements. The source and target NACs ensure that there is neither a primary attribute in the class diagram nor a primary key in the database model present when applying the rule. More formally, the depicted NACs are actually NAC schemata (see Def. 2.17), such that the NACs also forbid the cases when some of the specified variables are evaluated to the same values. Rule Association2Table creates an association between two classes and the corresponding table, together with two foreign keys, where the parameters an, cn1, cn2, t1 and t2 are used to set the names and types of the created nodes.

Fig. 3.9 shows a triple graph transformation step $G \xrightarrow{tr,m} H$ via rule tr = Subclass2Table, where we omitted the attribute values of the nodes and reduced the node types to the starting letters. The top line shows the rule with its left- and right-hand sides and the bottom line shows the given triple graph G and the resulting triple graph H. The effect of this step is the addition of a new subclass that is related to the existing table corresponding to the existing class.   △

Remark 3.10 (Possible extensions of the example). The example *CD2RDBM* provides a simplistic view on the general problem to define an object relational mapping. A possible extension would be to take into account multiplicities on associations. In that case, the TGG could be extended by an additional rule that would create an association with cardinality $1 - n$ (source multiplicity x..1 and destination

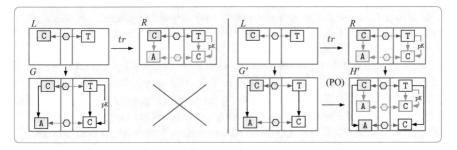

Fig. 3.10 Violation of NAC and satisfaction of NAC for rule `PrimaryAttr2Column`

multiplicity y..n) together with a single foreign key in the target domain (database model). The existing rule `Association2Table` would be refined to handle only the case for associations with cardinality $n - m$ (source multiplicity x..* and destination multiplicity y..*) using attribute conditions. The example also simplifies that data type names used for nodes of type `Attribute` (`datatype = t`) in class diagrams and for nodes of type `Column` (`type = t`) in database diagrams coincide. There are several ways to extend the TGG to define a mapping from certain data type names in one domain to differently named ones in the other domain. For example, this could be handled by specialising the rules for each pair of corresponding data type names or by using a constant structural fragment from which the name mapping information can be matched by the rules and used for the attribute assignments. Finally, one could consider composed primary keys in the database domain, which would require further extensions of the TGG.   △

Example 3.11 (Triple transformation with NACs). The left component of Fig. 3.10 shows a violation of the target NAC for rule `PrimaryAttr2Column`, whose target NAC forbids the presence of an existing primary key at the matched table. In its right component, the figure shows a NAC consistent transformation step, where no primary key is present and also the existing attribute is assumed to be not a primary one.   △

A *triple graph transformation system* $TGS = (TR)$ is based on triple graphs and a set of rules TR over them. A *triple graph grammar* $TGG = (TR, S)$ contains in addition a triple start graph S. For triple graph grammars, the generated language is given by all triple graphs G that can be derived from the start graph S via rules in TR. The source language $\mathcal{L}(TGG)^S$ contains all graphs that are the source component of a derived triple graph. Similarly, the target language $\mathcal{L}(TGG)^T$ contains all derivable target components. A triple graph grammar generates the language $\mathcal{L}(TGG)$ of integrated models, where each integrated model contains a source model and its corresponding target model. This language induces the model transformation relation MT_R that defines the set of all consistent pairs (G^S, G^T) of source and target models. Any other pair of models is seen to be inconsistent.

Definition 3.12 (Triple graph grammar and triple language). A triple graph grammar $TGG = (TR, S)$ consists of a set TR of triple rules and a triple graph S called triple start graph.

A language of triple graphs generated by TGG is given by $\mathcal{L}(TGG) = \{G \mid \exists$ triple transformation $S \overset{*}{\Rightarrow} G$ via rules in $TR\}$. The source language $\mathcal{L}(TGG)^S = \{G^S \mid (G^S \overset{s_G}{\hookleftarrow} G^C \overset{t_G}{\hookrightarrow} G^T) \in \mathcal{L}(TGG)\}$ contains all graphs that are the source component of a derived triple graph. Similarly, the target language $\mathcal{L}(TGG)^T = \{G^T \mid (G^S \overset{s_G}{\hookleftarrow} G^C \overset{t_G}{\hookrightarrow} G^T) \in \mathcal{L}(TGG)\}$ contains all derivable target components.

The model transformation relation $MT_R = \{(G^S, G^T) \in \mathcal{L}(TGG)^S \times \mathcal{L}(TGG)^T \mid \exists G = (G^S \leftarrow G^C \rightarrow G^T) \in \mathcal{L}(TGG)\}$ defines the set of all consistent pairs (G^S, G^T) of source and target models.   △

Example 3.13 (Triple language). The triple graph in Fig. 3.6 shows an instance triple graph $G = (G^S \leftarrow G^C \rightarrow G^T)$ of the triple language $\mathcal{L}(TGG)$ for the language CD2RDBM. It can be constructed via the transformation sequence $\emptyset \xrightarrow{\text{Class2Table}}$ $G_1 \xrightarrow{\text{Class2Table}} G_2 \xrightarrow{\text{SubClass2Table}} G_3 \xrightarrow{\text{Attribute2Column}} G_4 \xrightarrow{\text{PrimaryAttribute2Column}}$ $G_5 \xrightarrow{\text{Association2Table}} G_6$. The triple language contains all such triple graphs that can be created via the triple rules.   △

3.4 Model Transformation Based on TGGs

In Part III, we present the techniques and results for model transformation based on TGGs (see Def. 3.2). We cover different execution and analysis techniques in different transformation scenarios.

Part III starts with Chap. 7, which presents the automatic derivation of operational rules from a given TGG:

- operational rules for forward model transformations, i.e., transforming a model of the source language into a corresponding one of the target language
- operational rules for backward model transformations, i.e., transforming a model of the target language into a corresponding one of the source language
- operational rules for model integrations, i.e., taking a pair of existing source and target models and extending them to an integrated model that is consistent (if possible)
- operational rules for model synchronisation, i.e., taking and integrated model together with updates on both domains and propagating the changes from one domain to the other, including the handling of conflicts
- operational rules for consistency checking, i.e., checking whether a given integrated model is consistent

Furthermore, Chap. 7 presents appropriate execution algorithms for forward/ backward model transformations and model integrations. We focus on both the formal point of view based on category theory and the implementation point of view using an encoding of the formal control conditions via Boolean-valued attributes as a kind of caching structure and a flattening construction to enable the use of plain graph transformation tools. As the main result, we provide sufficient criteria for ensuring the fundamental composition and decomposition result for TGGs that is the basis for the results in Chapters 8 and 9. The decomposition result ensures that a triple graph transformation sequence can be decomposed into two operational sequences, where one is used for a kind of parsing of the inputs and the other performs the actual creation of the relevant elements for the output. Vice versa, the composition result ensures that two operational sequences can be composed if a certain consistency condition is satisfied.

Chap. 8 provides results for the analysis of model transformations and model integrations based on TGGs concerning the following three out of four properties of the dimension of functional aspects presented in Sect. 3.1, which concerns the reliability of the produced results:

1. *Syntactical Correctness:* For each model $M_S \in \mathcal{L}_S$ that is transformed by MT the resulting model M_T has to be syntactically correct, i.e., $M_T \in \mathcal{L}_T$.
2. *Completeness:* The model transformation MT can be performed on each model $M_S \in \mathcal{L}_S$. Additionally, it can be required that MT reaches all models $M_T \in \mathcal{L}_T$.
3. *Functional Behaviour:* For each source model $M_S \in \mathcal{L}_S$, the model transformation MT will always terminate and lead to the same resulting target model M_T.

As a general requirement, Chap. 8 assumes the validity of the (de-)composition result for TGGs, for which Theorem 7.21 in Chap. 7 provides a sufficient condition. The main results show that model transformations and integrations based on TGGs ensure syntactical correctness and completeness. Since some TGGs do not show functional behaviour for the derived operations, we provide sufficient conditions that ensure functional behaviour. These conditions are based on the notion of critical pairs and can be analysed statically using the tool AGG. We also show how the operational rules of a TGG can be extended by special application conditions to remove conflicts, which is used to improve efficiency of the execution and for showing functional behaviour in a more complex case. In addition, we show that model transformations based on TGGs are always information preserving in a weak sense and provide a sufficient condition for showing that they are completely information preserving, i.e., that the input can be reconstructed from the output without requiring any other information than the output model itself.

Chap. 9 presents the most complex case of model transformations—namely model synchronisation—where several different operations have to be combined. Model synchronisation is an important technique for keeping and gaining consistency of source and target models after changing one or both of them. We provide sufficient conditions and general results for ensuring

1. *Syntactical Correctness:* For each synchronisation of an integrated model $M \in \mathcal{L}(TG)$ with updates on the source and target domains, the result of the model synchronisation is a consistent integrated model $M' \in \mathcal{L}(TGG)$ together with corresponding updates on the source and target domains. Moreover, if no update is required, then the synchronisation preserves the inputs.

2. *Completeness:* The model synchronisation can be performed for any integrated model $M \in \mathcal{L}(TG)$.
3. *Invertibility:* The propagation of changes is symmetric in the following sense. Propagating changes from the first domain and then propagating the obtained changes back yields the initially given update on the first domain. If this property is required for a restricted set of updates only, we use the notion of weak invertibility.

In the basic case, model synchronisation is performed in a unidirectional way, i.e., an update on one domain is propagated to the corresponding domain. The more general cases handle concurrent updates on both domains, conflict resolution and nondeterminism.

Part II
$\mathcal{M}$-Adhesive Transformation Systems

This second part of the book presents the algebraic approach to graph transformation in the general framework of $\mathcal{M}$-adhesive categories, which instantiate to graphs of the form $G = (V, E, s, t)$ considered in Chap. 2, but also to a large variety of further types of graphs and other kinds of high-level structures, such as labelled graphs, typed graphs, hypergraphs and different kinds of low- and high-level Petri nets. The extension from graphs to high-level structures was introduced in [EHKP91a, EHKP91b], leading to the theory of high-level replacement (HLR) systems. In [EHPP04] the concept of HLR systems was joined to that of adhesive categories of Lack and Sobocinsky in [LS04], leading to the concepts of adhesive HLR categories used in [EEPT06] and $\mathcal{M}$-adhesive categories in this book, where all these concepts are introduced in Chap. 4. Moreover, this chapter includes an overview of different adhesive and HLR notions and several results concerning HLR properties, which are used in the general theories of Chapters 5 and 6 and for the construction of $\mathcal{M}$-adhesive categories. In fact, $\mathcal{M}$-adhesive categories and transformation systems constitute a suitable general framework for an abstract theory of graph and model transformations, which can be instantiated to various kinds of high-level structures, especially to those mentioned above. All the concepts and results—introduced for graph transformation in Chap. 2—are carefully presented and proven in Chap. 5 for $\mathcal{M}$-adhesive transformation systems and in Chap. 6 for multi-amalgamated transformations. Finally it is shown in Chap. 6 how multi-amalgamation can be used to define the semantics of elementary Petri nets.

Chapter 4
Adhesive and $\mathcal{M}$-Adhesive Categories

In this chapter, we introduce adhesive and $\mathcal{M}$-adhesive categories as the categorical foundation of graph and model transformations and present various constructions and properties. In Sect. 4.1, we introduce $\mathcal{M}$-adhesive categories based on the notion of van Kampen squares. Different versions of adhesive categories are compared to the one we use in this book in Sect. 4.2. In Sect. 4.3, we introduce some additional properties which are needed for results in $\mathcal{M}$-adhesive categories, as well as derive results that hold in any such category. A special variant of $\mathcal{M}$-adhesive categories using only finite objects is presented in Sect. 4.4. In Sect. 4.5, we show that $\mathcal{M}$-adhesive categories are closed under certain categorical constructions and analyse how far several of the additional properties are also preserved. This chapter is based on [EGH$^+$14, Gol10].

4.1 $\mathcal{M}$-Adhesive Categories

For the transformation of not only graphs, but also high-level structures such as Petri nets and algebraic specifications, high-level replacement (HLR) categories were established in [EHKP91a, EHKP91b], which require a list of so-called *HLR properties* to hold. They were based on a morphism class $\mathcal{M}$ used for the rule morphisms. This framework allowed a rich theory of transformations for all HLR categories, but the HLR properties were difficult and lengthy to verify for each category.

Adhesive categories were introduced in [LS04] as a categorical framework for deriving process congruences from reaction rules. They require a certain compatibility of pushouts and pullbacks, called the *van Kampen property*, for pushouts along monomorphisms in the considered category. Later, they were extended to quasiadhesive categories in [JLS07], where the van Kampen property has to hold only for pushouts along regular monomorphisms.

Adhesive categories behave well also for transformations, but interesting categories as typed attributed graphs are neither adhesive nor quasiadhesive. Combining adhesive and HLR categories leads to adhesive HLR categories in [EHPP04,

EPT04], where a subclass $\mathcal{M}$ of monomorphisms is considered and only pushouts over $\mathcal{M}$-morphisms have to fulfill the van Kampen property. They were slightly extended to weak adhesive HLR categories in [EEPT06], where a weaker version of the van Kampen property is sufficient to show the main results of graph and HLR transformations also for transformations in weak adhesive HLR categories. Not only many kinds of graphs, but also Petri nets and algebraic high-level nets are weak adhesive HLR categories, which allows to apply the theory to all these kinds of structures. In [EEPT06], the main theory including all the proofs for transformations in weak adhesive HLR categories can be found, while an introduction including motivation and examples for all the results is given in [PE07].

In this book, we use a slightly different version, the so-called $\mathcal{M}$-adhesive categories. The differences between several variants of adhesive categories are analysed in detail in the following section. Their main property is (a variant of) the van Kampen property, which is a special compatibility of pushouts and pullbacks in a commutative cube. The idea of a van Kampen square is that of a pushout which is stable under pullbacks, and, vice versa, where pullbacks are stable under combined pushouts and pullbacks.

Definition 4.1 (Van Kampen square). A commutative cube (2) with pushout (1) in the bottom face and where the back faces are pullbacks fulfills the *van Kampen property* if the following statement holds: the top face is a pushout if and only if the front faces are pullbacks.

A pushout (1) is a *van Kampen square* if the van Kampen property holds for all commutative cubes (2) with (1) in the bottom face.

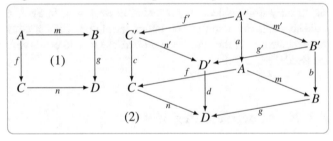

Given a morphism class $\mathcal{M}$, a pushout (1) with $m \in \mathcal{M}$ is an $\mathcal{M}$-*van Kampen square* if the van Kampen property holds for all commutative cubes (2) with (1) in the bottom face and $b, c, d \in \mathcal{M}$. $\triangle$

It might be expected that, at least in the category **Sets**, every pushout is a van Kampen square. Unfortunately, this is not true, but at least pushouts along monomorphisms are van Kampen squares in **Sets** and several other categories.

For an $\mathcal{M}$-adhesive category, we consider a category **C** together with a morphism class $\mathcal{M}$ of monomorphisms. We require pushouts along $\mathcal{M}$-morphisms to be $\mathcal{M}$-van Kampen squares, along with some rather technical conditions for the morphism class $\mathcal{M}$ called PO–PB compatibility, which are needed to ensure compatibility of $\mathcal{M}$ with pushouts and pullbacks.

Definition 4.2 (PO–PB compatibility). A morphism class $\mathcal{M}$ in a category **C** is called *PO–PB compatible* if

1. M is a class of monomorphisms, contains all identities, and is closed under composition ($f : A \to B \in M$, $g : B \to C \in M \Rightarrow g \circ f \in M$).
2. C has pushouts and pullbacks along M-morphisms, and M-morphisms are closed under pushouts and pullbacks. △

Remark 4.3. From Items 1 and 2, it follows that M contains all isomorphisms and is also closed under decomposition, i.e., $g \circ f \in M$, $g \in M$ implies $f \in M$. In fact, isomorphisms can be obtained by pushouts or pullbacks along identities, and the pullback of $(g \circ f, g)$ is (id, f) such that the pullback closure of M implies that $f \in M$ [Hei10]. △

Definition 4.4 (M-adhesive category). A category C with a PO–PB compatible morphism class M is called an M-*adhesive category* if pushouts in C along M-morphisms are M-van Kampen squares. △

Examples for M-adhesive categories are the categories **Sets** of sets, **Graphs** of graphs, **Graphs$_{TG}$** of typed graphs, **Hypergraphs** of hypergraphs, **ElemNets** of elementary Petri nets, and **PTNets** of place/transition nets, all together with the class M of injective morphisms, as well as the category **Specs** of algebraic specifications with the class M_{strict} of strict injective specification morphisms, the category **PTSys** of place/transition systems with the class M_{strict} of strict morphisms, and the category **AGraphs$_{ATG}$** of typed attributed graphs with the class M_{D-iso} of injective graph morphisms with isomorphic data part. The proof that **Sets** is an M-adhesive category is shown in [EEPT06], while the proofs for most of the other categories can be done using the Construction Theorem in the following Sect. 4.5.

4.2 Overview of Different Adhesive and HLR Notions

Several variants of HLR and adhesive categories have been introduced in the literature as categorical frameworks for graph transformation and HLR systems based on

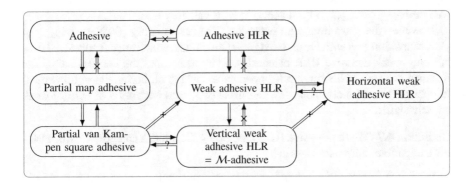

Fig. 4.1 Hierarchy of adhesive categories

the DPO approach. In this section, we compare and relate different relevant notions and build up a hierarchy between them, as shown in Fig. 4.1.

4.2.1 From Adhesive to $\mathcal{M}$-Adhesive Categories

Adhesive categories have been introduced by Lack and Sobocinski [LS04] as a categorical framework for graph transformation systems, which allows to verify the variety of axiomatic HLR properties required for the theory of high-level replacement systems in [EHKP91b]. Adhesive categories are based on the property that pushouts along monomorphisms are van Kampen squares.

Definition 4.5 (Adhesive category). **C** is an *adhesive category* if:

1. **C** has pushouts along monomorphisms.
2. **C** has pullbacks.
3. Pushouts in **C** along monomorphisms are van Kampen squares.   △

Important examples of adhesive categories are the categories **Sets** of sets, **Graphs** of graphs, and **Graphs$_{TG}$** of typed graphs, while the category **AGraphs$_{ATG}$** of typed attributed graphs is not adhesive. But in the latter case, pushouts along $\mathcal{M}$-morphisms are van Kampen squares for the class $\mathcal{M}_{mono-iso}$ of all monomorphisms which are isomorphic on the data type part. In fact, **AGraphs$_{ATG}$** is an adhesive HLR category.

Definition 4.6 (Adhesive HLR category). A category **C** with a PO–PB compatible morphism class $\mathcal{M}$ is called an *adhesive HLR category* if pushouts along $\mathcal{M}$-morphisms are van Kampen squares.   △

It can be shown that the class $\mathcal{M}_{mono}$ of all monomorphisms in an adhesive category **C** fulfills these properties [LS04], such that $(\mathbf{C}, \mathcal{M}_{mono})$ is also an adhesive HLR category, leading to the implication "adhesive implies adhesive HLR" in Fig. 4.1. This implication is proper, because **AGraphs$_{ATG}$** is not adhesive, but $(\mathbf{AGraphs_{ATG}}, \mathcal{M}_{mono-iso})$ is an adhesive HLR category [EEPT06].

However, there are important examples, like the category $(\mathbf{PTNets}, \mathcal{M}_{mono})$ of place/transition nets with the class $\mathcal{M}_{mono}$ of all monos, which are not adhesive HLR, but only weak adhesive HLR categories. This means that the corresponding van Kampen property holds for van Kampen cubes, where all horizontal or all vertical morphisms are $\mathcal{M}$-morphisms. We call these two cases horizontal or vertical weak adhesive HLR.

Definition 4.7 (Weak adhesive HLR category). Consider a category **C** with a PO–PB compatible morphism class $\mathcal{M}$.

- $(\mathbf{C}, \mathcal{M})$ is *horizontal weak adhesive HLR* if pushouts in **C** along $m \in \mathcal{M}$ are *horizontal weak van Kampen squares*, i.e., the van Kampen property holds for commutative cubes with $f, m \in \mathcal{M}$ (see Def. 4.1).

- (**C**, $\mathcal{M}$) is *vertical weak adhesive HLR* if pushouts in **C** along $m \in \mathcal{M}$ are *vertical weak van Kampen squares*, i.e., the van Kampen property holds for commutative cubes with $b, c, d \in \mathcal{M}$ (see Def. 4.1).
- (**C**, $\mathcal{M}$) is a *weak adhesive HLR category* if it is horizontal and vertical weak adhesive HLR. △

Remark 4.8. In the horizontal case, the closure of $\mathcal{M}$ under pushouts and pullbacks implies that all horizontal morphisms are in $\mathcal{M}$. Similarly, in the vertical case it follows that $a \in \mathcal{M}$. △

Using this definition, we have the implications on the right-hand side of Fig. 4.1 between the different variants of "adhesive HLR". The category (**PTNets**, $\mathcal{M}_{mono}$) shows that the implication from "adhesive HLR" to "weak adhesive HLR" is proper.

Recently, we recognised by inspection of the proofs for weak adhesive HLR categories that it is sufficient to consider vertical weak adhesive HLR categories in order to obtain all the important properties. For this reason, we use the short name $\mathcal{M}$-adhesive category for this important class of categories. Note, that this is in contrast to some other work like [BEGG10], where weak adhesive HLR categories are called $\mathcal{M}$-adhesive.

Fact 4.9. *An $\mathcal{M}$-adhesive category is a vertical weak adhesive HLR category.* △

Proof. This follows directly from Defs. 4.4 and 4.7. □

4.2.2 Partial Map and Partial Van Kampen Square Adhesive Categories

Another variant of adhesive categories has been introduced by Heindel [Hei10], called partial map adhesive categories. They are based on the requirement that pushouts along $\mathcal{M}$-morphisms are hereditary [Ken91]. Hereditary pushouts in a category **C** are those pushouts that are preserved by the embedding into the associated category $\mathbf{Par}_{\mathcal{M}}(\mathbf{C})$ of partial maps over **C**. Heindel has shown that hereditary pushouts can be characterised by a variant of van Kampen squares, called partial van Kampen squares, which are closely related to weak van Kampen squares in weak adhesive HLR categories. This leads to the new concept of partial map adhesive categories, which are equivalent to partial van Kampen square adhesive categories.

The concepts in this section are based on an admissible class $\mathcal{M}$ of monomorphisms according to [Hei10], which we call PB compatible in analogy to PO–PB compatibility in Def. 4.2.

Definition 4.10 (PB compatibility). A morphism class $\mathcal{M}$ in a category **C** is called *PB compatible* if

1. $\mathcal{M}$ is a class of monomorphisms, contains all identities, and is closed under composition ($f : A \rightarrow B \in \mathcal{M}, g : B \rightarrow C \in \mathcal{M} \Rightarrow g \circ f \in \mathcal{M}$).

2. $\mathbf{C}$ has pullbacks along M-morphisms, and M-morphisms are closed under pullbacks. △

A partial map category is constructed by the objects of a given category and spans of morphisms within this category.

Definition 4.11 (Partial map category). Given a category $\mathbf{C}$ with a PB-compatible morphism class M, the *partial map category* $\mathbf{Par}_M(\mathbf{C})$ of M-partial maps over $\mathbf{C}$ is defined as follows:

- The objects are all objects in $\mathbf{C}$.
- The morphisms from A to B are (isomorphism classes of) spans $(A \xleftarrow{m} A' \xrightarrow{f} B)$ with $f, m \in \mathbf{C}$, $m \in M$, called M-*partial maps* $(m, f) : A \to B$.
- The identities are identical spans and the composition of spans is defined by pullbacks. △

For any partial map category, we find an inclusion functor from its underlying category. If this functor preserves a pushout, this pushout is called hereditary. This leads to the definition of a partial map adhesive category, where all pushouts along M-morphisms are hereditary.

Definition 4.12 (Hereditary pushout). Given a category $\mathbf{C}$ with a PB-compatible morphism class M, the inclusion functor $I : \mathbf{C} \to \mathbf{Par}_M(\mathbf{C})$, called graphing functor in [Hei10], is defined by the identity on objects, and maps each morphism $f : A \to B$ in $\mathbf{C}$ to the M-partial map $I(f) = (id, f) : A \to B$ in $\mathbf{Par}_M(\mathbf{C})$.

A pushout in $\mathbf{C}$ is called *hereditary* if it is preserved by I. △

Definition 4.13 (Partial map adhesive category). A category $\mathbf{C}$ with a PB-compatible morphism class M is a *partial map adhesive category* if pushouts along M-morphisms exist and are hereditary. △

Remark 4.14. If $\mathbf{C}$ has pushouts along M-morphisms, a sufficient condition for $(\mathbf{C}, M)$ to be partial map adhesive is the existence of a cofree construction leading to a right adjoint functor R for the inclusion functor I. In this case, I is left adjoint and preserves all colimits, especially pushouts along M-morphisms, such that these pushouts are hereditary.

In the case of sets and monomorphisms, the partial map category $\mathbf{Par}_{M_{\mathrm{mono}}}(\mathbf{Sets})$ is isomorphic to the category of sets and partial functions, where $R(X)$ is the extension of a set X by one distinguished ("undefined") element. In this and several other examples, the construction of a right adjoint R is much easier than the explicit verification of the van Kampen property. △

For hereditary pushouts, we find an equivalent property using so-called partial van Kampen squares. These are closely related to vertical weak van Kampen squares in Def. 4.7, but the assumption and conclusion concerning $d \in M$ is different.

Definition 4.15 (Partial van Kampen square). Given a morphism class M, a commutative cube (2) with pushout (1) in the bottom face and where the back faces are pullbacks with $b, c \in M$ fulfills the *partial van Kampen property* if the following statement holds: top face is a pushout if and only if the front faces are pullbacks with $d \in M$.

A pushout (1) with $m \in M$ is a *partial van Kampen square* if the partial van Kampen property holds for all commutative cubes (2) with (1) in the bottom face.

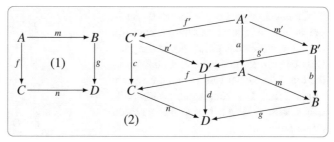

△

Definition 4.16 (Partial van Kampen square adhesive category). A category $\mathbf{C}$ with a PB-compatible morphism class M is a *partial van Kampen square adhesive category* if pushouts along M-morphisms exist and are partial van Kampen squares.

△

The following theorem shows that partial van Kampen square and partial map adhesive categories are in fact equivalent.

Theorem 4.17 (Equivalence of partial map and partial van Kampen square adhesive categories). *Given a category $\mathbf{C}$ with a PB compatible morphism class M such that $\mathbf{C}$ has pushouts along M-morphisms. Then pushouts along M-morphisms are hereditary if and only if they are partial van Kampen squares.* △

Proof. The proof is sketched in [Hei10] based on results in [Hei09].   □

Remark 4.18. As indicated in [Hei10], the statement and proof remains valid if "pushouts along M-morphisms" is replaced by arbitrary "pushouts".   △

In [Hei10], it is shown that adhesive categories are also partial map adhesive for the morphism class M_{mono}. Moreover, an example of a category **lSet** of list sets is given, which is partial map adhesive, but not adhesive. Together with Theorem 4.17 this leads to the implication and equivalence on the left-hand side of Fig. 4.1.

In the following, we will analyse the relationship between partial map adhesive categories and the different adhesive HLR notions. As a first step, it is shown in [Hei10] that PB compatibility already implies PO–PB compatibility in partial van Kampen square adhesive categories.

Theorem 4.19 (Equivalence of PB and PO–PB compatibility). *Given a partial van Kampen square adhesive category $(\mathbf{C}, M)$, we have that PB compatibility is equivalent to PO–PB compatibility.* △

Proof. It suffices to show that M is closed under pushouts in partial van Kampen square adhesive categories.

Consider a pushout with $m \in \mathcal{M}$ in the bottom of the commutative cube on the right, which has pullbacks in the back squares with $id, f \in \mathcal{M}$ and a pushout in the top. The partial van Kampen property implies that the front squares are pullbacks with $n \in \mathcal{M}$; hence $\mathcal{M}$ is closed under pushouts. $\qquad\square$

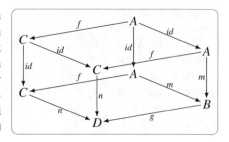

Using this result, we can show that partial van Kampen square adhesive categories are also $\mathcal{M}$-adhesive.

Theorem 4.20 (Partial van Kampen square adhesive categories are $\mathcal{M}$-adhesive). *Given a partial van Kampen square adhesive category* $(\mathbf{C}, \mathcal{M})$, $(\mathbf{C}, \mathcal{M})$ *is an $\mathcal{M}$-adhesive category.* $\qquad\qquad\qquad\qquad\triangle$

Proof. Given a partial van Kampen square adhesive category $(\mathbf{C}, \mathcal{M})$, $\mathcal{M}$ is already PO–PB compatible by Theorem 4.19. Moreover, pushouts along $\mathcal{M}$-morphisms satisfy the vertical weak van Kampen square property, because we only have to consider commutative cubes with $a, b, c, d \in \mathcal{M}$, in contrast to partial van Kampen squares, where the equivalence holds under the assumption that $a, b, c \in \mathcal{M}$. Hence, $(\mathbf{C}, \mathcal{M})$ is a vertical weak adhesive HLR and therefore an $\mathcal{M}$-adhesive category by definition. $\qquad\square$

By Theorem 4.20, we have the implication from "partial van Kampen square adhesive" to "$\mathcal{M}$-adhesive" in Fig. 4.1. Up to now it is open whether also the reverse direction holds.

In [Hei10] it is shown that the category **lSet** of list sets is partial map adhesive— and hence partial van Kampen square adhesive—but violates the property that pushouts over $\mathcal{M}$-morphisms are van Kampen squares. Therefore this category is not horizontal weak adhesive HLR, and also not weak adhesive HLR. This implies that there is no implication from "partial van Kampen square adhesive" to "weak adhesive HLR" in Fig. 4.1. It follows that there are no implications from "$\mathcal{M}$-adhesive" to "weak adhesive HLR" and to "horizontal weak adhesive HLR".

4.3 Results and Additional HLR Properties for $\mathcal{M}$-Adhesive Categories

In this section, we collect various results that hold in $\mathcal{M}$-adhesive categories and follow from the $\mathcal{M}$-van Kampen property, as well as additional HLR properties that are needed and have to be required. These results and properties are used to show the main theorems of graph transformation in [EEPT06] as well as various results in the following chapters.

4.3.1 Basic HLR Properties

In [EHKP91b], the following HLR properties were required for HLR categories. In the following, we call them basic HLR properties to distinguish them from the additional ones introduced later. All basic HLR properties are valid in $\mathcal{M}$-adhesive categories and can be proven using the $\mathcal{M}$-van Kampen property.

Definition 4.21 (Basic HLR properties). The following properties are called *basic HLR properties*:

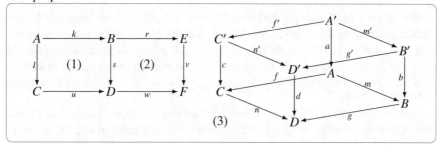

1. *Pushouts along $\mathcal{M}$-morphisms are pullbacks.* Given the above pushout (1) with $k \in \mathcal{M}$, then (1) is also a pullback.
2. *$\mathcal{M}$-pushout–pullback decomposition.* Given the above commutative diagram— where (1)+(2) is a pushout, (2) is a pullback, $w \in \mathcal{M}$, and ($l \in \mathcal{M}$ or $k \in \mathcal{M}$)—(1) and (2) are pushouts and also pullbacks.
3. *Cube pushout–pullback property.* Given the above commutative cube (3)—where all morphisms in the top and bottom faces are $\mathcal{M}$-morphisms, the top face is a pullback, and the front faces are pushouts—the following statement holds: the bottom face is a pullback if and only if the back faces of the cube are pushouts:
4. *Uniqueness of pushout complements.* Given $k : A \rightarrow B \in \mathcal{M}$ and $s : B \rightarrow D$, there is, up to isomorphism, at most one C with $l : A \rightarrow C$ and $u : C \rightarrow D$ such that (1) is a pushout.   △

All these HLR properties are valid in $\mathcal{M}$-adhesive categories.

Theorem 4.22 (HLR properties in $\mathcal{M}$-adhesive categories). *Given an $\mathcal{M}$-adhesive category $(\mathbf{C}, \mathcal{M})$, the following HLR properties are valid:*

1. *Pushouts along $\mathcal{M}$-morphisms are pullbacks,*
2. *$\mathcal{M}$-pushout–pullback decomposition,*
3. *Cube pushout–pullback property,*
4. *Uniqueness of pushout complements.*   △

Proof. See [EEPT06], where these properties are shown for weak adhesive HLR categories, but the proofs only use the vertical van Kampen property, i.e., are also valid in $\mathcal{M}$-adhesive categories.   □

4.3.2 Additional HLR Properties

The following additional HLR properties are essential to prove the main results for graph transformation and HLR systems in [EEPT06]. For the Parallelism Theorem (see Theorems 2.26 and 5.26), binary coproducts compatible with $\mathcal{M}$ are required in order to construct parallel rules. Initial pushouts are used in order to define the gluing condition and to show that consistency in the Embedding Theorem (see Theorems 2.37 and 5.34) is not only sufficient, but also necessary. In connection with the Concurrency Theorem (see Theorems 2.29 and 5.30) and for completeness of critical pairs (see Theorems 2.41 and 5.41), an $\mathcal{E}'$–$\mathcal{M}'$ pair factorisation is used such that the class $\mathcal{M}'$ satisfies the $\mathcal{M}$–$\mathcal{M}'$ pushout–pullback decomposition property. Moreover, a standard construction for $\mathcal{E}'$–$\mathcal{M}'$ pair factorisation uses an $\mathcal{E}$–$\mathcal{M}'$ factorisation of morphisms in **C**, where $\mathcal{E}'$ is constructed from $\mathcal{E}$ using binary coproducts. For the Amalgamation Theorem (see Theorems 2.34 and 6.17), we need effective pushouts.

As far as we know, these additional HLR properties cannot be concluded from the axioms of $\mathcal{M}$-adhesive categories; at least we do not know proofs for nontrivial classes $\mathcal{E}$, $\mathcal{E}'$, $\mathcal{M}$, and $\mathcal{M}'$.

Note that for $\mathcal{M}' = \mathcal{M}$, the $\mathcal{M}$–$\mathcal{M}'$ pushout–pullback decomposition property is the $\mathcal{M}$ pushout–pullback decomposition property, which is already valid in general $\mathcal{M}$-adhesive categories.

Definition 4.23 (Additional HLR properties). An $\mathcal{M}$-adhesive category $(\mathbf{C}, \mathcal{M})$ fulfills the *additional HLR properties* if all of the following items hold:

1. *Binary coproducts:* **C** has binary coproducts.
2. *$\mathcal{E}$–$\mathcal{M}$ factorisation:* Given a morphism class $\mathcal{E}$, for each $f : A \to B$ there is a factorisation over $e : A \to K \in \mathcal{E}$ and $m : K \to B \in \mathcal{M}$ such that $m \circ e = f$, and this factorisation is unique up to isomorphism.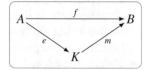
3. *$\mathcal{E}'$–$\mathcal{M}'$ pair factorisation:* Given a morphism class $\mathcal{M}'$ and a class of morphism pairs with common codomain $\mathcal{E}'$, for each pair of morphisms $f_1 : A_1 \to B$, $f_2 : A_2 \to B$ there is a factorisation over $e_1 : A_1 \to K$, $e_2 : A_2 \to K$, $m : K \to B$ with $(e_1, e_2) \in \mathcal{E}'$ and $m \in \mathcal{M}'$ such that $m \circ e_1 = f_1$ and $m \circ e_2 = f_2$.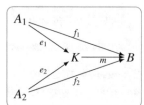
4. *Initial pushouts over $\mathcal{M}'$:* Given a morphism class $\mathcal{M}'$, for each $f : A \to D \in \mathcal{M}'$ there exists an initial pushout (1) with $b, c \in \mathcal{M}$. (1) is an initial pushout if the following condition holds: for all pushouts (2) with 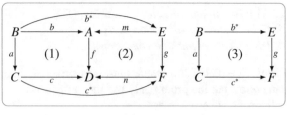 $m, n \in \mathcal{M}$ there exist unique morphisms $b^*, c^* \in \mathcal{M}$ such that $m \circ b^* = b$, $n \circ c^* = c$, and (3) is a pushout.

5. *Effective pushouts:* Given a pullback (4) and a pushout (5) with all morphisms being $\mathcal{M}$-morphisms, also the induced morphism $e : D \rightarrow D'$ is an $\mathcal{M}$-morphism.   △

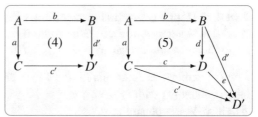

Remark 4.24. In the setting of effective pushouts, the morphism e has to be a monomorphism [LS05a]. But up to now we were not able to show that it is actually an $\mathcal{M}$-morphism if the class $\mathcal{M}$ does not contain all monomorphisms.   △

4.3.3 Finite Coproducts and $\mathcal{E}'-\mathcal{M}'$ Pair Factorisation

For the construction of coproducts, it often makes sense to use pushouts over $\mathcal{M}$-initial objects in the following sense.

Definition 4.25 ($\mathcal{M}$-initial object). An initial object I in $(\mathbf{C}, \mathcal{M})$ is called $\mathcal{M}$-*initial* if for each object $A \in \mathbf{C}$ the unique morphism $i_A : I \rightarrow A$ is in $\mathcal{M}$.   △

Note that if $(\mathbf{C}, \mathcal{M})$ has one $\mathcal{M}$-initial object then all initial objects are $\mathcal{M}$-initial due to $\mathcal{M}$ being closed under isomorphisms and composition.

In the $\mathcal{M}$-adhesive categories $(\mathbf{Sets}, \mathcal{M})$, $(\mathbf{Graphs}, \mathcal{M})$, $(\mathbf{Graphs_{TG}}, \mathcal{M})$, $(\mathbf{ElemNets}, \mathcal{M})$, and $(\mathbf{PTNets}, \mathcal{M})$ we have $\mathcal{M}$-initial objects defined by the empty set, empty graphs, and empty nets, respectively. But in $(\mathbf{AGraphs_{ATG}}, \mathcal{M})$, there is no $\mathcal{M}$-initial object. The initial attributed graph $(\varnothing, T_{DSIG})$ with term algebra T_{DSIG} of the data type signature $DSIG$ is not $\mathcal{M}$-initial because the data type part of the unique morphism $(\varnothing, T_{DSIG}) \rightarrow (G, D)$ is, in general, not an isomorphism.

The existence of an $\mathcal{M}$-initial object implies that we have finite coproducts.

Fact 4.26 (Existence of finite coproducts). *For each $\mathcal{M}$-adhesive category $(\mathbf{C}, \mathcal{M})$ with $\mathcal{M}$-initial object, $(\mathbf{C}, \mathcal{M})$ has finite coproducts, where the injections into coproducts are in $\mathcal{M}$.*   △

Proof. It suffices to show this for the binary case. The coproduct $A + B$ of A and B can be constructed by the pushout (1), which exists because of $i_A, i_B \in \mathcal{M}$. This also implies $in_A, in_B \in \mathcal{M}$, since $\mathcal{M}$-morphisms are closed under pushouts in $\mathcal{M}$-adhesive categories.   □

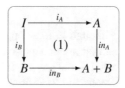

Note that an $\mathcal{M}$-adhesive category may still have coproducts even if it does not have an $\mathcal{M}$-initial object. For example, the $\mathcal{M}$-adhesive category $(\mathbf{AGraphs_{ATG}}, \mathcal{M})$ has no $\mathcal{M}$-initial object, but finite coproducts, as shown in [EEPT06].

For the construction of parallel rules in [EEPT06] the compatibility of the morphism class $\mathcal{M}$ with (finite) coproducts was required. In fact, finite coproducts (if they exist) are always compatible with $\mathcal{M}$ in $\mathcal{M}$-adhesive categories, as shown in [EHL10].

Fact 4.27 (Finite coproducts compatible with $\mathcal{M}$). *For each $\mathcal{M}$-adhesive category* $(\mathbf{C}, \mathcal{M})$ *with finite coproducts, finite coproducts are compatible with $\mathcal{M}$, i.e., $f_i \in \mathcal{M}$ for $i = 1, \ldots, n$ implies that $f_1 + \cdots + f_n \in \mathcal{M}$.* △

Proof. It suffices to show this for the binary case $n = 2$. For $f : A \rightarrow B \in \mathcal{M}$, we have pushout (1) with $(f + id_C) \in \mathcal{M}$, since $\mathcal{M}$-morphisms are closed under pushouts. Similarly, we have $(id_B + g) \in \mathcal{M}$ in pushout (2) for $g : C \rightarrow D$. Hence, $(f + g) = (id_B + g) \circ (f + id_C) \in \mathcal{M}$ by composition of $\mathcal{M}$-morphisms.

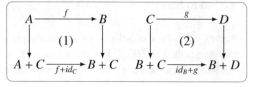

☐

Based on an $\mathcal{E}$–$\mathcal{M}'$ factorisation and binary coproducts, we obtain a standard construction for an $\mathcal{E}'$–$\mathcal{M}'$ pair factorisation with $\mathcal{E}'$ induced by $\mathcal{E}$.

Fact 4.28 (Construction of $\mathcal{E}'$–$\mathcal{M}'$ pair factorisation). *Given a category $\mathbf{C}$ with an $\mathcal{E}$–$\mathcal{M}'$ factorisation and binary coproducts, $\mathbf{C}$ has also an $\mathcal{E}'$–$\mathcal{M}'$ pair factorisation for the class $\mathcal{E}' = \{(e_A : A \rightarrow C, e_B : B \rightarrow C) \mid e_A, e_B \in \mathbf{C} \text{ with induced } e : A + B \rightarrow C \in \mathcal{E}\}$.* △

Proof. Given $f_A : A \rightarrow D$ and $f_B : B \rightarrow D$ with induced $f : A + B \rightarrow D$, we consider the $\mathcal{E}$–$\mathcal{M}'$ factorisation $f = m \circ e$ of f with $e \in \mathcal{E}$ and $m \in \mathcal{M}'$, and define $e_A = e \circ in_A$ and $e_B = e \circ in_B$.

Then $(e_A, e_B) \in \mathcal{E}'$ and $m \in \mathcal{M}'$ defines an $\mathcal{E}'$–$\mathcal{M}'$ pair factorisation of (f_A, f_B) which is unique up to isomorphism, since each other $\mathcal{E}'$–$\mathcal{M}'$ pair factorisation also leads to an $\mathcal{E}$–$\mathcal{M}'$ factorisation via the induced morphism in $\mathcal{E}$, and $\mathcal{E}$–$\mathcal{M}'$ factorisations are unique up to isomorphism. ☐

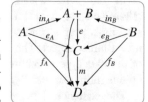

4.4 Finitary $\mathcal{M}$-Adhesive Categories

Although in most application areas of graph and model transformations only finite models are considered, the theory has been developed for general graphs, including also infinite graphs. It is implicitly assumed that the results can be restricted to finite graphs and to attributed graphs with a finite graph part, where the data algebra may be infinite. Obviously, not only **Sets** and **Graphs** are $\mathcal{M}$-adhesive categories, but also the full subcategories **Sets**$_{\text{fin}}$ of finite sets and **Graphs**$_{\text{fin}}$ of finite graphs. In this section, we consider the general restriction of an $\mathcal{M}$-adhesive category $(\mathbf{C}, \mathcal{M})$ to finite objects, leading to a category $(\mathbf{C}_{\text{fin}}, \mathcal{M}_{\text{fin}})$, where $\mathcal{M}_{\text{fin}}$ is the restriction of $\mathcal{M}$ to morphisms between finite objects. This section is based on [GBEG14].

4.4.1 Basic Notions of Finitary *M*-Adhesive Categories

Intuitively, we are interested in those objects where the graph or net part is finite. An object A is called finite if A has (up to isomorphism) only a finite number of *M*-subobjects, i.e., only finite many *M*-morphisms $m : A' \to A$ up to isomorphism.

Definition 4.29 (*M*-subobject and finite object). Given an object A in an *M*-adhesive category $(\mathbf{C}, M)$, an *M*-*subobject* of A is an iso-morphism class of morphisms $m : A' \to A \in M$, where $m_1 : A'_1 \to A$ and $m_2 : A'_2 \to A$ belong to the same *M*-subobject of A if there is an isomorphism $i : A'_1 \overset{\sim}{\to} A'_2$ with $m_1 = m_2 \circ i$. By $\mathrm{MSub}(A)$, we denote the set of all *M*-subobjects of A.

A is called *finite* if it has finitely many *M*-subobjects.   △

Finitary *M*-adhesive categories are *M*-adhesive categories with finite objects only. Note that the notion "finitary" depends on the class *M* of monomorphisms and "**C** is finitary" must not be confused with "**C** is finite" in the sense of having a finite number of objects and morphisms.

Definition 4.30 (Finitary *M*-adhesive category). An *M*-adhesive category $(\mathbf{C}, M)$ is called *finitary* if each object $A \in \mathbf{C}$ is finite.   △

In $(\mathbf{Sets}, M_{mono})$, the finite objects are the finite sets. Graphs in $(\mathbf{Graphs}, M_{mono})$ and $(\mathbf{Graphs_{TG}}, M_{mono})$ are finite if the node and edge sets have finite cardinality, while TG itself may be infinite. Petri nets in $(\mathbf{ElemNets}, M_{mono})$ and $(\mathbf{PTNets}, M_{mono})$ are finite if the number of places and transitions is finite. A typed attributed graph $G^T = ((G^E, D_G), type)$ in $(\mathbf{AGraphs_{ATG}}, M_{mono-iso})$ is finite if the graph part of G^T, i.e., all vertex and edge sets except the set V_D of data vertices generated from D_G, is finite, while the attributed type graph ATG or the data type part D_G may be infinite, because *M*-morphisms are isomorphisms on the data type part.

Finite *M*-intersections are a generalisation of pullbacks to an arbitrary but finite number of *M*-subobjects and, thus, a special case of limits.

Definition 4.31 (Finite *M*-intersection). Given an *M*-adhesive category $(\mathbf{C}, M)$ and morphisms $m_i : A_i \to B \in M$ with the same codomain object B and $i \in I$ for a finite set I, a *finite M-intersection* of $(m_i)_{i \in I}$ is an object A with morphisms $n_i : A \to A_i$, such that $m_i \circ n_i = m_j \circ n_j$ for all $i, j \in I$ and for each other object A' with morphisms $(n'_i : A' \to A_i)_{i \in I}$ with $m_i \circ n'_i = m_j \circ n'_j$ for $i, j \in I$ there is a unique morphism $a : A' \to A$ with $n_i \circ a = n'_i$ for all $i \in I$.

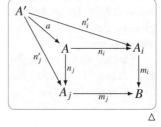

△

Remark 4.32. Note that finite *M*-intersections can be constructed by iterated pullbacks. Hence, they always exist in *M*-adhesive categories. Moreover, since pullbacks preserve *M*-morphisms, the morphisms n_i are also in *M*.   △

4.4.2 Additional HLR Properties in Finitary M-Adhesive Categories

In the case of a finitary M-adhesive category $(\mathbf{C}, M)$, we are able to show that the additional HLR properties from Def. 4.23 are valid for suitable classes $\mathcal{E}$ and $\mathcal{E}'$, and $M' = M$.

4.4.2.1 Binary Coproducts

For finitary M-adhesive categories with an M-initial object, we directly obtain binary coproducts.

Fact 4.33 (Binary coproducts). *Given a finitary M-adhesive category $(\mathbf{C}, M)$ with M-initial object, $\mathbf{C}$ has binary coproducts.* △

Proof. This follows directly from Fact 4.26.   □

4.4.2.2 $\mathcal{E}$–M Factorisation

The reason for the existence of an $\mathcal{E}$–M factorisation of morphisms in finitary M-adhesive categories is the fact that we only need finite intersections of M-subobjects, and no infinite intersections as required for general M-adhesive categories. Moreover, we fix the choice of the class $\mathcal{E}$ to extremal morphisms w. r. t. M.

Definition 4.34 (Extremal $\mathcal{E}$–M factorisation). Given an M-adhesive category $(\mathbf{C}, M)$, the class $\mathcal{E}$ of all *extremal morphisms* w. r. t. M is defined by $\mathcal{E} := \{e \in \mathbf{C} \mid \forall\, m, g \in \mathbf{C}, m \circ g = e : m \in M \Rightarrow m$ isomorphism$\}$.

For a morphism $f : A \to B$ in $\mathbf{C}$, an *extremal $\mathcal{E}$–M factorisation* of f is given by an object $\overline{B}$ and morphisms $e : A \to \overline{B} \in \mathcal{E}$ and $m : \overline{B} \to B \in M$ such that $m \circ e = f$.   △

Remark 4.35. In several example categories, the class $\mathcal{E}$ consists of all epimorphisms. But this is not necessarily the case for extremal morphisms w. r. t. M, as shown below.

If we require M to be the class of all monomorphisms and consider only epimorphisms e, g in the definition of $\mathcal{E}$, then $\mathcal{E}$ is the class of all extremal epimorphisms in the sense of [AHS90].   △

Fact 4.36 (Existence of extremal $\mathcal{E}$–M factorisation). *Given a finitary M-adhesive category $(\mathbf{C}, M)$, we can construct an extremal $\mathcal{E}$–M factorisation for each morphism in $\mathbf{C}$.*   △

Construction 4.37. For $f : A \rightarrow B$ consider all M-subobjects $m_i : B_i \rightarrow B$ such that there exists $e_i : A \rightarrow B_i$ with $f = m_i \circ e_i$, leading to a suitable finite index set I. Now $m: \overline{B} \rightarrow B$ is constructed as an M-intersection of $(m_i)_{i \in I}$ and $e : A \rightarrow \overline{B}$ is the induced unique morphism with $\overline{m_i} \circ e = e_i$ for all $i \in I$.   △

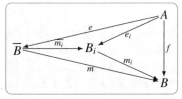

Proof. See Appendix B.2.1.   □

Fact 4.38 (Uniqueness of extremal $\mathcal{E}$–M factorisation). *In an M-adhesive category* (**C**, M), *extremal $\mathcal{E}$–M factorisations are unique up to isomorphism.*   △

Proof. See Appendix B.2.2.   □

In the categories (**Sets**, M), (**Graphs**, M), (**Graphs$_{TG}$**, M), (**ElemNets**, M), and (**PTNets**, M) with the classes M of all monomorphisms, the extremal $\mathcal{E}$–M factorisation is exactly the well-known epi–mono factorisation of morphisms which also works for infinite objects, because these categories have not only finite but also general intersections.

For (**AGraphs$_{ATG}$**, M), the extremal $\mathcal{E}$–M factorisation of $(f_G, f_D) : (G, D) \rightarrow (G', D')$ with finite (or infinite) G and G' is given by $(f_G, f_D) = (m_G, m_D) \circ (e_G, e_D)$, where e_G is an epimorphism, m_G a monomorphism, and m_D an isomorphism. In general, e_D, and hence also (e_G, e_D), is not an epimorphism, since e_D has to be essentially the same as f_D, because m_D is an isomorphism. This means that the class $\mathcal{E}$, which depends on M, is not necessarily a class of epimorphisms.

4.4.2.3 $\mathcal{E}'$–M Pair Factorisation

For finitary M-adhesive categories, we consider the special case $M' = M$ and use the extremal $\mathcal{E}$–M factorisation to construct an $\mathcal{E}'$–M pair factorisation in a standard way.

Fact 4.39 (Existence and uniqueness of $\mathcal{E}'$–M pair factorisation). *Given a finitary M-adhesive category* (**C**, M) *with an M-initial object (or finite coproducts), we can construct a unique $\mathcal{E}'$–M pair factorisation for each pair of morphisms in* **C** *with the same codomain, where $\mathcal{E}' = \{(e_A \colon A \rightarrow C, e_B \colon B \rightarrow C) \mid e_A, e_B \in$ **C** *with induced $e \colon A + B \rightarrow C \in \mathcal{E}\}$.*   △

Proof. This follows directly from Fact 4.28.   □

4.4.2.4 Initial Pushouts

As with the extremal $\mathcal{E}$–M factorisation, we are able to construct initial pushouts in finitary M-adhesive categories by finite M-intersections of M-subobjects.

Fact 4.40 (Initial pushouts). *A finitary $\mathcal{M}$-adhesive category has initial pushouts.*

$\triangle$

Construction 4.41. Given $m : L \rightarrow G$, we consider all those $\mathcal{M}$-subobjects $b_i : B_i \rightarrow L$ of L and $c_i : C_i \rightarrow G$ of G such that there is a pushout (P_i) over m. Since L and G are finite this leads to a finite index set I for all (P_i)
with $i \in I$. Now construct
$b : B \rightarrow L$ as the finite $\mathcal{M}$-
intersection of $(b_i)_{i \in I}$ and
$c : C \rightarrow G$ as the finite
$\mathcal{M}$-intersection of $(c_i)_{i \in I}$.

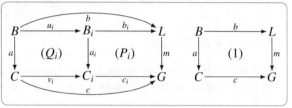

Then there is a unique $a : B \rightarrow C$ such that (Q_i) commutes for all $i \in I$ and the outer diagram (1) is the initial pushout over m.

$\triangle$

Proof. See Appendix B.2.3.

$\square$

4.4.2.5 Additional HLR Properties

The following theorem summarises that, except for effective pushouts, the additional HLR properties from Def. 4.23 are valid for all finitary $\mathcal{M}$-adhesive categories.

Theorem 4.42 (Additional HLR properties in finitary $\mathcal{M}$-adhesive categories).
Given a finitary $\mathcal{M}$-adhesive category $(\mathbf{C}, \mathcal{M})$, the following additional HLR properties hold:

1. $(\mathbf{C}, \mathcal{M})$ *has initial pushouts.*
2. $(\mathbf{C}, \mathcal{M})$ *has a unique extremal $\mathcal{E}$–$\mathcal{M}$ factorisation, where $\mathcal{E}$ is the class of all extremal morphisms w. r. t. $\mathcal{M}$.*

If $(\mathbf{C}, \mathcal{M})$ has an $\mathcal{M}$-initial object, we also have that:

3. $(\mathbf{C}, \mathcal{M})$ *has finite coproducts.*
4. $(\mathbf{C}, \mathcal{M})$ *has a unique $\mathcal{E}'$–$\mathcal{M}'$ pair factorisation for the classes $\mathcal{M}' = \mathcal{M}$ and $\mathcal{E}'$ induced by $\mathcal{E}$.*

$\triangle$

Proof. Item 1 follows from Fact 4.40, Item 2 follows from Facts 4.36 and 4.38, Item 3 follows from Fact 4.33, and Item 4 follows from Fact 4.39.

$\square$

For a concrete finitary $\mathcal{M}$-adhesive category, we still need to show that it has effective pushouts to ensure all additional HLR properties from Def. 4.23.

4.4.3 Finitary Restriction of $\mathcal{M}$-Adhesive Categories

In this subsection, we show that for any $\mathcal{M}$-adhesive category $(\mathbf{C}, \mathcal{M})$ the restriction $(\mathbf{C}_{\text{fin}}, \mathcal{M}_{\text{fin}})$ to finite objects is a finitary $\mathcal{M}$-adhesive category, where $\mathcal{M}_{\text{fin}}$ is the corresponding restriction of $\mathcal{M}$. In this case, the inclusion functor $I : \mathbf{C}_{\text{fin}} \to \mathbf{C}$ preserves $\mathcal{M}$-morphisms, such that finite objects in $\mathbf{C}_{\text{fin}}$ w. r. t. $\mathcal{M}_{\text{fin}}$ are exactly the finite objects in $\mathbf{C}$ w. r. t. $\mathcal{M}$.

Definition 4.43 (Finitary restriction of $\mathcal{M}$-adhesive category). Given an $\mathcal{M}$-adhesive category $(\mathbf{C}, \mathcal{M})$ the restriction to all finite objects of $(\mathbf{C}, \mathcal{M})$ defines the full subcategory $\mathbf{C}_{\text{fin}}$ of $\mathbf{C}$, and $(\mathbf{C}_{\text{fin}}, \mathcal{M}_{\text{fin}})$ with $\mathcal{M}_{\text{fin}} = \mathcal{M} \cap \mathbf{C}_{\text{fin}}$ is called *finitary restriction* of $(\mathbf{C}, \mathcal{M})$. △

Remark 4.44. Note that an object A in $\mathbf{C}$ is finite in $(\mathbf{C}, \mathcal{M})$ if and only if A is finite in $(\mathbf{C}_{\text{fin}}, \mathcal{M}_{\text{fin}})$. Even if $\mathcal{M}$ is the class of all monomorphisms in $\mathbf{C}$, $\mathcal{M}_{\text{fin}}$ is not necessarily the class of all monomorphisms in $\mathbf{C}_{\text{fin}}$. △

In order to prove that $(\mathbf{C}_{\text{fin}}, \mathcal{M}_{\text{fin}})$ is an $\mathcal{M}$-adhesive category, we show that the inclusion functor $I_{\text{fin}} : \mathbf{C}_{\text{fin}} \to \mathbf{C}$ creates and preserves pushouts and pullbacks along $\mathcal{M}_{\text{fin}}$ and $\mathcal{M}$, respectively.

Definition 4.45 (Creation and preservation of pushouts and pullbacks). Given an $\mathcal{M}$-adhesive category $(\mathbf{C}, \mathcal{M})$ and a full subcategory $\mathbf{C}'$ of $\mathbf{C}$ with $\mathcal{M}' = \mathcal{M} \cap \mathbf{C}'$, an inclusion functor $I : \mathbf{C}' \to \mathbf{C}$ *creates pushouts along* $\mathcal{M}$ if for each pushout (1) in $\mathbf{C}$ with $f, h \in \mathbf{C}'$ and $f \in \mathcal{M}'$ we have that $D \in \mathbf{C}'$ such that (1) is a pushout in $\mathbf{C}'$.

I *creates pullbacks along* $\mathcal{M}$ if for each pullback (1) in $\mathbf{C}$ with $g, k \in \mathbf{C}'$ and $g \in \mathcal{M}'$ we have that $A \in \mathbf{C}'$ such that (1) is a pullback in $\mathbf{C}'$.

I *preserves pushouts (pullbacks) along* $\mathcal{M}'$ if each pushout (pullback) (1) in $\mathbf{C}'$ with $f \in \mathcal{M}$ ($g \in \mathcal{M}'$) is also a pushout (pullback) in $\mathbf{C}$ with $f \in \mathcal{M}$ ($g \in \mathcal{M}$).

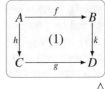

△

Fact 4.46 (Creation and preservation of pushouts and pullbacks). *Given an $\mathcal{M}$-adhesive category $(\mathbf{C}, \mathcal{M})$, the inclusion functor $I_{\text{fin}} : \mathbf{C}_{\text{fin}} \to \mathbf{C}$ creates pushouts and pullbacks along $\mathcal{M}$ and preserves pushouts and pullbacks along $\mathcal{M}_{\text{fin}}$.* △

Proof. 1. I_{fin} creates pushouts along $\mathcal{M}$. Given pushout (1) in $\mathbf{C}$ with $A, B, C \in \mathbf{C}_{\text{fin}}$ and $f \in \mathcal{M}$, also $g \in \mathcal{M}$. It remains to show that $D \in \mathbf{C}_{\text{fin}}$.

For any subobject $d : D' \to D \in \mathcal{M}$ we obtain subobjects $b : B' \to B \in \mathcal{M}$ and $c : C' \to C \in \mathcal{M}$ by pullback constructions in the front faces of the cube. Now we construct the back faces as pushouts with subobject $a : A' \to A \in \mathcal{M}$, and the $\mathcal{M}$-van Kampen property implies that the top face is a pushout.

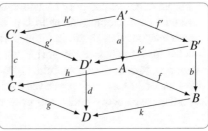

Consider the $\mathcal{M}$-subobject function $\Phi : \text{MSub}(D) \to \text{MSub}(B) \times \text{MSub}(C)$ defined by $\Phi([d]) = ([b], [c])$ in the construction above. Φ is injective since pushouts are unique up to isomorphism. $\text{MSub}(B)$ and $\text{MSub}(C)$ are finite, hence also $\text{MSub}(B) \times \text{MSub}(C)$ is finite, and injectivity of Φ implies that also $\text{MSub}(D)$ is finite. This means that $D \in \mathbf{C}_{\text{fin}}$.

2. I_{fin} creates pullbacks along $\mathcal{M}$. Given pullback (1) in $\mathbf{C}$ with $B, C, D \in \mathbf{C}_{\text{fin}}$ and $g \in \mathcal{M}$, also $f \in \mathcal{M}$. Moreover, each $\mathcal{M}$-subobject of A is also an $\mathcal{M}$-subobject of B, because $f \in \mathcal{M}$. Hence $B \in \mathbf{C}_{\text{fin}}$ implies that $A \in \mathbf{C}_{\text{fin}}$ and (1) is also a pullback in $\mathbf{C}_{\text{fin}}$ with $f \in \mathcal{M}_{\text{fin}}$.
3. I_{fin} preserves pushouts along $\mathcal{M}_{\text{fin}}$. Given pushout (1) in $\mathbf{C}_{\text{fin}}$ with $f \in \mathcal{M}_{\text{fin}}$, also $f \in \mathcal{M}$. Since I_{fin} creates pushouts along $\mathcal{M}$ by Item 1, the pushout (1′) of $f \in \mathcal{M}$ and h in $\mathbf{C}$ is also a pushout in $\mathbf{C}_{\text{fin}}$. By uniqueness of pushouts this means that (1) and (1′) are isomorphic and hence (1) is also a pushout in $\mathbf{C}$.
4. Similarly, we can show that I_{fin} preserves pullbacks along $\mathcal{M}_{\text{fin}}$ using the fact that I_{fin} creates pullbacks along $\mathcal{M}$, as shown in Item 2.   □

Using this result we are able to show that the finitary restriction of an $\mathcal{M}$-adhesive category leads to a finitary $\mathcal{M}$-adhesive category.

Theorem 4.47 (Finitary restriction). *The finitary restriction* $(\mathbf{C}_{\text{fin}}, \mathcal{M}_{\text{fin}})$ *of an $\mathcal{M}$-adhesive category* $(\mathbf{C}, \mathcal{M})$ *is a finitary $\mathcal{M}$-adhesive category.*   △

Proof. An object A in $\mathbf{C}$ is finite in $(\mathbf{C}, \mathcal{M})$ if and only if it is finite in $(\mathbf{C}_{\text{fin}}, \mathcal{M}_{\text{fin}})$. Hence, all objects in $(\mathbf{C}_{\text{fin}}, \mathcal{M}_{\text{fin}})$ are finite. Moreover, $\mathcal{M}_{\text{fin}}$ is a class of monomorphisms, contains all identities and is closed under composition, because this is valid for $\mathcal{M}$. $(\mathbf{C}_{\text{fin}}, \mathcal{M}_{\text{fin}})$ has pushouts along $\mathcal{M}_{\text{fin}}$ because $(\mathbf{C}, \mathcal{M})$ has pushouts along $\mathcal{M}$ and I_{fin} creates pushouts along $\mathcal{M}$ by Fact 4.46. This also implies that $\mathcal{M}_{\text{fin}}$ is preserved by pushouts along $\mathcal{M}_{\text{fin}}$ in $\mathbf{C}_{\text{fin}}$. Similarly, $(\mathbf{C}_{\text{fin}}, \mathcal{M}_{\text{fin}})$ has pullbacks along $\mathcal{M}_{\text{fin}}$ and $\mathcal{M}_{\text{fin}}$ is preserved by pullbacks along $\mathcal{M}_{\text{fin}}$ in $\mathbf{C}_{\text{fin}}$. Finally, the $\mathcal{M}$-van Kampen property of $(\mathbf{C}, \mathcal{M})$ implies that of $(\mathbf{C}_{\text{fin}}, \mathcal{M}_{\text{fin}})$ using the fact that I_{fin} preserves pushouts and pullbacks along $\mathcal{M}_{\text{fin}}$ and creates pushouts and pullbacks along $\mathcal{M}$.   □

A direct consequence of Theorem 4.47 is the fact that finitary restrictions of $(\mathbf{Sets}, \mathcal{M})$, $(\mathbf{Graphs}, \mathcal{M})$, $(\mathbf{Graphs}_{\text{TG}}, \mathcal{M})$, $(\mathbf{ElemNets}, \mathcal{M})$, $(\mathbf{PTNets}, \mathcal{M})$, and $(\mathbf{AGraphs}_{\text{ATG}}, \mathcal{M})$ are all finitary $\mathcal{M}$-adhesive categories satisfying not only the axioms of $\mathcal{M}$-adhesive categories, but also the additional HLR properties as stated in Theorem 4.42, except for effective pushouts.

For an adhesive category $\mathbf{C}$, which is based on the class of all monomorphisms, there may be monomorphisms in $\mathbf{C}_{\text{fin}}$ which are not monomorphisms in $\mathbf{C}$. Thus it is not clear whether the finite objects in $\mathbf{C}$ and $\mathbf{C}_{\text{fin}}$ are the same. This problem is avoided for $\mathcal{M}$-adhesive categories, where finitariness depends on $\mathcal{M}$. For adhesive categories, the restriction to finite objects leads to an adhesive category if the inclusion functor $I : \mathbf{C}_{\text{fin}} \to \mathbf{C}$ preserves monomorphisms. Currently, we are not aware of any adhesive category failing this property, or whether this can be shown in general.

4.4.4 Functorial Constructions of Finitary $\mathcal{M}$-Adhesive Categories

Similarly to general $\mathcal{M}$-adhesive categories, also finitary $\mathcal{M}$-adhesive categories are closed under product, slice, coslice, functor, and comma categories under suitable conditions [EEPT06]. It suffices to show this for functor and comma categories, because all others are special cases.

Fact 4.48 (Finitary functor category). *Given a finitary M-adhesive category* (**C**, M) *and a category* **X** *with a finite class of objects, also the functor category* (**Funct**(**X**, **C**), M_F) *is a finitary M-adhesive category, where M_F is the class of all M-functor transformations $t : F' \to F$, i.e., $t(X): F'(X) \to F(X) \in M$ for all objects X in* **X**. △

Proof. (**Funct**(**X**, **C**), M_F) is an M-adhesive category (see [EEPT06]) and it remains to show that each $F : \mathbf{X} \to \mathbf{C}$ is finite. W. l. o. g. we have objects $X_1, \ldots, X_n$ in **X**. We want to show that there are only finitely many M-functor transformations $t : F' \to F$ up to isomorphism. Since $F(X_k) \in \mathbf{C}$ and **C** is a finitary M-adhesive category we have $i_k \in \mathbb{N}$ different choices for $t(X_k) : F'(X_k) \to F(X_k) \in M$. Hence, altogether we have at most $i = i_1 \cdot \ldots \cdot i_n \in \mathbb{N}$ different $t : F' \to F$ up to isomorphism. □

For infinite, even discrete, **X**, the functor category is not finitary. For example, consider $\mathbf{C} = \mathbf{Sets}_{\text{fin}}$. The object $(2_i)_{i \in \mathbb{N}}$ with $2_i = \{1, 2\}$ has an infinite number of subobjects $(1_i)_{i \in \mathbb{N}}$ with $1_i = \{1\}$, because in each component $i \in \mathbb{N}$ we have two choices of injective functions $f_{1/2} : \{1\} \to \{1, 2\}$. Hence $(2_i)_{i \in \mathbb{N}}$ is not finite and **Funct**(**X**, **C**) is not finitary.

Fact 4.49 (Finitary comma category). *Given finitary M-adhesive categories* (**A**, M_1), (**B**, M_2) *and functors $F : \mathbf{A} \to \mathbf{C}$, $G : \mathbf{B} \to \mathbf{C}$, where F preserves pushouts along M_1 and G preserves pullbacks along M_2, the comma category* **ComCat**($F, G; I$) *with $M = (M_1 \times M_2) \cap$ **ComCat**($F, G; I$) is a finitary M-adhesive category.* △

Proof. **ComCat**($F, G; I$) is an M-adhesive category (see [EEPT06]). It remains to show that each object $(A, B, op) = \left[op^k : F(A) \to G(B) \right]_{k \in I}$ is finite.

By assumption, A and B are finite with a finite number of subobjects $m_{1,i} : A_i \to A \in M_1$ for $i \in I_1$ and $m_{2,j} : B_j \to B \in M_2$ for $j \in I_2$. Hence, we have at most $|I_1| \cdot |I_2|$ M-subobjects of (A, B, op), where for each i, j, k there is at most one $op_{i,j}^k$ such that diagram (1) commutes. This is due to the fact that G preserves

$$
\begin{array}{ccc}
F(A_i) & \xrightarrow{\;op_{i,j}^k\;} & G(B_j) \\
{\scriptstyle F(m_{1,i})}\big\downarrow & (1) & \big\downarrow{\scriptstyle G(m_{2,j})} \\
F(A) & \xrightarrow[\;op^k\;]{} & G(B)
\end{array}
$$

pullbacks along M_2, and therefore $G(m_{2,j})$ is a monomorphism in **C**. □

Remark 4.50. Note that I in **ComCat**($F, G; I$) is not required to be finite. △

4.5 Preservation of Additional HLR Properties

Similarly to the special case of finitary M-adhesive categories, also M-adhesive categories are closed under different categorical constructions. This means that we

can construct new $\mathcal{M}$-adhesive categories from given ones. In this section, we analyse how far also the additional HLR properties for $\mathcal{M}$-adhesive categories defined
in Def. 4.23 can be obtained from the categorical constructions if the underlying
$\mathcal{M}$-adhesive categories fulfill these properties. This work is based on [PEL08] and
extended to general comma categories and subcategories. Here, we only state the
results; the proofs can be found in Appendix B.4, and for examples, see [PEL08].

4.5.1 Binary Coproducts

In most cases, binary coproducts can be constructed in the underlying categories,
with some compatibility requirements for the preservation of binary coproducts.
Note that we do not have to analyse the compatibility of binary coproducts with $\mathcal{M}$,
as done in [PEL08], since this is a general result in $\mathcal{M}$-adhesive categories, as shown
in Fact 4.27.

Fact 4.51. *If the $\mathcal{M}$-adhesive categories $(\mathbf{C}, \mathcal{M}_1)$, $(\mathbf{D}, \mathcal{M}_2)$, and $(\mathbf{C}_j, \mathcal{M}_j)$ for $j \in \mathcal{J}$
have binary coproducts then also the following $\mathcal{M}$-adhesive categories have binary
coproducts:*

1. *the general comma category $(\mathbf{G}, (\times_{j\in\mathcal{J}} \mathcal{M}_j) \cap Mor_\mathbf{G})$, if for all $i \in I$ F_i preserves
   binary coproducts,*
2. *any full subcategory $(\mathbf{C}', \mathcal{M}_1|_{\mathbf{C}'})$ of $\mathbf{C}$, if*
   - *(i) the inclusion functor reflects binary coproducts or*
   - *(ii) $\mathbf{C}'$ has an initial object I and, in addition, we have general pushouts in $\mathbf{C}'$ or
     $i_A : I \to A \in \mathcal{M}$ for all $A \in \mathbf{C}'$,*
3. *the comma category $(\mathbf{F}, (\mathcal{M}_1 \times \mathcal{M}_2) \cap Mor_\mathbf{F})$, if $F : \mathbf{C} \to \mathbf{X}$ preserves binary
   coproducts,*
4. *the product category $(\mathbf{C} \times \mathbf{D}, \mathcal{M}_1 \times \mathcal{M}_2)$,*
5. *the slice category $(\mathbf{C}\backslash X, \mathcal{M}_1 \cap Mor_{\mathbf{C}\backslash X})$,*
6. *the coslice category $(X\backslash\mathbf{C}, \mathcal{M}_1 \cap Mor_{X\backslash\mathbf{C}})$, if $\mathbf{C}$ has general pushouts,*
7. *the functor category $([\mathbf{X}, \mathbf{C}], \mathcal{M}_1$-functor transformations).*

$\triangle$

Proof. See Appendix B.4.1. $\square$

4.5.1.1 Epi–$\mathcal{M}$ Factorisation

For epi–$\mathcal{M}$ factorisations, we obtain the same results as for $\mathcal{E}'$–$\mathcal{M}'$ pair factorisations by replacing the class of morphism pairs $\mathcal{E}'$ by the class of all epimorphisms
and $\mathcal{M}'$ by $\mathcal{M}$. We do not explicitly state these results here, but they can be easily
deduced from the results in the following subsection.

4.5.1.2 $\mathcal{E}'$–$\mathcal{M}'$ Pair Factorisation

For most of the categorical constructions, the $\mathcal{E}'$–$\mathcal{M}'$ pair factorisation from the underlying categories is preserved. But for functor categories, we need a stronger property, the $\mathcal{E}'$–$\mathcal{M}'$ *diagonal property*, for this result.

Definition 4.52 (Strong $\mathcal{E}'$–$\mathcal{M}'$ pair factorisation). An $\mathcal{E}'$–$\mathcal{M}'$ pair factorisation is called *strong* if the following $\mathcal{E}'$–$\mathcal{M}'$ diagonal property holds:

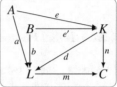

Given $(e, e') \in \mathcal{E}'$, $m \in \mathcal{M}'$, and morphisms a, b, n as shown in the following diagram, with $n \circ e = m \circ a$ and $n \circ e' = m \circ b$, there exists a unique $d : K \to L$ such that $m \circ d = n$, $d \circ e = a$, and $d \circ e' = b$. △

Fact 4.53. *In an $\mathcal{M}$-adhesive category $(\mathbf{C}, \mathcal{M})$, the following properties hold:*

1. *If $(\mathbf{C}, \mathcal{M})$ has a strong $\mathcal{E}'$–$\mathcal{M}'$ pair factorisation, then the $\mathcal{E}'$–$\mathcal{M}'$ pair factorisation is unique up to isomorphism.*

2. *A strong $\mathcal{E}'$–$\mathcal{M}'$ pair factorisation is functorial, i.e., given morphisms a, b, c, f_1, g_1, f_2, g_2 as shown in the right diagram with $c \circ f_1 = f_2 \circ a$ and $c \circ g_1 = g_2 \circ b$, and $\mathcal{E}'$–$\mathcal{M}'$ pair factorisations $((e_1, e_1'), m_1)$ and $((e_2, e_2'), m_2)$ of f_1, g_1 and f_2, g_2, respectively, there exists a unique $d : K_1 \to K_2$ such that $d \circ e_1 = e_2 \circ a$, $d \circ e_1' = e_2' \circ b$, and $c \circ m_1 = m_2 \circ d$.* △

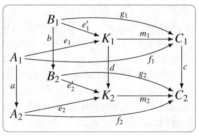

Proof. See [PEL08]. □

Fact 4.54. *Given $\mathcal{M}$-adhesive categories $(\mathbf{C}_j, \mathcal{M}_j)$, $(\mathbf{C}, \mathcal{M}_1)$, and $(\mathbf{D}, \mathcal{M}_2)$ with $\mathcal{E}_j'$–$\mathcal{M}_j'$, $\mathcal{E}_1'$–$\mathcal{M}_1'$, and $\mathcal{E}_2'$–$\mathcal{M}_2'$ pair factorisations, respectively, the following $\mathcal{M}$-adhesive categories have an $\mathcal{E}'$–$\mathcal{M}'$ pair factorisation and preserve strongness:*

1. *the general comma category $(\mathbf{G}, (\times_{j \in \mathcal{J}} \mathcal{M}_j) \cap Mor_\mathbf{G})$ with $\mathcal{M}' = (\times_{j \in \mathcal{J}} \mathcal{M}_j')$ and $\mathcal{E}' = \{((e_j,), (e_j')) \mid (e_j, e_j') \in \mathcal{E}_j'\} \cap (Mor_\mathbf{G} \times Mor_\mathbf{G})$, if $G_i(\mathcal{M}_{\ell_i}') \subseteq Isos$ for all $i \in I$,*

2. *any full subcategory $(\mathbf{C}', \mathcal{M}_1|_{\mathbf{C}'})$ of $\mathbf{C}$ with $\mathcal{M}' = \mathcal{M}_1'|_{\mathbf{C}'}$ and $\mathcal{E}' = \mathcal{E}_1'|_{(\mathbf{C}' \times \mathbf{C}')}$, if the inclusion functor reflects the $\mathcal{E}_1'$–$\mathcal{M}_1'$ pair factorisation,*

3. *the comma category $(\mathbf{F}, (\mathcal{M}_1 \times \mathcal{M}_2) \cap Mor_\mathbf{F})$ with $\mathcal{M}' = (\mathcal{M}_1' \times \mathcal{M}_2') \cap Mor_\mathbf{F}$ and $\mathcal{E}' = \{((e_1, e_2), (e_1', e_2')) \mid (e_1, e_1') \in \mathcal{E}_1', (e_2, e_2') \in \mathcal{E}_2'\} \cap (Mor_\mathbf{F} \times Mor_\mathbf{F})$, if $G(\mathcal{M}_2') \subseteq Isos$,*

4. *the product category $(\mathbf{C} \times \mathbf{D}, \mathcal{M}_1 \times \mathcal{M}_2)$ with $\mathcal{M}' = \mathcal{M}_1' \times \mathcal{M}_2'$ and $\mathcal{E}' = \{((e_1, e_2), (e_1', e_2')) \mid (e_1, e_1') \in \mathcal{E}_1', (e_2, e_2') \in \mathcal{E}_2'\}$,*

5. *the slice category $(\mathbf{C} \backslash X, \mathcal{M}_1 \cap Mor_{\mathbf{C} \backslash X})$ with $\mathcal{M}' = \mathcal{M}_1' \cap Mor_{\mathbf{C} \backslash X}$ and $\mathcal{E}' = \mathcal{E}_1' \cap (Mor_{\mathbf{C} \backslash X} \times Mor_{\mathbf{C} \backslash X})$,*

6. *the* coslice category $(X\backslash\mathbf{C}, \mathcal{M}_1 \cap Mor_{X\backslash\mathbf{C}})$ *with* $\mathcal{M}' = \mathcal{M}'_1 \cap Mor_{X\backslash\mathbf{C}}$ *and* $\mathcal{E}' = \mathcal{E}'_1 \cap (Mor_{X\backslash\mathbf{C}} \times Mor_{X\backslash\mathbf{C}})$, *if* $\mathcal{M}'_1$ *is a class of monomorphisms,*
7. *the* functor category $([\mathbf{X}, \mathbf{C}], \mathcal{M}_1\text{-}functor\ transformations)$ *with the class* $\mathcal{M}'$ *of all* $\mathcal{M}'_1$-*functor transformations and* $\mathcal{E}' = \{(e, e') \mid e, e'\ functor\ transformations,$ $(e(x), e'(x)) \in \mathcal{E}'_1\ for\ all\ x \in \mathbf{X}\}$, *if* $\mathcal{E}'_1$–$\mathcal{M}'_1$ *is a strong pair factorisation in* $\mathbf{C}$.   △

Proof. See Appendix B.4.2.   □

4.5.1.3 Initial Pushouts

In general, the construction of initial pushouts from the underlying categories is complicated since the existence of the boundary and context objects have to be ensured. In many cases, this is only possible under very strict limitations.

Fact 4.55. *If the M-adhesive categories* $(\mathbf{C}, \mathcal{M}_1)$, $(\mathbf{D}, \mathcal{M}_2)$, *and* $(\mathbf{C}_j, \mathcal{M}_j)$ *for* $j \in \mathcal{J}$ *have initial pushouts over* $\mathcal{M}'_1$, $\mathcal{M}'_2$, *and* $\mathcal{M}'_j$, *respectively, then also the following M-adhesive categories have initial pushouts over* $\mathcal{M}'$-*morphisms:*

1. *the* general comma category $(\mathbf{G}, (\times_{j\in\mathcal{J}} \mathcal{M}_j)\cap Mor_{\mathbf{G}})$ *with* $\mathcal{M}' = \times_{j\in\mathcal{J}} \mathcal{M}'_j\cap Mor_{\mathbf{G}}$, *if for all* $i \in I$ F_i *preserves pushouts along* $\mathcal{M}_{k_i}$-*morphisms and* $G_i(\mathcal{M}_{\ell_i}) \subseteq Isos$,
2. *any* full subcategory $(\mathbf{C}', \mathcal{M}_1|_{\mathbf{C}'})$ *of* $\mathbf{C}$ *with* $\mathcal{M}' = \mathcal{M}'_1|_{\mathbf{C}'}$, *if the inclusion functor reflects initial pushouts over* $\mathcal{M}'$-*morphisms,*
3. *the* comma category $(\mathbf{F}, (\mathcal{M}_1 \times \mathcal{M}_2)\cap Mor_{\mathbf{F}})$ *with* $\mathcal{M}' = \mathcal{M}'_1 \times \mathcal{M}'_2$, *if* F *preserves pushouts along* $\mathcal{M}_1$-*morphisms and* $G(\mathcal{M}_2) \subseteq Isos$,
4. *the* product category $(\mathbf{C} \times \mathbf{D}, \mathcal{M}_1 \times \mathcal{M}_2)$ *with* $\mathcal{M}' = \mathcal{M}'_1 \times \mathcal{M}'_2$,
5. *the* slice category $(\mathbf{C}\backslash X, \mathcal{M}_1 \cap Mor_{\mathbf{C}\backslash X})$ *with* $\mathcal{M}' = \mathcal{M}'_1 \cap Mor_{\mathbf{C}\backslash X}$,
6. *the* coslice category $(X\backslash\mathbf{C}, \mathcal{M}_1 \cap Mor_{X\backslash\mathbf{C}})$ *with* $\mathcal{M}' = \mathcal{M}'_1 \cap Mor_{X\backslash\mathbf{C}}$, *if for* $f : (A, a') \to (D, d') \in \mathcal{M}'$

   (i) *the initial pushout over* f *in* $\mathbf{C}$ *can be extended to a valid square in* $X\backslash\mathbf{C}$ *or*
   (ii) $a' : X \to A \in \mathcal{M}_1$ *and the pushout complement of* a' *and* f *in* $\mathbf{C}$ *exists,*

7. *the* functor category $([\mathbf{X}, \mathbf{C}], \mathcal{M}_1\text{-}functor\ transformations)$ *with* $\mathcal{M}' = \mathcal{M}'_1$-*functor transformations, if* $\mathbf{C}$ *has arbitrary limits and intersections of* $\mathcal{M}_1$-*subobjects.*
                                                                                                 △

Proof. See Appendix B.4.3.   □

4.5.1.4 Effective Pushouts

Effective pushouts are also preserved by categorical constructions. Using Rem. 4.24, we already know for the regarded situation that the induced morphism is a monomorphism. We only have to show that it is indeed an $\mathcal{M}$-morphism. This is the case if pullbacks, pushouts, and their induced morphisms are constructed componentwise.

Fact 4.56. *If the $\mathcal{M}$-adhesive categories* $(\mathbf{C}, \mathcal{M}_1)$, $(\mathbf{D}, \mathcal{M}_2)$, *and* $(\mathbf{C}_j, \mathcal{M}_j)$ *for* $j \in \mathcal{J}$ *have effective pushouts then also the following $\mathcal{M}$-adhesive categories have effective pushouts:*

1. *the* general comma category $(\mathbf{G}, (\times_{j \in \mathcal{J}} \mathcal{M}_j) \cap Mor_{\mathbf{G}})$,
2. *any* full subcategory $(\mathbf{C}', \mathcal{M}_1|_{\mathbf{C}'})$ *of* $\mathbf{C}$,
3. *the* comma category $(\mathbf{F}, (\mathcal{M}_1 \times \mathcal{M}_2) \cap Mor_{\mathbf{F}})$,
4. *the* product category $(\mathbf{C} \times \mathbf{D}, \mathcal{M}_1 \times \mathcal{M}_2)$,
5. *the* slice category $(\mathbf{C} \backslash X, \mathcal{M}_1 \cap Mor_{\mathbf{C} \backslash X})$,
6. *the* coslice category $(X \backslash \mathbf{C}, \mathcal{M}_1 \cap Mor_{X \backslash \mathbf{C}})$,
7. *the* functor category $([\mathbf{X}, \mathbf{C}], \mathcal{M}_1\text{-}functor\ transformations)$. △

Proof. See Appendix B.4.4. □

Chapter 5
$\mathcal{M}$-Adhesive Transformation Systems

In this chapter, we introduce $\mathcal{M}$-adhesive transformation systems based on the $\mathcal{M}$-adhesive categories introduced in Chap. 4. They describe a powerful framework for the definition of transformations of various models. By using a very general variant of application conditions we extend the expressive power of the transformations. For this chapter, we assume we have an $\mathcal{M}$-adhesive category with initial object, binary coproducts, an $\mathcal{E}$–$\mathcal{M}$ factorisation and an $\mathcal{E}'$–$\mathcal{M}$ pair factorisation (see Sect. 4.3).

In Sect. 5.1, we give a short introduction to conditions and constraints and define rules and transformations with application conditions. This is the generalisation of the theory presented in Sect. 2.2 for graphs to $\mathcal{M}$-adhesive categories. Various results for transformations with application conditions, which were motivated on an intuitive level in Sect. 2.3, are presented in the general setting of $\mathcal{M}$-adhesive transformation systems in Sect. 5.2. As a running example, we use a model of an elevator control and analyse its behaviour. Application conditions induce additional dependencies for transformation steps. In Sect. 5.3, we show that the notion of equivalence for transformation sequences with application conditions can be described and analysed adequately using the notion of permutation equivalence, which generalises the notion of switch equivalence. This chapter is based on [EGH+14, EGH+12, Gol10, HCE14].

5.1 Rules and Transformations with Application Conditions

Nested conditions were introduced in [HP05, HP09] to express properties of objects in a category. They are expressively equivalent to first-order formulas on graphs. Later, we will use them to express application conditions for rules to increase the expressiveness of transformations. We only present the general theory for $\mathcal{M}$-adhesive categories in this section; for examples of conditions and their constructions, see Sect. 2.2.

Basically, a condition describes the existence or nonexistence of a certain structure for an object.

Definition 5.1 (Condition). A *(nested) condition ac* over an object P is of the form

- $ac = \text{true}$,
- $ac = \exists\,(a, ac')$, where $a : P \rightarrow C$ is a morphism and ac' is a condition over C,
- $ac = \neg ac'$, where ac' is a condition over P,
- $ac = \wedge_{i \in I} ac_i$, where $(ac_i)_{i \in I}$ with an index set I are conditions over P, or
- $ac = \vee_{i \in I} ac_i$, where $(ac_i)_{i \in I}$ with an index set I are conditions over P.

Moreover, false abbreviates $\neg$true, $\exists\,a$ abbreviates $\exists\,(a, \text{true})$, and $\forall\,(a, ac)$ abbreviates $\neg\,\exists\,(a, \neg ac)$.   △

A condition is satisfied by a morphism into an object if the required structure exists, which can be verified by the existence of suitable morphisms.

Definition 5.2 (Satisfaction of condition). Given a condition ac over P a morphism $p : P \rightarrow G$ satisfies ac, written $p \models ac$, if

- $ac = \text{true}$,
- $ac = \exists\,(a, ac')$ and there exists a morphism $q \in \mathcal{M}$ with $q \circ a = p$ and $q \models ac'$,
- $ac = \neg ac'$ and $p \not\models ac'$,
- $ac = \wedge_{i \in I} ac_i$ and $\forall\, i \in I : p \models ac_i$, or
- $ac = \vee_{i \in I} ac_i$ and $\exists\, i \in I : p \models ac_i$.

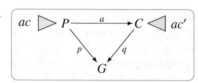

Two conditions ac and ac' over P are *semantically equivalent*, denoted by $ac \cong ac'$, if $p \models ac \Leftrightarrow p \models ac'$ for all morphisms p with domain P.   △

As shown in [HP09, EHL10], conditions can be shifted over morphisms into equivalent conditions over the codomain. For this shift construction, all $\mathcal{E}'$-overlappings of the codomain of the shift morphism and the codomain of the condition morphism have to be collected.

Definition 5.3 (Shift over morphism). Given a condition ac over P and a morphism $b : P \rightarrow P'$, $\text{Shift}(b, ac)$ is a condition over P' defined by

- $\text{Shift}(b, ac) = \text{true}$ if $ac = \text{true}$,
- $\text{Shift}(b, ac) = \vee_{(a',b') \in \mathcal{F}}\,\exists\,(a', \text{Shift}(b', ac'))$ if $ac = \exists\,(a, ac')$ and $\mathcal{F} = \{(a', b') \in \mathcal{E}' \mid b' \in \mathcal{M}, b' \circ a = a' \circ b\}$,
- $\text{Shift}(b, ac) = \neg\text{Shift}(b, ac')$ if $ac = \neg ac'$,
- $\text{Shift}(b, ac) = \wedge_{i \in I}\text{Shift}(b, ac_i)$ if $ac = \wedge_{i \in I} ac_i$, or
- $\text{Shift}(b, ac) = \vee_{i \in I}\text{Shift}(b, ac_i)$ if $ac = \vee_{i \in I} ac_i$.   △

Fact 5.4. *Given a condition ac over P and morphisms $b : P \rightarrow P'$, $b' : P' \rightarrow P''$, and $p : P' \rightarrow G$,*

- $p \models \text{Shift}(b, ac)$ *if and only if* $p \circ b \models ac$ *and*
- $\text{Shift}(b', \text{Shift}(b, ac)) \cong \text{Shift}(b' \circ b, ac)$.

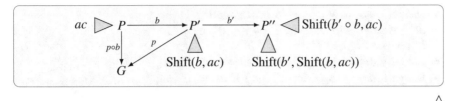

$\triangle$

Proof. We show the first statement by structural induction.

Basis. The equivalence holds trivially for the condition ac = true.

Induction step. Consider a condition $ac = \exists\,(a, ac')$ with the corresponding shift construction, $\text{Shift}(b, ac)$, along morphism b.

"$\Rightarrow$" Suppose $p \models \text{Shift}(b, ac)$, i.e., there exists $(a', b') \in \mathcal{F}$ such that $p \models \exists\,(a', \text{Shift}(b', ac'))$. This means that there exists a morphism $q : C' \to G \in \mathcal{M}$ with $q \circ a' = p$ and $q \models \text{Shift}(b', ac')$. By induction hypothesis, $q \circ b' \models ac'$, i.e., we have a morphism $q \circ b' \in \mathcal{M}$ with $q \circ b' \circ a = q \circ a' \circ b = p \circ b$ and $p \circ b \models \exists\,(a, ac') = ac$.

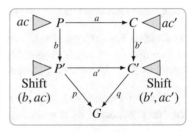

"$\Leftarrow$" If $p \circ b \models \exists\,(a, ac')$ then there exists $q \in \mathcal{M}$ with $p \circ b = q \circ a$ and $q \models ac'$. Now consider the $\mathcal{E}'$–$\mathcal{M}$ pair factorisation $((a', b'), q')$ of p and q. Since $q, q' \in \mathcal{M}$, by $\mathcal{M}$-decomposition it follows that $b' \in \mathcal{M}$, and $q' \circ a' \circ b = p \circ b = q \circ a = q' \circ b' \circ a$ implies that $a' \circ b = b' \circ a$, i.e., $(a', b') \in \mathcal{F}$. By induction hypothesis, $q = q' \circ b' \models ac'$

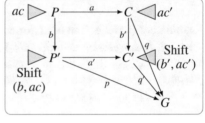

implies that $q' \models \text{Shift}(b', ac')$, and therefore $p \models \exists\,(a', \text{Shift}(b', ac'))$, i.e., $p \models \text{Shift}(b, ac)$.

Similarly, this holds for composed conditions.

The second statement follows directly from the first one: for a morphism $q : P'' \to G$, we have that $q \models \text{Shift}(b' \circ b, ac) \Leftrightarrow q \circ b' \circ b \models ac \Leftrightarrow q \circ b' \models \text{Shift}(b, ac) \Leftrightarrow q \models \text{Shift}(b', \text{Shift}(b, ac))$. □

As with the shift construction, we can also merge a condition over a morphism. The difference lies in different $\mathcal{M}$-morphisms to be required, with $a' \in \mathcal{M}$ instead of $b' \in \mathcal{M}$. Additionally, b' has to be from a distinguished morphism class $\mathcal{O}$ of match morphisms.

Definition 5.5 (Merge over morphism). Given a condition ac over P and a morphism $b : P \to P'$, $\text{Merge}(b, ac)$ is a condition over P' defined by

• $\text{Merge}(b, ac)$ = true if ac = true,

- $\text{Merge}(b, ac) = \bigvee_{(a',b')\in\mathcal{F}'} \exists\,(a', \text{Merge}(b',$ $ac'))$ if $ac = \exists\,(a, ac')$ and $\mathcal{F}' = \{(a',b') \in$ $\mathcal{E}' \mid a' \in \mathcal{M}, b' \in \mathcal{O}, b' \circ a = a' \circ b\}$,
- $\text{Merge}(b, ac) = \neg\text{Merge}(b, ac')$ if $ac = \neg ac'$,
- $\text{Merge}(b, ac) = \bigwedge_{i\in I}\text{Merge}(b, ac_i)$ if $ac = \bigwedge_{i\in I} ac_i$,

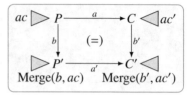

- $\text{Merge}(b, ac) = \bigvee_{i\in I}\text{Merge}(b, ac_i)$ if $ac = \bigvee_{i\in I} ac_i$. △

The merge construction is used to specify schemata of conditions, which are inspired by a construction proposed in [KHM06] for negative application conditions. An AC schema consists of the disjunction of all merges of a condition along $\mathcal{E}$-morphisms starting from its domain.

Definition 5.6 (AC schema). Given a condition ac over P and the set $\mathcal{E}_P = \{e \in \mathcal{E} \mid \text{dom}(e) = P\}$ of all $\mathcal{E}$-morphisms with domain P, the *AC schema* $\overline{ac}$ over P is given by $\overline{ac} = \bigvee_{f\in\mathcal{E}_P} \exists\,(f, \text{Merge}(f, ac))$. △

The satisfaction of an AC schema by a morphism only depends on the satisfaction of one component of the corresponding $\mathcal{E}$–$\mathcal{M}$ factorisation of the morphism. To prove this, we first show a slightly more general result for disjunctions over the set $\mathcal{E}_P$.

Lemma 5.7 (Satisfaction of disjunction). *Consider the set $\mathcal{E}_P = \{e \in \mathcal{E} \mid \text{dom}(e) = P\}$ of all $\mathcal{E}$-morphisms with domain P, a condition $ac = \bigvee_{f\in\mathcal{E}_P} \exists\,(f, ac'_f)$ over P, and a morphism $p : P \to G$ with an $\mathcal{E}$–$\mathcal{M}$ factorisation $m \circ e = p$. Then $p \models ac$ if and only if $m \models ac'_e$.* △

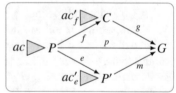

Proof. The following equivalences prove this:

$$p \models ac \Leftrightarrow \exists\,f \in \mathcal{E}_P : p \models \exists\,(f, ac'_f) \qquad\qquad \text{(Def. 5.2)}$$
$$\Leftrightarrow \exists\,f \in \mathcal{E}_P, g \in \mathcal{M} : g \circ f = p \wedge g \models ac'_f \quad \text{(Def. 5.2)}$$
$$\Leftrightarrow m \models ac'_e \wedge m \circ e = p \qquad\qquad\qquad \text{(Def. 4.23)} \qquad \square$$

Fact 5.8 (AC schema satisfaction). *Given an AC schema $\overline{ac}$ over P and a morphism $p\colon P \to G$ with an $\mathcal{E}$–$\mathcal{M}$ factorisation $m \circ e = p$, $p \models \overline{ac}$ if and only if $m \models \text{Merge}(e, ac)$.* △

Proof. By Def. 5.6 we have that $\overline{ac} = \bigvee_{f\in\mathcal{E}_P} \exists\,(f, \text{Merge}(f, ac))$. We can directly apply Lem. 5.7, leading to the required result. □

Remark 5.9. If $p : P \to G$ is an $\mathcal{M}$-morphism, then the satisfaction of an AC schema coincides with classical satisfaction, because the factorisation is trivially $p = p \circ id$. △

In contrast to conditions, constraints describe global requirements for objects. They can be interpreted as conditions over the initial object, which means that a constraint $\exists (i_C, \text{true})$ with the initial morphism i_C into C is valid for an object G if there exists a morphism $c : C \rightarrow G$. This constraint expresses that the existence of C as a part of G is required.

Definition 5.10 (Constraint). Given an initial object A, a condition ac over A is called a *constraint*. △

The satisfaction of a constraint is that of the corresponding conditions, adapted to the special case of a condition over an initial object.

Definition 5.11 (Satisfaction of constraint). Given a constraint ac (over the initial object A), an object G *satisfies* ac, written $G \models ac$, if

- $ac = \text{true}$,
- $ac = \exists (i_C, ac')$ and there exists a morphism $c \in \mathcal{M}$ with $c \models ac'$,
- $ac = \neg ac'$ and $G \not\models ac'$,
- $ac = \wedge_{i \in I} ac_i$ and $\forall i \in I : G \models ac_i$, or
- $ac = \vee_{i \in I} ac_i$ and $\exists i \in I : G \models ac_i$. △

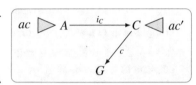

In [EEPT06], transformation systems based on a categorical foundation using weak adhesive HLR categories were introduced which can be instantiated to various graphs and graph-like structures. In addition, application conditions extend the standard approach of transformations. Here, we present the theory of transformations in $\mathcal{M}$-adhesive categories for rules with application conditions in general.

A rule is a general description of local changes that may occur in objects of the transformation system. Mainly, it consists of some deletion part and some construction part, defined by the rule morphisms l and r, respectively. In addition, an application condition restricts the application of this rule to certain objects.

Definition 5.12 (Rule). A *rule* $p = (L \xleftarrow{l} K \xrightarrow{r} R, ac)$ consists of objects L, K, and R, called left-hand side, gluing, and right-hand side, respectively, two morphisms l and r with $l, r \in \mathcal{M}$, and a condition ac over L, called *application condition*. △

A transformation describes the application of a rule to an object via a match. It can only be applied if the match satisfies the application condition.

Definition 5.13 (Transformation). Given a rule $p = (L \xleftarrow{l} K \xrightarrow{r} R, ac)$, an object G, and a morphism $m : L \rightarrow G$, called match, such that $m \models ac$, a *direct transformation* $G \xRightarrow{p,m} H$ from G to an object H is given by the pushouts (1) and (2).

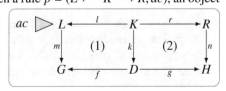

A sequence of direct transformations is called a *transformation*. △

Remark 5.14. Note that for the construction of the pushout (1) we have to construct the pushout complement of $m \circ l$, which is only possible if the so-called gluing condition is satisfied (see [EEPT06]).   △

In analogy to the application condition over L, which is a pre-application condition, it is also possible to define post-application conditions over the right-hand side R of a rule. Since these application conditions over R can be translated into equivalent application conditions over L (and vice versa), we can restrict our rules to application conditions over L.

Definition 5.15 (Shift over rule). Given a rule $p = (L \xleftarrow{l} K \xrightarrow{r} R, ac)$ and a condition ac_R over R, $L(p, ac_R)$ is a condition over L defined by

- $L(p, ac_R) = $ true if $ac_R = $ true,
- $L(p, ac_R) = \exists (b, L(p^*, ac'_R))$ if $ac_R = \exists (a, ac'_R)$, $a \circ r$ has a pushout complement (1), and $p^* = (Y \xleftarrow{l^*} Z \xrightarrow{r^*} X)$ is the derived rule by constructing pushout (2); $L(p, \exists (a, ac'_R)) = $ false otherwise,

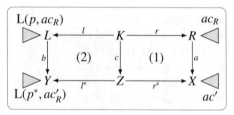

- $L(p, ac_R) = \neg L(p, ac'_R)$ if $ac_R = \neg ac'_R$,
- $L(p, ac_R) = \wedge_{i \in I} L(p, ac_{R,i})$ if $ac_R = \wedge_{i \in I} ac_{R,i}$, or
- $L(p, ac_R) = \vee_{i \in I} L(p, ac_{R,i})$ if $ac_R = \vee_{i \in I} ac_{R,i}$.

Dually, for a condition ac_L over L we define $R(p, ac_L) = L(\overline{p}^{-1}, ac_L)$, where the *inverse rule* $\overline{p}^{-1}$ without application conditions is defined by $\overline{p}^{-1} = (R \xleftarrow{r} K \xrightarrow{l} L)$.   △

Fact 5.16. *Given a transformation $G \xRightarrow{p,m} H$ via a rule $p = (L \xleftarrow{l} K \xrightarrow{r} R, ac)$ and a condition ac_R over R, $m \models L(p, ac_R)$ if and only if $n \models ac_R$.*

Dually, for a condition ac_L over L we have that $m \models ac_L$ if and only if $n \models R(p, ac_L)$.

Moreover, for any transformation $G' \xRightarrow{m', p'} H'$ we have that $m' \models \text{Shift}(m, L(p, ac_R))$ if and only if $m' \models L(p', \text{Shift}(n, ac_R))$ for $p' = (G \xleftarrow{f} D \xrightarrow{g} H)$.   △

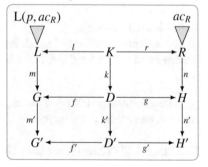

Proof. We show the first statement by structural induction.

Basis. The equivalence holds trivially for the condition $ac_R = $ true.

Induction step. Consider a condition $ac_R = \exists (a, ac'_R)$.

Case 1: $a \circ r$ has a pushout complement (1) and $L(p, ac_R) = \exists (b, L(p^*, ac'_R))$ from the construction.

"$\Rightarrow$" Suppose $m \models L(p, ac_R)$, i.e., there exists $q \in M$ with $q \circ b = m$ and $q \models L(p^*, ac'_R)$. Now construct pullback (3′) and obtain the induced morphism c'; and by M-pushout–pullback decomposition (Theorem 4.22), both (2′) and (3′) are pushouts. By uniqueness of pushout complements (Theorem 4.22), (2′) and (2) are equivalent such that there exists a pushout (3) equivalent to (3′) and a decomposition of $G \xrightarrow{p,m} H$ into pushouts (1)–(4). By PO–

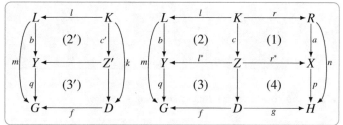

PB compatibility of M, $q \in M$ implies that $p \in M$, and by induction hypothesis, $p \models ac'_R$, i.e., $n \models ac_R$.

"$\Leftarrow$" If $n \models ac_R$, there exists $p \in M$ with $p \circ a = n$ and $p \models ac'_R$. As above, we can decompose $G \xrightarrow{p,m} H$ into pushouts (1)–(4), with $q \in M$, $q \circ b = m$, and $q \models L(p_*, ac'_R)$ such that $m \models L(p, ac_R)$.

Case 2: $a \circ r$ has no pushout complement and $L(p, ac_R)$ = false. Suppose $n \models ac_R$; then we have, by the construction above, a pushout complement (1) of $a \circ r$, which is a contradiction. Therefore, $n \not\models ac_R$ and $m \not\models$ false.

Similarly, this holds for composed conditions.

The second statement follows from the dual constructions.

For the third statement, we have that $m' \models \text{Shift}(m, L(p, ac_R)) \Leftrightarrow m' \circ m \models L(p, ac_R) \Leftrightarrow n' \circ n \models ac_R \Leftrightarrow n' \models \text{Shift}(n, ac_R) \Leftrightarrow m' \models L(p', \text{Shift}(n, ac_R))$, which follows from the first statement, Fact 5.4, and the composition of pushouts. □

A set of rules constitutes an M-adhesive transformation system, and combined with a start object an M-adhesive grammar. The language of such a grammar contains all objects derivable from the start object.

Definition 5.17 (M-adhesive transformation system and grammar). An M-adhesive transformation system $AS = (\mathbf{C}, M, P)$ consists of an M-adhesive category $(\mathbf{C}, M)$ and a set of rules P.

An M-adhesive grammar $AG = (AS, S)$ consists of an M-adhesive transformation system AS and a start object S.

The *language L* of an M-adhesive grammar AG is defined by

$$L = \{G \mid \exists \text{ transformation } S \xrightarrow{*} G \text{ via } P\}. \qquad \triangle$$

*Example 5.18 (*Elevator*).* Now we introduce our running example for this chapter, an elevator control. The type of control we model is used in buildings where the elevator transports people from or to one main stop; in our example this is the lowest floor. This situation occurs, for example, in apartment buildings or multistory car parks. Each floor in the building is equipped with a call button. Such *external* call requests are served by the elevator only if it is in downwards mode in order to transport people to the main stop. When inside the elevator, *internal* stop requests

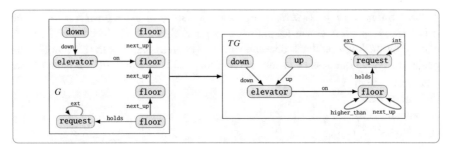

Fig. 5.1 The type graph TG and a model G of Elevator

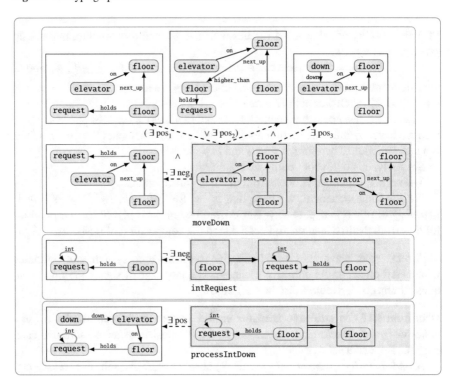

Fig. 5.2 Three rules for processing internal requests in the Elevator example

for each floor can be delivered. These are served as soon as the elevator car reaches
the requested floor, both in upwards and in downwards mode. As long as there are
remaining requests in the running direction, the direction of the elevator car is not
changed. If the elevator car arrives at a floor, all requests for this floor are deleted.

We model this system using typed graphs (see Sect. 2.1). In the right of Fig. 5.1,
the type graph TG for our elevator example is depicted. This type graph expresses
that an elevator car of type elevator exists, which can be on a specific floor.

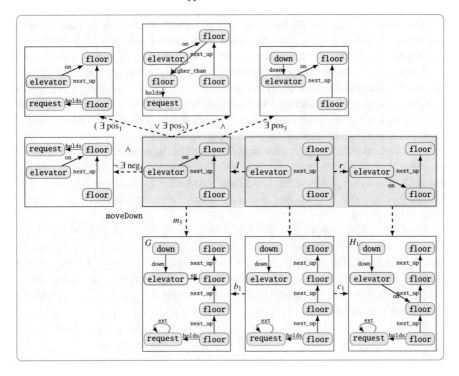

Fig. 5.3 Application of the rule moveDown

Moreover, the elevator can be in **up**wards or **down**wards mode. Floors are connected by next_up edges expressing which floor is directly above another floor. Moreover, higher_than edges express that a floor is arranged higher in the building than another floor. Each floor can hold requests of two different types. The first type is an external request expressing that an external call for the elevator car on this floor is given. The second type is an internal request expressing that an internal call from within the elevator car is given for stopping it on this floor.

In the left of Fig. 5.1, a graph G typed over this type graph TG is shown, describing a four story building, where the elevator car is on the second floor in downwards mode with an external call request on the ground floor. Note that G contains altogether six higher_than edges from each floor which is higher than another floor (corresponding to the transitive closure of opposite edges of next_up); these are not depicted.

In Fig. 5.2, three rules are shown modelling part of the elevator control for internal stop requests. Note that only the left- and right-hand sides of the rules are depicted—the gluing consists of all nodes and edges occurring in both L and R. The morphisms map the obvious elements by type and position and are therefore not explicitly marked.

At the top of Fig. 5.2, we have the rule `moveDown` describing that the elevator car moves down one floor. The combined application condition on L consists of three positive application conditions ($\exists \, \text{pos}_i$ for $i = 1, 2, 3$) and a negative one ($\neg \, \exists \, \text{neg}_1$). This combined application condition states that some request has to be present on the next lower floor (pos_1) or some other lower floor (pos_2), no request should be present on the elevator floor (neg_1), and the elevator car is in downwards mode (pos_3). Note that both pos_1 and pos_2 are necessary because the satisfaction of application conditions depends on injective morphisms.

As a second rule, `intRequest` describes that an internal stop request is made on some floor under the condition that no internal request is already given for this floor. The third rule `processIntDown` describes that an internal stop request is processed for a specific floor under the condition that the elevator is on this floor and in downwards mode.

In Fig. 5.3, the application of the rule `moveDown` to the graph G from Fig. 5.1 is shown, where also the gluing graph of the rule is explicitly depicted. Note that the match m_1 satisfies the application condition, because $m_1 \models \exists \, \text{pos}_2$ for the request on the lowest floor (remember the implicit `higher_than` edges between the floors), $m_1 \models \exists \, \text{pos}_3$ since the elevator is in downwards mode, and $m_1 \models \neg \, \exists \, \text{neg}_1$ with no request on the current floor.   △

5.2 Results for Transformations with Application Conditions

In this section, we present the main important results for M-adhesive transformation systems for rules with application conditions, generalising the corresponding well-known theorems for rules without application conditions [EEPT06] and with negative application conditions [Lam10]. The intuition and motivation for these results has already been given in Sect. 2.3—here we now state the full definitions, results and proofs. Note that the Local Church–Rosser, Parallelism, Concurrency, Embedding, Extension and Local Confluence Theorems are stated and proven in [EGH+14, EGH+12] for the case of rules with application conditions; in addition, we present new examples.

Most of the proofs are based on the corresponding statements for rules without application conditions and Facts 5.4 and 5.16, stating that application conditions can be shifted over morphisms and rules. The idea is the following: We switch from transformations with application conditions to the corresponding transformations without application conditions, use the results for transformations without application conditions,

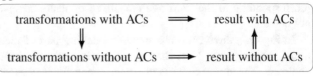

and lift the results without application conditions to the corresponding ones with application conditions.

In the following, let $p_i = (L_i \xleftarrow{l_i} K_i \xrightarrow{r_i} R_i, ac_i)$ be a rule with a left application condition and $\overline{p}_i = (L_i \xleftarrow{l_i} K_i \xrightarrow{r_i} R_i)$ the underlying plain rule for $i \in \mathbb{N}$. For every direct transformation $G \xRightarrow{p_i, m_i} H$ via such a rule p_i, there is a direct transformation $G \xRightarrow{\overline{p}_i, m_i} H$ via the underlying plain rule $\overline{p}_i$, called the underlying plain transformation.

5.2.1 Local Church–Rosser and Parallelism Theorem

The Local Church–Rosser Theorem is concerned with parallel and sequential independence of direct transformations. We study under what conditions two direct transformations applied to the same graph can be applied in arbitrary order, leading to the same result.

For parallel independence of two direct transformations $H_1 \xLeftarrow{p_1} G \xRightarrow{p_2} H_2$, the first obvious condition is that the underlying plain transformations have to be parallel independent. In addition, we have to require that the matches of p_2 and p_1 in H_1 and H_2, respectively, satisfy the application conditions of the corresponding rule.

Definition 5.19 (Parallel independence). Two direct transformations $G \xRightarrow{p_1, m_1} H_1$ and $G \xRightarrow{p_2, m_2} H_2$ are *parallel independent* if there are morphisms $d_{12} : L_1 \to D_2$ and $d_{21} : L_2 \to D_1$ such that $f_2 \circ d_{12} = m_1$, $f_1 \circ d_{21} = m_2$, $g_2 \circ D_{12} \models ac_1$, and $g_1 \circ d_{21} \models ac_2$.

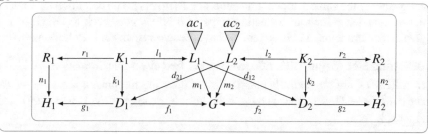

$\triangle$

Example 5.20. The pair $H_1 \xLeftarrow{\texttt{moveDown},m_1} G \xRightarrow{\texttt{intRequest},m_2} H_2$ of direct transformations in Fig. 5.4 is parallel dependent. The left rule application is the one already considered in Fig. 5.3, while m_2 matches the floor in the left-hand side of $\texttt{intRequest}$ to the current floor with the elevator. The morphisms d_{12} and d_{21} exist such that $f_1 \circ d_{21} = m_2$, $f_2 \circ d_{12} = m_1$, and $m_2' = g_1 \circ d_{21} \models \neg \exists \text{ neg}$. But $m_1' = g_2 \circ d_{12} \not\models \neg \exists \text{ neg}_1$, since the rule $\texttt{intRequest}$ added a request at the current floor with the elevator forbidding the application of the rule $\texttt{moveDown}$. Therefore, the transformations are parallel dependent. Note that the underlying plain transformations are parallel independent.

For sequential independence of transformations $G \xRightarrow{p_1} H_1 \xRightarrow{p_2} G'$ we need the sequential independence of the underlying plain rules. Moreover, the match of p_2 in

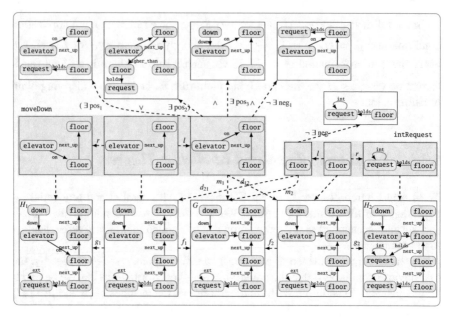

Fig. 5.4 Parallel dependent transformations

G has to satisfy its corresponding application condition and the co-match of p_1 to G' has to satisfy the shifted application condition $R(p_1, ac_1)$. The definition of sequential independence for transformation steps with NACs goes back to [HHT96] for graph transformation, and is generalised to adhesive systems in [LEOP08, Lam10].

Definition 5.21 (Sequential independence). Two direct transformations $G \xRightarrow{p_1, m_1} H_1 \xRightarrow{p_2, m_2} G'$ are *sequentially independent* if there are morphisms $d_{12} : R_1 \to D_2$ and $d_{21} : L_2 \to D_1$ such that $f_2 \circ d_{12} = n_1$, $g_1 \circ d_{21} = m_2$, $g_2 \circ d_{12} \models R(p_1, ac_1)$, and $f_1 \circ d_{21} \models ac_2$.

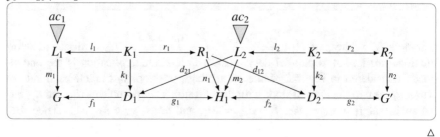

$\triangle$

Example 5.22. The sequence $H_2 \xRightarrow{\text{processIntDown}, m_0} G \xRightarrow{\text{intRequest}, m_3} H_3$ of direct transformations in Fig. 5.5 is sequentially independent. Note that m_0 matches the floor in the left-hand side of processIntDown to the current floor with the elevator, while m_3 matches the floor of the left-hand side of intRequest to the floor one

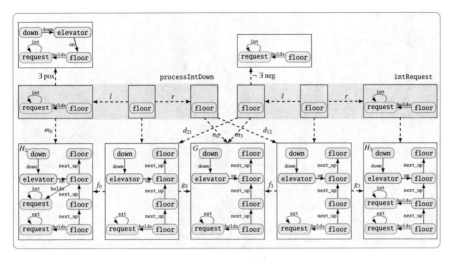

Fig. 5.5 Sequentially independent transformations

level down. The morphisms d_{12} and d_{21} exist such that $g_0 \circ d_{21} = m_3$, $f_3 \circ d_{12} = n_0$, $g_3 \circ d_{12} \models R(\text{processIntDown}, \exists \text{ pos})$, and $f_0 \circ d_{21} \models \neg \exists \text{ neg}$. △

In the case of parallel independence of two direct transformations via rules p_1 and p_2, the parallel rule $p_1 + p_2$ can be defined by binary coproducts of the components of the rules.

Definition 5.23 (Parallel rule). Given rules $p_1 = (L_1 \xleftarrow{l_1} K_1 \xrightarrow{r_1} R_1, ac_1)$ and $p_2 = (L_2 \xleftarrow{l_2} K_2 \xrightarrow{r_2} R_2, ac_2)$, the *parallel rule* $p_1 + p_2 = (L_1 + L_2 \xleftarrow{l_1+l_2} K_1 + K_2 \xrightarrow{r_1+r_2} R_1 + R_2, ac)$ is defined by the componentwise coproducts of the left-hand sides, glueings, and right-hand sides including the morphisms, and $ac = \text{Shift}(i_{L_1},$ $ac_1) \wedge L(p_1 + p_2, \text{Shift}(i_{R_1}, R(p_1, ac_1))) \wedge \text{Shift}(i_{L_2}, ac_2)$ $\wedge L(p_1 + p_2, \text{Shift}(i_{R_2}, R(p_2, ac_2)))$.

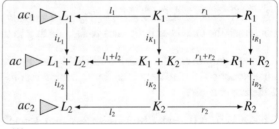

A direct transformation via a parallel rule is called *parallel transformation*. △

The parallel rule is well defined and, in particular, its morphisms are actually $\mathcal{M}$-morphisms.

Fact 5.24. *The morphisms $l_1 + l_2 : K_1 + K_2 \to L_1 + L_2$ and $r_1 + r_2 : K_1 + K_2 \to R_1 + R_2$ are in $\mathcal{M}$.* △

Proof. This follows directly from Fact 4.27. □

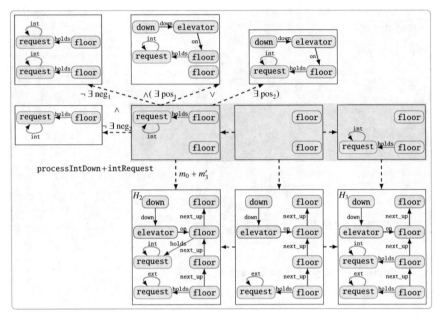

Fig. 5.6 Parallel rule and transformation

Example 5.25. In the upper row of Fig. 5.6, the parallel rule `processIntDown` + `intRequest` is shown. We have only depicted the relevant application conditions and left out those which should not appear in a valid model, for example, graphs with two floors holding the same request. The application conditions describe for various overlappings of the two floors that there is an elevator in downwards mode at the upper floor and no internal request at the lower one. The application $H_2 \xrightarrow{\text{processIntDown+intRequest},m_0+m'_3} H_3$ of this parallel rule is shown in Fig. 5.6 and combines the effects of both rules to H_2, leading to the graph H_3.   △

Now we present the Local Church–Rosser and Parallelism Theorem, which is an abstraction of Theorem 2.26, by replacing graphs by objects from a suitable $\mathcal{M}$-adhesive category.

Theorem 5.26 (Local Church–Rosser and Parallelism Theorem). *Given two*

parallel independent direct transformations $G \xrightarrow{p_1,m_1} H_1$ *and* $G \xrightarrow{p_2,m_2} H_2$*, there is an object* G' *together with direct transformations* $H_1 \xrightarrow{p_2,m'_2} G'$ *and* $H_2 \xrightarrow{p_1,m'_1} G'$ *such that* $G \xrightarrow{p_1,m_1} H_1 \xrightarrow{p_2,m'_2} G'$ *and* $G \xrightarrow{p_2,m_2} H_2 \xrightarrow{p_1,m'_1} G'$ *are sequentially independent.*

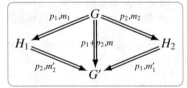

Given two sequentially independent direct transformations $G \xrightarrow{p_1,m_1} H_1 \xrightarrow{p_2,m'_2}$
G', there is an object H_2 together with direct transformations $G \xrightarrow{p_2,m_2} H_2 \xrightarrow{p_1,m'_1} G'$
such that $G \xrightarrow{p_1,m_1} H_1$ *and* $G \xrightarrow{p_2,m_2} H_2$ *are parallel independent.*

In any case of independence, there is a parallel transformation $G \xrightarrow{p_1+p_2,m} G'$,
and, vice versa, a direct transformation $G \xrightarrow{p_1+p_2,m} G'$ *via the parallel rule $p_1 + p_2$*
can be sequentialised both ways.   △

Proof. Consider the parallel independent direct transformations $G \xrightarrow{p_1,m_1} H_1$ and
$G \xrightarrow{p_2,m_2} H_2$. Then the underlying plain transformations $G \xrightarrow{\overline{p_1},m_1} H_1$ and $G \xrightarrow{\overline{p_2},m_2} H_2$ are also parallel independent.

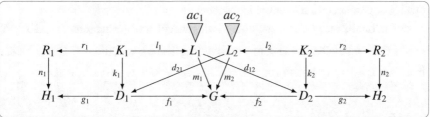

By the Local Church–Rosser Theorem without application conditions [EEPT06], there are an object G' and plain direct transformations $H_1 \xrightarrow{\overline{p_2},m'_2} G' \xleftarrow{\overline{p_1},m'_1} H_2$ such that the plain transformations $G \xrightarrow{\overline{p_1},m_1} H_1 \xrightarrow{\overline{p_2},m'_2} G'$ and $G \xrightarrow{\overline{p_2},m_2} H_2 \xrightarrow{\overline{p_1},m'_1} G'$ are sequentially independent.

Since the two direct transformations are parallel independent, there are morphisms $d_{12} : L_1 \to D_2$ and $d_{21} : L_2 \to D_1$ such that $m_1 = f_2 \circ d_{12}$ and $m_2 = f_1 \circ d_{21}$. Moreover, we have that $m'_1 = g_2 \circ d_{12}$ and $m'_2 = g_1 \circ d_{21}$ from the proof of the plain Local Church–Rosser Theorem. By assumption, $g_2 \circ d_{12} = m'_1 \models ac_1$ and $g_1 \circ d_{21} = m'_2 \models ac_2$; therefore the transformations $H_2 \xrightarrow{p_1,m'_1} G'$ and $H_1 \xrightarrow{p_2,m'_2} G'$ are well defined.

The plain transformations $G \xrightarrow{\overline{p_1},m_1} H_1 \xrightarrow{\overline{p_2},m'_2} G'$ are sequentially independent, with morphisms $d_{21} : L_2 \to D_1$ such that $g_1 \circ d_{21} = m'_2$ and $d'_{12} : R_1 \to D'_2$ such that $f'_2 \circ d'_{12} = n_1$.

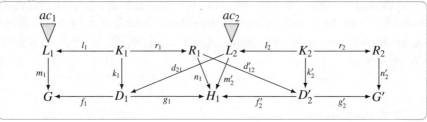

By precondition, $f_1 \circ d_{21} = m_2 \models ac_2$. The proof of the plain Local Church–Rosser Theorem shows that the diagrams (1)–(4) are pushouts, and therefore $g_2 \circ d_{12} \models ac_1 \cong L(R(p_1, ac_1))$ implies that $g'_2 \circ d'_{12} \models R(p_1, ac_1)$ using Fact 5.16. Therefore, also the transformations $G \xrightarrow{p_1, m_1} H_1 \xrightarrow{p_2, m'_2} G'$ are sequentially independent.

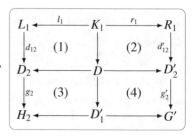

Similarly, the transformations $G \xrightarrow{p_2, m_2} H_2 \xrightarrow{p_1, m'_1} G'$ can be shown to be sequentially independent.

The second statement follows from the first one by using the inverse rule and duality of parallel and sequential independence.

For parallelism, given sequentially independent transformations $G \xrightarrow{p_1, m_1} H_1 \xrightarrow{p_2, m'_2} G'$, the Parallelism Theorem without application conditions [EEPT06] states that there is a parallel transformation $G \xrightarrow{\overline{p_1 + p_2}, m} G'$ with $m_1 = m \circ i_{L_1}$ and $n'_2 = n \circ i_{R_2}$.

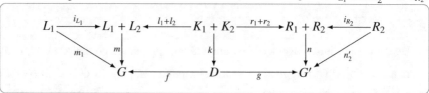

By assumption, $m_1 \models ac_1$ and $m'_2 \models ac_2$. Using Facts 5.4 and 5.16 and Def. 5.23, we have that $m_1 \models ac_1 \Leftrightarrow m \models \text{Shift}(i_{L_1}, ac_1)$ and $m'_2 \models ac_2 \Leftrightarrow n'_2 = n \circ i_{R_2} \models R(p_2, ac_2) \Leftrightarrow n \models \text{Shift}(i_{R_2}, R(p_2, ac_2)) \Leftrightarrow L(p_1 + p_2, \text{Shift}(i_{R_2}, R(p_2, ac_2)))$. Similarly, the sequentially independent transformation $G \xrightarrow{p_2, m_2} H_2 \xrightarrow{p_1, m'_1} G'$ implies that $m \models \text{Shift}(i_{L_2}, ac_2)$ and $m \models L(p_1 + p_2, \text{Shift}(i_{R_1}, R(p_1, ac_1)))$. Thus, $m \models ac$, i.e., the parallel transformation satisfies the application condition.

Vice versa, let $G \xrightarrow{p_1 + p_2, m} G'$ be a parallel transformation. Then there is an underlying plain parallel transformation and, by the Parallelism Theorem without application conditions [EEPT06], there is a sequentially independent direct transformation $G \xrightarrow{\overline{p_1}, m_1} H_1 \xrightarrow{\overline{p_2}, m'_2} G'$ with $m_1 = m \circ i_{L_1}$ and $n'_2 = n \circ i_{R_2}$. By assumption, $m \models ac_L$. From the equivalences above it follows that $m_1 \models ac_1$ and $m'_2 \models ac_2$, i.e., the direct transformations satisfy the application conditions. Similarly, this holds for the sequentially independent direct transformation $G \xrightarrow{p_2, m_2} H_2 \xrightarrow{p_1, m'_1} G'$ with $m_2 \models ac_2$ and $m'_1 \models ac_1$. $\square$

5.2.2 Concurrency Theorem

Sequentially dependent transformations $G \xrightarrow{p_1} H \xrightarrow{p_2} G'$ cannot be combined using the parallel rule. Instead, we use an E-dependency relation and construct an E-concurrent rule $p_1 *_E p_2$ for p_1 and p_2. The resulting E-concurrent transformation $G \xrightarrow{p_1 *_E p_2}$ leads to the same result G' as the E-related transformations $G \xrightarrow{p_1} H \xrightarrow{p_2} G'$. The connection between E-related and E-concurrent transformations is established in the Concurrency Theorem.

The construction of an E-concurrent rule is based on an E-dependency relation which guarantees the existence of suitable pushout complements. The application condition of the E-concurrent rule guarantees that, whenever it is applicable, the rules p_1 and, afterwards, p_2 can be applied.

Definition 5.27 (Concurrent rule). Given rules $p_1 = (L_1 \xleftarrow{l_1} K_1 \xrightarrow{r_1} R_1, ac_1)$ and $p_2 = (L_2 \xleftarrow{l_2} K_2 \xrightarrow{r_2} R_2, ac_2)$, an object E with morphisms $e_1 : R_1 \to E$ and $e_2 : L_2 \to E$ such that $(e_1, e_2) \in \mathcal{E}'$ is an E-dependency relation of p_1 and p_2 if the pushout complements (1) of $e_1 \circ r_1$ and (2) of $e_2 \circ l_2$ exist.

Given an E-dependency relation (E, e_1, e_2) of p_1 and p_2 the E-concurrent rule $p_1 *_E p_2 = (L \xleftarrow{s_1 \circ w_1} K \xrightarrow{t_2 \circ w_2} R, ac)$ is constructed by pushouts (1)–(4) and pullback (5), with $ac = \text{Shift}(u_1, ac_1) \wedge L(p^*, \text{Shift}(e_2, ac_2))$ and $p^* = (L \xleftarrow{s_1} C_1 \xrightarrow{t_1} E)$.

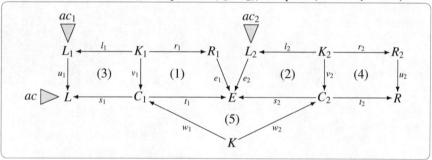

A sequence $G \xrightarrow{p_1, m_1} H \xrightarrow{p_2, m_2} G'$ is called E-related if there exist $h : E \to H$, $c_1 : C_1 \to D_1$, and $c_2 : C_2 \to D_2$ such that $h \circ e_1 = n_1$, $h \circ e_2 = m_2$, $c_1 \circ v_1 = k_1$, $c_2 \circ v_2 = k_2$, and (6) and (7) are pushouts.

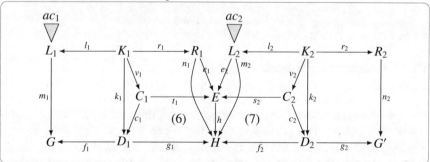

A direct transformation via an E-concurrent rule is called E-concurrent transformation. △

Example 5.28. In Fig. 5.7, a sequentially dependent transformation sequence is shown applying first the rule inRequest, followed by the rule moveDown. Note

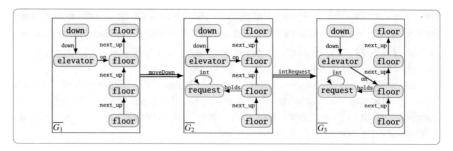

Fig. 5.7 A sequentially dependent transformation

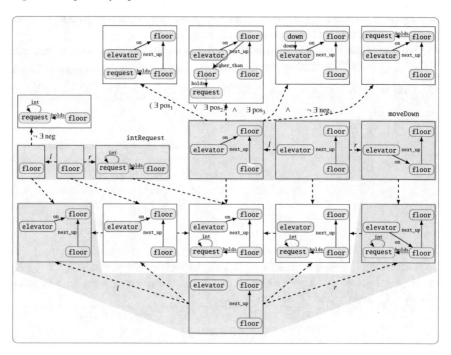

Fig. 5.8 *E*-concurrent rule construction

that these two rules cannot be switched because $\overline{G_1}$ does not fulfill the application condition $\exists\,\mathrm{pos}_1 \lor \exists\,\mathrm{pos}_2$. The corresponding *E*-dependency relation and the construction of the *E*-concurrent rule is depicted in Fig. 5.8. In this case, *E* results from overlapping the floor of the right-hand side of the rule intRequest with the lower floor of the left-hand side of the rule moveDown. The resulting rule is shown in the upper part of Fig. 5.9. Note that the application condition is only depicted for those graphs that may actually occur in valid models. Moreover, the translation $\mathrm{L}(p^*, \mathrm{Shift}(e_2,\ \exists\,\mathrm{pos}_1 \lor \exists\,\mathrm{pos}_2))$ evaluates to true and is therefore ignored. The

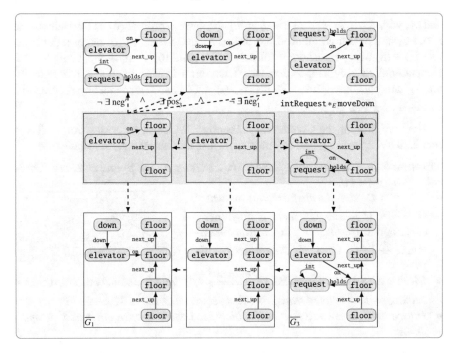

Fig. 5.9 The application of the E-concurrent rule

application of the E-concurrent rule to $\overline{G_1}$ is shown in Fig. 5.9, leading to the E-concurrent transformation $\overline{G_1} \xRightarrow{\text{intRequest}*_E\text{moveDown}} \overline{G_3}$. △

For a given transformation sequence $G \xRightarrow{p_1} H \xRightarrow{p_2} G'$, we are able to compute the corresponding E-dependency relation using the $\mathcal{E}'-\mathcal{M}$ pair factorisation.

Fact 5.29. *For a transformation* $G \xRightarrow{p_1,m_1} H \xRightarrow{p_2,m_2} G'$, *there is an E-dependency relation E such that* $G \xRightarrow{p_1,m_1} H \xRightarrow{p_2,m_2} G'$ *is E-related.* △

Proof. Given a transformation $G \xRightarrow{p_1,m_1} H \xRightarrow{p_2,m_2} H'$ with co-match n_1 for the first direct transformation, let $(e_1, e_2) \in \mathcal{E}'$, $h \in \mathcal{M}$ be an $\mathcal{E}'-\mathcal{M}$ pair factorisation of n_1

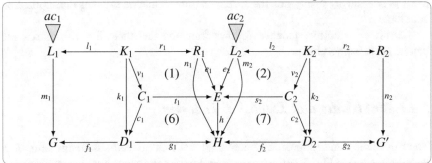

and m_2 with $h \circ e_1 = n_1$ and $h \circ e_2 = m_2$. Now we construct (6) as the pullback of h and g_1 and obtain an induced morphism v_1. From $h \in \mathcal{M}$, (6) being a pullback, and $(1) + (6)$ being a pushout it follows that (1) and (6) are pushouts using the $\mathcal{M}$-pushout–pullback decomposition (see Theorem 4.22). Analogously, diagrams (2) and (7) are pushouts. Thus, E with $(e_1, e_2) \in \mathcal{E}'$ is an E-dependency relation and $G \xrightarrow{p_1, m_1} H \xrightarrow{p_2, m_2} G'$ is E-related. $\square$

Now we present the Concurrency Theorem, which is an abstraction of Theorem 2.29 by replacing graphs by objects from a suitable $\mathcal{M}$-adhesive category.

Theorem 5.30 (Concurrency Theorem). *For rules p_1 and p_2 and an E-concurrent rule $p_1 *_E p_2$ we have:*

- *Given an E-related transformation sequence $G \xrightarrow{p_1, m_1} H \xrightarrow{p_2, m_2} G'$, there is a* synthesis construction *leading to a direct transformation $G \xrightarrow{p_1 *_E p_2, m} G'$ via the E-concurrent rule $p_1 *_E p_2$.*

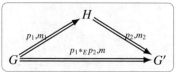

- *Given a direct transformation $G \xrightarrow{p_1 *_E p_2, m} G'$, there is an* analysis construction *leading to an E-related transformation sequence $G \xrightarrow{p_1, m_1} H \xrightarrow{p_2, m_2} G'$.*
- *The synthesis and analysis constructions are inverse to each other up to isomorphism.* $\triangle$

Proof. Let $G \xrightarrow{p_1, m_1} H \xrightarrow{p_2, m_2} G'$ be E-related. Then the underlying plain transformation is E-related and, by the Concurrency Theorem without application conditions [EEPT06], there is an E-concurrent transformation $G \xrightarrow{\overline{p_1 *_E p_2}, m} G'$ with $m \circ u_1 = m_1$.

By assumption, $m_1 \models ac_1$ and $m_2 \models ac_2$. From Facts 5.4 and 5.16 it follows that $m_1 \models ac_1 \wedge m_2 \models ac_2 \Leftrightarrow m \models \text{Shift}(u_1, ac_1) \wedge h \models \text{Shift}(e_2, ac_2) \Leftrightarrow m \models \text{Shift}(u_1, ac_1) \wedge m \models L(p^*, \text{Shift}(e_2, ac_2)) \Leftrightarrow m \models \text{Shift}(u_1, ac_1) \wedge L(p*, \text{Shift}(e_2, ac_2)) = ac$. Thus, $m \models ac$, i.e., the E-concurrent transformation satisfies the application condition.

Vice versa, let $G \xrightarrow{p, m} G'$ be an E-concurrent transformation. Then the underlying plain direct transformation is E-concurrent and, by the Concurrency Theorem without application conditions [EEPT06], there is an E-related transformation $G \xrightarrow{\overline{p_1}, m_1} H \xrightarrow{\overline{p_2}, m_2} G'$. By assumption, $m \models ac$. As shown above, this is equivalent to $m_1 \models ac_1$ and $m_2 \models ac_2$, i.e., the E-related transformation satisfies the application conditions.

The bijective correspondence follows from the fact that all constructions are unique up to isomorphism. $\square$

5.2.3 Embedding and Extension Theorem

The Embedding and Extension Theorem allows us to extend a transformation to a larger context (see Fig. 5.10). An extension diagram describes how a transformation

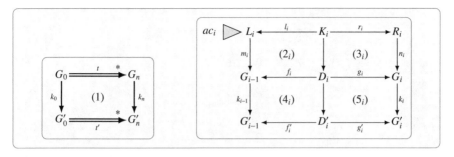

Fig. 5.10 Embedding and extension: sequence (left) and intermediate step (right)

$t : G_0 \overset{*}{\Rightarrow} G_n$ can be extended to a transformation $t' : G'_0 \overset{*}{\Rightarrow} G'_n$ via the same rules and an extension morphism $k_0 : G_0 \to G'_0$. For each rule application and transformation step, we have two double pushout diagrams (2_i)–(5_i), where the rule p_i is applied to both G_{i-1} and G'_{i-1}.

A sufficient and necessary condition for this extension is the consistency of an extension morphism. It is based on the notions of derived span and derived application condition of a transformation. The derived span describes the combined changes of a transformation by condensing it into a single span. The derived application condition summarises all application conditions of a transformation into a single one.

Definition 5.31 (Derived span and application condition). Given a transformation $t : G_0 \overset{*}{\Rightarrow} G_n$ via rules $p_1, \ldots, p_n$, the *derived span* $der(t)$ is inductively defined by

$$der(t) = \begin{cases} G_0 \overset{f_1}{\leftarrow} D_1 \overset{g_1}{\to} G_1 & \text{for } t : G_0 \overset{p_1,m_1}{\Longrightarrow} G_1 \\ G_0 \overset{d'_0 \circ d}{\leftarrow} D \overset{g_n \circ d_n}{\longrightarrow} G_n & \text{for } t : G_0 \overset{*}{\Rightarrow} G_{n-1} \overset{p_n,m_n}{\Longrightarrow} G_n \text{ with pullback } (PB) \text{ and} \\ & der(G_0 \overset{*}{\Rightarrow} G_{n-1}) = (G_0 \overset{d'_0}{\leftarrow} D' \overset{d'_{n-1}}{\to} G_{n-1}) \end{cases}$$

$$G_0 \overset{d'_0}{\longleftarrow} D' \overset{d'_{n-1}}{\longrightarrow} G_{n-1} \overset{f_n}{\longleftarrow} D_n \overset{g_n}{\longrightarrow} G_n$$
$$(PB)$$
$$d \searrow \qquad \swarrow d_n$$
$$D$$

Moreover, the *derived application condition* $ac(t)$ is defined by

$$ac(t) = \begin{cases} \text{Shift}(m_1, ac_1) & \text{for } t : G_0 \overset{p_1,m_1}{\Longrightarrow} G_1 \\ ac(G_0 \overset{*}{\Rightarrow} G_{n-1}) & \text{for } t : G_0 \overset{*}{\Rightarrow} G_{n-1} \overset{p_n,m_n}{\Longrightarrow} G_n \\ \wedge L(p_n^*, \text{Shift}(m_n, ac_n)) & \text{with } p_n^* = der(G_0 \overset{*}{\Rightarrow} G_{n-1}) \end{cases} \qquad \triangle$$

Example 5.32. Consider the transformation $\overline{G_1} \overset{\text{intRequest}}{\Longrightarrow} \overline{G_2} \overset{\text{moveDown}}{\Longrightarrow} \overline{G_3} \overset{\text{processIntDown}}{\Longrightarrow} \overline{G_4}$, where the first part of the transformation is shown in Fig. 5.9, while the last direct transformation step deletes the request on the second lowest floor. The derived span of this transformation and its derived application condition

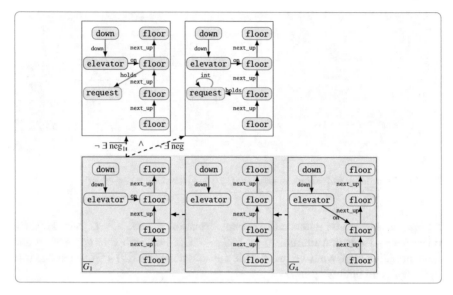

Fig. 5.11 The derived span and derived application condition of $\overline{G_1} \overset{*}{\Rightarrow} \overline{G_4}$

are shown in Fig. 5.11 and combine all changes applied in the single transformation steps. Note that in the translation of the derived application condition, several parts evaluate to true and are therefore not depicted.   △

The notion of consistency combines that of boundary consistency, ensuring the preservation of certain structures, and AC consistency, ensuring that the application conditions hold.

Definition 5.33 (Consistency). Given a transformation $t : G_0 \overset{*}{\Rightarrow} G_n$ with derived span $der(t) = (G_0 \overset{d_0^*}{\longleftarrow} D \overset{d_n^*}{\longrightarrow} G_n)$ and derived application condition $ac(t)$, and a morphism $k_0 : G_0 \to G_0' \in \mathcal{M}'$:

- k_0 is *boundary-consistent* with respect to t if there exists an initial pushout (6) over k_0 (see Def. 4.23) and a morphism $b^* \in \mathcal{M}$ with $d_0^* \circ b^* = b$.

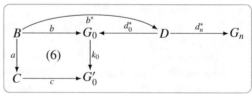

- k_0 is *AC-consistent* with respect to t if $k_0 \models ac(t)$.
- k_0 is *consistent* with respect to t if k_0 is both boundary- and AC-consistent with respect to t.   △

The consistency condition is sufficient and necessary for the construction of extension diagrams, provided that we have initial pushouts over $\mathcal{M}'$-morphisms. Moreover, we are able to give a direct construction of G_n' in the extension diagram, which avoids constructing the complete transformation $t' : G_0' \overset{*}{\Rightarrow} G_n'$.

Theorem 5.34 (Embedding and Extension Theorem). *Given a transformation* $t : G_0 \overset{*}{\Rightarrow} G_n$ *and a morphism* $k_0 : G_0 \to G_0' \in \mathcal{M}'$ *which is consistent with respect to* t, *there is an extension diagram over* t *and* k_0.

Vice versa, given a transformation $t : G_0 \overset{*}{\Rightarrow} G_n$ *with an extension diagram* (1) *and initial pushout* (6) *over* $k_0 : G_0 \to G_0' \in \mathcal{M}'$ *as above, we have that:*

1. k_0 *is consistent with respect to* $t : G_0 \overset{*}{\Rightarrow} G_n$.

2. There is a rule $p^* = (der(t), ac(t))$ *leading to a direct transformation* $G_0' \overset{p^*}{\Rightarrow} G_n'$.
3. G_n' *is the pushout of* C *and* G_n *along* B, *i.e.,* $G_n' = G_n +_B C$.   △

Proof. Let $t : G_0 \overset{*}{\Rightarrow} G_n$ be a transformation and $k_0 : G_0 \to G_0' \in \mathcal{M}'$ consistent with respect to t. Then k_0 is boundary-consistent with respect to the underlying plain transformation $\bar{t}$ and, by the Embedding Theorem for rules without application conditions [EEPT06], there is a plain extension diagram over $\bar{t}$ and k_0.

By assumption, $k_0 \models ac(t)$. It remains to show that the application condition ac_i is fulfilled for each single transformation step in the extension diagram, i.e., $k_{i-1} \circ m_i \models ac_i$ for $i = 1, \ldots, n$. This is proven by induction over the number of direct transformation steps n.

Basis. For a transformation $t : G_0 \Rightarrow G_0$ of length 0, $k_0 \models ac(t) = \text{true}$. For a transformation $t : G_0 \overset{p_1, m_1}{\Longrightarrow} G_1$ of length 1, $k_0 \models ac(t) = \text{Shift}(m_1, ac_1)$ if and only if $k_0 \circ m_1 \models ac_1$.

Induction hypothesis. For a transformation $t : G_0 \overset{*}{\Rightarrow} G_i$ of length $i \geq 1$, $k_0 \models ac(t) \Leftrightarrow k_{j-1} \circ m_j \models ac_j$ for $j = 1, \ldots, i$.

Induction step. Consider now the transformation $t : G_0 \overset{*}{\Rightarrow} G_i \Rightarrow G_{i+1}$. Then we have that:

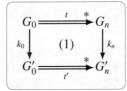

$$k_0 \models ac(G_0 \overset{*}{\Rightarrow} G_{i+1})$$
$$\Leftrightarrow k_0 \models ac(G_0 \overset{*}{\Rightarrow} G_i) \wedge \text{L}(der(G_0 \overset{*}{\Rightarrow} G_i), \text{Shift}(m_{i+1}, ac_{i+1})) \quad \text{Def. 5.31}$$
$$\Leftrightarrow k_0 \models ac(G_0 \overset{*}{\Rightarrow} G_i) \wedge k_i \models \text{Shift}(m_{i+1}, ac_{i+1}) \quad \text{Fact 5.16}$$
$$\Leftrightarrow k_0 \models ac(G_0 \overset{*}{\Rightarrow} G_i) \wedge k_i \circ m_{i+1} \models ac_{i+1} \quad \text{Fact 5.4}$$
$$\Leftrightarrow k_{j-1} \circ m_j \models ac_j \text{ for } j = 1, \ldots, i+1 \quad \text{Induction hypothesis}$$

This means that the resulting plain extension diagram is actually valid for all direct transformation steps, i.e., it is an extension diagram over t and k_0.

Vice versa, let $t : G_0 \overset{*}{\Rightarrow} G_n$ be a transformation with an extension diagram (1) and initial pushout (6) over $k_0 \in \mathcal{M}'$. By the Extension Theorem for rules without application conditions [EEPT06], k_0 is boundary-consistent with respect to the underlying plain transformation $\bar{t}$ with morphism $b^* : B \to D$ such that $d_0^* \circ b^* = b$. By assumption,

(1) is an extension diagram, i.e., $t' : G_0' \overset{*}{\Rightarrow} G_n'$ is a transformation via the rules $p_1, \ldots, p_n$ with $k_{i-1} \circ m_i \models ac_i$ for $i = 1, \ldots, n$. This means that $k_0 \models ac(t)$, and hence it is consistent with respect to t. Moreover, Items 2 and 3 are valid. which

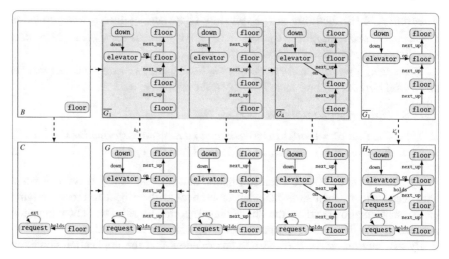

Fig. 5.12 The embedding of $\overline{G_1}$ into G and H_2

follows directly from the corresponding results for the underlying plain transformation.   □

Example 5.35. We can embed the graph $G_0 = \overline{G_1}$ from Fig. 5.9 into the larger context graph $G'_0 = G$ from Fig. 5.3, where G contains an additional external request on the lowest floor. The boundary B contains only this lowest floor, while the context graph C adds the additional request to this floor, as shown in the left of Fig. 5.12. Since this floor is not deleted, the extension morphism k_0 is boundary-consistent. Moreover, it is AC-consistent—because there is no request on the second or third floor, the derived application condition is fulfilled. Therefore, we have consistency and can construct the transformation $G'_0 \stackrel{*}{\Rightarrow} G'_3 = H_1$, where H_1 is constructed as the pushout of $\overline{G_4}$ and C along B.

In contrast, the morphism $k'_0 : \overline{G_1} \to H_2$ in the right of Fig. 5.12 is not consistent, because it does not satisfy the application condition $\neg \exists \overline{neg_1}$.   △

5.2.4 Critical Pairs and Local Confluence Theorem

Confluence is the property ensuring the functional behaviour of transformation systems. A system is confluent if whenever an object G can be transformed into two objects H_1 and H_2, these can be transformed into a common object G'. A slightly weaker property is local confluence, where we only ask this property for direct transformations of G into H_1 and H_2. Confluence co-incides with local conflu-

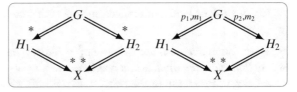

ence when the given transformation system is terminating, as shown in [New42].

For parallel independent transformations, the Local Church–Rosser Theorem ensures confluence (see, e.g., [EEPT06]). In general, however, not all pairs of direct transformations are parallel independent. It remains to analyse the parallel dependent ones.

The intuition behind using critical pairs to check (local) confluence is that we do not have to study all possible cases of parallel dependent pairs of rule applications, but only some minimal ones which are built by gluing the left-hand sides of these pairs. A *critical pair* for the rules p_1, p_2 is a pair of parallel dependent transformations, $P_1 \xLeftarrow{p_1,o_1} K \xRightarrow{p_2,o_2} P_2$, where o_1 and o_2 are in $\mathcal{E}'$. A completeness lemma, showing that every pair of parallel dependent rule applications embeds a critical pair, justifies why it is enough to consider the confluence of these cases.

As shown in [Plu93, Plu05], the confluence of all critical pairs is not a sufficient condition for the local confluence of a general transformation system, as it is in the case of term rewriting. Instead, a stronger notion of confluence is needed, called *strict confluence*. In [Plu05, EEPT06] it is shown for plain rules that strict confluence of all critical pairs implies the local confluence of a transformation system. In the following we show how to extend these results to rules with application conditions, which considerably complicate the confluence analysis. We introduce a new notion of *strict AC confluence* of critical pairs for an adequate handling of application conditions, leading to local confluence of the transformation system.

First, we present a new simple but weak notion of critical pairs. We know that all pairs of rule applications are potentially nonconfluent, even if their underlying plain transformations are parallel independent. Thus, we define as weak critical pairs all the minimal contexts of all pairs of plain rule applications.

Definition 5.36 (Weak critical pair). Given rules $p_1 = (\overline{p}_1, ac_1)$ and $p_2 = (\overline{p}_2, ac_2)$, a pair $P_1 \xLeftarrow{\overline{p}_1,o_1} K \xRightarrow{\overline{p}_2,o_2} P_2$ of plain transformations is a *weak critical pair* for (p_1, p_2), if $(o_1, o_2) \in \mathcal{E}'$.

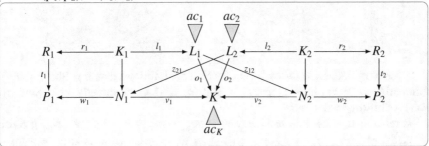

Every weak critical pair induces an application condition $ac_K = ac_K^E \wedge ac_K^C$ on K with

- *extension application condition:* $ac_K^E = \text{Shift}(o_1, ac_1) \wedge \text{Shift}(o_2, ac_2)$ and
- *conflict-inducing application condition:* $ac_K^C = \neg(ac_{z_{21}} \wedge ac_{z_{12}})$, with
  if $(\exists z_{12} : v_2 \circ z_{12} = o_1$ then $ac_{z_{12}} = L(p_2^*, \text{Shift}(w_2 \circ z_{12}, ac_1))$ else $ac_{z_{12}} = $ false,
  $$\text{with } p_2^* = (K \xleftarrow{v_2} N_2 \xrightarrow{w_2} P_2)$$
  if $(\exists z_{21} : v_1 \circ z_{21} = o_2$ then $ac_{z_{21}} = L(p_1^*, \text{Shift}(w_1 \circ z_{21}, ac_2))$ else $ac_{z_{21}} = $ false,
  $$\text{with } p_1^* = (K \xleftarrow{v_1} N_1 \xrightarrow{w_1} P_1) \qquad \triangle$$

The two application conditions ac_K^E and ac_K^C are used to characterise the extensions of K that may give rise to a confluence conflict. If a morphism $m : K \to G$ models ac_K^E then $m \circ o_1$ and $m \circ o_2$ are two matches of p_1 and p_2, respectively, satisfying their associated application conditions. If these two plain transformations are parallel independent then ac_K^C is precisely the condition that ensures that the two transformations with application conditions are parallel dependent.

We can prove that each pair of parallel dependent transformations is an extension of a weak critical pair.

Fact 5.37. *For each pair of parallel dependent direct transformations*
$H_1 \xLeftarrow{p_1, m_1} G \xRightarrow{p_2, m_2} H_2$ *there is a weak critical pair* $P_1 \xLeftarrow{\overline{p}_1, o_1} K \xRightarrow{\overline{p}_2, o_2} P_2$ *with induced application condition ac_K and morphism $m : K \to G \in \mathcal{M}$ with $m \models ac_K$, leading to extension diagrams (1) and (2).* $\qquad \triangle$

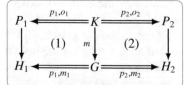

Proof. Consider the parallel dependent transformations $H_1 \xLeftarrow{p, m_1} G \xRightarrow{p_2, m_2} H_2$. For m_1 and m_2 there exists an $\mathcal{E}'$–$\mathcal{M}$ pair factorisation with an object K and morphisms $m \in \mathcal{M}$, $(o_1, o_2) \in \mathcal{E}'$ such that $m_1 = m \circ o_1$ and $m_2 = m \circ o_2$. Using the Restriction Theorem without application conditions [EEPT06] we obtain transformation $K \xRightarrow{\overline{p}_1, o_1} P_1$ and $K \xRightarrow{\overline{p}_2, o_2} P_i$, leading to the required plain extension diagrams.

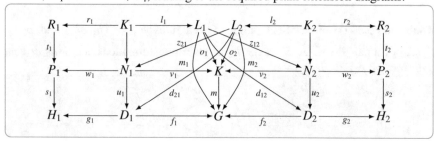

By assumption, $m_1 = m \circ o_1 \models ac_1$. By Fact 5.4 it follows that $m \models \text{Shift}(o_1, ac_1)$. Similarly, $m_2 \models ac_2$ implies that $m \models \text{Shift}(o_2, ac_2)$. Consequently, $m \models \text{Shift}(o_1, ac_1) \wedge \text{Shift}(o_2, ac_2) = ac_K^E$.

It remains to show that $m \models ac_K^C = \neg(ac_{z_{21}} \wedge ac_{z_{12}}) = \neg ac_{z_{21}} \vee \neg ac_{z_{12}})$. Since $H_1 \xLeftarrow{p, m_1} G \xRightarrow{p_2, m_2} H_2$ are parallel dependent, we have at least one of the following cases:

1. $\nexists d_{12} : f_2 \circ d_{12} = m_1$. Then also $\nexists z_{12} : v_2 \circ z_{12} = o_1$, because otherwise we could define $d_{12} = u_2 \circ z_{12}$. By definition, $ac_{z_{12}} = $ false. This means that $m \not\models ac_{z_{12}}$, i.e., $m \models ac_K^C$.

2. $\exists d_{12} : f_2 \circ d_{12} = m_1$, but $g_2 \circ d_{12} \not\models ac_1$. If $\nexists z_{12} : v_2 \circ z_{12} = o_1$ then $ac_{z_{12}} = $ false and $m \models ac_K^C$ as in Case 1. Otherwise, $ac_{z_{12}} = L(p_2^*, \text{Shift}(w_2 \circ z_{12}, ac_1))$. Using Facts 5.4 and 5.16, we have that $m \models L(p_2^*, \text{Shift}(w_2 \circ z_{12}, ac_1)) \Leftrightarrow s_2 \models \text{Shift}(w_2 \circ z_{12}, ac_1) \Leftrightarrow s_2 \circ w_2 \circ z_{12} \models ac_1$. From $f_2 \in \mathcal{M}$ and $f_2 \circ d_{12} = m_1 = m \circ o_1 = m \circ v_2 \circ z_{12} = f_2 \circ u_2 \circ z_{12}$ it follows that $u_2 \circ z_{12} = d_{12}$. Thus, using $g_2 \circ d_{12} = g_2 \circ u_2 \circ z_{12} = s_2 \circ w_2 \circ z_{12}$ it follows that $m \not\models ac_{z_{12}}$, i.e., $m \models ac_K^C$.

3. $\nexists d_{21} : f_1 \circ d_{21} = m_2$. Similarly to Case 1, $ac_{z_{21}} = $ false and $m \models ac_K^C$.

4. $\exists d_{21} : f_1 \circ d_{21} = m_2$, but $g_1 \circ d_{21} \not\models ac_2$. Similarly to Case 2, either $\nexists z_{21} : v_1 \circ z_{21} = o_2$ and $ac_{z_{21}} = $ false, or $ac_{z_{21}} = L(p_1^*, \text{Shift}(w_1 \circ z_{21}, ac_2))$ and $g_1 \circ d_{21} \not\models ac_2$ implies that $m \not\models ac_{z_{21}}$. In both cases, it follows that $m \models ac_K^C$. $\square$

It can be shown that the conflict-inducing application condition ac_K^C is characteristic for parallel dependency.

Fact 5.38. *Consider a pair of transformations* $H_1 \overset{p_1,m_1}{\Longleftarrow} G \overset{p_2,m_2}{\Longrightarrow} H_2$ *embedding a weak critical pair* $P_1 \overset{\overline{p}_1,o_1}{\Longleftarrow} K \overset{\overline{p}_2,o_2}{\Longrightarrow} P_2$ *with morphism* $m \in \mathcal{M}$ *and* $m \models ac_K^E$. *Then we have that* $H_1 \overset{p_1,m_1}{\Longleftarrow} G \overset{p_2,m_2}{\Longrightarrow} H_2$ *is parallel dependent if and only if* $m \models ac_K^C$. $\triangle$

Proof. "$\Rightarrow$". This follows from Fact 5.37.

"$\Leftarrow$". If the pair of underlying plain transformations is parallel dependent, then also $H_1 \overset{p_1,m_1}{\Longleftarrow} G \overset{p_2,m_2}{\Longrightarrow} H_2$ is parallel dependent. Otherwise, we have morphisms d_{12} and d_{21} with $f_2 \circ d_{12} = m_1 = m \circ o_1$ and $f_1 \circ d_{21} = m_2 = m \circ o_2$. Since N_2 is a pullback object we obtain a unique morphism z_{12} with $v_2 \circ z_{12} = o_1$ and $u_2 \circ z_{12} = d_{12}$. Similarly, the pullback object N_1 induces a unique morphism z_{21} with $v_1 \circ z_{21} = o_2$ and $u_1 \circ z_{21} = d_{21}$.

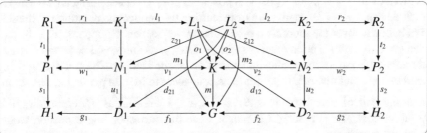

As shown in the proof of Fact 5.37, in this case $m \models ac_{z_{12}} \Leftrightarrow g_2 \circ d_{12} \models ac_1$ and $m \models ac_{z_{21}} \Leftrightarrow g_1 \circ d_{21} \models ac_2$. By assumption, $m \models ac_K^C$ and, by definition of ac_K^C, $m \not\models ac_{z_{12}}$ or $m \not\models ac_{z_{21}}$. If $m \not\models ac_{z_{12}}$ then $g_2 \circ d_{12} \not\models ac_1$, and similarly if $m \not\models ac_{z_{21}}$ then $g_1 \circ d_{21} \not\models ac_2$. Therefore, at least one of the application conditions is not fulfilled and the pair $H_1 \overset{p_1,m_1}{\Longleftarrow} G \overset{p_2,m_2}{\Longrightarrow} H_2$ is parallel dependent. $\square$

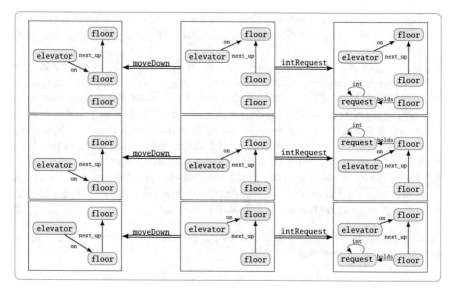

Fig. 5.13 Three weak critical pairs for moveDown and intRequest

Example 5.39. For the rules moveDown and intRequest, three weak critical pairs exist as shown in Fig. 5.13. As analysed in Ex. 5.20, the pair $H_1 \xLeftarrow{\text{moveDown},m_1} G \xRightarrow{\text{intRequest},m_2} H_2$ of direct transformations in Fig. 5.4 is parallel dependent. The second weak critical pair can be embedded into this pair of transformations. Note that this critical pair actually consists of plain transformations, but not valid transformations with application conditions, because $\exists\, \text{pos}_3$ is not fulfilled by o_1. K coincides with the left-hand side of the rule moveDown. The application condition ac_K is given by $ac_K = ac_K^E = (\exists\, \text{pos}_1 \vee \exists\, \text{pos}_2) \wedge \exists\, \text{pos}_3 \wedge \neg\, \exists\, \text{neg}_1$. Actually, we had to include neg' : $K \rightarrow P_2$, stemming from shifting the application condition of the rule intRequest over the morphism o_2. But if any request on the upper floor is forbidden ($\neg\, \exists\, \text{neg}_1$), so is an internal request ($\neg\, \exists\, \text{neg}'$), meaning $\neg\, \exists\, \text{neg}_1 \Rightarrow \neg\, \exists\, \text{neg}'$; therefore this part of the application condition can be ignored. The conflict-inducing application condition ac_K^C turns out to be equivalent to true. In particular, this means that any pair of transformations $H_1 \xLeftarrow{\text{moveDown}} G \xRightarrow{\text{intRequest}} H_2$ embedding this weak critical pair is parallel dependent since the corresponding extension would trivially satisfy ac_K^C.   △

Not every weak critical pair may be embedded in a parallel dependent pair of rule applications. Weak critical pairs without an extension m satisfying ac_K are useless for checking local confluence, because no extension of parallel dependent transformation exists. Therefore, we extend our notion of critical pair in the sense that they are also complete and each of them is embedded in at least one case of parallel dependence. In particular, a critical pair is a weak critical pair such that there is at least one extension satisfying ac_K.

Definition 5.40 (Critical pair). Given rules $p_1 = (\overline{p}_1, ac_1)$ and $p_2 = (\overline{p}_2, ac_2)$, a weak critical pair $P_1 \xLeftarrow{\overline{p}_1, o_1} K \xRightarrow{\overline{p}_2, o_2} P_2$ is a *critical pair* if there exists an extension of the pair via a morphism $m : K \to G \in \mathcal{M}$ such that $m \circ o_1 = m_1$, $m \circ o_2 = m_2$, and $m \models ac_K$. △

Note that this new notion of critical pairs is different from the one for rules with negative application conditions in [Lam10]. The main difference is that critical pairs in our sense may disregard the application conditions. In the case that all application conditions are negative application conditions, the above notion of critical pairs does not coincide with the notion defined in [Lam10], although they are in some sense equivalent. In [Lam10], so-called produce–forbid critical pairs may contain, in addition to an overlap of the left-hand sides of the rules, a part of the corresponding negative application conditions. In our notion, this additional part would be included in ac_K^C.

Also, critical pairs as defined above are complete. Moreover, the converse property also holds, in the sense that each critical pair can be extended to a pair of parallel dependent rule applications.

Theorem 5.41 (Completeness Theorem). *For each pair of parallel dependent direct transformations $H_1 \xLeftarrow{p_1, m_1} G \xRightarrow{p_2, m_2} H_2$ there is a critical pair $P_1 \xLeftarrow{\overline{p}_1, o_1} K \xRightarrow{\overline{p}_2, o_2} P_2$ and a morphism $m : K \to G \in \mathcal{M}$ with $m \models ac_K$, leading to extension diagrams (1) and (2).*

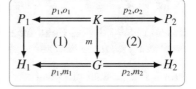

Moreover, for each critical pair $P_1 \xLeftarrow{\overline{p}_1, o_1} K \xRightarrow{\overline{p}_2, o_2} P_2$ for (p_1, p_2) there is a parallel dependent pair $H_1 \xLeftarrow{p_1, m_1} G \xRightarrow{p_2, m_2} H_2$ and a morphism $m : K \to G \in \mathcal{M}$ such that $m \models ac_K$, leading to to extension diagrams (1) and (2). △

Proof. By Fact 5.37, for parallel dependent $H_1 \xLeftarrow{p_1, m_1} G \xRightarrow{p_2, m_2} H_2$ there is a weak critical pair $P_1 \xLeftarrow{\overline{p}_1, o_1} K \xRightarrow{\overline{p}_2, o_2} P_2$ with $m \models ac_K$, leading to extension diagrams (1) and (2). Since this extension exists, the weak critical pair is indeed a critical pair.

Conversely, it follows from the definition of critical pair that each critical pair can be extended to a pair of parallel dependent transformations. □

Example 5.42. Due to Theorem 5.41, the weak critical pair in Ex. 5.39 is actually a critical pair. In particular, the parallel dependent transformations depicted in Fig. 5.4 satisfy ac_K. △

In order to show local confluence, we have to require that all critical pairs be confluent. However, even in the case of plain graph transformation rules, this is not sufficient to show local confluence [Plu93, Plu05]. For transformations without application conditions, the strict confluence of critical pairs ensures that the confluent transformations of the critical pair can be extended to the original pair of transformations. However, if we consider application conditions we may be unable to apply some of these rules if the corresponding matches of the extensions fail to satisfy the application conditions. Therefore, we compute a derived application condition $ac(\bar{t})$ (see Def. 5.31), collecting all application conditions in the transformation, which is used in the notion of strict AC confluence.

Definition 5.43 (Strict AC confluence). A critical pair $P_1 \xLeftarrow{\bar{p}_1,o_1} K \xRightarrow{\bar{p}_2,o_2} P_2$ with induced application conditions ac_K is strictly AC-confluent if it is

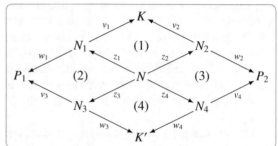

1. confluent without application conditions, i.e., there are plain transformations $P_1 \overset{*}{\Rightarrow} K'$ and $P_2 \overset{*}{\Rightarrow} K'$,
2. strict, i.e., given derived spans $der(P_i \xRightarrow{\bar{p}_i,o_i} K_i) = (K \xleftarrow{v_i} N_i \xrightarrow{w_i} P_i)$ and $der(P_i \overset{*}{\Rightarrow} K') = (P_i \xleftarrow{v_{i+2}} N_{i+2} \xrightarrow{w_{i+2}} K')$ for $i = 1, 2$ and pullback (1), there exist morphisms z_3, z_4 such that diagrams (2), (3), and (4) commute, and

3. for $\bar{t}_i : K \xRightarrow{\bar{p}_i,o_i} P_i \overset{*}{\Rightarrow} K'$ it holds that $ac_K \Rightarrow ac(\bar{t}_i)$ for $i = 1, 2$.   △

Theorem 5.44 (Local Confluence Theorem). *A transformation system is locally confluent if all its critical pairs are strictly AC-confluent.*   △

Proof. For parallel independent direct transformations Theorem 5.26 implies that we have local confluence.

For a parallel dependent pair of transformations $H_1 \xLeftarrow{p_1,m_1} G \xRightarrow{p_2,m_2} H_2$ we find a critical pair $P_1 \xLeftarrow{\bar{p}_1,o_1} K \xRightarrow{\bar{p}_2,o_2} P_2$ and embedding diagrams (5) and (6) with a morphism $m \in \mathcal{M}$ such that $m \models ac_K$ (Theorem 5.41). By assumption, this critical pair is confluent without application conditions, i.e., we have diagram (7). The Local Confluence Theorem without application conditions (see [EEPT06]) implies that we have corresponding extensions t'_1 of t_1 and t'_2 of t_2 in diagrams (8) and (9).

Since the extension diagrams exist, the Extension Theorem without application conditions (see [EEPT06]) implies that m is boundary-consistent w.r.t. t_1 and t_2 as well as $\bar{t}_1 : K \xRightarrow{p_1,o_1} P_1 \Rightarrow_{t_1} K'$ and $\bar{t}_2 : K \xRightarrow{p_2,o_2} P_2 \Rightarrow_{t_2} K'$.

It remains to show that all direct transformations in $\bar{t}'_1 : G \xRightarrow{p_1,m_1} H_1 \overset{t'_1}{\Rightarrow} G'$ and $\bar{t}'_2 : G \xRightarrow{p_2,m_2} H_2 \overset{t'_2}{\Rightarrow} G'$ satisfy their corresponding application conditions. This means that we have to show that m is AC-consistent w.r.t. $\bar{t}_1$ and $\bar{t}_2$, i.e., $m \models$

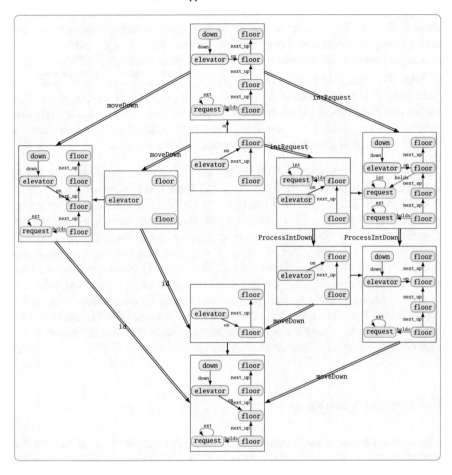

Fig. 5.14 Strict AC confluence of the critical pair in Examples 5.39 and 5.42

$ac(\bar{t}_1)$ and $m \models ac(\bar{t}_2)$. By AC confluence, especially AC compatibility, we have that $ac_K \Rightarrow ac(\bar{t}_1)$ and $ac_K \Rightarrow ac(\bar{t}_2)$. Since $m \models ac_K$ it follows that $m \models ac(\bar{t}_1)$ and $m \models ac(\bar{t}_2)$. Now Theorem 5.34 implies that t_1' and t_2' as well as $\bar{t}_1$ and $\bar{t}_2$ are valid transformations, even with application conditions.   □

Example 5.45. Using Fig. 5.14, we want to show that the critical pair in Examples 5.39 and 5.42 is strictly AC-confluent. First of all, when applying first the rule processIntDown and then the rule moveDown to P_2 as rules without application conditions, we obtain $K' = P_1$ as a confluent model (see Fig. 5.14). This is a strict solution, since it deletes none of the two floors and the edge between them nor the elevator—these four elements are the only structures preserved by the critical pair. To show AC compatibility, we have to analyse the transformations $\bar{t}_1 : K \xrightarrow{\text{moveDown}} P_1 = K'$ and $\bar{t}_2 : K \xrightarrow{\text{intRequest}} P_2 \xrightarrow{\text{processIntDown}}$

$P_3 \overset{\text{moveDown}}{\Longrightarrow} K'$. Since $ac(\bar{t}_1) = ac_K$ we have that $ac_K \Rightarrow ac(\bar{t})$. Similarly, we have $ac(\bar{t}_2) = ac_K \wedge \exists\, \text{pos}' \wedge \neg\, \exists\, \text{neg}'$ with $\text{pos}' = \text{pos}_3$ and, as explained in Ex. 5.39, $\neg\, \exists\, negac_1 \Rightarrow \text{neg}'$. Therefore $ac_K \Rightarrow ac(\bar{t}_2)$ and the critical pair is AC-compatible.

Now Theorem 5.44 implies that the pair $H_1 \overset{\text{moveDown},m_1}{\Longleftarrow} G \overset{\text{intRequest},m_2}{\Longrightarrow} H_2$ from Fig. 5.4 is locally confluent, as shown in the outer diagram of Fig. 5.14. This means that if the elevator is in downwards mode with a request on the lowest floor, we can first process a new internal request on the actual floor and then continue moving downwards instead of moving downwards immediately.   △

Remark 5.46. As shown in [HP09], in the case of graphs, application conditions are expressively equivalent to first-order graph formulas. This means that the satisfiability problem for application conditions is undecidable and, as a consequence, constructing the set of critical pairs for a transformation system with arbitrary application conditions would be a noncomputable problem. Similarly, showing logical consequence, and in particular AC compatibility, and therefore showing strict confluence, is also undecidable. However, in [OEP08, Pen08], techniques are presented to tackle the satisfiability and the deduction problems in practice. Obviously, this kind of techniques would be important in our context for computing critical pairs. Nevertheless, it must be taken into account that, as shown in [Plu05], checking local confluence for terminating graph transformation systems is undecidable, even in the case of rules without application conditions.   △

5.3 Process Analysis

This section presents general techniques for the analysis of processes of $\mathcal{M}$-adhesive transformation systems, i.e., of equivalence classes of executions differing only for the interleaving of the same transformation steps. The main problem in this context is to analyse whether a sequence of transformation steps can be rearranged in order to generate all possible equivalent executions, or some specific and possibly better ones. We define processes of $\mathcal{M}$-adhesive transformation systems based on subobject transformation systems inspired by processes for Petri nets [RE96] and adhesive rewriting systems [BCH+06]. For this purpose, we use the concept of *permutation equivalence* [Her09, HCE14] for transformation systems with negative application conditions (NACs) in $\mathcal{M}$-adhesive categories. Permutation equivalence is coarser than switch equivalence with NACs and has interesting applications in the area of business processes [BHE09b, BHG11]. This section is based on [Her09, HCE14].

In the main results of this section, we show that processes represent equivalence classes of permutation-equivalent transformation sequences. Moreover, they can be analysed efficiently by complete firing sequences of a Petri net, which can be constructed effectively as a dependency net of a given transformation sequence. Most constructions and results are illustrated by a case study of a typed attributed graph transformation system. Tool support for the analysis is available by the tool AGT-M [HCEK10, BHE09b], based on Wolfram Mathematica. This section is based

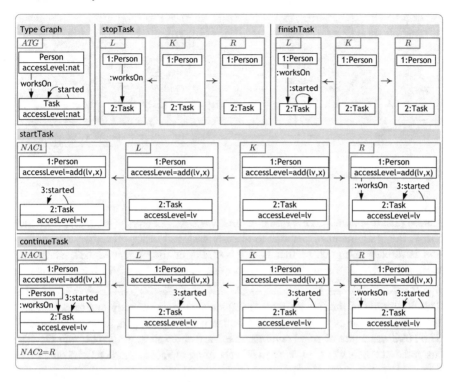

Fig. 5.15 Typed attributed graph transformation system GTS

on [CHS08, Her09, HCEK10, HCE14]. We present all results for transformation systems with NACs as a special kind of general application condition.

General Assumption: We generally assume $\mathcal{M}$-adhesive transformation systems with effective pushouts (see Def. 4.23.5) and $\mathcal{E}$–$\mathcal{M}$ factorisation for the morphism class O of the matches.

Example 5.47 (Typed attributed graph transformation system). As a running example for the analysis of permutation equivalence, we use the following typed attributed graph transformation system with the match morphism class O containing all morphisms that are injective on the graph part, i.e., possibly noninjective on data values. The *type graph ATG* specifies persons and tasks: a task is active if it has a ":started" loop, and it can be assigned to a person with a ":worksOn" edge. Moreover, the attribute "accessLevel" specifies the required access level of tasks and the allowed maximal access level of persons. Rule "startTask" is used to start a task, where the access level of the task can be at most equal to the access level of the considered person and the NAC schema ensures that the task is not started already. Rules "stopTask" and "finishTask" removes the assignment of a person, where "finishTask" additionally deletes the marker ":started" to specify that the task has been completed. Finally, rule "continueTask" assigns an already

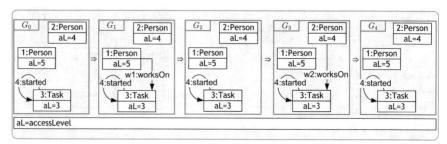

Fig. 5.16 Transformation sequence d of GTS

started task to a person. This rule contains two NAC schemata (see Def. 5.6) which forbid the assignment of persons to already assigned tasks—if either another person is already assigned to that task ("NAC1") or the person himself is already assigned ("NAC2"). Fig. 5.16 shows a NAC-consistent transformation sequence $d = (G_0 \xoverset{continueTask,f_1}{\Longrightarrow} G_1 \xoverset{stopTask,f_2}{\Longrightarrow} G_2 \xoverset{continueTask,f_3}{\Longrightarrow} G_3 \xoverset{stopTask,f_4}{\Longrightarrow} G_4)$ of GTS. The first graph of the transformation sequence contains exactly one task, which is first assigned to node "1:Person", and then, after being stopped, to node "2:Person". The NAC schemata of rule "*continueTask*" are checked at graphs G_0 and G_2. The instantiated NACs $n' : L \to N'$ with N' according to Fact 5.8 and Def. 5.5 contain an edge of type *worksOn*. Since G_0 and G_2 do not contain an edge of this type there is no embedding q from N' into these graphs such that the NAC schemata are satisfied by the matches. Therefore, the transformation sequence is NAC-consistent, because the remaining steps do not involve NACs.   △

5.3.1 Permutation Equivalence

The classical theory of the DPO approach introduces an equivalence among transformation sequences, called *switch equivalence*, that relates the sequences that differ only in the order in which independent transformation steps are performed. More precisely, two sequences are switch-equivalent if each of them can be obtained from the other by repeatedly exchanging consecutive transformation steps that are *sequentially independent* (see Def. 5.21).

Definition 5.48 (Switch equivalence for transformation sequences). Let $d = (d_1; \ldots; d_k; d_{k+1}; \ldots; d_n)$ be a transformation sequence, where d_k and d_{k+1} are two sequentially independent transformation steps, and let d' be obtained from d by switching them according to the Local Church–Rosser Theorem (Theorem 5.26). Then, d' is a *switching of* d, written $d \overset{sw}{\sim} d'$. The *switch equivalence*, denoted by

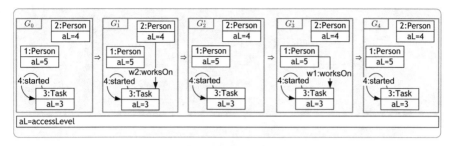

Fig. 5.17 Permutation-equivalent transformation sequence d' of GTS

$\overset{sw}{\approx}$, is the smallest equivalence on transformation sequences containing both $\overset{sw}{\sim}$ and the isomorphism relation $\cong$.[1] $\triangle$

In our opinion, however, the switch equivalence for NAC-consistent sequences is too restrictive, for the following reason. Suppose that $d_1; d_2$ are sequentially independent without considering application conditions, but that after the switching $d'_2; d'_1$ is not NAC-consistent. Then either d'_2 does not satisfy the NACs, which means that d_2 can fire after d_1 because d_1 deletes some resource that would represent a forbidden context for d_2; or the NACs of d'_1 are not satisfied, because d_2 creates a resource that matches (part of) a NAC of the transformation rule of d_1. In both cases, we argue that there is no information flow from d_1 to d_2, and therefore that there is no conceptual obstacle to the possibility that the two steps occur in the opposite order (even if not consecutively) in another equivalent transformation sequence.

These considerations justify the following definition of *permutation equivalence* [Her09, HCE14] for NAC-consistent transformation sequences, which is coarser than the corresponding switch equivalence in the sense that it equates more sequences.

Definition 5.49 (Permutation equivalence of transformation sequences). Two NAC-consistent transformation sequences d and d' are *permutation-equivalent*, written $d \overset{\pi}{\approx} d'$, if, disregarding the NACs, they are switch-equivalent as per Def. 5.48. The equivalence class π-$Equ(d)$ of all permutation-equivalent transformation sequences of d is given by π-$Equ(d) = \{d' \mid d' \overset{\pi}{\approx} d\}$. $\triangle$

Example 5.50 (Permutation equivalence). Fig. 5.17 shows a NAC-consistent transformation sequence $d' = (G_0 \xrightarrow{continueTask, f'_1} G'_1 \xrightarrow{stopTask, f'_2} G'_2 \xrightarrow{continueTask, f'_3} G'_3 \xrightarrow{stopTask, f'_4} G_4)$, which is permutation-equivalent to the transformation sequence d of Fig. 5.16, by performing the following switchings of steps disregarding NACs (we denote by $(d'_i; d'_j)$ the result of switching $(d_j; d_i)$):

---

[1] Informally, transformation sequences d and d' are isomorphic ($d \cong d'$) if they have the same length and there are isomorphisms between the corresponding objects of d and d' compatible with the involved morphisms.

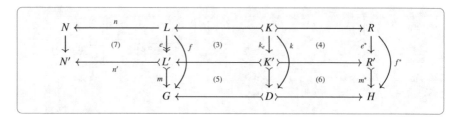

Fig. 5.18 Construction of instantiated rules and transformation steps

$(d_2; d_3), (d_1; d_3'), (d_2'; d_4), (d_1'; d_4')$. The equivalent transformation sequences are not switch-equivalent with NACs, because there is no pair of independent consecutive transformation steps in any of the transformation sequences.   △

While general matches for $\mathcal{M}$-adhesive transformation systems lead to extended concepts for NACs and NAC satisfaction, we now show that we can reduce the analysis of a concrete given transformation sequence to the case of $\mathcal{M}$-matches by instantiating the rules and transformation diagrams along the given matches. Note in particular that, for transformation steps along $\mathcal{M}$-matches, the instantiated transformation steps coincide with the given ones.

Definition 5.51 (Instantiated rules and transformation sequences). Let $G \overset{p,f}{\Longrightarrow} H$ be a NAC-consistent transformation step via a rule $p = ((L \hookleftarrow K \hookrightarrow R), \mathbf{N})$ with NAC schemata $\mathbf{N}$. Let $f = m \circ e$ be the extremal $\mathcal{E}$–$\mathcal{M}$ factorisation of match f. The *instantiated transformation step* is given by $G \overset{p',m}{\Longrightarrow} H$ with *instantiated rule p' derived via e* and constructed as follows according to Fig. 5.18. Construct pullback (PB) (5) leading to pushouts (POs) (3) and (5) by PB splitting and $\mathcal{M}$-pushout–pullback decomposition (see Def. 4.21). Construct PO (4) leading to PO (6) by PO splitting. Instantiate each NAC schema $n : L \to N$ in $\mathbf{N}$ along morphism e (square (7) according to Fact 5.8 and Def. 5.5), leading to new NACs $n' : L' \hookrightarrow N'$. Let $\mathbf{N}'$ be the new set of NACs consisting of all NACs $n' : L' \hookrightarrow N'$ obtained from all $n \in \mathbf{N}$. The *instantiated rule* is given by $p' = ((L' \hookleftarrow K' \hookrightarrow R'), \mathbf{N}')$ and the *instantiated transformation step* is defined by $G \overset{p',m}{\Longrightarrow} H$ with $m \in \mathcal{M}$ via DPO diagram $((5) + (6))$.

Let d be a transformation sequence; then the instantiated transformation sequence d_I is derived by instantiating each transformation step as defined above.   △

Example 5.52 (Instantiation of transformation sequence). The instantiation of the transformation sequence d in Fig. 5.16 via rules of Fig. 5.15 is performed according to Def. 5.51. We derive an instantiated transformation sequence d_I. By definition, the lower line of the DPO diagrams coincides with the one of d in Fig. 5.16. The instantiated rules for the four steps are depicted in Figs. 5.19 and 5.20 (rules "stop1", "stop2", "cont1", and "cont2") and they are used in the following sections for the analysis of permutation equivalence.   △

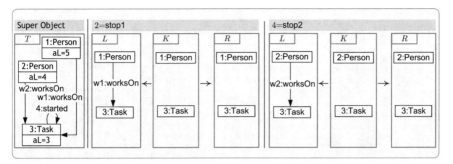

Fig. 5.19 Super object T and two rules of process $Prc(d)$

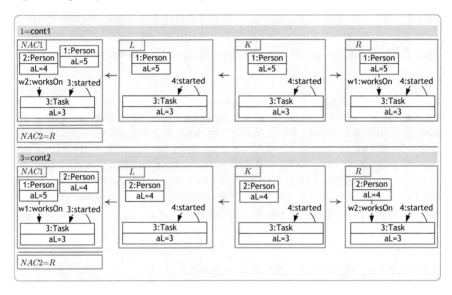

Fig. 5.20 Further rules of STS $STS(d)$

Fact 5.53 (Reduction of permutation equivalence for general matches to M-matches). *Two transformation sequences d and d' with general matches are permutation-equivalent if and only if their instantiated transformation sequences d_I and d'_I with M-matches are permutation-equivalent, i.e., $d \stackrel{\pi}{\approx} d' \Leftrightarrow d_I \stackrel{\pi}{\approx} d'_I$.* △

Proof (Idea). The full proof (see [HCE14]) first shows that switch equivalence disregarding NACs is implied for both directions using Def. 5.48. In a second step, we showed that the transformation sequences are additionally NAC consistent. Therefore, $d \stackrel{\pi}{\approx} d' \Leftrightarrow d_I \stackrel{\pi}{\approx} d'_I$. □

Remark 5.54 (Permutation equivalence for general matches). By the above fact, we can base our analysis techniques in the following on the derived transformation se-

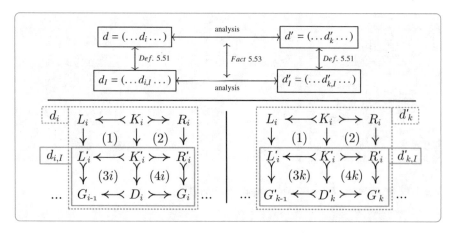

Fig. 5.21 Correspondence between transformation sequences and their instantiations

quences with $\mathcal{M}$-morphisms only as depicted in Fig. 5.21. Given a transformation sequence d, we first instantiate d according to Def. 5.51, such that the lower transformation diagrams form a new transformation sequence d_I with $\mathcal{M}$-matches only and all NAC morphisms $n' : L' \to N'$ are $\mathcal{M}$-morphisms. Thereafter, we can analyse permutation equivalence for d_I and derive the analysis results for d via Fact 5.53. In particular, the derived permutation-equivalent transformation sequences d'_I of d_I can be composed with the upper DPO diagrams of the instantiation, leading to permutation-equivalent transformation sequences d' of d.   △

5.3.2 Subobject Transformation Systems

We now present subobject transformation systems (STSs) as a formal framework for the concurrent semantics of $\mathcal{M}$-adhesive transformation systems. This concept generalises the notion of elementary nets, which form the category of process nets for P/T Petri nets, in the way that STSs form the category of process transformation systems for $\mathcal{M}$-adhesive transformation systems. Subobject transformation systems are essentially double pushout transformation systems over the lattice of subobjects $\mathbf{Sub}(T)_{\mathcal{M}}$ of a given object T of an $\mathcal{M}$-adhesive category $\mathbf{C}$. By $|\mathbf{C}|$ we denote the class of objects of $\mathbf{C}$.

Definition 5.55 (Category of $\mathcal{M}$-subobjects).   Let T be an object of an $\mathcal{M}$-adhesive category $\mathbf{C}$. Given two $\mathcal{M}$-morphisms $a : A \hookrightarrow T$ and $a' : A' \hookrightarrow T$, they are *equivalent* if there exists an isomorphism $\phi : A \to A'$ such that $a = a' \circ \phi$. An *$\mathcal{M}$-subobject* $[a : A \hookrightarrow T]$ of T is an equivalence class of $\mathcal{M}$-morphisms with target T. The *category of $\mathcal{M}$-subobjects of T*, denoted by $\mathbf{Sub}_{\mathcal{M}}(T)$, has the $\mathcal{M}$-subobjects of T as objects. Furthermore, there is an arrow from $[a : A \hookrightarrow T]$ to

$[b : B \hookrightarrow T]$ if there exists a morphism $f : A \to B$ such that $a = b \circ f$; in this case f is an $\mathcal{M}$-morphism and it is unique up to isomorphism (therefore $\mathbf{Sub}_{\mathcal{M}}(T)$ is a partial order), and we write $[a : A \hookrightarrow T] \subseteq [b : B \hookrightarrow T]$.

Usually we will denote an $\mathcal{M}$-subobject $[a : A \hookrightarrow T]$ simply by A, leaving the $\mathcal{M}$-morphism a implicit, and correspondingly we write $A \subseteq B$ if $[a : A \hookrightarrow T] \subseteq [b : B \hookrightarrow T]$ and denote the corresponding embedding by $f : A \hookrightarrow B$. △

If $\mathcal{M}$ is the class of all monomorphisms of $\mathbf{C}$, as for adhesive categories, then $\mathbf{Sub}_{\mathcal{M}}(T)$ for $T \in |\mathbf{C}|$ is the standard category of subobjects of T. The following notions of "intersection" and "union" will be used in the definition of direct derivations of an STS.

Definition 5.56 (Intersection and union in $\mathbf{Sub}_{\mathcal{M}}(T)$). Let $A, B \in |\mathbf{Sub}_{\mathcal{M}}(T)|$ be two $\mathcal{M}$-subobjects, with $T \in |\mathbf{C}|$. The product of A and B in $\mathbf{Sub}_{\mathcal{M}}(T)$ is called their *intersection*, denoted by $A \cap B$. The coproduct of A and B in $\mathbf{Sub}_{\mathcal{M}}(T)$ is called *union*, denoted by $A \cup B$. △

Since pushouts are not effective in general, we require this property by our general assumption. As shown in [HCE14] for $\mathcal{M}$-adhesive transformation systems based on [LS04], intersections and unions exist, and $\mathbf{Sub}_{\mathcal{M}}(T)$ is a distributive lattice for any $T \in \mathbf{C}$.

Remark 5.57 (Unions in $\mathbf{Sub}_{\mathcal{M}}(T)$ for $(\mathbf{AGraphs}_{ATG}, \mathcal{M})$). The $\mathcal{M}$-adhesive category $(\mathbf{AGraphs}_{ATG}, \mathcal{M})$ has effective pushouts, because by commutativity of the diagram in item 5 of Def. 4.23, the morphism d is an isomorphism on the data part. Therefore, the union $A \cup B$ of two $\mathcal{M}$-subobjects A and B can be constructed as the pushout over the intersection $A \cap B$ in $\mathbf{C}$. △

Definition 5.58 (STS with NACs). A *subobject transformation system (STS) with NACs* $S = (T, P, \pi)$ over an $\mathcal{M}$-adhesive category $\mathbf{C}$ with effective unions consists of a super object $T \in \mathbf{C}$, a set of rule names P—also called productions—and a function π, which maps each rule name $q \in P$ to a *rule with negative application conditions* $((L, K, R), \mathbf{N})$, where L, K, and R are objects in $\mathbf{Sub}_{\mathcal{M}}(T)$, $K \subseteq L$, $K \subseteq R$ and its NACs $\mathbf{N}$ are given by $\mathbf{N} = (N, \nu)$, consisting of a set N of names for the NACs together with a function ν mapping each NAC name $i \in N$ to a *NAC* $\nu(i)$, which is given by a subobject $\nu(i) = N_i \in \mathbf{Sub}_{\mathcal{M}}(T)$ with $L \subseteq N_i \subseteq T$. The short notation $\mathbf{N}[i]$ refers to a NAC N_i of rule p with $\nu(i) = N_i$. △

Direct derivations $(G \overset{q}{\Rightarrow} G')$ with NACs in an STS correspond to transformation steps with NACs in an $\mathcal{M}$-adhesive TS, but the construction is simplified, because morphisms between two subobjects are unique. There is no need for pattern matching, and for this reason, we use the notion of derivations within an STS in contrast to transformation sequences in an $\mathcal{M}$-adhesive TS, and we use names $\{p_1, \ldots, p_n\}$ for rules in an $\mathcal{M}$-adhesive TS and $\{q_1, \ldots, q_n\}$ for rules in an STS.

Definition 5.59 (Direct derivations in an STS). Let $S = (T, P, \pi)$ be a subobject transformation system with NACs, let $\pi(q) = ((L, K, R), \mathbf{N})$ be a production with

NACs, and let $G \in |\mathbf{Sub}_{\mathcal{M}}(T)|$. Then there is a *direct derivation disregarding NACs* from G to G' using q if $G' \in |\mathbf{Sub}_{\mathcal{M}}(T)|$ and there is an object $D \in \mathbf{Sub}_{\mathcal{M}}(T)$ such that:

$$(i) \quad L \cup D = G; \qquad (ii) \quad L \cap D = K;$$
$$(iii) \quad D \cup R = G', \text{ and } (iv) \quad D \cap R = K.$$

We say that there is a *direct derivation with NACs* from G to G' using q, if in addition to all the conditions above it also holds that $\mathbf{N}[i] \not\subseteq G$ for each $\mathbf{N}[i]$ in $\mathbf{N}$. In both cases we write $G \overset{q}{\Rightarrow} G'$. $\triangle$

Given a transformation sequence d with matches in $\mathcal{M}$, we can construct its corresponding STS, which we will use for the analysis of its processes in a similar way to that presented for adhesive systems in [BCH$^+$06].

Definition 5.60 (STS of a transformation sequence with $\mathcal{M}$-matches). Let $d = (G_0 \overset{p_1,m_1}{\Longrightarrow} \dots \overset{p_n,m_n}{\Longrightarrow} G_n)$ be a NAC-consistent transformation sequence in an $\mathcal{M}$-adhesive TS with matches in $\mathcal{M}$. The *STS with NACs generated by* d is given by $STS(d) = (T, P, \pi)$ and its components are constructed as follows. T is an arbitrarily chosen but fixed colimit of the sequence of DPO diagrams given by d; $P = \{i \mid 0 < i \leq n\}$ is a set of natural numbers that contains a canonical rule occurrence name for each rule occurrence in d. For each $k \in P$, $\pi(k)$ is defined as $\pi(k) = ((L_k, K_k, R_k), \mathbf{N}_k)$, where each component X of a production p_k ($X \in \{L_k, K_k, R_k\}$) is regarded as a subobject of T via the natural embedding $in_T(X)$. Furthermore, for each $k \in \{1, \dots, n\}$ the NACs $\mathbf{N}_k = (N_k, \nu)$ are constructed as follows. Let $J_{\mathbf{N}_k}$ be the set of subobjects of T which are possible images of NACs of production $(p_k, \mathbf{N}_k)$, with respect to the match $in_T : L_k \hookrightarrow T$; namely,

$$J_{\mathbf{N}_k} = \{[j : N \hookrightarrow T] \in \mathbf{Sub}_{\mathcal{M}}(T) \mid \exists (n : L_k \hookrightarrow N) \in \mathbf{N}_k \wedge j \circ n = in_T(L_k)\}.$$

Then the NAC names N_k are given by $N_k = \{i \mid 0 < i \leq |J_{\mathbf{N}_k}|\}$ and the function ν is an arbitrary but fixed bijective function $\nu : N_k \to J_{\mathbf{N}_k}$ mapping NAC names to corresponding subobjects. $\triangle$

When analysing permutation equivalence in concrete case studies we consider only transformation sequences such that the colimit object T is finite, i.e., has finitely many $\mathcal{M}$-subobjects, in order to ensure termination. Finiteness is guaranteed if each rule of TS has finite left- and right-hand sides, and if the start object of the transformation sequence is finite. For typed attributed graphs, this means that T is finite on the structural part, but the carrier sets of the data algebra for the attribution component may by infinite ($\mathcal{M}$-morphisms in $\mathbf{AGraphs}_{ATG}$ are isomorphisms on the data part).

Remark 5.61. Note that during the construction of $STS(d)$ the set of instantiated NACs for a NAC of a rule p applied in d may be empty, which means that the NAC n cannot be found within T. This would be the case for rule *continueTask* if we replace the variable lv within the NACs by the constant 4, i.e., the NAC pattern would never be present in the transformation sequence. Furthermore, if we require T to be finite, the sets of NACs in $STS(d)$ are finite. $\triangle$

Example 5.62 (Derived STS STS(d)). For the transformation sequence in Fig. 5.16 the construction leads to the STS as shown in Figs. 5.19 and 5.20. The transformation sequence d involves the rules "continueTask" and "stopTask", and thus the derived STS contains the rule occurrences "cont1", "cont2", "stop1" and "stop2". △

Deterministic processes for DPO graph transformation systems are introduced in [CMR96] and characterised as occurrence grammars in [Bal00]: these concepts generalise the corresponding notions for Petri nets [Rei85], and are generalised further in [BCH$^+$06] to adhesive transformation systems. By Def. 5.60, we generalise these notions and define the process of a transformation sequence d in an $\mathcal{M}$-adhesive transformation system. The process consists of the STS derived from d together with an embedding v relating the STS with the $\mathcal{M}$-adhesive TS of the given transformation sequence.

Definition 5.63 (Process of a transformation sequence with NACs). Let $d = (G_0 \xrightarrow{q_1,m_1} \cdots \xrightarrow{q_n,m_n} G_n)$ be a NAC-consistent transformation sequence in an $\mathcal{M}$-adhesive transformation system TS $= (P_{TS}, \pi_{TS})$. The *process* $Prc(d) = (STS(d), \mu)$ of d consists of the derived STS $STS(d) = (T, P, \pi)$ of d together with the mapping $\mu : STS(d) \rightarrow TS$ given by $\mu : P \rightarrow P_{TS}, \mu(i) = q_i$ for each step i of d. △

Note that the mapping μ induces a function $\mu_\pi : \pi(P) \rightarrow \pi_{TS}(P_{TS})$ mapping each rule in $STS(d)$ to the corresponding rule in TS, where $\mu_\pi(\pi(q)) = \pi_{TS}(\mu(q))$. Given the process $Prc(d) = ((T, P, \pi), \mu)$ of a derivation d, often we will denote by $seq(d) \in P^*$ the sequence of production names of $Prc(d)$ that corresponds to the order in which productions are applied in d; from the canonical choice of production names in P (see Def. 5.60) it follows that $seq(d) = (1, 2, \ldots, n)$, where n is the length of d.

The notion of processes for transformation sequences corresponds to the notion of processes for Petri nets given by an occurrence net together with a Petri net morphism into the system Petri net. Moreover, as shown in [CHS08] the process construction yields a *pure* STS, meaning that no rule deletes and produces again the same part of a subobject, i.e., $L \cap R = K$. This terminology is borrowed from the theory of elementary net systems, where a system which does not contain transitions with a self-loop is called "pure". Therefore, the class of pure STSs can be seen as a generalisation of elementary nets to the setting of $\mathcal{M}$-adhesive transformation systems, and thus as a generalisation of the Petri net class of occurrence nets.

The following relations between the rules of an STS with NACs specify the possible dependencies among them: the first four relations are discussed in [CHS08], while the last two are introduced in [Her09, HCE14].

Definition 5.64 (Relations on rules). Let q_1 and q_2 be two rules in an STS $S = (T, P, \pi)$ with $\pi(q_i) = ((L_i, K_i, R_i), \mathbf{N}_i)$ for $i \in \{1, 2\}$. The relations on rules are defined on P as shown in Table 5.1. △

In words, $q_1 <_{rc} q_2$ (read: "q_1 causes q_2 by read causality") if q_1 produces an element which is used but not consumed by q_2. Analogously, $q_1 <_{wc} q_2$ (read: "q_1 causes q_2 by write causality") if q_1 produces an element which is consumed by

Table 5.1 Relations on rules in an STS

Name	Notation	Condition
Read Causality	$q_1 <_{rc} q_2$	$R_1 \cap K_2 \not\subseteq K_1$
Write Causality	$q_1 <_{wc} q_2$	$R_1 \cap L_2 \not\subseteq K_1 \cup K_2$
Deactivation	$q_1 <_d q_2$	$K_1 \cap L_2 \not\subseteq K_2$
Independence	$q_1 \diamond q_2$	$(L_1 \cup R_1) \cap (L_2 \cup R_2) \subseteq K_1 \cap K_2$
Weak NAC Enabling	$q_1 <_{wen[i]} q_2$	$0 < i \le |N_2| \wedge L_1 \cap N_2[i] \not\subseteq K_1 \cup L_2$
Weak NAC Disabling	$q_1 <_{wdn[i]} q_2$	$0 < i \le |N_1| \wedge N_1[i] \cap R_2 \not\subseteq L_1 \cup K_2$

Table 5.2 Relations on rules in the example STS

cont1 $<_{wc}$ stop1	stop1 $<_{wen[1]}$ cont1	stop2 $<_{wen[2]}$ cont1	cont1 $<_{wdn[1]}$ cont1	cont2 $<_{wdn[2]}$ cont2
cont2 $<_{wc}$ stop2	stop1 $<_{wen[1]}$ cont2	stop2 $<_{wen[2]}$ cont2	cont2 $<_{wdn[1]}$ cont1	cont1 $<_{wdn[2]}$ cont2

q_2 and $q_1 <_d q_2$ (read: "q_1 is deactivated by q_2") precisely when q_1 preserves an element which is consumed by q_2, meaning that q_1 is not applicable afterwards. Furthermore $q_1 \diamond q_2$ if they overlap only on items that are preserved by both. Finally, $q_1 <_{wen[i]} q_2$ (read: "q_1 weakly enables q_2 at i") if q_1 deletes a piece of the NAC $N[i]$ of q_2; instead $q_1 <_{wdn[i]} q_2$ ("q_2 weakly disables q_1 at i") if q_2 produces a piece of the NAC $N[i]$ of q_1. It is worth stressing that the relations introduced above are not transitive in general.

Example 5.65 (Relations of an STS). The rules of $STS(d)$ in Ex. 5.62 are related by the dependencies listed in Table 5.2. $\triangle$

Definition 5.66 (STS-switch equivalence of sequences disregarding NACs). Let $S = (T, P, \pi)$ be an *STS*, let d be a derivation in S disregarding NACs and let $s = \langle q_1, \ldots, q_n \rangle$ be its corresponding sequence of rule occurrence names. If $q_k \diamond q_{k+1}$, then the sequence $s' = \langle q_1, \ldots, q_{k+1}, q_k, \ldots, q_n \rangle$ is *STS-switch-equivalent* to the sequence s, written $s \overset{sw}{\sim}_S s'$. Switch equivalence $\overset{sw}{\approx}_S$ of rule sequences is the transitive closure of $\overset{sw}{\sim}_S$. $\triangle$

In order to characterise the set of possible permutations of transformation steps of a given transformation sequence, we now define suitable conditions for permutations of rule occurrences. We call rule sequences s of a derived STS $STS(d)$ *legal sequences* if they are switch-equivalent without NACs to the sequence of rules $seq(d)$ of d and if the following condition concerning NACs holds: For every NAC $N[i]$ of a rule q_k, either there is a rule which *deletes* part of $N[i]$ and is applied *before* q_k, or there is a rule which *produces* part of $N[i]$ and is applied *after* q_{k-1}. In both cases, $N[i]$ cannot be present when applying q_k, because the STS $STS(d)$ is a sort of "unfolding" of the transformation sequence, and every subobject is created at most once and deleted at most once (see [CHS08]). Note that the first condition already ensures that each rule name in P occurs exactly once in a legal sequence s.

Definition 5.67 (Legal sequence). Let $d = (d_1; \ldots; d_n)$ be a NAC-consistent transformation sequence in an $\mathcal{M}$-adhesive TS, and let $STS(d) = (T, P, \pi_N)$ be its derived STS. A sequence $s = \langle q_1; \ldots; q_n \rangle$ of rule names of P is *locally legal at position* $k \in \{1, \ldots, n\}$ *with respect to* d if the following conditions hold:

1. $s \overset{sw}{\approx}_{STS(d)} seq(d)$
2. $\forall$ NAC $N_k[i]$ of $q_k : \left(\begin{array}{l} \exists \, e \in \{1, \ldots, k-1\} : q_e <_{wen[i]} q_k \text{ or} \\ \exists \, l \in \{k, \ldots, n\} : q_k <_{wdn[i]} q_l. \end{array} \right)$

A sequence s of rule names is *legal with respect to* d, if it is locally legal at all positions $k \in \{1, ..., n\}$ with respect to d. △

Definition 5.68 (STS equivalence of rule sequences). Let d be a NAC-consistent transformation sequence of an $\mathcal{M}$-adhesive TS and let $Prc(d) = (STS(d), \mu)$ be its derived process. Two sequences s, s' of rule names in $STS(d)$ are STS-equivalent, written $s \approx_{STS(d)} s'$, if they are legal sequences with respect to d. The set of all STS-equivalent sequences of $Prc(d)$ is given by $Seq(d) = \{s \mid s \approx_{STS(d)} seq(d)\}$. Moreover, the specified class of transformation sequences of $Seq(d)$ is given by $Trafo(s) = [trafo_{STS(d)}(s)]_\cong$ for single sequences and $Trafo(Seq(d)) = \bigcup_{s \in Seq(d)} Trafo(s)$ for the complete set. △

Theorem 5.69 (Characterisation of permutation equivalence based on STSs). *Given the process $Prc(d)$ of a NAC-consistent transformation sequence d.*

1. *The class of permutation-equivalent transformation sequences of d coincides with the set of derived transformation sequences of the process $Prc(d)$ of d:* $\pi\text{-}Equ(d) = Trafo(Seq(d))$
2. *The mapping $Trafo$ defines a bijective correspondence between STS-equivalent sequences of rule names and permutation-equivalent transformation sequences:* $Trafo : Seq(d) \xrightarrow{\sim} (\pi\text{-}Equ(d))/_\cong$ △

Proof (Idea). Let d be a NAC-consistent transformation sequence in an $\mathcal{M}$-adhesive TS and let $Prc(d) = (S, \mu)$ be the process of d with $S = (T, P, \pi)$. We have to show that each STS-equivalent rule sequence s' of $seq(d)$ in S defines a permutation-equivalent transformation sequence $trafo_{STS(d)}(s')$ of d; and vice versa, for each permutation-equivalent transformation sequence d' of d there is an STS-equivalent rule sequence s' of $seq(d)$ in S such that $d' \cong trafo_{STS(d)}(s')$.

$$\forall \, s' \in P^* : s' \approx_{STS(d)} seq(d) \Rightarrow trafo_{STS(d)}(s') \overset{\pi}{\approx} d \qquad (1)$$
$$\forall \, d' : \quad d' \overset{\pi}{\approx} d \Rightarrow \exists \, s'. s' \approx_{STS(d)} seq(d) \wedge trafo_{STS(d)}(s') \cong d' \qquad (2)$$

The proof is based on Thm. 1 in [Her09], which concerns the results (1) and (2) for the case of adhesive transformation systems with NACs and monomorphic matches and is extended to the case of $\mathcal{M}$-adhesive transformation systems in [HCE14]. By Def. 5.68 we have that $d' \in Trafo(Prc(d))$ is equivalent to $d' \cong trafo_{STS(d)}(s')$ and $s' \approx_{STS(d)} seq(d)$. Using (1) and (2) above together with Def. 5.49, we derive $\pi\text{-}Equ(d) = Trafo(Prc(d))$. □

According to Theorem 5.69, the construction of the process $Prc(d)$ of a transformation sequence d specifies the equivalence class of all transformation sequences which are permutation-equivalent to d. In the next section, we present an efficient analysis technique for processes based on Petri nets.

5.3.3 Analysis Based on Petri Nets

In order to efficiently analyse the process of a transformation sequence, we present the construction of its *dependency net*, given by a P/T Petri net which specifies only the dependencies between the transformation steps. All details about the internal structure of the objects and the transformation rules are excluded. The names of the generated places of the dependency net are composed of constant symbols and numbers, where constant symbols s are denoted by s. We use the monoidal notation of P/T Petri nets according to [MM90] and ISO/IEC 15909-1:2004 [ISO04], which is equivalent to the classical notation of P/T Petri nets [Rei85].

Definition 5.70 (Dependency net $DNet$ of a transformation sequence). Let d be a NAC-consistent transformation sequence of an $\mathcal{M}$-adhesive TS, let $STS(d) = (T, P, \pi)$ be the generated STS of d and let $s = seq(d) = \langle q_1, \ldots, q_n \rangle$ be the sequence of rule names in $STS(d)$ according to the steps in d. The dependency net of d is given by the following marked Petri net $DNet(d) = (Net, M)$, $Net = (PL, TR, pre, post)$:

- $TR = P = \{i \mid 1 \le i \le |P|\}$
- $PL = \{p(q) \mid q \in TR\} \cup \{p(q' <_x q) \mid q, q' \in TR, x \in \{rc, wc, d\}, q' <_x q\}$
  $\cup \{p(q, N[i]) \mid q \in TR, \pi(q) = ((L_q, K_q, R_q), \mathbf{N}), 0 < i \le |\mathbf{N}|, q \not<_{wdn[i]} q\}$

- $pre(q) = p(q) \oplus \underset{\substack{q' <_x q \\ x \in \{rc,wc,d\}}}{\sum} p(q' <_x q) \oplus \underset{\substack{q' <_{wdn[i]} q \\ q' \ne q}}{\sum} p(q', N[i]) \oplus \underset{p(q,N[i]) \in PL}{\sum} p(q, N[i])$

- $post(q) = \underset{\substack{q <_x q' \\ x \in \{rc,wc,d\}}}{\sum} p(q <_x q') \oplus \underset{q <_{wen[i]} q'}{\sum} p(q', N[i]) \oplus \underset{p(q,N[i]) \in PL}{\sum} p(q, N[i])$

- $M = \underset{q \in TR}{\sum} p(q) \oplus \underset{\substack{q' <_{wdn[i]} q \\ p(q',N[i]) \in PL}}{\sum} p(q', N[i])$ $\hfill \triangle$

Fig. 5.22 shows how the dependency net is constructed algorithmically. The construction steps are performed in the order in which they appear in the table. Each step is shown as a rule, where gray lines and plus signs mark the elements to be

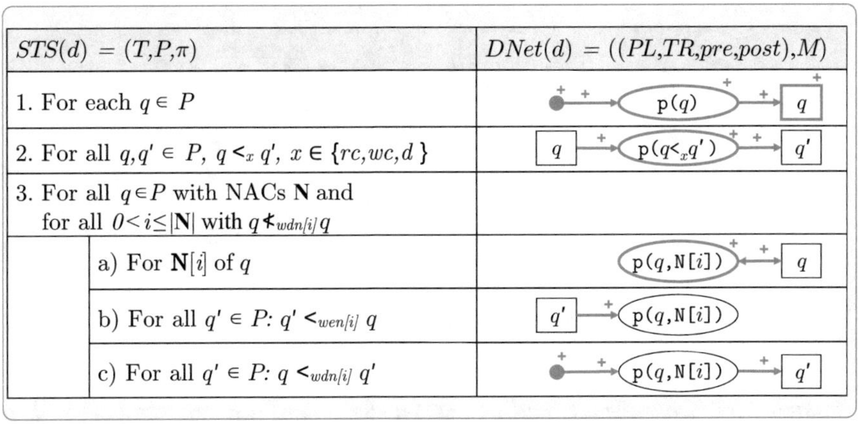

$STS(d) = (T,P,\pi)$	$DNet(d) = ((PL,TR,pre,post),M)$
1. For each $q \in P$	$p(q)$ q
2. For all $q,q' \in P$, $q <_x q'$, $x \in \{rc,wc,d\}$	q $p(q<_xq')$ q'
3. For each $q \in P$ with NACs $\mathbf{N}$ and for all $0 < i \le |\mathbf{N}|$ with $q \nleq_{wdn[i]} q$	
a) For $\mathbf{N}[i]$ of q	$p(q,\mathbf{N}[i])$ q
b) For all $q' \in P$: $q' <_{wen[i]} q$	q' $p(q,\mathbf{N}[i])$
c) For all $q' \in P$: $q <_{wdn[i]} q'$	$p(q,\mathbf{N}[i])$ q'

Fig. 5.22 Visualisation of the construction of the Petri net

inserted. The matched context that is preserved by a rule is marked by black lines, e.g., in Step 2 the new place "$p(q <_x q')$" is inserted between the already existing transitions q and q'. The tokens of the initial marking of the net are represented by bullets that are connected to their places via arcs. In the first step, each rule q of the STS is encoded as a transition and it is connected to a marked place, which prevents the transition from firing more than once. In Step 2, between each pair of transitions in each of the relations $<_{rc}$, $<_{wc}$ and $<_d$, a new place is created in order to enforce the corresponding dependency. The rest of the construction is concerned with places which correspond to NACs and can contain several tokens in general. Each token in such a place represents *the absence* of a piece of the NAC; therefore if the place is empty, the NAC is complete.

In this case, by Step (3a) the transition cannot fire. Consistently with this intuition, if $q' <_{wen[i]} q$, i.e., transition q' consumes part of the NAC $\mathbf{N}[i]$ of q, then by Step (3b) q' produces a token in the place corresponding to $\mathbf{N}[i]$. Symmetrically, if $q <_{wdn[i]} q'$, i.e., q' produces part of NAC $\mathbf{N}[i]$ of q, then by Step (3c) q' consumes a token from the place corresponding to $\mathbf{N}[i]$. Notice that each item of a NAC is either already in the start graph of the transformation sequence or produced by a single rule. If a rule generates part of one of its NACs, say $\mathbf{N}[i]$ ($q <_{wdn[i]} q$), then by the acyclicity of $Prc(d)$ the NAC $\mathbf{N}[i]$ cannot be completed before the firing of q: therefore we ignore it in the third step of the construction of the dependency net. Examples of such weakly self-disabling rules are rules ($1 = cont1$) and ($3 = cont2$) in Fig. 5.20, where the specific NACs coincide with the right-hand sides of the rules ($NAC2 = R$).

Note that the constructed net in general is not a safe one, because the places for the NACs can contain several tokens. Nevertheless it is a bounded P/T net. The bound is the maximum of 1 and the maximal number of adjacent edges at a NAC place minus 2.

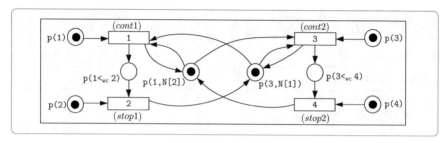

Fig. 5.23 Dependency net *DNet*(*d*) as Petri net

Example 5.71 (Dependency net). Consider the transformation sequence *d* in Fig. 5.16 from Ex. 5.47 and its derived STS in Ex. 5.62. The marked Petri net in Fig. 5.23 is the dependency net *DNet*(*d*) according to Def. 5.70. The places encoding the write causality relation are "$p(1 <_{wc} 2)$" and "$p(3 <_{wc} 4)$". For the NAC dependencies we have the places p(1,N[2]) for the second instantiated NAC in the first transformation step of *d* and p(3,N[1]) for the third transformation step and its first instantiated NAC. The other two instantiated NACs are not considered, because the corresponding rules are weakly self-disabling ($q <_{wdn[i]} q$). At the beginning, transitions 1 and 2 (*cont*1 and *cont*2) are enabled. The firing sequences according to the transformation sequences *d* and *d'* in Figs. 5.16 and 5.17 can be executed and they are the only complete firing sequences of this net. Thus, the net specifies exactly the transformation sequences which are permutation-equivalent to *d*.   △

We now show that we can exploit the constructed Petri net *DNet*(*d*) to characterise STS equivalence of sequences of rule occurrences by Fact 5.73. Note that according to Def. 5.70 each sequence *s* of rule names in the STS of *Prc*(*d*) can be interpreted as a sequence of transitions in the derived marked Petri net *DNet*(*d*), and vice versa. This correspondence allows us to transfer the results of the analysis of the dependency net back to the STS. Notice that the construction of the dependency net (Def. 5.70) ensures that each transition can fire at most once by construction.

Definition 5.72 (Transition-complete firing sequences). A firing sequence of a Petri net is called *transition-complete* if each transition of the net occurs exactly once. The set of transition-complete firing sequences of a dependency net *DNet*(*d*) is denoted by *FSeq*(*DNet*(*d*)).   △

Fact 5.73 (Characterisation of STS equivalence based on Petri nets). *Given the process Prc*(*d*) *and the dependency net DNet*(*d*) *of a NAC-consistent transformation sequence d of an M-adhesive transformation system with M-matches, the class of STS-equivalent sequences of seq*(*d*) *coincides with the set of transition-complete firing sequences in the dependency net DNet*(*d*), *i.e., Seq*(*d*) = *FSeq*(*DNet*(*d*)).*   △

Remark 5.74 (Bijective correspondence). Analogously to Theorem 5.69, there is also a bijective correspondence between STS sequences and transition-complete

firing sequences, which is in this case directly given by the identity function
$id : Seq(d) \xrightarrow{\sim} FSeq(DNet(d))$.   △

Proof (Idea). The proof (see [HCE14]) shows that $s \approx_{STS(d)} seq(d)$ **iff** s is a
transition-complete firing sequence of $DNet(d)$. Direction "⇒" uses the property
that s is a legal sequence with respect to d in $STS(d)$ and thus s is a permutation of
$seq(d)$ and each transition occurs exactly once in s. For each transition, we can ensure its activation using Def. 5.70. Direction "⇐" starts with the transition-complete
firing sequence s of $DNet(d)$ and shows that s is a legal sequence with respect to d
in $STS(d)$, i.e., that the two conditions in Def. 5.67 hold.   □

In order to solve the challenge of computing the set of all permutation-equivalent
transformation sequences for a given one, we can now combine the presented results, leading to our forth main result by Theorem 5.75 below, where we show that
the analysis of permutation equivalence can be completely performed on the dependency net $DNet(d)$.

Theorem 5.75 (Analysis of permutation equivalence based on Petri nets). *Given
the process $Prc(d)$ and the dependency net $DNet(d)$ of a NAC-consistent transformation sequence d.*

1. *The class of permutation-equivalent transformation sequences of d coincides
   with the set of derived transformation sequences using $DNet(d)$:*
   $\pi\text{-}Equ(d) = Trafo(FSeq(DNet(d)))$.
2. *The mapping Trafo according to Def. 5.68 defines a bijective correspondence
   between transition-complete firing sequences and permutation-equivalent transformation sequences:*
   $Trafo : FSeq(DNet(d)) \xrightarrow{\sim} (\pi\text{-}Equ(d))/_{\cong}$.   △

Proof. By combining the characterisations of Theorem 5.69 and Fact 5.73 we
derive the equality $\pi\text{-}Equ(d) = Trafo(FSeq(DNet(d)))$, and the bijection $Trafo :$
$FSeq(DNet(d)) \xrightarrow{\sim} (\pi\text{-}Equ(d))/_{\cong}$ is given by $Trafo : Seq(d) \xrightarrow{\sim} (\pi\text{-}Equ(d))/_{\cong}$ of Theorem 5.69 with $Seq(d) = FSeq(DNet(d))$ in Fact 5.73.   □

Remark 5.76 (Analysis of permutation equivalence). We now describe how the presented results can be used for an efficient analysis of permutation equivalence, i.e.,
for the generation of the complete set of permutation-equivalent transformation sequences for a given one and for checking permutation equivalence of specific ones.
Given a NAC-consistent transformation sequence with general matches and NAC
schemata, we can first reduce the analysis problem to the derived instantiated transformation sequence with $\mathcal{M}$-matches and standard NACs according to Fact 5.53 and
Rem. 5.54. According to Theorem 5.75, we can perform the analysis of permutation
equivalence based on Petri nets by first constructing the dependency net $DNet(d)$.
For the generation of all permutation-equivalent sequences, we construct the complete reachability graph of $DNet(d)$, where each path specifies one permutation-
equivalent transformation sequence up to isomorphism. If only specific reorderings
of the transformation steps shall be checked, then the corresponding firing sequences
are checked for being executable in $DNet(d)$.   △

Another computational model closely related to transformation systems with NACs are Petri nets with inhibitor arcs (or *inhibitor nets*) [JK95, BP99, KK04, BBCP04]. In such nets, a transition cannot fire if there are tokens on its *inhibitor places*, i.e., on the places that are linked to it with inhibitor arcs.[2] Therefore these places play a role conceptually similar to NACs'.

Note that the proposed notion of permutation equivalence would be original also in the framework of inhibitor nets. In fact, if we encode the system of Ex. 5.47 into an inhibitor net (by forgetting the graphical structure), the standard semantics for such nets would not consider equivalent the firing sequences corresponding to the two transformation sequences d of Fig. 5.16 and d' of Fig. 5.17.

---

[2] For simplicity we consider only the case of unweighted inhibitor arcs.

Chapter 6
Multi-amalgamated Transformations

In this chapter, we introduce amalgamated transformations. An amalgamated rule is based on a kernel rule, which defines a fixed part of the match, and multi rules, which extend this fixed match. From a kernel and a multi rule, a complement rule can be constructed which characterises the effect of the multi rule exceeding the kernel rule. If multiple rules can be applied using the same kernel rule, as a first main result the Multi-amalgamation Theorem states that a bundle of s-amalgamable transformations is equivalent to a corresponding amalgamated transformation. An interaction scheme is defined by a kernel rule and available multi rules, leading to a bundle of multi rules that specifies in addition how often each multi rule is applied. Amalgamated rules are in general standard rules in M-adhesive transformation systems; thus all the results follow. In addition, we are able to refine parallel independence of amalgamated rules based on the induced multi rules. If we extend an interaction scheme as large as possible we can describe the transformation for an unknown number of matches, which otherwise would have to be defined by an infinite number of rules. This leads to maximal matchings, which are useful for defining the semantics of models. For this chapter, we require an M-adhesive category with binary coproducts as well as initial and effective pushouts (see Sect. 4.3). The theoretical results in this chapter are based on [GHE14].

In Sect. 6.1, kernel, multi and complement rules are presented. In Sect. 6.2, we introduce amalgamated rules and transformations and show some important results in Sect. 6.3. In Sect. 6.4, we define interaction schemes and maximal matching and use these concepts for the firing semantics of elementary Petri nets modelled by typed graphs using amalgamation. This chapter is based on [EGH$^+$14, Gol11].

6.1 Kernel Rules, Multi Rules, and Complement Rules

In the following, a *bundle* represents a family of morphisms or transformation steps with the same domain, which means that a *bundle* of things always starts at the same object.

A kernel morphism describes how a smaller rule, the kernel rule, is embedded into a larger rule, the multi rule. The multi rule has its name because it can be applied multiple times for a given kernel rule match, as described later. We need some more technical preconditions to make sure that the embeddings of the L-, K-, and R-components as well as the application conditions are consistent and allow us to construct a complement rule.

Definition 6.1 (Kernel morphism). Given rules $p_0 = (L_0 \xleftarrow{l_0} K_0 \xrightarrow{r_0} R_0, ac_0)$ and $p_1 = (L_1 \xleftarrow{l_1} K_1 \xrightarrow{r_1} R_1, ac_1)$, a kernel morphism $s_1 : p_0 \rightarrow p_1$, $s_1 = (s_{1,L}, s_{1,K}, s_{1,R})$ consists of M-morphisms $s_{1,L} : L_0 \rightarrow L_1$, $s_{1,K} : K_0 \rightarrow K_1$, and $s_{1,R} : R_0 \rightarrow$

R_1 such that the diagrams (1_1) and (2_1) are pullbacks, (1_1) has a pushout complement $(1'_1)$ for $s_{1,L} \circ l_0$, and $ac_1 \Rightarrow \text{Shift}(s_{1,L}, ac_0)$. In this case, p_0 is called *kernel rule* and p_1 *multi rule*.

ac_0 and ac_1 are *complement-compatible* w.r.t. s_1 if there is some application condition ac'_1 on the pushout complement L_{10} such that $ac_1 \cong \text{Shift}(s_{1,L}, ac_0) \wedge L(p_1^*,$ $\text{Shift}(v_1, ac'_1))$ for the pushout (3_1) and $p_1^* = (L_1 \xleftarrow{u_1} L_{10} \xrightarrow{v_1} E_1)$. $\triangle$

Remark 6.2. The complement-compatibility makes sure that there is a decomposition of ac_1 into parts on L_0 and L_{10}. The latter are used later for the application conditions of the complement rule, which ensure the equivalence of the composition. $\triangle$

Example 6.3. To explain the concept of amalgamation, in our example we model a small transformation system for switching the direction of edges in labeled graphs, where we have different labels for edges—black and dotted ones. The kernel rule p_0 is depicted in Fig. 6.1. It selects a node with a black loop, deletes this loop, and adds a dotted loop, all of this if no dotted loop is already present. The matches are defined by the numbers at the nodes and can be induced for the edges by their position.

In Fig. 6.2, two multi rules p_1 and p_2 are shown which extend the rule p_0 and in addition reverse an edge if no backward edge is present. They also inherit the application condition of p_0, forbidding a dotted loop at the selected node. There is a kernel morphism $s_1 : p_0 \rightarrow p_1$, as shown in the top of Fig. 6.2, with pullbacks (1_1), (2_1) and pushout complement $(1'_1)$. Similarly, there is a kernel morphism $s_2 : p_0 \rightarrow p_2$, as shown in the bottom of Fig. 6.2, with pullbacks (1_2), (2_2) and pushout complement $(1'_2)$.

For the application conditions, it holds that $ac_1 = \text{Shift}(s_{1,L}, ac_0) \wedge \neg \exists a_1 \cong \text{Shift}(s_{1,L}, ac_0) \wedge L(p_1^*, \text{Shift}(v_1, \neg \exists a'_1))$ with a'_1 as shown in the left of Fig. 6.3. We have that $\text{Shift}(v_1, \neg \exists a'_1) = \neg \exists a_{11}$, because square $(*)$ is the only possible commuting square leading to morphism (a_{11}, b_{11}) being jointly surjective and b_{11}

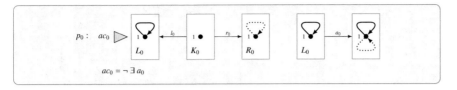

Fig. 6.1 The kernel rule p_0 deleting a loop at a node

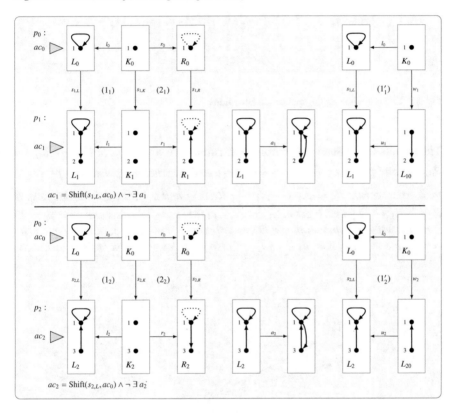

Fig. 6.2 The multi rules p_1 and p_2 describing the reversion of an edge

being injective. $L(p_1^*, \neg \exists a_{11}) = \neg \exists a_1$, as shown by the two pushout squares (PO_1) and (PO_2) in the middle of Fig. 6.3. Thus $ac_1' = \neg \exists a_1'$, and ac_0 and ac_1 are complement-compatible w. r. t. s_1. Similarly, it can be shown that ac_0 and ac_2 are complement-compatible w. r. t. s_2. △

For a given kernel morphism, the complement rule is the remainder of the multi rule after the application of the kernel rule, i.e., it describes what the multi rule does in addition to the kernel rule.

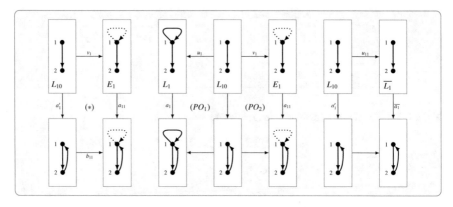

Fig. 6.3 Constructions for the application conditions

Theorem 6.4 (Existence of complement rule). *Given rules $p_0 = (L_0 \xleftarrow{l_0} K_0 \xrightarrow{r_0} R_0, ac_0)$ and $p_1 = (L_1 \xleftarrow{l_1} K_1 \xrightarrow{r_1} R_1, ac_1)$, and a kernel morphism $s_1 : p_0 \to p_1$, there exists a rule $\overline{p_1} = (\overline{L_1} \xleftarrow{\overline{l_1}} \overline{K_1} \xrightarrow{\overline{r_1}} \overline{R_1}, \overline{ac_1})$ and a jointly epimorphic cospan $R_0 \xrightarrow{e_{11}} E_1 \xleftarrow{e_{12}} \overline{L_1}$ such that the E_1-concurrent rule $p_0 *_{E_1} \overline{p_1}$ exists and $p_1 = p_0 *_{E_1} \overline{p_1}$ for rules without application conditions. Moreover, if ac_0 and ac_1 are complement-compatible w. r. t. s_1 then $p_1 \cong p_0 *_{E_1} \overline{p_1}$ also for rules with application conditions.*

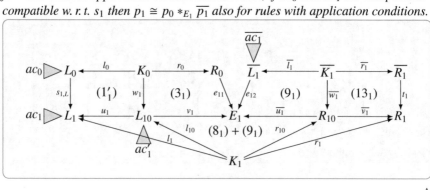

$\triangle$

Proof. First, we consider the construction without application conditions. Since s_1 is a kernel morphism the following diagrams (1_1) and (2_1) are pullbacks and we have a pushout complement $(1'_1)$ for $s_{1,L} \circ l_0$. Now construct the pushout (3_1) and the initial pushout (4_1) over $s_{1,R}$ with $b_1, c_1 \in \mathcal{M}$.

Consider P_1 as the pullback object of r_0 and b_1, and the pushout (5_1) where we obtain an induced morphism $s_{13} : S_1 \to R_0$ with $s_{13} \circ s_{12} = b_1$, $s_{13} \circ s_{11} = r_0$, and $s_{13} \in \mathcal{M}$ by effective pushouts.

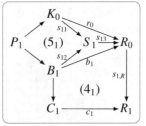

Since (1_1) is a pullback, Lem. B.1 implies that there is a unique morphism $l_{10} : K_1 \to L_{10}$ with $l_{10} \circ s_{1,K} = w_1$, $u_1 \circ l_{10} = l_1$, and $l_{10} \in \mathcal{M}$, and we can construct pushouts (6_1)–(9_1) as a decomposition of pushout (3_1), which leads to $\overline{L_1}$ and $\overline{K_1}$ of the complement rule, and with $(7_1) + (9_1)$ being a pushout, e_{11} and e_{12} are jointly epimorphic.

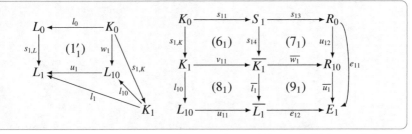

The pushout (4_1) can be decomposed into pushouts (10_1) and (11_1), obtaining the right-hand side $\overline{R_1}$ of the complement rule, while pullback (2_1) can be decomposed into pushout (6_1) and square (12_1), which is a pullback by Lem. B.2.

$$
\begin{array}{ccccc}
B_1 & \xrightarrow{s_{12}} & S_1 & \xrightarrow{s_{13}} & R_0 \\
\downarrow & (10_1)\ u_{13}\downarrow & (11_1) & & \downarrow s_{1,R} \\
C_1 & \longrightarrow & R_1 & \xrightarrow{t_1} & R_1
\end{array}
\qquad
\begin{array}{ccccc}
K_0 & \xrightarrow{s_{11}} & S_1 & \xrightarrow{s_{13}} & R_0 \\
s_{1,K}\downarrow & (6_1) & s_{14}\downarrow & (12_1) & \downarrow s_{1,R} \\
K_1 & \xrightarrow{v_{11}} & \overline{K_1} & \xrightarrow{v_{12}} & R_1
\end{array}
$$

Now Lem. B.1 implies that there is a unique morphism $\overline{r_1} : \overline{K_1} \to \overline{R_1}$ with $\overline{r_1} \circ s_{14} = u_{13}$, $t_1 \circ \overline{r_1} = v_{12}$, and $\overline{r_1} \in \mathcal{M}$. With pushout (7_1) there is a unique morphism $\overline{v_1} : R_{10} \to \overline{R_1}$ and by pushout decomposition of $(11_1) = (7_1) + (13_1)$ square (13_1) is a pushout.

Moreover, $(8_1) + (9_1)$ as a pushout over $\mathcal{M}$-morphisms is also a pullback which completes the construction of the rule, and $p_1 = p_0 *_{E_1} \overline{p_1}$ for rules without application conditions.

For the application conditions, suppose $ac_1 \cong \mathrm{Shift}(s_{1,L}, ac_0) \wedge \mathrm{L}(p_1^*, \mathrm{Shift}(v_1, ac_1'))$ for $p_1^* = (L_1 \xleftarrow{u_1} L_{10} \xrightarrow{v_1} E_1)$ with $v_1 = e_{12} \circ u_{11}$ and ac_1' over L_{10}. Now define $\overline{ac_1} = \mathrm{Shift}(u_{11}, ac_1')$, which is an application condition on $\overline{L_1}$.

We have to show that $(p_1, ac_{p_0 *_{E_1} \overline{p_1}}) \cong (p_1, ac_1)$. By construction of the E_1-concurrent rule we have that $L(p_1^*, \mathrm{Shift}(e_{12}, \overline{ac_1})) \cong L(p_1^*, \mathrm{Shift}(e_{12}, \mathrm{Shift}(u_{11}, ac_1')))$ $\cong L(p_1^*, \mathrm{Shift}(e_{12} \circ u_{11}, ac_1')) \cong L(p_1^*, \mathrm{Shift}(v_1, ac_1'))$. It follows that $ac_{p_0 *_{E_1} \overline{p_1}} \cong$ $\mathrm{Shift}(s_{1,L}, ac_0) \wedge L(p_1^*, \mathrm{Shift}(e_{12}, \overline{ac_1})) \cong \mathrm{Shift}(s_{1,L}, ac_0) \wedge L(p_1^*, \mathrm{Shift}(v_1, ac_1')) \cong ac_1$.

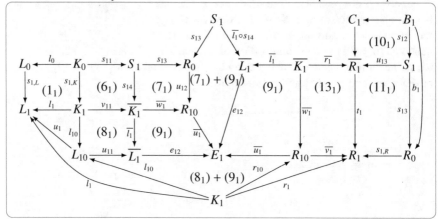

□

Remark 6.5. Note that by construction the interface K_0 of the kernel rule has to be preserved in the complement rule. The construction of $\overline{p_1}$ is not unique w. r. t. the property $p_1 = p_0 *_{E_1} \overline{p_1}$, since other choices for S_1 with $\mathcal{M}$-morphisms s_{11} and s_{13} also lead to a well-defined construction. In particular, one could choose $S_1 = R_0$, leading to $\overline{p_1} = E_1 \xleftarrow{\overline{u_1}} R_{10} \xrightarrow{\overline{v_1}} R_1$. Our choice represents the smallest possible complement, which should be preferred in most application areas.   △

Definition 6.6 (Complement rule). Given rules $p_0 = (L_0 \xleftarrow{l_0} K_0 \xrightarrow{r_0} R_0, ac_0)$ and $p_1 = (L_1 \xleftarrow{l_1} K_1 \xrightarrow{r_1} R_1, ac_1)$, and a kernel morphism $s_1 : p_0 \to p_1$ such that ac_0 and ac_1 are complement-compatible w. r. t. s_1, the rule $\overline{p_1} = (\overline{L_1} \xleftarrow{\overline{l_1}} \overline{K_1} \xrightarrow{\overline{r_1}} \overline{R_1}, \overline{ac_1})$ constructed in Theorem 6.4 is called *complement rule* (of s_1).

If we choose $\overline{ac_1} = \mathrm{true}$, this leads to the *weak complement rule* (of s_1) $\overline{p_1} =$ $(\overline{L_1} \xleftarrow{\overline{l_1}} \overline{K_1} \xrightarrow{\overline{r_1}} \overline{R_1}, \mathrm{true})$, which is defined even if ac_0 and ac_1 are not complement-compatible.   △

Example 6.7. Consider the kernel morphism s_1 depicted in Fig. 6.2. Using Theorem 6.4 we obtain the complement rule depicted in the top row of Fig. 6.4 with the application condition $\overline{ac_1} = \neg \exists \overline{a_1}$ constructed in the right of Fig. 6.3. The diagrams in Fig. 6.5 show the complete construction as done in the proof. Similarly, we obtain a complement rule for the kernel morphism $s_2 : p_0 \to p_2$ in Fig. 6.2, which is shown in the bottom row of Fig. 6.4.   △

Each direct transformation via a multi rule can be decomposed into a direct transformation via the kernel rule followed by a direct transformation via the (weak) complement rule.

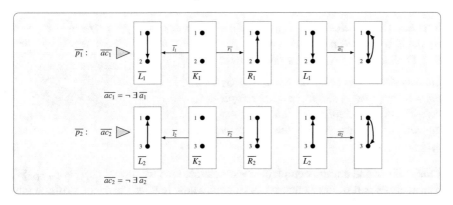

Fig. 6.4 The complement rules for the kernel morphisms

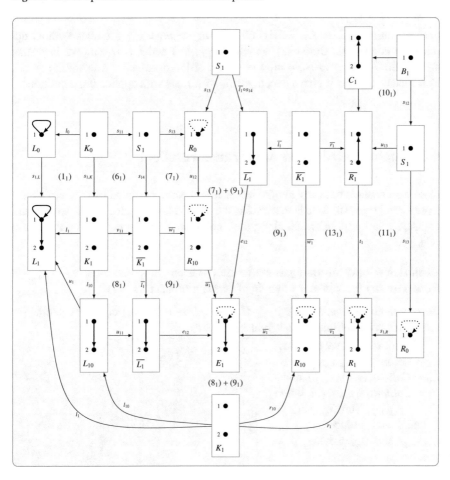

Fig. 6.5 The construction of the complement rule for the kernel morphism s_1

Fact 6.8. *Given rules* $p_0 = (L_0 \xleftarrow{l_0} K_0 \xrightarrow{r_0} R_0, ac_0)$ *and* $p_1 = (L_1 \xleftarrow{l_1} K_1 \xrightarrow{r_1} R_1, ac_1)$, *a kernel morphism* $s_1 : p_0 \to p_1$, *and a direct transformation* $t_1 : G \xRightarrow{p_1,m_1} G_1$, t_1 *can be decomposed into the transformation* $G \xRightarrow{p_0,m_0} G_0 \xRightarrow{\overline{p_1},\overline{m_1}} G_1$ *with* $m_0 = m_1 \circ s_{1,L}$ *using either the weak complement rule* $\overline{p_1}$ *or the complement rule* $\overline{p_1}$ *if* ac_0 *and* ac_1 *are complement-compatible with respect to* s_1.

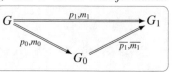

△

Proof. If ac_0 and ac_1 are complement-compatible then we have that $p_1 \cong p_0 *_{E_1} \overline{p_1}$. The analysis part of the Concurrency Theorem now implies the decomposition into $G \xRightarrow{p_0,m_0} G_0 \xRightarrow{\overline{p_1},\overline{m_1}} G_1$ with $m_0 = m_1 \circ s_{1,L}$.

If ac_0 and ac_1 are not complement-compatible we can apply the analysis part of the Concurrency Theorem without application conditions leading to a decomposition into $G \xRightarrow{p_0,m_0} G_0 \xRightarrow{\overline{p_1},\overline{m_1}} G_1$ with $m_0 = m_1 \circ s_{1,L}$ for rules without application conditions. Since $ac_1 \Rightarrow \text{Shift}(s_{1,L}, ac_0)$ and $m_1 \models ac_1$ we have that $m_1 \models \text{Shift}(s_{1,L}, ac_0) \Leftrightarrow m_0 = m_1 \circ s_{1,L} \models ac_0$. Moreover, $\overline{ac_1} = true$ and $\overline{m_1} \models \overline{ac_1}$. This means that this is also a decomposition for rules with application conditions. □

6.2 Amalgamated Rules and Transformations

Now we consider not only single kernel morphisms, but bundles of them over a fixed kernel rule. The idea is to combine the multi rules of such a bundle to an amalgamated rule by gluing them along their common elements defined by the kernel rule.

Definition 6.9 (Multi-amalgamated rule). Given rules $p_i = (L_i \xleftarrow{l_i} K_i \xrightarrow{r_i} R_i, ac_i)$ for $i = 0, \ldots, n$ and a bundle of kernel morphisms $s = (s_i : p_0 \to p_i)_{i=1,\ldots,n}$, the *(multi-)amalgamated rule* $\tilde{p}_s = (\tilde{L}_s \xleftarrow{\tilde{l}_s} \tilde{K}_s \xrightarrow{\tilde{r}_s} \tilde{R}_s, \tilde{ac}_s)$ is constructed as the componentwise colimit of the kernel morphisms.

This means that we construct $\tilde{L}_s = Colimit((s_{i,L})_{i=1,\ldots,n})$, $\tilde{K}_s = Colimit((s_{i,K})_{i=1,\ldots,n})$, and $\tilde{R}_s = Colimit((s_{i,R})_{i=1,\ldots,n})$, with $\tilde{ac}_s = \bigwedge_{i=1,\ldots,n} \text{Shift}(t_{i,L}, ac_i)$, and $\tilde{l}_s$ and $\tilde{r}_s$ are induced by $(t_{i,L} \circ l_i)_{i=1,\ldots,n}$ and $(t_{i,R} \circ r_i)_{i=1,\ldots,n}$, respectively. △

This definition is well-defined. Moreover, if the application conditions of the kernel morphisms are complement-compatible, this also holds for the application

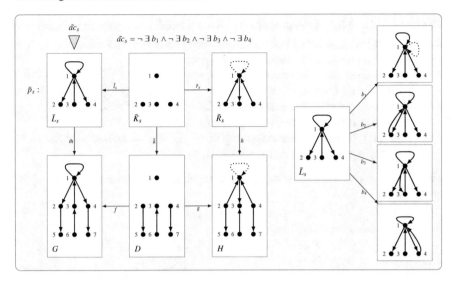

Fig. 6.6 An amalgamated transformation

condition of the amalgamated rule with respect to the morphisms from the original kernel and multi rules.

Fact 6.10. *The amalgamated rule as defined in Def. 6.9 is well defined and we have kernel morphisms* $t_i = (t_{i,L}, t_{i,K}, t_{i,R}) : p_i \rightarrow \tilde{p}_s$ *for* $i = 0, 1, \ldots, n$. *If* ac_0 *and* ac_i *are complement-compatible w. r. t.* s_i *for all* $i = 1, \ldots, n$ *then also* ac_i *and* $\tilde{ac}_s$ *as well as* ac_0 *and* $\tilde{ac}_s$ *are complement compatible w. r. t.* t_i *and* t_0, *respectively.* △

Proof. See Appendix B.5.1. □

The application of an amalgamated rule yields an amalgamated transformation.

Definition 6.11 (Amalgamated transformation). The application of an amalgamated rule to a graph G is called an *amalgamated transformation*. △

Example 6.12. Consider the bundle $s = (s_1, s_2, s_3 = s_1)$ of the kernel morphisms depicted in Fig. 6.2. The corresponding amalgamated rule $\tilde{p}_s$ is shown in the top row of Fig. 6.6. This amalgamated rule can be applied to the graph G, leading to the amalgamated transformation depicted in Fig. 6.6, where the application condition $\tilde{ac}_s$ is obviously fulfilled by the match $\tilde{m}$. △

If we have a bundle of direct transformations of an object G, where for each transformation one of the multi rules is applied, we want to analyse if the amalgamated rule is applicable to G combining all the single transformation steps. These transformations are compatible, i.e., multi-amalgamable, if the matches agree on the kernel rules, and are independent outside.

Definition 6.13 (Multi-amalgamable). Given a bundle of kernel morphisms $s = (s_i : p_0 \to p_i)_{i=1,...,n}$, a bundle of direct transformations steps $(G \overset{p_i,m_i}{\Longrightarrow} G_i)_{i=1,...n}$ is *s-multi-amalgamable*, or in short *s-amalgamable*, if

- it has *consistent matches*, i.e., $m_i \circ s_{i,L} = m_j \circ s_{j,L} =: m_0$ for all $i, j = 1, \ldots, n$ and
- it has *weakly independent matches*, i.e., for all $i \neq j$ consider the pushout complements

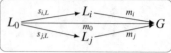

(1'$_i$) and (1'$_j$) for which there exist morphisms $p_{ij} : L_{i0} \to D_j$ and $p_{ji} : L_{j0} \to D_i$ such that $f_j \circ p_{ij} = m_i \circ u_i$ and $f_i \circ p_{ji} = m_j \circ u_j$.

Moreover, if ac_0 and ac_i are complement-compatible we require $g_j \circ p_{ij} \models ac'_i$ for all $j \neq i$.

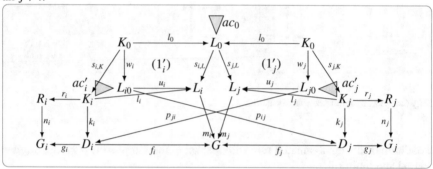

$\triangle$

As with to the characterisation of parallel independence in [EEPT06], we can give a set-theoretical characterisation of weak independence.

Fact 6.14. *For graphs and other set-based structures, weakly independent matches means that*

$$m_i(L_i) \cap m_j(L_j) \subseteq m_0(L_0) \cup (m_i(l_i(K_i)) \cap m_j(l_j(K_j)))$$

for all $i \neq j = 1, \ldots, n$, i.e., the elements in the intersection of the matches m_i and m_j are either preserved by both transformations, or are also matched by m_0.

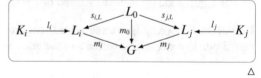

$\triangle$

Proof. We have to prove the equivalence of $m_i(L_i) \cap m_j(L_j) \subseteq m_0(L_0) \cup (m_i(l_i(K_i)) \cap m_j(l_j(K_j)))$ for all $i \neq j = 1, \ldots, n$ with the definition of weakly independent matches.

"$\Leftarrow$" Let $x = m_i(y_i) = m_j(y_j)$, and suppose $x \notin m_0(L_0)$. Since (1'$_i$) is a pushout we have that $y_i = u_i(z_i) \in u_i(L_{i0} \backslash w_i(K_0))$, and $x = m_i(u_i(z_i)) = f_j(p_i(z_i)) = m_j(y_j)$, and by pushout properties $y_j \in l_j(K_j)$ and $x \in m_j(l_j(K_j))$. Similarly, $x \in m_i(l_i(K_i))$.

"$\Rightarrow$" For $x \in L_{i0}$, $x = w_i(k)$ define $p_{ij}(x) = k_j(s_{j,K}(k))$; then $f_j(p_{ij}(x)) = f_j(k_j(s_{j,K}(k))) = m_j(l_j(s_{j,K}(k))) = m_j(s_{j,L}(l_0(k))) = m_i(s_{i,L}(l_0(k))) = m_i(n_i(w_i(k))) = m_i(u_i(x))$. Otherwise, $x \notin w_i(K_0)$, i.e., $u_i(x) \notin s_{i,L}(L_0)$, and we define $p_{ij}(x) = y$ with $f_j(y) = m_i(u_i(x))$. This y exists, because either $m_i(u_i(x)) \notin m_j(L_j)$ or $m_i(u_i(x)) \in$

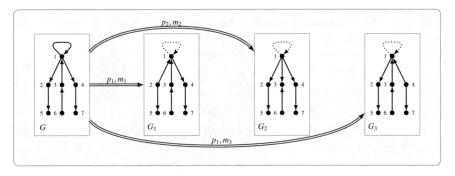

Fig. 6.7 An s-amalgamable bundle of direct transformations

$m_j(L_j)$ and then $m_i(u_i(x)) \in m_j(l_j(K_j))$, and in both cases $m_i(u_i(x)) \in f_j(D_j)$. Similarly, we can define p_{ji} with the required property.   □

Example 6.15. Consider the bundle $s = (s_1, s_2, s_3 = s_1)$ of kernel morphisms from Ex. 6.12. For the graph G given in Fig. 6.6 we find matches $m_0 : L_0 \to G$, $m_1 : L_1 \to G$, $m_2 : L_2 \to G$, and $m_3 : L_1 \to G$ mapping all nodes from the left-hand side to their corresponding nodes in G, except for m_3 mapping node 2 in L_1 to node 4 in G. For all these matches, the corresponding application conditions are fulfilled and we can apply the rules p_1, p_2, p_1, respectively, leading to the bundle of direct transformations depicted in Fig. 6.7. This bundle is s-amalgamable, because the matches m_1, m_2, and m_3 agree on the match m_0, and are weakly independent, because they only overlap in m_0.   △

For an s-amalgamable bundle of direct transformations, each single transformation step can be decomposed into an application of the kernel rule followed by an application of the (weak) complement rule, as shown in Fact 6.8. Moreover, all kernel rule applications lead to the same object, and the following applications of the complement rules are parallel independent.

Fact 6.16. *Given a bundle of kernel morphisms* $s = (s_i : p_0 \to p_i)_{i=1,...,n}$ *and an s-amalgamable bundle of direct transformations* $(G \xrightarrow{p_i,m_i} G_i)_{i=1,...,n}$, *each direct transformation* $G \xrightarrow{p_i,m_i} G_i$ *can be decomposed into a transformation* $G \xrightarrow{p_0,m_0} G_0 \xrightarrow{\overline{p_i},\overline{m_i}} G_i$, *where* $\overline{p_i}$ *is the (weak) complement rule of* s_i. *Moreover, the transformations* $G_0 \xrightarrow{\overline{p_i},\overline{m_i}} G_i$ *are pairwise parallel independent.*   △

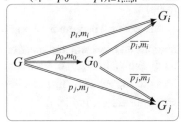

Proof. See Appendix B.5.2.   □

If a bundle of direct transformations of an object G is s-amalgamable we can apply the amalgamated rule directly to G, leading to a parallel execution of all the changes done by the single transformation steps.

Theorem 6.17 (Multi-amalgamation Theorem). *Consider a bundle of kernel morphisms $s = (s_i : p_0 \to p_i)_{i=1,\dots,n}$.*

1. Synthesis. *Given an s-amalgamable bundle of direct transformations $(G \overset{p_i,m_i}{\Longrightarrow} G_i)_{i=1,\dots,n}$, there is an amalgamated transformation $G \overset{\tilde{p}_s,\tilde{m}}{\Longrightarrow} H$ and transformations $G_i \overset{q_i}{\Longrightarrow} H$ over the complement rules q_i of the kernel morphisms $t_i : p_i \to \tilde{p}_s$ such that $G \overset{p_i,m_i}{\Longrightarrow} G_i \overset{q_i}{\Longrightarrow} H$ is a decomposition of $G \overset{\tilde{p}_s,\tilde{m}}{\Longrightarrow} H$.*

2. Analysis. *Given an amalgamated transformation $G \overset{\tilde{p}_s,\tilde{m}}{\Longrightarrow} H$, there are s_i-related transformations $G \overset{p_i,m_i}{\Longrightarrow} G_i \overset{q_i}{\Longrightarrow} H$ for $i = 1,\dots,n$ such that the bundle $(G \overset{p_i,m_i}{\Longrightarrow} G_i)_{i=1,\dots,n}$ is s-amalgamable.*

3. Bijective Correspondence. *The synthesis and analysis constructions are inverse to each other up to isomorphism.* △

Proof. See Appendix B.5.3.   □

Remark 6.18. Note that q_i can be constructed as the amalgamated rule of the kernel morphisms $(p_{K_0} \to \overline{p_j})_{j \neq i}$, where $p_{K_0} = (K_0 \overset{id_{K_0}}{\longleftarrow} K_0 \overset{id_{K_0}}{\longrightarrow} K_0, \text{true}))$ and $\overline{p_j}$ is the complement rule of p_j.

For $n = 2$, the Multi-amalgamation Theorem specialises to the Amalgamation Theorem in [BFH87, EGH$^+$14] for rules without application conditions. Moreover, if p_0 is the empty rule, this is the Parallelism Theorem in [EHL10], since the transformations are parallel independent for an empty kernel match.   △

Example 6.19. As already observed in Ex. 6.15, the transformations $G \overset{p_1,m_1}{\Longrightarrow} G_1$, $G \overset{p_2,m_2}{\Longrightarrow} G_2$, and $G \overset{p_1,m_3}{\Longrightarrow} G_3$ shown in Fig. 6.7 are s-amalgamable for the bundle $s = (s_1, s_2, s_3 = s_1)$ of kernel morphisms. Applying Fact 6.16, we can decompose these transformations into a transformation $G \overset{p_0,m_0}{\Longrightarrow} G_0$ followed by transformations $G_0 \overset{\overline{p_1},\overline{m_1}}{\Longrightarrow} G_1$, $G_0 \overset{\overline{p_2},\overline{m_2}}{\Longrightarrow} G_2$, and $G_0 \overset{\overline{p_1},\overline{m_3}}{\Longrightarrow} G_3$ via the complement rules, which are pairwise parallel independent. These transformations are depicted in Fig. 6.8.

Moreover, Theorem 6.17 implies that we obtain for this bundle of direct transformations an amalgamated transformation $G \overset{\tilde{p}_s,\tilde{m}}{\Longrightarrow} H$, which is the transformation already shown in Fig. 6.6. Vice versa, the analysis of this amalgamated transformation leads to the s-amalgamable bundle of transformations $G \overset{p_1,m_1}{\Longrightarrow} G_1, G \overset{p_2,m_2}{\Longrightarrow} G_2$, and $G \overset{p_1,m_3}{\Longrightarrow} G_3$ in Fig. 6.7.   △

For an $\mathcal{M}$-adhesive transformation system with amalgamation we define a set of kernel morphisms and allow all kinds of amalgamated transformations using bundles from this set.

Definition 6.20 ($\mathcal{M}$-adhesive grammar with amalgamation). An $\mathcal{M}$-adhesive transformation system with amalgamation $ASA = (\mathbf{C}, \mathcal{M}, P, S)$ is an $\mathcal{M}$-adhesive

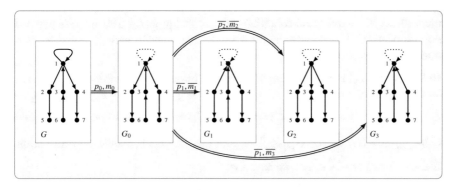

Fig. 6.8 The decomposition of the s-amalgamable bundle

transformation system $(\mathbf{C}, \mathcal{M}, P)$ with a set of kernel morphisms S between rules in P.

An $\mathcal{M}$-*adhesive grammar with amalgamation* $AGA = (ASA, S_0)$ consists of an $\mathcal{M}$-adhesive transformation system with amalgamation ASA and a start object S_0.

The language L of an $\mathcal{M}$-adhesive grammar with amalgamation AGA is defined by

$$L = \{G \mid \exists \text{ amalgamated transformation } S_0 \stackrel{*}{\Rightarrow} G\},$$

where all amalgamated rules over arbitrary bundles of kernel morphisms in S are allowed to be used.   △

Remark 6.21. Note that by including the kernel morphism $id_p : p \rightarrow p$ for a rule p into the set S the transformation $G \stackrel{p,m}{\Longrightarrow} H$ is also an amalgamated transformation for this kernel morphism as the only one considered in the bundle.   △

6.3 Results for Amalgamated Transformations

Since amalgamated rules are normal rules in an $\mathcal{M}$-adhesive transformation system with only a special way of constructing them, we obtain all the results from Sect. 5.2 also for amalgamated transformations. Especially for parallel independence, we can analyse this property in more detail to connect the result to the underlying kernel and multi rules.

6.3.1 Parallel Independence of Amalgamated Transformations

The parallel independence of two amalgamated transformations of the same object can be reduced to the parallel independence of the involved transformations via the multi rules if the application conditions are handled properly. This leads to two new notions of parallel independence for amalgamated transformations and bundles of transformations.

Definition 6.22 (Parallel amalgamation and bundle independence). Given two bundles of kernel morphisms $s = (s_i : p_0 \to p_i)_{i=1,\dots,n}$ and $s' = (s'_j : p'_0 \to p'_j)_{j=1,\dots,n'}$, and two bundles of s- and s'-amalgamable transformations $(G \xRightarrow{p_i,m_i} G_i)_{i=1,\dots,n}$ and $(G \xRightarrow{p'_j,m'_j} G'_j)_{j=1,\dots,n'}$ leading to the amalgamated transformations $G \xRightarrow{\tilde{p}_s,\tilde{m}} H$ and $G \xRightarrow{\tilde{p}_{s'},\tilde{m}'} H'$, we have that

- $G \xRightarrow{\tilde{p}_s,\tilde{m}} H$ and $G \xRightarrow{\tilde{p}_{s'},\tilde{m}'} H'$ are *parallel amalgamation independent* if they are parallel independent, i.e., there are morphisms $\tilde{r}_s$ and $\tilde{r}_{s'}$ with $f \circ \tilde{r}_{s'} = \tilde{m}'$, $f' \circ \tilde{r}_s = \tilde{m}$, $g \circ \tilde{r}_{s'} \models \tilde{a}c_{s'}$, and $g' \circ \tilde{r}_s \models \tilde{a}c_s$, and in addition we have that $g_i \circ d_i \circ \tilde{r}_{s'} \models \mathrm{Shift}(t'_{j,L}, ac'_j)$ and $g'_j \circ d'_j \circ \tilde{r}_s \models \mathrm{Shift}(t_{i,L}, ac_i)$ for all $i = 1,\dots,n$, $j = 1,\dots,n'$.

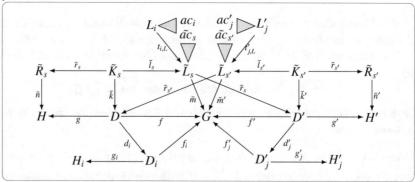

- $(G \xRightarrow{p_i,m_i} G_i)_{i=1,\dots,n}$ and $(G \xRightarrow{p'_j,m'_j} G'_j)_{j=1,\dots,n'}$ are *parallel bundle independent* if they are pairwise parallel independent for all i, j, i.e., there are morphisms r_{ij} and r'_{ji} with $f'_j \circ r_{ij} = m_i$, $f_i \circ r'_{ji} = m'_j$, $g'_j \circ r_{ij} \models ac_i$, and $g_i \circ r'_{ji} \models ac'_j$, and in addition we have for the induced morphisms $\tilde{r}_s : \tilde{L}_s \to D'$ and $\tilde{r}_{s'} : \tilde{L}_{s'} \to D$ that $g \circ \tilde{r}_{s'} \models \tilde{a}c_{s'}$ and $g' \circ \tilde{r}_s \models \tilde{a}c_s$.

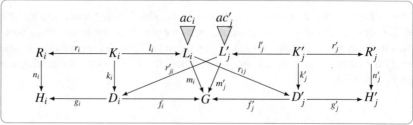

$\triangle$

Remark 6.23. Note that all objects and morphisms in the above diagrams originate from the construction in the proof of Theorem 6.17 and the parallel independence.

$\triangle$

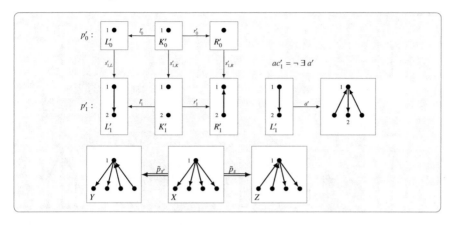

Fig. 6.9 A counterexample for parallel independence of amalgamated transformations

Two amalgamated transformations are parallel amalgamation independent if and only if the corresponding bundles of transformations are parallel bundle independent.

Theorem 6.24 (Characterisation of parallel independence). *Given two bundles of kernel morphisms $s = (s_i : p_0 \rightarrow p_i)_{i=1,\ldots,n}$ and $s' = (s'_j : p'_0 \rightarrow p'_j)_{j=1,\ldots,n'}$, and two bundles of s- and s'-amalgamable transformations $(G \xrightarrow{p_i,m_i} G_i)_{i=1,\ldots,n}$ and $(G \xrightarrow{p'_j,m'_j} G'_j)_{j=1,\ldots,n'}$ leading to the amalgamated transformations $G \xrightarrow{\tilde{p}_s,\tilde{m}} H$ and $G \xrightarrow{\tilde{p}_{s'},\tilde{m}'} H'$, the following holds: $(G \xrightarrow{p_i,m_i} G_i)_{i=1,\ldots,n}$ and $(G \xrightarrow{p'_j,m'_j} G'_j)_{j=1,\ldots,n'}$ are parallel bundle independent if and only if $G \xrightarrow{\tilde{p}_s,\tilde{m}} H$ and $G \xrightarrow{\tilde{p}_{s'},\tilde{m}'} H'$ are parallel amalgamation independent.* △

Proof. See Appendix B.5.4.   □

Remark 6.25. Note that the additional verification of the application conditions is necessary because the common effect of all rule applications may invalidate the amalgamated application condition, although the single applications of the multi rules behave well. For example, consider the kernel morphism s'_1 in Fig. 6.9, where the bundles $s = (s'_1, s'_1)$ and $s' = (s'_1, s'_1)$ are applied to the graph X. Although all pairs of applications of the rule p'_1 to X are pairwise parallel independent, the amalgamated transformations are not parallel independent because they invalidate the application condition.

Similarly, a positive condition may be fulfilled for the amalgamated rule, but not for all single multi rules.   △

Given two amalgamated rules, the parallel rule can be constructed as an amalgamated rule using some componentwise coproduct constructions of the kernel and multi rules.

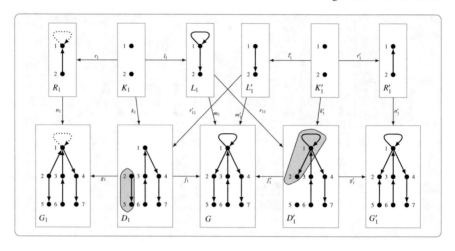

Fig. 6.10 Parallel independence of the transformations $G \xrightarrow{p_1,m_1} G_1$ and $G \xrightarrow{p'_1,m'_1} G'_1$

Fact 6.26. *Given two bundles of kernel morphisms* $s = (s_i : p_0 \to p_i)_{i=1,...,n}$ *and* $s' = (s'_j : p'_0 \to p'_j)_{j=1,...,n'}$ *leading to amalgamated rules* $\tilde{p}_s$ *and* $\tilde{p}_{s'}$, *respectively, the parallel rule* $\tilde{p}_s + \tilde{p}_{s'}$ *is constructed by* $\tilde{p}_s + \tilde{p}_{s'} = \tilde{p}_t$ *as the amalgamated rule of the bundle of kernel morphisms* $t = (t_i : p_0+p'_0 \to p_i+p'_0, t'_j : p_0+p'_0 \to p_0+p'_j)$. △

Proof. This follows directly from the general construction of colimits and their compatibility. □

Example 6.27. Consider the amalgamated transformation $G \xrightarrow{\tilde{p}_s,\tilde{m}} H$ in Fig. 6.6 and the bundle of kernel morphisms $s' = (s'_1)$ using the kernel morphism depicted in Fig. 6.9. The amalgamated rule $\tilde{p}_{s'}$ can also be applied to G via match $\tilde{m}'$ matching the nodes 1 and 2 in L'_1 to the nodes 2 and 5 in G, respectively. This results in the amalgamated transformation $G \xrightarrow{\tilde{p}_{s'},\tilde{m}'} G'_1$.

For the analysis of parallel amalgamation independence, we first analyse the pairwise parallel independence of the transformations $G \xrightarrow{p_i,m_i} G_i$ and $G \xrightarrow{p'_1,m'_1} G'_1$ for $i = 1,2,3$, with $m'_1 = \tilde{m}'$. This is done exemplarily for $i = 1$ in Fig. 6.10, where we do not show the application conditions. The morphisms r_{11} and r'_{11} are marked in their corresponding domains D'_1 and D_1, leading to $f'_1 \circ r_{11} = m_1$ and $f_1 \circ r'_{11} = m'_1$. Moreover, $g \circ r'_{11} \models ac'_1$, because there are no ingoing edges into node 2, and $g' \circ r_{11} \models ac_1$, because there is no dotted loop at node 1 and no reverse edge. Thus, both transformations are parallel independent, and this follows analogously for $i = 2,3$. Moreover, the induced morphism $\tilde{r}_{s'} : \tilde{L}_{s'} = L'_1 \to D$ leads to $g \circ \tilde{r}_{s'} \models \tilde{ac}_{s'} = ac'_1$. In the other direction, $\tilde{r}_s : \tilde{L}_s \to D' = D'_1$ ensures that $g'_1 \circ \tilde{r}_s \models \tilde{ac}_s$. Thus, the two bundles are parallel bundle independent and, using Theorem 6.24, it follows that $G \xrightarrow{\tilde{p}_s,\tilde{m}} H$ and $G \xrightarrow{\tilde{p}_{s'},\tilde{m}'} H'$ are parallel amalgamation independent.

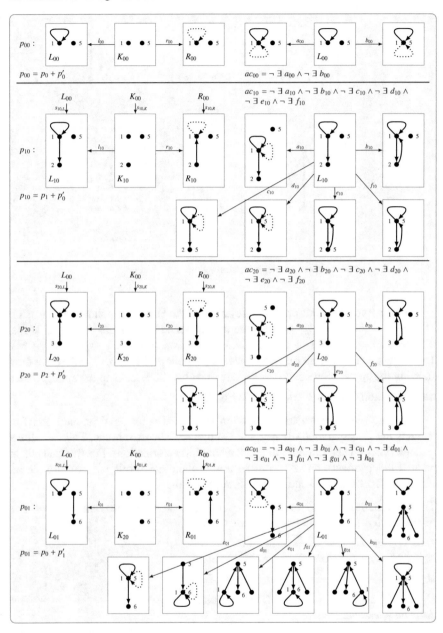

Fig. 6.11 The kernel morphisms leading to the parallel rule

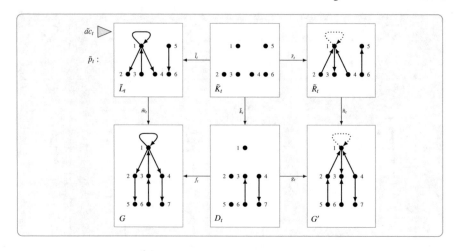

Fig. 6.12 A parallel amalgamated graph transformation

The construction of this parallel rule according to Fact 6.26 is shown in Fig. 6.11. The parallel rule $\tilde{p}_s + \tilde{p}_{s'} = \tilde{p}_t$ is the amalgamated rule of the bundle of kernel morphisms $t = (s_{10} = s_1 + id_{p'_0}, s_{20} = s_2 + id_{p'_0}, s_{10} = s_1 + id_{p'_0}, s_{01} = id_{p_0} + s'_1)$. The corresponding parallel rule is depicted in the top of Fig. 6.12, where we omit showing the application condition ac_t due to its length. It leads to the amalgamated transformation $G \xrightarrow{\tilde{p}_t, \tilde{m}_t} G'$ depicted in Fig. 6.12.   △

As in any $\mathcal{M}$-adhesive transformation system, also for amalgamated transformations the Local Church–Rosser and Parallelism Theorem holds. This is a direct instantiation of Theorem 2.26 to amalgamated transformations. For the analysis of parallel independence and the construction of the parallel rule we may use the results from Theorem 6.24 and Fact 6.26, respectively.

Theorem 6.28 (Local Church–Rosser and Parallelism Theorem). *Given two parallel independent amalgamated transformations $G \xrightarrow{\tilde{p}_s} H_1$ and $G \xrightarrow{\tilde{p}_{s'}} H_2$, there is an object G' together with direct transformations $H_1 \xrightarrow{\tilde{p}_{s'}} G'$ and $H_2 \xrightarrow{\tilde{p}_s} G'$ such that $G \xrightarrow{\tilde{p}_s} H_1 \xrightarrow{\tilde{p}_{s'}} G'$ and $G \xrightarrow{\tilde{p}_{s'}} H_2 \xrightarrow{\tilde{p}_s} G'$ are sequentially independent.*

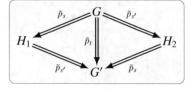

Given two sequentially independent direct transformations $G \xrightarrow{\tilde{p}_s} H_1 \xrightarrow{\tilde{p}_{s'}} G'$, there is an object H_2 with direct transformations $G \xrightarrow{\tilde{p}_{s'}} H_2 \xrightarrow{\tilde{p}_s} G'$ such that $G \xrightarrow{\tilde{p}_s} H_1$ and $G \xrightarrow{\tilde{p}_{s'}} H_2$ are parallel independent.

In any case of independence, there is a parallel transformation $G \xrightarrow{\tilde{p}_t} G'$ via the parallel rule $\tilde{p}_s + \tilde{p}_{s'} = \tilde{p}_t$ and, vice versa, a direct transformation $G \xrightarrow{\tilde{p}_t} G'$ can be sequentialised both ways.   △

Proof. This follows directly from Theorem 5.26, where all transformations are amalgamated transformations.   □

Example 6.29. In addition to the results from Ex. 6.27, from Theorem 6.28 we obtain amalgamated transformations $H \xrightarrow{\bar{p}_{s'}} G'$ and $H' \xrightarrow{\bar{p}_{s}} G'$, with $G \xrightarrow{\bar{p}_{s}} G \xrightarrow{\bar{p}_{s'}} G'$ and $G \xrightarrow{\bar{p}_{s'}} H' \xrightarrow{\bar{p}_{s}} G'$ being sequentially independent transformation sequences.   △

6.3.2 Other Results for Amalgamated Transformations

For $\mathcal{M}$-adhesive transformation systems with amalgamation, also the other results stated in Sect. 5.2 are valid for amalgamated transformations. But additional results for the analysis of the results for amalgamated rules based on the underlying kernel and multi rules are future work:

- For the Concurrency Theorem, two amalgamated rules leading to parallel dependent amalgamated transformations can be combined to an E-concurrent rule and the corresponding transformation. It would be interesting to analyse if this E-concurrent rule could be constructed as an amalgamated rule based on the underlying kernel and multi rules.
- For the Embedding and Extension Theorem, an amalgamated rule can be embedded if the embedding morphism is consistent. Most likely, consistency w. r. t. an amalgamated transformation can be formulated as a consistency property w. r. t. the bundle of transformations.
- For the Local Confluence Theorem, if all critical pairs depending on all available amalgamated rules are strictly AC-confluent then the $\mathcal{M}$-adhesive transformation system with amalgamation is locally confluent. It would be interesting to find a new notion of critical pairs depending not on the amalgamated rules, but on the kernel morphisms. For arbitrary amalgamated rules, any bundle of kernel morphisms had to be analysed. It would be more efficient if some kinds of minimal bundles were sufficient for constructing all critical pairs or dependent transformations of the $\mathcal{M}$-adhesive transformation system with amalgamation.

6.4 Interaction Schemes and Maximal Matchings

For many interesting application areas, including the operational semantics of Petri nets and statcharts, we do not want to define the matches for the multi rules explicitly, but to obtain them dependent on the object to be transformed. In this case, only an interaction scheme is given, which defines a set of kernel morphisms but does not include a count of how often each multi rule is used in the bundle leading to the amalgamated rule.

Definition 6.30 (Interaction scheme). A kernel rule p_0 and a set of multi rules $\{p_1, \ldots, p_k\}$ with kernel morphisms $s_i : p_0 \to p_i$ form an *interaction scheme is =* $\{s_1, \ldots, s_k\}$.   △

When given an interaction scheme, we want to apply as many rules occurring in the interaction scheme as often as possible over a certain kernel rule match. Here we consider two different possible maximal matchings: maximal weakly independent and maximal weakly disjoint matchings. For maximal weakly independent matchings, we require the matchings of the multi rules to be weakly independent to ensure that the resulting bundle of transformations is amalgamable. This is the minimal requirement to meet the definition. In addition, for maximal weakly disjoint matchings the matches of the multi rules should be disjoint up to the kernel rule match. This variant is preferred for implementation, because it eases the computation of additional matches when we can rule out model parts that were already matched.

Definition 6.31 (Maximal weakly independent matching). Given an object G and an interaction scheme $is = \{s_1, \ldots, s_k\}$, a maximal weakly independent matching $m = (m_0, m_1, \ldots, m_n)$ is defined as follows:

1. Set $i = 0$. Choose a kernel matching $m_0 : L_0 \to G$ such that $G \xRightarrow{p_0, m_0} G_0$ is a valid transformation.
2. As long as possible: Increase i, choose a multi rule $\hat{p}_i = p_j$ with $j \in \{1, \ldots, k\}$, and find a match $m_i : L_j \to G$ such that $m_i \circ s_{j,L} = m_0$, $G \xRightarrow{p_j, m_i} G_i$ is a valid transformation, the matches $m_1, \ldots, m_i$ are weakly independent, and $m_i \neq m_\ell$ for all $\ell = 1, \ldots, i - 1$.
3. If no more valid matches for any rule in the interaction scheme can be found, return $m = (m_0, m_1, \ldots, m_n)$.

The maximal weakly independent matching leads to a bundle of kernel morphisms $s = (s_i : p_0 \to \hat{p}_i)$ and an s-amalgamable bundle of direct transformations $G \xRightarrow{\hat{p}_i, m_i}$ G_i.   △

Definition 6.32 (Maximal weakly disjoint matching). Given an object G and an interaction scheme $is = \{s_1, \ldots, s_k\}$, a maximal weakly disjoint matching $m = (m_0, m_1, \ldots, m_n)$ is defined as follows:

1. Set $i = 0$. Choose a kernel matching $m_0 : L_0 \to G$ such that $G \xRightarrow{p_0, m_0} G_0$ is a valid transformation.
2. As long as possible: Increase i, choose a multi rule $\hat{p}_i = p_j$ with $j \in \{1, \ldots, k\}$, and find a match $m_i : L_j \to G$ such that $m_i \circ s_{j,L} = m_0$, $G \xRightarrow{p_j, m_i} G_i$ is a valid transformation, the matches $m_1, \ldots, m_i$ are weakly independent, and $m_i \neq m_\ell$ and the square $(P_{i\ell})$ is a pullback for all $\ell = 1, \ldots, i - 1$.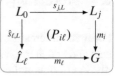
3. If no more valid matches for any rule in the interaction scheme can be found, return $m = (m_0, m_1, \ldots, m_n)$.

The maximal weakly disjoint matching leads to a bundle of kernel morphisms $s = (s_i : p_0 \rightarrow \hat{p}_i)$ and an s-amalgamable bundle of direct transformations $G \overset{\hat{p}_i, m_i}{\Longrightarrow} G_i$. $\triangle$

Note that for maximal weakly disjoint matchings, the pullback requirement already implies the existence of the morphisms for the weakly independent matches. Only the property for the application conditions has to be checked in addition.

Fact 6.33. *Given an object G, a bundle of kernel morphisms $s = (s_1, \ldots, s_n)$, and matches $m_1, \ldots, m_n$ leading to a bundle of direct transformations $G \overset{p_i, m_i}{\Longrightarrow} G_i$ such that $m_i \circ s_{i,L} = m_0$ and square (P_{ij}) is a pullback for all $i \neq j$, the bundle $G \overset{p_i, m_i}{\Longrightarrow} G_i$ is s-amalgamable for transformations without application conditions.* $\triangle$

Proof. By construction, the matches m_i agree on the match m_0 of the kernel rule. It remains to show that they are weakly independent.

Given the transformations $G \overset{p_i, m_i}{\Longrightarrow} G_i$ with pushouts (20_i) and (21_i), consider the following cube, where the bottom face is pushout (20_i), the back right face is pullback (1_i), and the front right face is pullback (P_{ij}). Now construct the pullback of f_i and m_j as the front left face, and from $m_j \circ s_{j,L} \circ l_0 = m_i \circ s_{i,L} \circ l_0 = m_i \circ l_i \circ s_{i,K} = f_i \circ k_i \circ s_{i,K}$ we obtain a morphism p with $\hat{f} \circ p = s_{j,L} \circ l_0$ and $\hat{m} \circ p = k_i \circ s_{i,K}$.

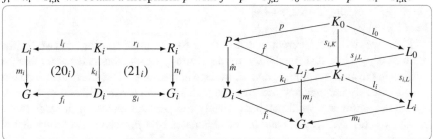

From pullback composition and decomposition of the right and left faces it follows that also the back left face is a pullback. Now the $\mathcal{M}$-van Kampen property can be applied, leading to a pushout in the top face. Since pushout complements are unique up to isomorphism, we can substitute the top face by pushout $(1'_i)$ with $P \cong L_{j0}$. Thus we have found the morphism $p_{ji} := \hat{m}$ with $f_i \circ p_{ji} = m_j \circ u_i$. This construction can be applied for all pairs i, j leading to weakly independent matches without application conditions. $\square$

This fact leads to a set-theoretical characterisation of maximal weakly disjoint matchings.

Fact 6.34. *For graphs and graph-based structures, valid matches $m_0, m_1, \ldots, m_n$ with $m_i \circ s_{i,L} = m_0$ for all $i = 1, \ldots, n$ form a maximal weakly disjoint matching without application conditions if and only if $m_i(L_i) \cap m_j(L_j) = m_0(L_0)$.* $\triangle$

Proof. Valid matches means that the transformations $G \overset{p_i, m_i}{\Longrightarrow}$ are well defined. In graphs and graph-like structures, (P_{ij}) is a pullback if and only if $m_i(L_i) \cap m_j(L_j) = m_0(L_0)$. Then Fact 6.33 implies that the matches form a maximal weakly disjoint matching without application conditions. $\square$

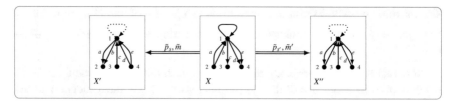

Fig. 6.13 Application of an amalgamated rule via maximal matchings

Example 6.35. Consider the interaction scheme $is = (s_1, s_2)$ defined by the kernel morphisms s_1 and s_2 in Fig. 6.2, the graph X depicted in the middle of Fig. 6.13, and the kernel rule match m_0 mapping the node 1 in L_0 to the node 1 in X.

If we choose maximal weakly independent matchings, the construction works as follows, defining the following matches, where f is the edge from 1 to 2 in L_1 and g the reverse edge in L_2:

$i = 1 : \hat{p}_1 = p_1, m_1 : 2 \mapsto 3, f \mapsto c,$
$i = 2 : \hat{p}_2 = p_1, m_2 : 2 \mapsto 4, f \mapsto d,$
$i = 3 : \hat{p}_3 = p_2, m_3 : 3 \mapsto 2, g \mapsto a,$
$i = 4 : \hat{p}_4 = p_1, m_4 : 2 \mapsto 4, f \mapsto e,$
$i = 5 : \hat{p}_5 = p_2, m_5 : 3 \mapsto 2, g \mapsto b.$

Thus, we find five different matches, three for the multi rule p_1 and two for the multi rule p_2. Note that in addition to the overlapping m_0, the matches m_3 and m_5 overlap in the node 2, while m_2 and m_4 overlap in the node 4. But since these matches are still weakly independent, because the nodes 2 and 4 are not deleted by the rule applications, this is a valid maximal weakly independent matching. It leads to the bundle $s = (s_1, s_1, s_1, s_2, s_2)$ and the amalgamated rule $\tilde{p}_s$, which can be applied to X, leading to the amalgamated transformation $X \xRightarrow{\tilde{p}_s, \tilde{m}} X'$ as shown in the left of Fig. 6.13.

If we choose maximal weakly disjoint matchings instead, the matches m_4 and m_5 are no longer valid because they overlap with m_2 and m_3, respectively, in more than the match m_0. Thus we obtain the maximal weakly disjoint matching (m_0, m_1, m_2, m_3), the corresponding bundle $s' = (s_1, s_1, s_2)$ leading to the amalgamated rule $\tilde{p}_{s'}$ and the amalgamated transformation $X \xRightarrow{\tilde{p}_{s'}, \tilde{m}'} X''$ depicted in the right of Fig. 6.13. Note that this matching is not unique; also, (m_0, m_1, m_2, m_4) could have been chosen as a maximal weakly disjoint matching.   △

6.4.1 Main Results for Amalgamated Transformations Based on Maximal Matchings

If we allow applying amalgamated rules only via maximal matchings, the main results from Sect. 5.2 do not hold instantly as is the case for arbitrary matchings. The main problem is that the amalgamated transformations obtained from the constructions are in general not applied via maximal matchings. The analysis and definition of properties ensuring these results is future work:

- The Local Church–Rosser Theorem guarantees that for parallel independent amalgamated transformations $G \xrightarrow{\tilde{p}_s} H_1$ and $G \xrightarrow{\tilde{p}_{s'}} H_2$ via maximal matchings there exist transformations $H_1 \xrightarrow{\tilde{p}_{s'}} G'$ and $H_2 \xrightarrow{\tilde{p}_s} G'$. But in general, these resulting transformations will not be via maximal matchings, since $\tilde{p}_{s'}$ or $\tilde{p}_s$ may create new matchings for s or s', respectively. Thus, we have to find properties that make sure that no new matches, or at least no new disjoint matches, are created.
- For the Parallelism Theorem, the property of maximal weakly independent matchings is transferred to the application of the parallel rule, as shown below.
- For the Concurrency Theorem, we have to formulate results concerning the construction of an E-concurrent rule as an amalgamated rule based on the underlying kernel and multi rules before relating the results to maximal matchings.
- For the Embedding and Extension Theorem, embedding an object G with a maximal matching into a larger context G' in general enables more matches, i.e., the application of the amalgamated rule to G' may not be maximal. We need to define properties to restrict the embedding to certain parts outside the matches of the multi rules to ensure that the same matchings are maximal in G and G'.
- For the Local Confluence Theorem, maximal matchings may actually lead to fewer critical pairs if we have additional information about the objects to be transformed, since some conflicting transformations may not occur at all due to maximal matchings.

In case of parallel independent transformations, the property of a maximal weakly independent matching is transferred to the application of the parallel rule. Note that for maximal weakly disjoint matchings, we have to require in addition that the matches of the two amalgamated transformations not overlap.

Theorem 6.36 (Parallelism of maximal weakly independent matchings). *Given parallel independent amalgamated transformations* $G \xrightarrow{\tilde{p}_s, \tilde{m}} H_1$ *and* $G \xrightarrow{\tilde{p}_{s'}, \tilde{m}'} H_2$ *leading to the induced transformations* $G \xrightarrow{\tilde{p}_t, \tilde{m}_t} G'$ *via the parallel rule* $\tilde{p}_t = \tilde{p}_s + \tilde{p}_{s'}$, *the following holds: if* $G \xrightarrow{\tilde{p}_s, \tilde{m}} H_1$ *and* $G \xrightarrow{\tilde{p}_{s'}, \tilde{m}'} H_2$ *are transformations via maximal weakly independent matchings then also* $G \xrightarrow{\tilde{p}_t, \tilde{m}_t} G'$ *is a transformation via a maximal weakly independent matching.*   △

Proof. Consider parallel independent amalgamated transformations $G \xrightarrow{\tilde{p}_s, \tilde{m}} H_1$ and $G \xrightarrow{\tilde{p}_{s'}, \tilde{m}'} H_2$ via maximal weakly independent matchings $(m_0, m_1, \ldots, m_n)$ with $\tilde{m} \circ t_{i,L} = m_i$ and $(m'_0, m'_1, \ldots, m'_{n'})$ with $\tilde{m}' \circ t'_{j,L} = m'_j$, respectively. Then we have the matching $m = ([m_0, m'_0], ([m_i, m'_0])_{i=1,\ldots,n}, ([m_0, m'_j])_{j=1,\ldots,n'})$ for the parallel transformation $G \xrightarrow{\tilde{p}_t, \tilde{m}_t} G'$, with $[m_i, m'_0] \circ (s_{i,L} + id_{L_0}) = [m_0, m'_0]$ and $[m_0, m'_j] \circ (id_{L_0} + s'_{j,L}) = [m_0, m'_0]$. We have to show the maximality of m.

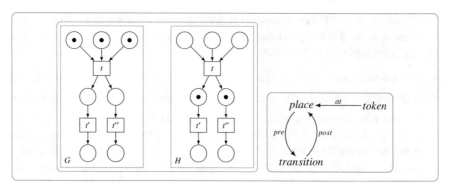

Fig. 6.14 The firing semantics and type graph for elementary Petri nets

Suppose m is not maximal. This means that there is, w.l.o.g., some match $\hat{m}$: $L_k + L'_0 \to G$ such that $\hat{m} \circ (s_{k,L} + id_{L'_0}) = [m_0, m'_0]$ and $\hat{m} \neq [m_i, m'_0]$ for all $i = 1, \dots, n$ such that $(m, \hat{m})$ is also weakly independent. Then we find a match $\hat{m}_k := \hat{m} \circ i_{L_k}$ for the rule p_k with $\hat{m}_k \circ s_{k,L} = m_0$ and $\hat{m}_k \neq m_i$ for all i. It follows that $(m_0, m_1, \dots, m_n, \hat{m}_k)$ are also weakly independent, which is a contradiction to the maximality of $(m_0, m_1, \dots, m_n)$.   □

6.4.2 Semantics for Elementary Nets

As a concrete and more complex example, we use amalgamation and maximal matchings to model the operational semantics of elementary Petri nets. Using amalgamation allows the description of a semantical step in an unknown surrounding with only one interaction scheme. We do not need specific rules for each occurring situation as is the case for Petri nets with standard graph transformation.

In the following, we present a semantics for the firing behaviour of elementary Petri nets using graph transformation and amalgamation. Elementary Petri nets are nets where at most one token is allowed on each place. A transition t is activated if there is a token on each pre-place of t and all post-places of t are token-free. In this case, the transition may fire, leading to the follower marking where the tokens on all the pre-places of t are deleted and at all post-places of t a token appears. An example is depicted on the right of Fig. 6.14, where the transition t in the elementary Petri net G is activated on the left and the follower marking is depicted on the right, leading to the elementary Petri net H.

We model these nets by typed graphs. The type graph is depicted on the left of Fig. 6.14 and consists simply of places, transitions, the corresponding pre- and post-arcs, and tokens attached to their places. For the following examples, we use the well-known concrete syntax of Petri nets, modelling a place by a circle, a transition by a rectangle, and a token by a small filled circle placed on its place.

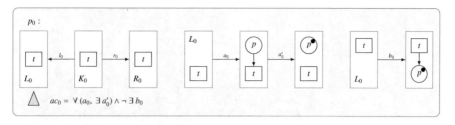

Fig. 6.15 The kernel rule selecting an activated transition

In Figs. 6.15 and 6.16, three rules, p_0, p_1, and p_2, are shown, which, combined as an amalgamated rule with maximal weakly disjoint matchings, will result in a firing step of the net. The rule p_0 in Fig. 6.15 selects a transition t which is not changed at all. But note that the application condition restricts this rule to be only applicable if there is no empty pre-place of t and we have only empty post-places. This means that the transition t is activated in the elementary net. The rule p_1 describes the firing of a pre-place, where the token on this place is deleted. It only inherits the application condition of p_0 to guarantee a kernel morphism $s_1 : p_0 \rightarrow p_1$, as shown at the top of Fig. 6.16. s_1 is indeed a kernel morphism because (1) and (2) are pullbacks and (3) is the required pushout complement. ac_0 and ac_1 are complement-compatible w. r. t. s_1 with $ac_1' = $ true. Similarly, rule p_2 describes the firing of a post-place, where a token is added on this place. Again, there is a kernel morphism $s_2 : p_0 \rightarrow p_2$, as shown in the bottom of Fig. 6.16 with pullbacks (1′) and (2)′, (1′) is already a pushout, and ac_0 and ac_2 are complement-compatible w. r. t. s_2 with $ac_2' = $ true.

Theorem 6.37 (Equivalence of amalgamated transformation and firing step).
Using the interaction scheme is $= \{s_1 : p_0 \rightarrow p_1, s_2 : p_0 \rightarrow p_2\}$ of the rules defined in Figs. 6.15 and 6.16 with maximal weakly disjoint matchings, the derived amalgamated transformations are equivalent to the firing steps of elementary Petri nets. △

For the multi rules in Fig. 6.16, the complement rules are the rules p_1 and p_2 themselves but with empty application condition true, because they contain everything which is done in addition to p_0, including the connection with K_0, while the application condition is already ensured by p_0.

Now consider the interaction scheme $is = \{s_1, s_2\}$ leading to the bundle of kernel morphisms $s = (s_1, s_1, s_1, s_2, s_2)$. The construction of the corresponding amalgamated rule $\tilde{p}_s$ is shown in Fig. 6.17 without application conditions. This amalgamated rule can be applied to the elementary Petri net G as depicted in Fig. 6.18, leading to the amalgamated transformation $G \xRightarrow{\tilde{p}_s, \tilde{m}} H$.

Moreover, we can find a bundle of transformations $G \xRightarrow{m_1, p_1} G_1, G \xRightarrow{m_2, p_1} G_2,$ $G \xRightarrow{m_3, p_1} G_3, G \xRightarrow{m_4, p_2} G_4,$ and $G \xRightarrow{m_5, p_2} G_5$ with the resulting nets depicted in Fig. 6.19 and matches $m_0 : t \mapsto t, m_1 : p_1 \mapsto q_1, m_2 : p_1 \mapsto q_2, m_3 : p_1 \mapsto q_3, m_4 : p_2 \mapsto q_4,$ and $m_3 : p_2 \mapsto q_5$. This bundle is s-amalgamable, because it has

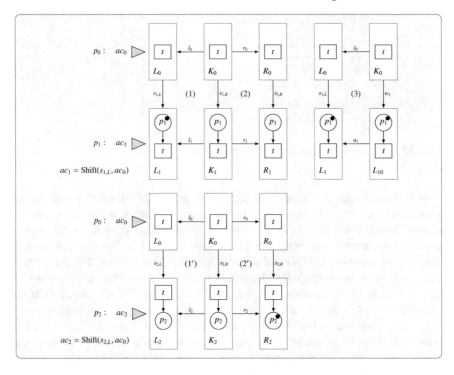

Fig. 6.16 The multi rules describing the handling of each place

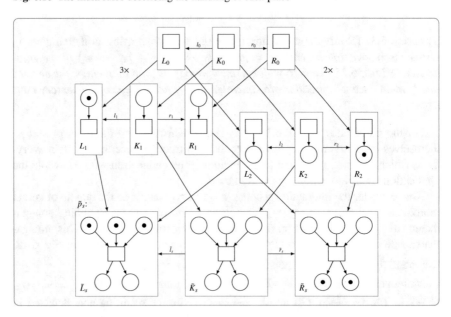

Fig. 6.17 The construction of the amalgamated rule

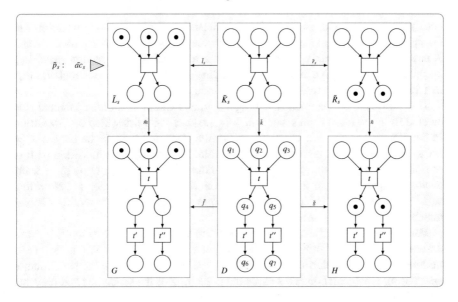

Fig. 6.18 An amalgamated transformation

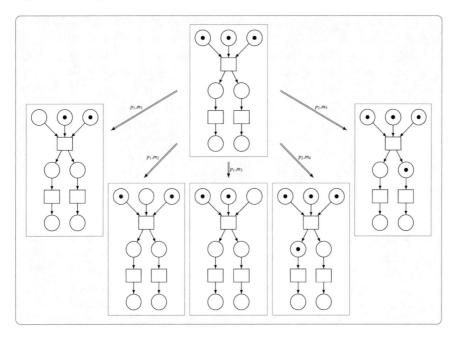

Fig. 6.19 An s-amalgamable transformation bundle

consistent matches with m_0 matching the transition t from p_0 to the transition t in G, and all matches are weakly independent; they only overlap in L_0. $(m_0, \ldots, m_5)$ is both a maximal weakly independent and a maximal weakly disjoint matching, because no other match can be found extending the kernel rule match, and all these matches are disjoint up to the selected transition t.

If we always use maximal matchings, any application of an amalgamated rule created from the interaction scheme $is = \{s_1, s_2\}$ is a valid firing step of a transition in the elementary net. For example, to fire the transition t' in G the bundle $s' = (s_1, s_2)$ leads to the required amalgamated rule. In general, for a transition with m pre- and n post-arcs, the corresponding bundle $s = ((s_1)_{i=1,\ldots,m}, (s_2)_{j=1,\ldots,n})$ leads to the amalgamated rule firing this transition via a maximal matching. Note that each maximal weakly independent matching is already a maximal weakly disjoint matching due to the net structure.

For elementary Petri nets we only need one kernel rule and two multi rules to describe the complete firing semantics for all well-defined nets. We neither need infinite many rules, which are difficult to analyse, nor any control or helper structure when using amalgamation. This eases the modelling of the semantics and prevents errors [GHE14].

Part III
Model Transformation Based on Triple Graph Grammars

This third part presents model transformation, model integration and model synchronisation based on triple graph grammars. Following up on the informal introduction to model transformation in Chap. 3 of Part I, we present the formal theory of graph transformation based on triple graph grammars. In Chap. 7, we give the foundations of triple graph grammars leading to model transformation and model integration. It is important to note that transformation and integration are based on operational rules, which can be generated automatically from the triple graph grammar rules. A flattening construction allows us to show the equivalence of model transformations based on triple graph grammars and plain graph grammars. In Chap. 8, we present several analysis techniques for model transformations, which are supported by tools discussed in Part IV. Important properties, which are analysed in Chap. 8, include correctness and completeness, functional behaviour and information preservation, as well as conflict resolution and optimisation. In Chap. 9, model transformation techniques are applied to model synchronisation, which is an important technique for gaining and keeping consistency of source and target models after changing one or both of them. This leads to unidirectional and concurrent model synchronisation, respectively.

Chapter 7
Model Transformation and Model Integration

In this chapter, we describe the formal framework for model transformation and model integration based on triple graph grammars. For this purpose, we use triple graph transformation systems as introduced in Chap. 3 and show in Sect. 7.1 that they instantiate the general framework of $\mathcal{M}$-adhesive transformation systems presented in Chap. 5. This ensures that all results for $\mathcal{M}$-adhesive transformation systems hold for the specific case of triple graph transformation systems. A triple graph grammar is a constructive specification of a language of integrated models, which are specified by their underlying abstract syntax graphs. Based on this general concept, we first derive a transformation system for forward model transformations, which are defined in Sect. 7.3. In Sect. 7.4, we introduce forward translation rules as an alternative to forward rules and show the equivalence of model transformations based on either forward or forward translation rules. The concept of forward translation rules simplifies the control mechanism for executing model transformations. In addition to that, it offers improved capabilities for analysis and execution, which we will study in detail in Chap. 8. Model integration is a technique to integrate two given models—one from the source and one from the target language. In Sect. 7.5, we present model integration based on TGGs and show formally that this concept is closely related to model transformations. The last section (Sect. 7.6) of this chapter relates the presented concepts based on TGGs with standard model transformations based on plain graph grammars. The chapter is based on [Her11, HEGO14, EEE$^+$07, EEH08c, HHK10, GEH11].

7.1 Triple Graphs form an $\mathcal{M}$-adhesive Category

A triple graph is an integrated model $G = (G^S \leftarrow G^C \rightarrow G^T)$ containing a source model G^S from the source language, a target model G^T from the target language, and explicit correspondences between them specified via a correspondence model G^C.

The category of triple graphs can be constructed from underlying $\mathcal{M}$-adhesive categories of typed attributed graphs.

Fact 7.1 (Construction of categories of triple graphs). *The category of typed attributed triple graphs in Def. 3.4 and the base categories for triple graphs in Def. 3.5 can be constructed as follows.*

- *The category* **TrGraphs** *of triple graphs and triple graph morphisms can be constructed as functor category* [**X**, **Graphs**] *over the category* **Graphs** *of graphs with schema category* $\mathbf{X} = (\{S, C, T\}, \{s \colon C \to S, t \colon C \to T, id_S, id_C, id_T\})$.
- *The category* **ATrGraphs** *of attributed triple graphs can be constructed as the functor category* [**X**, **AGraphs**] *over the category* **AGraphs** *of attributed graphs with the same schema category as above.*
- *The category* **TrGraphs**$_{TG}$ *of typed triple graphs (or* **ATrGraphs**$_{ATG}$ *of typed attributed triple graphs) for a given triple graph TG in* **TrGraphs** *(or ATG in* **ATrGraphs***) can be constructed as slice category* **TrGraphs**$\backslash TG$ *over* **TrGraphs** *(or* **ATrGraphs**$\backslash ATG$ *over* **ATrGraphs***).* △

Proof. A triple graph $G = (G^S \xleftarrow{s_G} G^C \xrightarrow{t_G} G^T)$ is represented by a functor $\mathcal{G} \colon \mathbf{X} \to$ **Graphs** with $\mathcal{G}(S) = G^S, \mathcal{G}(C) = G^C, \mathcal{G}(T) = G^T$ and $\mathcal{G}(s) = s_G, \mathcal{G}(t) = t_G$. The compatibility condition for triple graph morphisms follows from the compatibility condition of functor transformations that form the morphisms in a functor category. The typing and attribution extensions are compatible with the construction of the functor category. □

Theorem 7.2 (Category of triple graphs is $\mathcal{M}$-adhesive). *The categories* **TrGraphs**, **TrGraphs**$_{TG}$, **ATrGraphs**, *and* **ATrGraphs**$_{ATG}$ *are $\mathcal{M}$-adhesive.* △

Proof. Using Fact 7.1, we derive the categories as functor and slice constructions over $\mathcal{M}$-adhesive categories (**Graphs**, $\mathcal{M}$) and (**AGraphs**, $\mathcal{M}$) and can apply Theorem B.13 to derive that the constructed categories are again $\mathcal{M}$-adhesive categories. □

By Theorem 7.2, we can conclude that the results in Chapters 4 to 6 for $\mathcal{M}$-adhesive categories hold for triple graphs and triple graph transformations. In particular, we will apply the theory and analysis for critical pairs and confluence (see Sect. 5.2.4 in Chap. 5) in Chap. 8 for analysing functional behaviour and information preservation. Using Theorem 7.2, we derive the classes of $\mathcal{M}$-morphisms for the different kinds of triple graphs as constructions from the $\mathcal{M}$-adhesive categories **Graphs** and **AGraphs**. The class of $\mathcal{M}$-morphisms is given by all triple graph morphisms that are injective on the graph part and isomorphisms on the data part.

From the application point of view a model transformation should be injective on the structural part, i.e., the transformation rules are applied along matches that do not identify structural elements. Thus, the translation of each element is explicitly specified and there is no confusion. But it would be too restrictive to require injectivity of the matches also on the data and variable nodes, because we must allow two different variables to be mapped to the same data value. For this reason we introduce the notion of almost injective matches, which requires matches to be injective except for the data value nodes. This way, attribute values can still be specified as terms within a rule and matched noninjectively to the same value. For the rest of this chapter, we generally require almost injective matching for the transformation sequences. Moreover, we require that application conditions contain almost injective morphisms only. For the constructions, we can assume without loss of generality that each almost injective morphism $f = (f_G, f_D): G \to H$ is given by $f = (inc_G, f_D)$, i.e., f_G is an inclusion.

Definition 7.3 (Almost injective match). An attributed triple graph morphism $m : L \to G$ is called *almost injective* if it is noninjective at most for the set of variables and data values.   △

Remark 7.4 (Restriction of application conditions to almost injective internal morphisms). The internal morphisms of an application condition are not restricted by definition. If we consider almost injective matches, it is sufficient to use almost injective morphisms for the internal morphisms of an application condition. The reason is that those internal morphisms that are not almost injective can never be completed with a compatible mediating morphism $q: P \to G$. Thus, they are not relevant and do not restrict the applicability of the rule. Therefore, we use application conditions with almost injective internal morphisms only. Note that this restriction is compatible with the notion of AC schemata. Given an AC schema $\overline{ac}$ (see Def. 5.6), where ac contains almost injective internal morphisms only, the induced application condition $\overline{ac}$ also contains almost injective internal morphisms only due to the merge construction (see Def. 5.5).   △

Following the discussion and explanations for almost injective morphisms above, we generally require almost injective morphisms for TGGs as stated by our general assumption below.

Remark 7.5 (General assumption). The formal results in this chapter are presented for TGGs that are executed via almost injective matches and where all internal morphisms of application conditions are almost injective.   △

7.2 Derivation of Operational Rules

The operational rules for executing forward and backward model transformations are derived from the set of triple rules of a given TGG. This process requires us to split the application conditions of the triple rules and to distribute them to the corresponding derived rules. For this reason, we need to specify a restriction of application conditions, which ensures that the split can be performed. This restriction, however, is not problematic from an application point of view, which we will also see in our running example.

Definition 7.6 (Special application conditions). Given a triple rule $tr : L \to R$, an application condition $ac = \exists (a, ac')$ over L with $a : L \to P$ is an

- *S-application condition* if a^C, a^T are identities, i.e., $P^C = L^C$, $P^T = L^T$, and ac' is an S-application condition over P,
- *S-extending application condition* if a^S is an identity, i.e., $P^S = L^S$, and ac' is an S-extending application condition over P.

- *T-application condition* if a^S, a^C are identities, i.e., $P^S = L^S$, $P^C = L^C$, and ac' is a T-application condition over P,
- *T-extending application condition* if a^T is an identity, i.e., $P^T = L^T$, and ac' is a T-extending application condition over P,

- *ST-application condition* if a^C is an identity, i.e., $P^C = L^C$, and ac' is an ST-application condition over P.

Moreover, true is an S- (S-extending, ST-, T-, T-extending) application condition, and if ac, ac_i are S- (S-extending, ST-, T-, T-extending) application conditions so are $\neg ac$, $\wedge_{i \in I} ac_i$, and $\vee_{i \in I} ac_i$.   △

During the generation of the operational rules, each application condition ac of a triple rule $tr \in TR$ has to be transferred to the operational rules. This transfer is achieved by decomposing ac into (1) a part on the source rule and an S-extending application condition for the forward rule, or (2) a part on the target rule and a T-extending application condition for the backward rule, or (3) an ST-condition for the source–target rules and an empty remainder for the model integration rules.

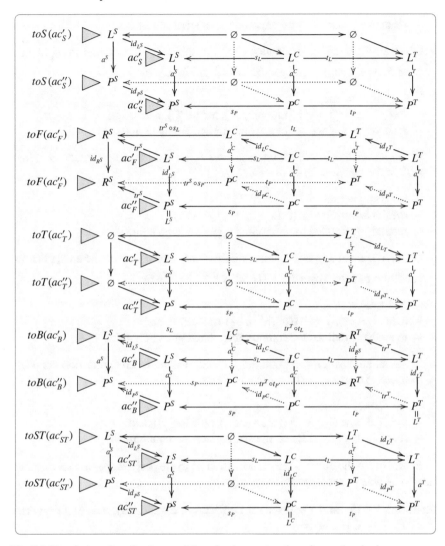

Fig. 7.1 Translation of application conditions for the construction of operational rules

Definition 7.7 (Translated application condition). Consider a triple rule $tr = (tr: L \to R, ac)$; then an application condition ac can be translated as follows, according to Fig. 7.1. The data component for each inclusion $\emptyset \to X$ for a triple graph X is given by an identity, i.e., the construction does not change the data component.

- Given an application condition ac'_S over L, we define an application condition $toS(ac'_S)$ over $L_S = (L^S \leftarrow \emptyset \to \emptyset)$ by

  - $toS(\text{true}) = \text{true}$,
  - $toS(\exists (a, ac''_S)) = \exists ((a^S, id_\emptyset, id_\emptyset), toS(ac''_S))$, and

– recursively defined for composed application conditions.

- Given an S-extending application condition ac'_F over L, we define an application condition $toF(ac'_F)$ over $L_F = (R^S \xleftarrow{tr^S \circ s_L} L^C \xrightarrow{t_L} L^T)$ by

  – $toF(\text{true}) = \text{true}$,
  – $toF(\exists\,(a, ac''_F)) = \exists\,((id_{R^S}, a^C, a^T), toF(ac''_F))$, and
  – recursively defined for composed application conditions.

- Given an application condition ac'_T over L, we define an application condition $toT(ac'_T)$ over $L_T = (\emptyset \leftarrow \emptyset \rightarrow L^T)$ by

  – $toT(\text{true}) = \text{true}$,
  – $toT(\exists\,(a, ac''_T)) = \exists\,((id_\emptyset, id_\emptyset, a^T), toS(ac''_T))$, and
  – recursively defined for composed application conditions.

- Given a T-extending application condition ac'_B over L, we define an application condition $toB(ac'_B)$ over $L_B = (L^S \xleftarrow{s_L} L^C \xrightarrow{tr^T \circ t_L} R^T)$ by

  – $toB(\text{true}) = \text{true}$,
  – $toB(\exists\,(a, ac''_B)) = \exists\,((a^S, a^C, id_{R^T}), toB(ac''_B))$, and
  – recursively defined for composed application conditions.

- Given an application condition ac'_{ST} over L, we define an application condition $toST(ac'_{ST})$ over $L_{ST} = (L^S \leftarrow \emptyset \rightarrow L^T)$ by

  – $toST(\text{true}) = \text{true}$,
  – $toST(\exists\,(a, ac''_{ST})) = \exists\,((a^S, id_\emptyset, a^T), toST(ac''_{ST}))$, and
  – recursively defined for composed application conditions.   △

In order to assign an application condition ac to the derived operational rules, we have to be able to decompose it properly.

Definition 7.8 (S- and T-consistent application conditions). Given a triple rule $tr = (tr: L \rightarrow R, ac)$, ac is

- *S-consistent* if it can be decomposed into $ac \cong ac'_S \wedge ac'_F$ such that $ac'_S \cong$ Shift$((id_{L^S}, \emptyset_{L^C}, \emptyset_{L^T}), toS(ac'_S))$ and ac'_F is an S-extending application condition.
- *T-consistent* if it can be decomposed into $ac \cong ac'_T \wedge ac'_B$ such that $ac'_T \cong$ Shift$((\emptyset_{L^S}, \emptyset_{L^C}, id_{L^T}), toT(ac'_T))$ and ac'_B is a T-extending application condition.
- *ST-consistent* if it can be decomposed into $ac \cong ac'_{ST}$ such that $ac'_{ST} \cong$ Shift$((id_{L^S}, \emptyset_{L^C}, id_{L^T}), toST(ac'_{ST}))$.

If ac is S-consistent (T-consistent, ST-consistent), we also say that tr is S-consistent (T-consistent, ST-consistent).   △

The consistency conditions for application conditions in Def. 7.8 can be checked as follows. First of all, for an S-consistent application condition we require that we be able to split it into a conjunction of an application condition ac'_S that concerns the source component only and an application condition ac'_F that does not restrict source structures. Thus, a modeller should use application conditions that are already of this form. Furthermore, we require that $ac'_S \cong \mathrm{Shift}((id_{L^S}, \varnothing_{L^C}, \varnothing_{L^T}), toS(ac'_S))$. By Lem. 7.9 we show that our general assumption (see Rem. 7.5) already ensures this condition. This means that the additional condition is guaranteed by using almost injective morphisms for matches and application conditions.

Lemma 7.9 (Validity of S-consistency for almost injective morphisms). *Let ac be an S-application condition for a triple rule $tr = (L \to R)$ and $ac' = \mathrm{Shift}((id_{L^S}, \varnothing_{L^C}, \varnothing_{L^T}), toS(ac))$. Let further all morphisms in ac be almost injective. Then, $ac' \equiv ac$ for almost injective matches, i.e.: for each almost injective morphism $m\colon L \to G$ it holds that $m \models ac \Leftrightarrow m \models ac'$.* △

Proof. See Appendix B.6.1. □

Example 7.10 (Consistent application conditions). Two of the triple rules in Fig. 3.8 described in Ex. 3.9 contain application conditions. The application condition of rule `Association2ForeignKey` is both, a T-application condition and an S-extending application condition. The rule `PrimaryAttr2Column` contains a conjunction of two NACs, where `NAC1` is an application condition that is an S-application condition and a T-extending application condition and `NAC2` is an application condition that is a T-application condition and an S-extending application condition. Therefore, all application conditions are S-consistent and T-consistent, because they can be decomposed as required using $ac \cong (ac \wedge \mathrm{true})$. Moreover, all application conditions are ST-consistent, because they are each either S- or T-consistent. △

From a triple rule, we can derive a source rule tr_S and a target rule tr_T, which specify the changes made by this rule in the source and target components, respectively. Similarly, we derive a source–target rule tr_{ST} specifying the changes made by this rule in the source and target components synchronously. Additionally, we derive the forward rule tr_F describing the changes made by the rule to the correspondence and target parts, the backward rule tr_B concerning the correspondence and source parts, and the integration rule tr_I concerning the correspondence parts. Intuitively, these rules require that their counterparts (source, target, and source–target, respectively) have been applied already. Intuitively, source rules are used to parse a given source model in order to control the actual forward model transformation via forward rules from source to target models. Vice versa, target rules are used to parse a given target model in order to control the actual backward model transformation via backward rules from target to source models. Source–target rules are used to parse a given pair of source and target models in order to control the actual model integration via integration rules, yielding a fully integrated model. Technically, the source rules recreate the given source model, such that the matches can be used to induce partial matches for the forward rules that transform the model into the corresponding target model.

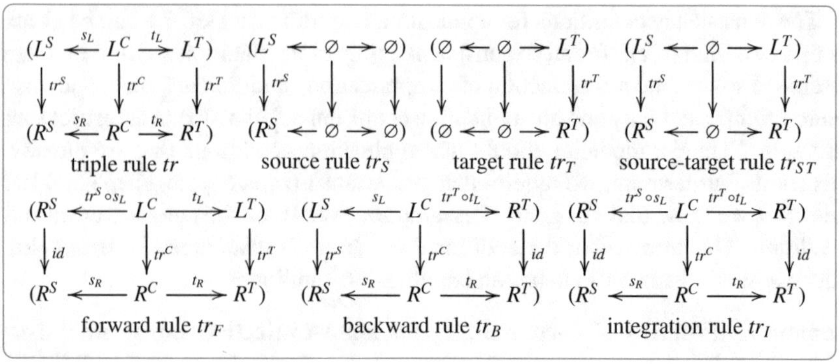

$$(L^S \xleftarrow{s_L} L^C \xrightarrow{t_L} L^T) \quad (L^S \leftarrow \varnothing \rightarrow \varnothing) \quad (\varnothing \leftarrow \varnothing \rightarrow L^T) \quad (L^S \leftarrow \varnothing \rightarrow L^T)$$

$$\downarrow tr^S \quad \downarrow tr^C \quad \downarrow tr^T \qquad \downarrow tr^S \quad\downarrow\quad\downarrow \qquad \downarrow\quad\downarrow\quad \downarrow tr^T \qquad \downarrow tr^S \quad\downarrow\quad \downarrow tr^T$$

$$(R^S \xleftarrow{s_R} R^C \xrightarrow{t_R} R^T) \quad (R^S \leftarrow \varnothing \rightarrow \varnothing) \quad (\varnothing \leftarrow \varnothing \rightarrow R^T) \quad (R^S \leftarrow \varnothing \rightarrow R^T)$$

$$\text{triple rule } tr \qquad\qquad \text{source rule } tr_S \qquad\qquad \text{target rule } tr_T \qquad\quad \text{source-target rule } tr_{ST}$$

$$(R^S \xleftarrow{tr^S \circ s_L} L^C \xrightarrow{t_L} L^T) \quad (L^S \xleftarrow{s_L} L^C \xrightarrow{tr^T \circ t_L} R^T) \quad (R^S \xleftarrow{tr^S \circ s_L} L^C \xrightarrow{tr^T \circ t_L} R^T)$$

$$\downarrow id \quad \downarrow tr^C \quad \downarrow tr^T \qquad \downarrow tr^S \quad \downarrow tr^C \quad \downarrow id \qquad \downarrow id \quad \downarrow tr^C \quad \downarrow id$$

$$(R^S \xleftarrow{s_R} R^C \xrightarrow{t_R} R^T) \quad (R^S \xleftarrow{s_R} R^C \xrightarrow{t_R} R^T) \quad (R^S \xleftarrow{s_R} R^C \xrightarrow{t_R} R^T)$$

$$\text{forward rule } tr_F \qquad\qquad \text{backward rule } tr_B \qquad\qquad \text{integration rule } tr_I$$

operational rule	left-hand side	right-hand side	rule morphism
source rule $tr_S : L_S \rightarrow R_S$	$L_S = (L^S \leftarrow \varnothing \rightarrow \varnothing)$	$R_S = (R^S \leftarrow \varnothing \rightarrow \varnothing)$	$tr_S = (tr^S, \varnothing, \varnothing)$
target rule $tr_T : L_T \rightarrow R_T$	$L_T = (\varnothing \leftarrow \varnothing \rightarrow L^T)$	$R_T = (\varnothing \leftarrow \varnothing \rightarrow R^T)$	$tr_T = (\varnothing, \varnothing, tr^T)$
forward rule $tr_F : L_F \rightarrow R_F$	$L_F = (R^S \xleftarrow{tr_S \circ s_L} L^C \xrightarrow{t_L} L^T)$	$R_F = R$	$tr_F = (id_R^S, tr^C, tr^T)$
backward rule $tr_B : L_B \rightarrow R_B$	$L_B = (L^S \xleftarrow{s_L} L^C \xrightarrow{tr^T \circ t_L} R^T)$	$R_B = R$	$tr_B = (tr^S, tr^C, id_R^T)$
source–target rule $tr_{ST} : L_{ST} \rightarrow R_{ST}$	$L_{ST} = (L^S \leftarrow \varnothing \rightarrow L^T)$	$R_S = (R^S \leftarrow \varnothing \rightarrow R^T)$	$tr_S = (tr^S, \varnothing, tr^T)$
integration rule $tr_I : L_I \rightarrow R_I$	$L_I = (R^S \xleftarrow{tr_S \circ s_L} L^C \xrightarrow{tr_T \circ t_L} R^T)$	$R_I = R$	$tr_I = (id_R^S, tr^C, id_R^T)$

Fig. 7.2 The main components of derived operational rules for model transformation

Definition 7.11 (Derived operational rules without application conditions).
Given a triple rule $tr = (tr: L \rightarrow R, ac)$, we derive its operational rules tr_S, tr_T, tr_{ST}, tr_F, tr_B, tr_I without application conditions according to Fig. 7.2. The data component for each inclusion $\varnothing \rightarrow X$ for a triple graph X is given by an identity, i.e., the construction does not change the data component.   △

We combine the translated application conditions with the derived rules without application conditions, leading to the derived rules of a triple rule with application conditions.

Definition 7.12 (Derived operational rules). Given a triple rule $tr = (tr: L \rightarrow R, ac)$, we derive its operational rules without application conditions according to Def. 7.11. If tr contains an application condition ac of L, then the derivation is extended as defined below, requiring additional consistency conditions. If ac is an S-consistent application condition, we derive $ac \cong ac_S' \wedge ac_F'$ and we obtain the source rule $tr_S = (tr_S, ac_S)$ with $ac_S = toS(ac_S')$ and the forward rule

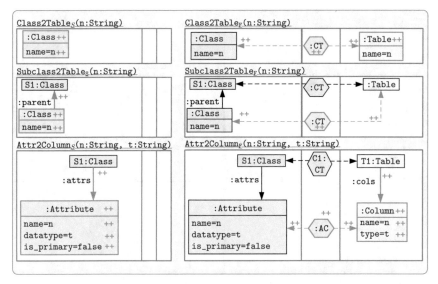

Fig. 7.3 Derived source rules (left) and forward rules (right) for *CD2RDBM*

$tr_F = (tr_F, ac_F)$ with $ac_F = toF(ac'_F)$. If ac is a T-consistent application condition, we derive $ac \cong ac'_T \wedge ac'_B$ and we obtain the target rule $tr_T = (tr_T, ac_T)$ with $ac_T = toT(ac'_T)$ and the backward rule $tr_B = (tr_B, ac_B)$ with $ac_B = toB(ac' B)$. If ac is an ST-application condition, we obtain the source–target rule $tr_{ST} = (tr_{ST}, ac_{ST})$ with $ac_{ST} = toST(ac)$ and the integration rule $tr_I = (tr_I, true)$. By TR_S, TR_T, TR_{ST}, TR_F, TR_B and TR_I, we denote the sets of all source, target, source–target, forward, backward and integration rules derived from *TR*.   △

Remark 7.13 (Symmetry of forward and backward case). According to Def. 7.12, the definition of operational rules shows symmetries. Source rules are symmetric to target rules and forward rules are symmetric to backward rules. For this reason, all further constructions can be presented based on source and forward rules and the symmetric constructions and results for target and backward rules follow immediately.   △

Example 7.14 (Derived operational rules). The derived operational source and forward rules for the model transformation *CD2RDBM* are depicted in Fig. 7.3. The derived operational target and backward rules follow by the symmetry in Def. 7.12. Intuitively, the source rules are obtained by deleting all elements in the correspondence and target components, including the components of the application conditions. The forward rules are obtained by removing all double plus signs in the source component and by removing the application conditions on the source component. The derived operational source–target and integration rules for the model transformation *CD2RDBM* are depicted in Figs. 7.4 and 7.5. Intuitively, the source–target rules are obtained by deleting all elements in the correspondence component, in-

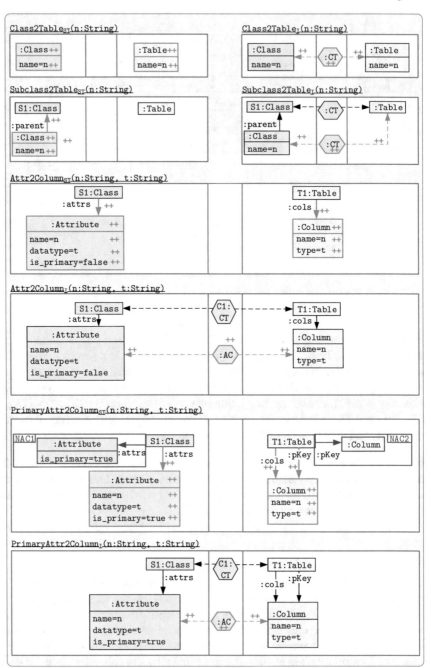

Fig. 7.4 Derived source–target and integration rules for *CD2RDBM* (part 1)

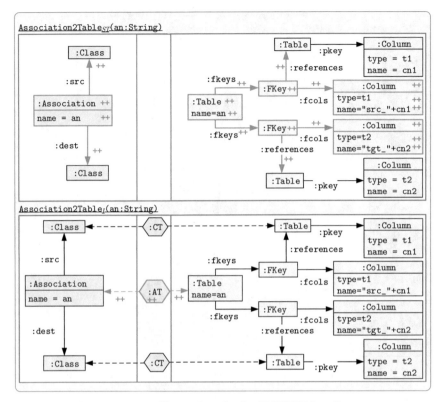

Fig. 7.5 Derived source–target and integration rules for *CD2RDBM* (part 2)

cluding the components of the application conditions. The integration rules are obtained by removing all double plus signs in the source and target components and by removing the application conditions on the source and target components.   △

In the following, we want to split each triple rule into the corresponding source and forward rules. Each triple rule without application conditions is the E-concurrent rule of its source and forward rules as well as the E-concurrent rule of its target and backward rules (see also the proof for Thm. 1 in [EEE$^+$07]). Moreover, each triple rule without application conditions is the E-concurrent rule of its source–target and integration rules (see also the proof for Lem. 1 in [EEH08a]).

Fact 7.15 (Splitting of triple rule without application condition). *Given a triple rule $tr = (tr: L \rightarrow R)$ without application conditions, we have that $tr = tr_S *_{E_1} tr_F = tr_T *_{E_2} tr_B = tr_{ST} *_{E_3} tr_I$ with E_1, E_2 and E_3 being the domains of tr_F, tr_B and tr_I, respectively.*   △

Proof. At first, we show that $tr = tr_S *_{E_1} tr_F$ where $E_1 = L_F$. Triple graph morphisms $e_{1,F}$ and $e_{2,F}$ are obtained from pushouts $(1F)$ and $(2F)$ in **ATrGraphs**$_{ATG}$ (see Fig. 7.6). Using $tr = tr_F \circ (tr^S, id, id)$ we obtain $tr_S *_E tr_F = tr$. Symmetrically, we derive that $tr = tr_T *_{E_2} tr_B$ where $E_2 = L_B$ by exchanging tr_S with tr_T

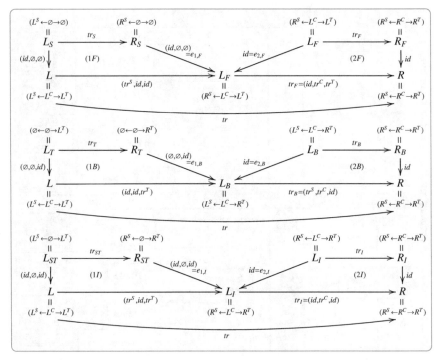

Fig. 7.6 Splitting of triple rules via forward, backward and integration rules (from top to bottom)

and tr_F with tr_B. Analogously, we derive that $tr = tr_{ST} *_{E_3} tr_I$ where $E_3 = L_I$ by exchanging tr_S with tr_{ST} and tr_F with tr_I. In that case, $e_{1,I} = (id, \varnothing, id)$, $e_{2,I} = id$, and $tr = tr_I \circ (tr^S, id, tr^T)$.   □

In case of S-consistency, each triple rule is the E-concurrent rule of its source and forward rules. Similarly, in case of T-consistency, each triple rule is the E-concurrent rule of its target and backward rules (see also Fact 5.17 in [Gol11]). Moreover, in case of an ST-application condition, each triple rule is the E-concurrent rule of its source–target and its integration rule. This correspondence is made explicit in Prop. 7.16 below and used to show the composition and decomposition theorem for TGGs in Theorem 7.21, which builds the basis for the correctness and completeness properties in Sect. 8.1.

Proposition 7.16 (Splitting of triple rule with application condition). *Given a triple rule $tr = (tr: L \to R, ac)$ with S-consistent ac, $tr = tr_S *_{E_1} tr_F$ with $E_1 = L_F$. Dually, if ac is T-consistent we have that $tr = tr_T *_{E_2} tr_B$ with $E_2 = L_B$. If ac is ST-consistent, we have that $tr = tr_{ST} *_{E_3} tr_I$ with $E_3 = L_I$.*   △

Proof. By Fact 7.15 we know that this holds for triple rules without application conditions. It remains to show the property for the application conditions. By Def. 5.27, the application condition ac^* of $(tr_S *_{L_F} tr_F)$ is given by $ac^* =$

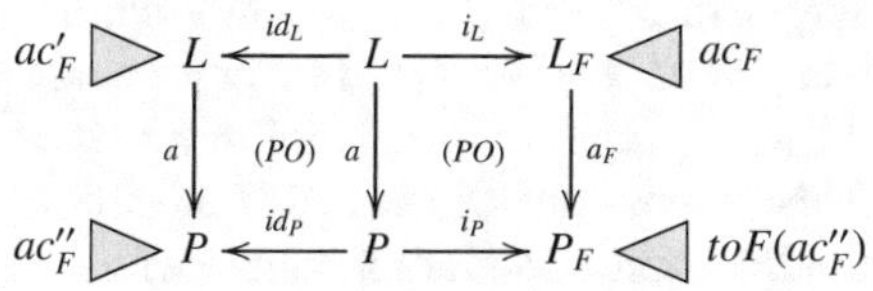

$$i_L = (tr^S, id_{L^C}, id_{L^T})$$
$$i_P = (tr^S, id_{P^C}, id_{P^T})$$
$$a = (id_{L^S}, a^C, a^T)$$
$$a_F = (id_{R^S}, a^C, a^T)$$

Fig. 7.7 Obtaining P as pushout complement from left shift construction

Shift$(u_1, ac_1) \wedge$ L$(p^*, $ Shift$(e_2, ac_2))$ with $u_1 = (id_{L^S}, \varnothing_{L^C}, \varnothing_{L^T})$, $ac_1 = ac_S$, $p^* = (L \overset{id_L}{\longleftarrow} L \xrightarrow{(tr^S, id_{L^C}, id_{L^T})} L_F)$, $e_2 = e_{2,F} = id_{L_F}$ and $ac_2 = ac_F$. Thus, we have that $ac^* =$ Shift$((id_{L^S}, \varnothing_{L^C}, \varnothing_{L^T}), ac_S) \wedge$ L$((L \overset{id_L}{\longleftarrow} L \xrightarrow{(tr^S, id_{L^C}, id_{L^T})} L_F),$ Shift$(id_{E_1}, ac_F))$. We show that $ac = ac'_S \wedge ac'_F \cong ac^*$ in two steps:

1. $ac'_S \cong$ Shift$((id_{L^S}, \varnothing_{L^C}, \varnothing_{L^T}), ac_S)$. Since $ac_S = toS(ac'_S)$ and ac is S-consistent, we can conclude directly that $ac'_S \cong$ Shift$((id_{L^S}, \varnothing_{L^C}, \varnothing_{L^T}), toS(ac'_S))$.
2. $ac'_F \cong$ L$((L \overset{id_L}{\longleftarrow} L \xrightarrow{(tr^S, id_{L^C}, id_{L^T})} L_F),$ Shift$(id_{L_F}, ac_F))$. Shift$(id_{L_F}, ac_F) \cong ac_F$ by Item 1 of Fact 5.4, and therefore $ac'_F \cong$ L$((L \overset{id_L}{\longleftarrow} L \xrightarrow{(tr^S, id_{L^C}, id_{L^T})} L_F), ac_F)$. With $ac_F = toF(ac'_F)$ this is obvious for $ac'_F =$ true. Consider $ac'_F = \exists\,(a, ac''_F)$ with L$((L^S \leftarrow P^C \rightarrow P^T) \rightarrow (R^S \leftarrow P^C \rightarrow P^T), toF(ac''_F)) \cong ac''_F$. Then $(P^S = L^S \overset{s_P}{\longleftarrow} P^C \overset{t_P}{\rightarrow} P^T)$ is the pushout complement constructed for the left shift construction (see Fig. 7.7). Thus, we have that L$((L \xrightarrow{(tr^S, id_{L^C}, id_{L^T})} L_F), toF(\exists\,(a, ac''_F))) \cong$ L$((L \overset{id_L}{\longleftarrow} L \xrightarrow{(tr^S, id_{L^C}, id_{L^T})} L_F), \exists\,((id_{R^S}, a^C, a^T), toF(ac''_F))) \cong \exists\,((id_{L^S}, a^C, a^T),$ L$((P \overset{id_P}{\longleftarrow} (L^S \leftarrow P^C \rightarrow P^T) \rightarrow (R^S \leftarrow P^C \rightarrow P^T)), toF(ac''_F)) \cong \exists\,(a, ac''_F) = ac'_F$. This can be recursively done, leading to the result that indeed L$((L \overset{id_L}{\longleftarrow} L \xrightarrow{(tr^S, id_{L^C}, id_{L^T})} L_F),$ Shift$(id_{L_F}, ac_F)) \cong ac'_F$.

It follows that $ac \cong ac'_S \wedge ac'_F \cong$ Shift$((id_{L^S}, \varnothing_{L^C}, \varnothing_{L^T}), ac_S) \wedge$ L$((L \xrightarrow{(tr^S, id_{L^C}, id_{L^T})} E_1),$ Shift$(id_{E_1}, ac_F))$.

Dually, we can obtain the result for the splitting into target and backward rules for a T-consistent application condition $ac \cong ac'_T \wedge ac'_B \cong$ Shift$((\varnothing_{L^S}, \varnothing_{L^C}, id_{L^T}), ac^T) \wedge$ L$((L \xrightarrow{(id_{L^S}, id_{L^C}, tr^T)} E_2),$ Shift$(id_{E_2}, ac_B))$.

Analogously, we can obtain the result for the splitting into source–target and integration rules for an ST-consistent application condition ac. We show that Shift$((id_{L^S}, \varnothing_{L^C}, id_{L^T}), ac_{ST}) \cong ac$. With $ac_{ST} = toST(ac)$ this is obviously true for $ac =$ true. Consider $ac = \exists\,(a, ac'')$ and suppose Shift$((id_{P^S}, \varnothing_{L^C}, id_{P^T}), toST(ac'')) = ac''$. Then we have that $(P^S \overset{s_P}{\longleftarrow} P^C = L^C \overset{t_P}{\rightarrow} P^T)$ is the only square that we have to consider in the shift construction; for the correspondence component, (C) is the only jointly epimorphic extension we have to consider because all morphisms in the application conditions are identities in the correspondence component. For any square (1) with a monomorphism b^S and (b^S, c^S) jointly epimorphic, it follows that b^S is an epimorphism, i. e., $P^S \cong Q^S$. For any square (2) with a monomorphism b^T and (b^T, c^T) jointly epimorphic, it follows that b^T is an epimorphism, i. e., $P^T \cong Q^T$. This means that (S) and (T) are the only epimorphic extensions that we obtain in the source and target components. It follows that $\text{Shift}((id_{L^S}, \varnothing_{L^C}, id_{L^T}), toST(\;\exists\,(a, ac''))) \cong \exists\,(a, \text{Shift}((id_{P^S}, \varnothing_{L^S}, id_{P^T}), toST(ac''))$ $\cong \exists\,(a, ac''_{ST}) = ac'_{ST}$. This can be recursively done, leading to the result that indeed $\text{Shift}((id_{L^S}, \varnothing_{L^C}, id_{P^T}), ac_{ST}) \cong ac$.

$$
\begin{array}{ccccccc}
L^S \xrightarrow{id_{L^S}} L^S & \quad & L^T \xrightarrow{id_{L^T}} L^T & \quad & L^S \xrightarrow{id_{L^S}} L^S & \quad \varnothing \longrightarrow \varnothing \quad & L^T \xrightarrow{id_{L^T}} L^T \\
a^S \downarrow \;\; (1) \;\; \downarrow c^S & & a^T \downarrow \;\; (2) \;\; \downarrow c^T & & a^S \downarrow \;\; (S) \;\; \downarrow a^S & \downarrow \;\; (C) \;\; \downarrow & a^T \downarrow \;\; (T) \;\; \downarrow a^T \\
P^S \xrightarrow[b^S]{} Q^S & & P^T \xrightarrow[b^T]{} Q^T & & P^S \xrightarrow[id_{P^S}]{} P^S & L^C \xrightarrow[id_{L^C}]{} L^C & P^T \xrightarrow[id_{P^T}]{} P^T
\end{array}
$$

Therefore, $ac \cong \text{Shift}((id_{L^S}, \varnothing_{L^C}, id_{L^T}), ac_{ST})$. $\square$

7.3 Model Transformation Based on Forward Rules

In order to perform model transformations that are compatible with the consistency specification of a TGG, the triple rules of the TGG are used to generate operational rules. These operational rules have to be executed in a controlled way. This section presents the execution of forward transformations using the control condition *source consistency* and backward transformations using the control condition *target consistency*. In Sect. 7.4, we use extended operational rules for simplifying and improving both, analysis and execution techniques.

The general idea of model transformations based on TGGs from source to target models is to take the given source model and apply forward rules in order to complete the missing elements in the correspondence and target components. This process has to be driven by a suitable control condition, which ensures termination, correctness and completeness of the transformation with respect to the triple language $\mathcal{L}(TGG)$ generated by the TGG. This control condition has been formalised by the notion of source consistency in [EEE$^+$07, EEHP09, GEH11]. As we show in Chap. 8, source consistency ensures syntactical correctness and completeness. In combination with an additional static condition on the rules, termination is ensured as well.

Example 7.17 (Inconsistent transformation sequence via forward rules). The forward transformation sequence $G_0 \xRightarrow{tr_{1,F}, m_{1,F}} G_1 \xRightarrow{tr_{2,F}, m_{2,F}} G_2$ in Fig. 7.8 is inconsistent. Types are abbreviated by the first letter. The two nodes of type Class are translated into two nodes of type Table using the forward rule Class2Table$_F$ for each step. However, the edge of type parent between the two nodes of type Class

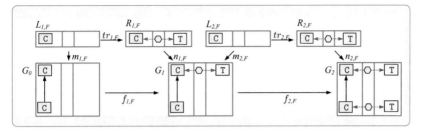

Fig. 7.8 Inconsistent forward sequence

was not handled by one of the two transformation steps. Still, we could execute further forward transformation steps via rules Class2Table$_F$ or SubClass2Table$_F$. However, there is no extension of this sequence that would yield a triple graph $G_n \in \mathcal{L}(TGG)$. Thus, the result will never be consistent with the given TGG. Intuitively, each further step would need to effectively translate again a node of type Class. △

As illustrated by Ex. 7.17, matches for forward rules have to be consistent with the given source model. The main idea of source consistency is the following: Let us consider a source model $G^S \in \mathcal{L}(TG^S)$. If there is a target model $G^T \in \mathcal{L}(TG^T)$, such that there is a triple graph $G = (G^S \leftarrow G^C \rightarrow G^T)$ of the triple language $\mathcal{L}(TGG)$, we know that G^T corresponds to G^S according to the TGG. Otherwise, there is no consistent model transformation sequence starting with G^S. The challenge is to compute such a triple graph $G \in \mathcal{L}(TGG)$, if it exists. By definition, $G \in \mathcal{L}(TGG)$ means that there is a triple sequence $\varnothing \xRightarrow{tr^*} G$ via triple rules $tr \in TR$. It turns out that this sequence can be decomposed into a source sequence $s_S = \langle \varnothing \xRightarrow{tr_S^*} G_{n,0} = G_S \rangle$ with $G_S = (G^S \leftarrow \varnothing \rightarrow \varnothing)$ via source rules $tr_S \in TR_S$ and a forward sequence $s_F = \langle G_S \xRightarrow{tr_F^*} G_{n,n} = G \rangle$ via forward rules $tr_F \in TR_F$. In addition to that, the sequences correspond stepwise to each other concerning the applied rules and the used matches. The exact correspondence is characterised by the notion of source consistency. The remaining challenge is to compute possible source sequences from a given source model, which intuitively means to parse the source model. In Sect. 7.4, we provide an efficient technique for this purpose.

In a first step, we analyse how a triple transformation sequence can be decomposed into a transformation applying first the source rules followed by a sequence of forward rules. Match consistency of the decomposed transformation means that the co-matches of the source rules define the source part of the matches of the forward rules. This notion provides the basis for the actual control condition source consistency for forward model transformations. Note that triple transformation sequences always satisfy the application conditions of the corresponding rules.

Definition 7.18 (Source and match consistency). Given a sequence $(tr_i)_{i=1,\dots,n}$ of triple rules with S-consistent application conditions leading to corresponding sequences $(tr_{i,S})_{i=1,\dots,n}$ and $(tr_{i,F})_{i=1,\dots,n}$ of source and forward rules. A triple trans-

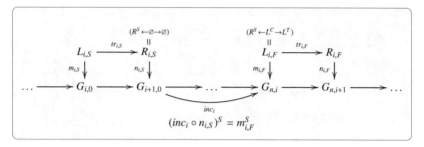

Fig. 7.9 Match and source consistency conditions

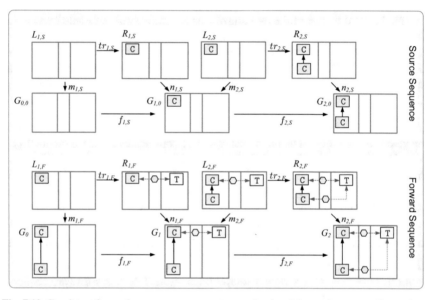

Fig. 7.10 Consistent forward sequence: source sequence (top) and forward sequence (bottom)

formation sequence $\emptyset = G_{0,0} \overset{tr_S^*}{\Longrightarrow} G_{n,0} \overset{tr_F^*}{\Longrightarrow} G_{n,n}$ via first $tr_{1,S}, \ldots, tr_{n,S}$ and then $tr_{1,F}, \ldots, tr_{n,F}$ with matches $m_{i,S}$ and $m_{i,F}$ and co-matches $n_{i,S}$ and $n_{i,F}$, respectively, is *match consistent* if the source component of the match $m_{i,F}$ is uniquely defined by the co-match $n_{i,S}$ (see Fig. 7.9).

A triple transformation $G_{n,0} \overset{tr_F^*}{\Longrightarrow} G_{n,n}$ is called *source consistent* if there is a match consistent sequence $G_{0,0} \overset{tr_S^*}{\Longrightarrow} G_{n,0} \overset{tr_F^*}{\Longrightarrow} G_{n,n}$.   △

Example 7.19 (Consistent transformation sequence via forward rules). The forward transformation sequence $G_0 \overset{tr_{1,F},m_{1,F}}{\Longrightarrow} G_1 \overset{tr_{2,F},m_{2,F}}{\Longrightarrow} G_2$ in the bottom of Fig. 7.10 is source consistent. The compatible source sequence $G_{0,0} = \emptyset \overset{tr_{1,S},m_{1,S}}{\Longrightarrow} G_{1,0} \overset{tr_{2,S},m_{2,S}}{\Longrightarrow} G_{2,0} = G_0$ provides co-matches $n_{i,S}$ that induce the source component $m_{i,F}^S$ of the forward matches $m_{i,F}$ of the forward sequence. Intuitively, each forward step is triggered by a corresponding source step until the given source model G_0^S is completely constructed by the source rules. In fact, the first source step creates the upper node of type Class and the second source step creates the remaining node and edge. △

As mentioned in the beginning of this section, the triple rules $tr_i \in TR$ generate the language of consistent integrated models $\mathcal{L}(TGG)$. Therefore, is is important that there is a compatibility between the transformation sequences via the triple rules $tr_i \in TR$ and the consistent transformation sequences via forward rules. This compatibility is expressed by the decomposition and composition property. The main idea is that we can split a transformation $G_0 \overset{tr_1}{\Longrightarrow} G_1 \Rightarrow \ldots \overset{tr_n}{\Longrightarrow} G_n$ via triple rules $tr_i \in TR$ into transformations $G_0 \overset{tr_{1,S}}{\Longrightarrow} G_0' \overset{tr_{1,F}}{\Longrightarrow} G_1 \Rightarrow \ldots \overset{tr_{n,S}}{\Longrightarrow} G_{n-1}' \overset{tr_{n,F}}{\Longrightarrow} G_n$.

Definition 7.20 (Decomposition and composition (forward case)). A TGG satisfies the *decomposition and composition property for forward sequences* if the following holds:

1. **Decomposition:** For each triple transformation sequence
   (1) $G_0 \overset{tr_1}{\Longrightarrow} G_1 \Rightarrow \ldots \overset{tr_n}{\Longrightarrow} G_n$ via rules in TR
   there is a corresponding match consistent triple transformation sequence
   (2) $G_0 = G_{0,0} \overset{tr_{1,S}}{\Longrightarrow} G_{1,0} \Rightarrow \ldots \overset{tr_{n,S}}{\Longrightarrow} G_{n,0} \overset{tr_{1,F}}{\Longrightarrow} G_{n,1} \Rightarrow \ldots \overset{tr_{n,F}}{\Longrightarrow} G_{n,n} = G_n$
   via rules in TR_S and TR_F.
2. **Composition:** For each match consistent triple transformation sequence (2) as above there is a canonical triple transformation sequence (1) as above.
3. **Bijective Correspondence:** Composition and decomposition are inverse to each other. △

A sufficient condition for the (de)composition property of TGGs in the forward case is that the application conditions of triple rules are S-consistent application conditions as stated by Theorem 7.21 [EEE$^+$07, EHS09, GEH11] below. The proof shows that a decomposition into a match consistent transformation can be found in general, but the composition of match consistent transformations into transformations via the corresponding triple rules requires the additional condition. The investigation of further sufficient criteria to ensure the (de)composition property forms a future research topic for TGGs.

Theorem 7.21 (Decomposition and composition). *For triple transformation sequences with S-consistent application conditions, the decomposition and composition property for forward sequences holds.* △

Proof. At first, we concern only triple rules without ACs (see Fig. 7.11).

1. *Decomposition:* Given the TGT sequence (1) $G_0 \overset{tr_1}{\Longrightarrow} G_1 \Rightarrow \ldots \overset{tr_n}{\Longrightarrow} G_n$ we first consider the case $n = 1$. A TGT step $G_0 \overset{tr_1}{\Longrightarrow} G_1$ can be decomposed uniquely into a match consistent TGT sequence $G_0 = G_{0,0} \overset{tr_{1,S}}{\Longrightarrow} G_{1,0} \overset{tr_{1,F}}{\Longrightarrow} G_{1,1} = G_1$.

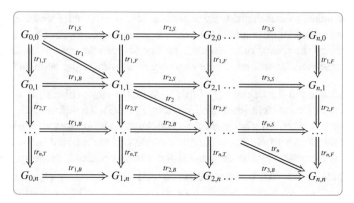

Fig. 7.11 Composition and decomposition of triple sequences

By Fact 7.15 we have shown that tr_1 can be represented as E-concurrent rule $tr_1 = tr_{1,S} *_E tr_{1,F}$. Using the Concurrency Theorem the TGT step $G_0 \stackrel{tr_1}{\Longrightarrow} G_1$ can be decomposed uniquely into an E-related sequence as given above. In this special case, an E-relation is equivalent to the fact that the S-components of the co-match of $G_{0,0} \stackrel{tr_{1,S}}{\Longrightarrow} G_{1,0}$ and the match of $G_{1,0} \stackrel{tr_{1,F}}{\Longrightarrow} G_{1,1}$ coincide, which corresponds exactly to match consistency. Using this construction for $i = 1, \dots, n$ the transformation sequence (1) can be decomposed canonically to an intermediate version between (1) and (2) called (1.5): $G_0 = G_{0,0} \stackrel{tr_{1,S}}{\Longrightarrow} G_{1,0} \stackrel{tr_{1,F}}{\Longrightarrow} G_{1,1} \stackrel{tr_{2,S}}{\Longrightarrow} G_{2,1} \stackrel{tr_{2,F}}{\Longrightarrow} G_{2,2} \Rightarrow \dots \stackrel{tr_{n,S}}{\Longrightarrow} G_{n,n1} \stackrel{tr_{n,F}}{\Longrightarrow} G_{n,n}$, where each subsequence $G_{i1,i1} \stackrel{tr_{i,S}}{\Longrightarrow} G_{i,i1} \stackrel{tr_{i,F}}{\Longrightarrow} G_{i,i}$ is match consistent. Moreover, $G_{1,0} \stackrel{tr_{1,F}}{\Longrightarrow} G_{1,1} \stackrel{tr_{2,S}}{\Longrightarrow} G_{2,1}$ is sequentially independent, because we have a morphism $d: L_2 \to G_{1,0}$, with $L_2 = (L_2^S \leftarrow \varnothing \to \varnothing)$ and $d = (m_2^S, \varnothing, \varnothing)$. The morphism $m_2: L_2 \to G_{1,1}$ is the match of $G_{1,1} \stackrel{tr_{2,S}}{\Longrightarrow} G_{2,1}$, because the S-components of $G_{1,0}$ and $G_{1,1}$ are equal according to the forward rule $tr_{1,F}$. Now the Local Church–Rosser Theorem (Theorem 5.26) leads to an equivalent sequentially independent sequence $G_{1,0} \stackrel{tr_{2,S}}{\Longrightarrow} G_{2,0} \stackrel{tr_{1,F}}{\Longrightarrow} G_{2,1}$ such that $G_{0,0} \stackrel{tr_{1,S}}{\Longrightarrow} G_{1,0} \stackrel{tr_{2,S}}{\Longrightarrow} G_{2,0} \stackrel{tr_{1,F}}{\Longrightarrow} G_{2,1} \stackrel{tr_{2,F}}{\Longrightarrow} G_{2,2}$ is match consistent. The iteration of this shift between $tr_{i,F}$ and $tr_{j,S}$ leads to a shift-equivalent transformation sequence (2) $G_0 = G_{0,0} \stackrel{tr_{1,S}}{\Longrightarrow} G_{1,0} \Rightarrow \dots \stackrel{tr_{n,S}}{\Longrightarrow} G_{n,0} \stackrel{tr_{1,F}}{\Longrightarrow} G_{n,1} \Rightarrow \dots \stackrel{tr_{n,F}}{\Longrightarrow} G_{n,n} = G_n$, which is still match consistent.

2. *Composition:* Vice versa, each match consistent transformation sequence (2) leads to a canonical sequence (1.5) by inverse shift equivalence, where each subsequence as above is match consistent. In fact, match consistency of (2) implies that the corresponding subsequences are sequentially independent in order to allow inverse shifts in an order opposite to that in Item 1, using again the Local Church–Rosser Theorem. Match consistent subsequences of (1.5) are E-related,

as discussed in Item 1, which allows to apply the Concurrency Theorem to obtain the TGT sequence (1).

3. *Bijective Correspondence:* The bijective correspondence of composition and decomposition is a direct consequence of the bijective correspondence in the Local Church–Rosser and the Concurrency Theorem, where the bijective correspondence for the Local Church–Rosser Theorem is not explicitly formulated in Theorem 5.26, but is a direct consequence of the proof in analogy to Theorem 5.30.

Now we consider the case of triple rules with ACs. We use the facts that $tr_i = tr_{i,S} *_{E_i} tr_{i,F}$, as shown in Prop. 7.16, and that the transformations via $tr_{i,S}$ and $tr_{j,F}$ are sequentially independent for $i > j$, as shown above for rules without application conditions. This result can be extended to triple rules with application conditions, as shown in the following.

It suffices to show that the transformations $G_{1,0} \xrightarrow{tr_{1,F},m_1} G_{1,1} \xrightarrow{tr_{2,S},m_2} G_{2,1}$ are sequentially independent. From the sequential independence without application conditions we obtain morphisms $i : R_{1,F} \rightarrow G_{1,1}$ with $i = n_1$ and $j : L_{2,S} \rightarrow G_{1,0}$ with $g_1 \circ j = m_2$.

It remains to show the compatibility with the application conditions (see Fig. 7.12):

- $j \models ac_{2,S}$ (see Fig. 7.12(a)): $ac_{2,S} = toS(ac'_{2,S})$. For $ac'_{2,S} = $ true, also $ac_{2,S} = $ true and therefore $j \models ac_{2,S}$. Suppose $ac'_{2,S} = \exists (a, ac''_{2,S})$, leading to $ac_{2,S} = \exists ((a^S, id_\varnothing, id_\varnothing), toS(ac''_{2,S}))$. Moreover, $tr_{1,F}$ is a forward rule, i.e., it does not change the source component and $G^S_{1,1} = G^S_{1,0}$. We know that $m_2 = g_1 \circ j \models ac_{2,S}$, which means that there exists $p : P \rightarrow G_{1,1}$ with $p \circ a = g_1 \circ j$, $p \models toS(ac''_{2,S})$, and $p^C = \varnothing$, $p^T = \varnothing$. Then there exists $q : P \rightarrow G_{1,0}$ with $q = (p^S, \varnothing, \varnothing)$, $q \circ a = (p^S \circ a^S, \varnothing, \varnothing) = j$, and $q \models toS(ac''_{2,S})$ because all objects occurring in $toS(ac''_{2,S})$ have empty correspondence and target components. This means that $j \models ac_{2,S}$ for this case, and can be shown recursively for composed $ac_{2,S}$.

- $g_2 \circ n_1 \models ac_R := R(tr_{1F}, ac_{1F})$ (see Fig. 7.12(b)): $ac_{1,F} = toF(ac'_{1,F})$, where $ac'_{1,F}$ is an S-extending application condition. For $ac'_{1,F} = $ true also $ac_{1,F} = $ true and $ac_R = $ true, therefore $g_2 \circ n_1 \models ac_R$. Now suppose $ac'_{1,F} = \exists (a, ac''_{1,F})$, leading to $ac_{1,F} = \exists ((id_{R^S_1}, a^C, a^T), toF(ac''_{1,F}))$ and $ac_R = \exists ((id_{R^S_1}, b^C, b^T), ac'_R)$ by componentwise pushout construction for the right shift with $ac'_R = R(u, toF(ac''_{1,F}))$. Moreover, $tr_{2,S}$ is a source rule, which means that g^C_2 and g^T_2 are identities. From the shift property of application conditions we know that $n_1 \models ac_R$, using $m_1 \models ac_{1,F}$. This means that there is a morphism $p : P \rightarrow G_{1,1}$ with $p \circ a = n_1$, $p \models ac'_R$, and $p^S = n^S_1$. It follows that $g_2 \circ p \circ a = g_2 \circ n_1$ and $g_2 \circ p = (g^S_2 \circ p^S, p^C, p^T) \models ac'_R$, because it only differs from p in the S-component, which is identical in all objects occurring in ac'_R. This means that $g_2 \circ n_1 \models ac_R = \exists (a, ac'_R)$, and can be shown recursively for composed ac_R. □

Remark 7.22 (Composition and decomposition for backward case). For each TGT sequence $G_0 \xRightarrow{tr^*} G_n$ there is also a corresponding match consistent backward TGT sequence $G_0 = G_{00} \xRightarrow{tr_{1,T}} G_{01} \Rightarrow \ldots \xRightarrow{tr_{n,T}} G_{0n} \xRightarrow{tr_{1,F}} G_{1n} \Rightarrow \ldots \xRightarrow{tr_{n,F}} G_{nn} = G_n$

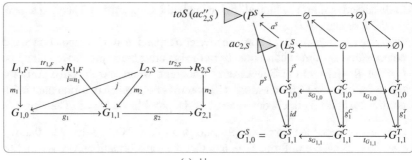

(a) $j \models ac_{2,S}$

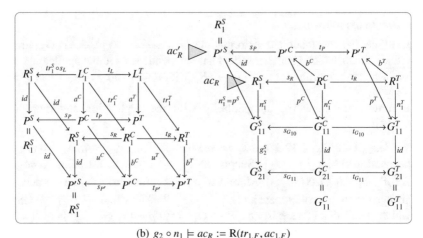

(b) $g_2 \circ n_1 \models ac_R := \mathrm{R}(tr_{1,F}, ac_{1,F})$

Fig. 7.12 Constructions for showing compatibility with the application conditions

based on target and backward rules, leading to a backward model transformation $MT_B : \mathcal{L}(TG^T) \Rightarrow \mathcal{L}(TG^S)$ with similar results as in the forward case. △

Based on source consistent forward transformations we define model transformations, where we assume that the start graph of the given TGG is the empty graph.

Definition 7.23 (Model transformation based on forward rules). A *(forward) model transformation sequence* $(G^S, G_0 \overset{tr_F^*}{\Longrightarrow} G_n, G^T)$ is given by a source graph G^S, a target graph G^T, and a source consistent forward transformation $G_0 \overset{tr_F^*}{\Longrightarrow} G_n$ with $G_0 = (G^S \overset{\varnothing}{\longleftarrow} \varnothing \overset{\varnothing}{\longrightarrow} \varnothing)$ and $G_n^T = G^T$.

A *(forward) model transformation* $MT_F : VL_S \Rightarrow VL_T$ is defined by all (forward) model transformation sequences. △

Example 7.24 (Model transformation sequence). Fig. 7.13 shows the resulting triple graph G of a transformation sequence $G_0 \overset{tr_F^*}{\Longrightarrow} G_n$ via forward rules. This sequence

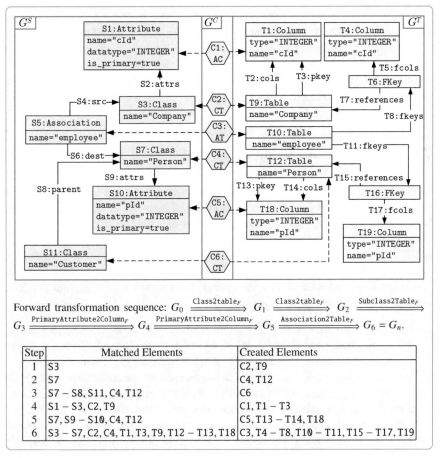

Fig. 7.13 Forward model transformation for *CD2RDBM*: result (top) and forward transformation sequence (bottom)

specifies a model transformation sequence $(G^S, G_0 \xRightarrow{tr_F^*} G_n, G^T)$ from a source model G^S to a target model G^T via forward rules, where $G_0 = (G^S \leftarrow \varnothing \rightarrow \varnothing)$. The table in the bottom of the figure shows the corresponding steps with numbers for matched and created elements. According to the numbers of the elements, the correspondence and target components are completely created during the forward model transformation sequence. Moreover, there is a source sequence $\varnothing \xRightarrow{tr_S^*} G_0$ such that $\varnothing \xRightarrow{tr_S^*} G_0 \xRightarrow{tr_F^*} G_n$ is match consistent. The co-matches of the source steps are given by the numbers for the source elements in the forward matches. Indeed, we can inspect the figure and conclude that the sequence $\varnothing \xRightarrow{tr_S^*} G_0 \xRightarrow{tr_F^*} G_n$ is match consistent, because the source elements of each forward match are created by the corresponding source rule applications in $\varnothing \xRightarrow{tr_S^*} G_0$.   △

Similar to the structuring concepts for plain graph grammars [KKvT10], techniques for rule refinement and control structures have been introduced for TGGs to improve the development process concerning usability and maintainability. We refer you to [ASLS14] for an overview. In the next section, we present how the control condition "source consistency" can be encoded by additional attributes, which is used for analysis and implementation purposes.

7.4 Model Transformation Based on Forward Translation Rules

While the concept of model transformations based on source consistent forward sequences in Sect. 7.3 provides an abstract formal basis for model transformations based on TGGs, this section presents a possible encoding of the control condition source consistency by extending the forward rules. This concept has been introduced in [HEOG10b] and extended in [HEGO10, HEGO14]. The main idea is to extend the operational forward rules with additional markers for keeping track of the elements that have been translated during the execution of a forward transformation. This concept achieves the following goals.

1. *Simplification of execution:* The complex and descriptive control condition source consistency that is based on the existence and compatibility of a source sequence is replaced by a constructive check of marker values of the resulting graph.
2. *Improvement of formal analysis:* The general results for confluence analysis for $\mathcal{M}$-adhesive transformation systems based on critical pair analysis cannot be applied directly to systems with forward rules and would need to be accompanied with additional techniques in order to capture the effect of the control condition source consistency. The encoding of this condition within the triple graphs enables us to apply the critical pair analysis directly as performed in Sect. 8.2.
3. *Implementation of the approach:* The resulting execution strategy for model transformations based on forward translation rules has a constructive nature that can be implemented as an extension to existing graph transformation engines. It has been used for realising the implementation in the tool HenshinTGG and it has a close correspondence to the pointer structures that are used in execution algorithms of other TGG tools.

The main idea is to extend the source component of the triple graph by additional Boolean-valued attributes that specify for each element whether it has been already translated. The main result in this section shows that model transformations based on source consistent forward TGT sequences are equivalent to those based on complete forward translation TGT sequences, as stated by Fact 7.36. The control condition source consistency is ensured by the completeness of forward translation TGT sequences, which are based on the generated forward translation rules. For this reason, the check of source consistency for forward TGT sequences is reduced to a check for whether all translation attributes are set to "**T**", which ensures that the model is completely translated. Note that the encoding via translation attributes

requires the general assumption (see Rem. 7.5), which states that matches and application conditions are based on almost injective morphisms. This ensures that the
translation markers are independent from each other, because the morphisms do not
identify structural elements.

In many practical applications, model transformations are required to preserve
the source model in order to use database-driven model repositories. For this reason, we have presented in [HEOG10a] how the translation attributes can be exter-
nalised using the concept of triple graphs with interfaces. The translation attributes
are equivalently replaced by external pointer structures such that the model transformation can be performed without any modification of the source model. This
concept corresponds to the transformation algorithm in [SK08], which uses a separate set of translated elements. Furthermore, it shows how the source sequence for
a source consistent forward sequence in the previous section can be computed by
an implementation with additional pointer structures, as explained at the end of this
section.

7.4.1 Translation Attributes

The extension of forward rules to forward translation rules is based on new attributes
that control the translation process to ensure the source consistency condition. For
each node, edge and attribute of a graph a new attribute is created and labelled with
the prefix "tr". Given an attributed graph $AG = (G, D)$ and a family of subsets
$M \subseteq G$ for nodes and edges, we call AG' a graph with translation attributes over AG
if it extends AG with one Boolean-valued attribute $\mathtt{tr\_}x$ for each element x (node or
edge) in M and one Boolean-valued attribute $\mathtt{tr\_}x\mathtt{\_}a$ for each attribute associated
to such an element x in M. In order to distinguish between a triple rule tr and the
prefix $\mathtt{tr}$ of a translation attribute, we use a different font shape (typewriter). The
family M together with all these additional translation attributes is denoted by Att_M.

Definition 7.25 (Family with translation attributes). Given an attributed graph
$AG = (G, D)$ (see Def. 2.4), we denote by $|G| = (V_G^G, V_G^D, E_G^G, E_G^{NA}, E_G^{EA})$
the underlying family of sets containing all nodes and edges. Let $M \subseteq$
$|G|$ with $(V_M^G, V_M^D, E_M^G, E_M^{NA}, E_M^{EA})$; then a *family with translation attributes* for
(G, M) extends M by additional translation attributes and is given by $Att_M =$
$(V_M^G, V_M^D, E_M^G, E^{NA}, E^{EA})$ with:

- $E^{NA} = E_M^{NA} \cup \{\mathtt{tr\_}x \mid x \in V_M^G\} \cup \{\mathtt{tr\_}x\mathtt{\_}a \mid a \in E_M^{NA}, src_G^{NA}(a) = x \in V_G^G\}$,
- $E^{EA} = E_M^{EA} \cup \{\mathtt{tr\_}x \mid x \in E_M^G\} \cup \{\mathtt{tr\_}x\mathtt{\_}a \mid a \in E_M^{EA}, src_G^{EA}(a) = x \in E_G^G\}$. $\quad\triangle$

Definition 7.26 (Graph with translation attributes). Given an attributed graph
$AG = (G, D)$ (see Def. 2.4) and a family of subsets $M \subseteq |G|$ with $\{\mathbf{T}, \mathbf{F}\} \subseteq V_M^D$, let
Att_M be a family with translation attributes for (G, M) according to Def. 7.25. Then
$AG' = (G', D)$ is a *graph with translation attributes* over AG, where the domains $|G'|$

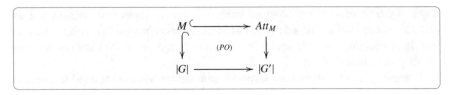

Fig. 7.14 Triple graph with translation attributes: construction

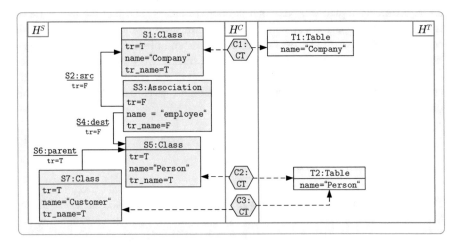

Fig. 7.15 Triple graph with translation attributes: example

of G' are given by the gluing via the pushout of $|G|$ and Att_M over M (see Fig. 7.14) and the source and target functions of G' are defined as follows:

- $src_{G'}^G = src_G^G, trg_{G'}^G = trg_G^G,$

- $src_{G'}^X(z) = \begin{cases} src_G^X(z) & z \in E_G^X \\ x & z = \text{tr\_}x \text{ or } z = \text{tr\_}x\_a \end{cases}$ for $X \in \{NA, EA\},$

- $trg_{G'}^X(z) = \begin{cases} trg_G^X(z) & z \in E_G^X \\ \mathbf{T} \text{ or } \mathbf{F} & z = \text{tr\_}x \text{ or } z = \text{tr\_}x\_a \end{cases}$ for $X \in \{NA, EA\}.$

Att_M^v, where $v = \mathbf{T}$ or $v = \mathbf{F}$, denotes a family with translation attributes where all attributes are set to v. Moreover, we denote by $AG \oplus Att_M$ that AG is extended by the translation attributes in Att_M, i.e., $AG \oplus Att_M = (G', D)$ for $AG' = (G', D)$, as defined above. Analogously, we use the notion $AG \oplus Att_M^v$ for translation attributes with value v and we use the short notation $Att^v(AG) := AG \oplus Att_{|G|}^v$. △

Example 7.27 (Triple graph with translation attributes). Fig. 7.15 shows the triple graph $H = (H^S \leftarrow H^C \rightarrow H^T)$ which is extended by some translation attributes in the source component. The translation attributes with value "**T**" indicate that the owning elements have been translated during a model transformation sequence using forward translation rules, which are defined in Def. 7.29 hereafter. The remaining elements (edges S2, S4, node S3 and the attribute name of S3) in the source component are still marked with translation attributes set to "**F**". These elements can still be matched for a continuation of the translation and may be translated at later steps.   △

The concept of forward translation rules, which we have introduced in [HEOG10b], extends the construction of forward rules by additional translation attributes in the source component. As described in Ex. 7.27, the translation attributes are used to keep track of the elements that have been translated so far. Since triple rules may create new attributes for existing nodes by definition, we also have to keep track of the translation of the attributes. The separate handling of nodes and their attributes is used, e.g., in synchronisation scenarios [HEO⁺11a]. At the beginning, the source model of a model transformation sequence is extended by translation attributes that are all set to "**F**" and, step by step, they are set to "**T**" when their containing elements are translated by a forward translation rule.

7.4.2 Execution via Forward Translation Rules

The extension of forward rules to forward translation rules integrates additional translation attributes on the source component that keep track of the elements that have been translated during the execution of a forward transformation.

The application conditions of a forward translation rule are derived from the forward rule by adding translation attributes with value **T** to all additional elements that are not contained in the left hand side L_{FT} of the forward translation rule tr_{FT}. Therefore, we introduce the construction $tExt$ in Def. 7.28 below that extends an application condition with additional translation attributes, which are set to the value **T**. The third argument X of this construction specifies the triple components that are extended. This enables us to use this construction for several kinds of operational translation rules in this book, such as the consistency creating rules in Def. 7.44.

Definition 7.28 (T-Extension of application conditions). Given an application condition ac over P, a triple graph with translation attributes P' (extended premise graph) and a subset of triple components $X \subseteq \{S, C, T\}$, the T-extension $tExt(ac, P', X)$ of ac is given by

- $tExt(\text{true}, P', X) = \text{true}$,
- $tExt(\neg(ac'), P', X) = \neg(tExt(ac', P', X))$,
- $tExt(ac_1 \wedge ac_2, P', X) = tExt(ac_1, P', X) \wedge tExt(ac_2, P', X)$,
- $tExt(ac_1 \vee ac_2, P', X) = tExt(ac_1, P', X) \vee tExt(ac_2, P', X)$,
- $tExt(\exists (a = (inc_P, a_D): P \rightarrow C, ac'), P', X) = \exists (a_E: P' \rightarrow C', tExt(ac', C', X))$
  with $C' = P' +_P C \oplus \cup_{x \in X}(Att^{\mathbf{T}}_{C^x \setminus P^x})$ and $a_E = (inc_{P_E}, a_D)$ given by the algebra homomorphism a_D on the data part and the inclusion inc_{P_E} on the graph part (derived from inc_P).   △

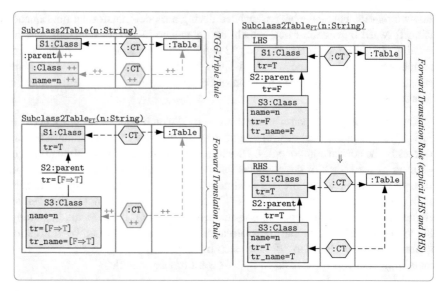

Fig. 7.16 Forward translation rule Subclass2Table$_{FT}$($n : String$)

Definition 7.29 (Forward translation rule). Consider a triple rule $tr = (tr: L \rightarrow R, ac)$ with S-consistent application condition. Let $tr_S = (tr_S: L_S \rightarrow R_S, ac_S)$ be the derived source rule and $tr_F = (tr_F: L_F \rightarrow R_F, ac_F)$ be the derived forward rule with $ac \cong ac_S \wedge ac_F$. The *forward translation rule* of tr is given by $tr_{FT} = (tr_{FT}: L_{FT} \xleftarrow{l_{FT}} K_{FT} \xrightarrow{r_{FT}} R_{FT}, ac_{FT})$, defined as follows:

- $L_{FT} = L_F \oplus Att_{L_S}^{\mathbf{T}} \oplus Att_{R_S \setminus L_S}^{\mathbf{F}}$,
- $K_{FT} = L_F \oplus Att_{L_S}^{\mathbf{T}}$,
- $R_{FT} = R_F \oplus Att_{L_S}^{\mathbf{T}} \oplus Att_{R_S \setminus L_S}^{\mathbf{T}} = R_F \oplus Att_{R_S}^{\mathbf{T}}$,
- l_{FT} and r_{FT} are the induced inclusions,
- $ac_{FT} = tExt(ac, L_{FT}, \{S\})$.

Given a set of triple rules TR, we denote by TR_{FT} the set of all tr_{FT} with $tr \in TR$.   △

Remark 7.30 (Construction of application conditions). The construction of the application condition for a forward translation rule tr_{FT} starts with the left hand side L_{FT} that contains translation attributes and adds additional translation attributes recursively for each new element in the premise and conclusion graphs P_{FT} and C_{FT}. Note that initially L_{FT} plays the role of the first premise P_{FT} of a nested application condition. Note further that $(P_{FT} +_P C)$ is the union of P_{FT} and C with shared P (constructed as a pushout) and for an S-extending application condition ac the forward translation application condition ac_{FT} does not contain any additional translation attributes because $C^S = P^S$ for all contained morphisms $a: P \rightarrow C$.   △

Example 7.31 (Derived forward translation rules). The rule "Subclass2Table$_{FT}$" in Fig. 7.16 is the derived forward translation rule of the triple rule "Subclass2Table"

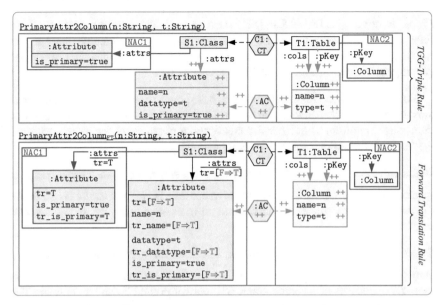

Fig. 7.17 Forward translation rule with NACs `PrimaryAttribute2Column`$_{FT}(n : String)$

in Fig. 3.8. Note that we abbreviate "`tr_x`" for an item (node or edge) x by "`tr`" and "`tr_x_a`" by "`tr_type(a)`" in the figures to increase readability. The compact notation of forward translation rules specifies the modification of translation attributes by "[**F** ⇒ **T**]", meaning that the attribute is matched with the value "**F**" and set to "**T**" during the transformation step. The detailed complete notation of a forward translation rule is shown on the right of Fig. 7.16 for "*Subclass2Table*$_{FT}$".

Fig. 7.17 shows the forward translation rule with NACs "*PrimaryAttr2Column*$_{FT}$" derived from the triple rule "*PrimaryAttr2Column*" in Fig. 3.8. According to Def. 7.29 the source elements of the triple rule are extended by translation attributes and changed by the rule from "**F**" to "**T**" if the owning elements are created by the triple rule. Furthermore, the forward translation rule contains both, the source and the target NACs of the triple rule, where the NAC only elements in the source NACs are extended by translation attributes set to "**T**". Thus, a source NAC concerns only elements that have been translated so far. △

Since forward translation rules are deleting attribution edges only, each NAC-consistent match is applicable according to Fact 7.32 below, which was first presented in [HEGO10]. Note that in the general case of deleting rules the additional gluing condition has to be checked [EEPT06]. This ensures, in particular, that edges do not become dangling due to the deletion of nodes.

Fact 7.32 (Gluing condition for forward translation rules). *Let* tr_{FT} *be a forward translation rule and* $m_{FT} : L_{FT} \to G$ *be an almost injective match; then the gluing condition is satisfied, i.e., there is the transformation step* $G \overset{tr_{FT}, m_{FT}}{\Longrightarrow} H$. △

Proof. According to Def. 9.8 in [EEPT06] we need to check that $DP \cup IP \subseteq GP$. First of all, by the restriction of the match, the set IP may only contain data elements which are in GP. Furthermore, the set DP does only contain nodes. The rule is only deleting on attribution edges, and thus $DP \cup IP \subseteq GP$. $\square$

Now, we define model transformations based on forward translation rules via almost injective matches in a similar way as for forward rules in Def. 7.23. We replace the control condition source consistency of the forward sequence by requiring that the forward translation sequence be complete.

Definition 7.33 (Complete forward translation sequence). A forward translation sequence $G_0 \overset{tr_{FT}^*}{\Longrightarrow} G_n$ with almost injective matches is called *complete* if no further forward translation rule is applicable and all translation attributes in G_n are set to true ("**T**").   △

Definition 7.34 (Model transformation based on forward translation rules). A *model transformation sequence* $(G^S, G_0' \overset{tr_{FT}^*}{\Longrightarrow} G_n', G^T)$ based on forward translation rules TR_{FT} consists of a source graph G^S, a target graph G^T, and a complete TGT sequence $G_0' \overset{tr_{FT}^*}{\Longrightarrow} G_n'$ typed over $TG' = TG \oplus Att_{|TG^S|}^{\mathbf{F}} \oplus Att_{|TG^S|}^{\mathbf{T}}$ based on TR_{FT} with $G_0' = (Att^{\mathbf{F}}(G^S) \leftarrow \varnothing \rightarrow \varnothing)$ and $G_n' = (Att^{\mathbf{T}}(G^S) \leftarrow G^C \rightarrow G^T)$.

A *model transformation* $MT : \mathcal{L}(TG^S) \Rightarrow \mathcal{L}(TG^T)$ based on TR_{FT} is defined by all model transformation sequences as above with $G^S \in \mathcal{L}(TG^S)$ and $G^T \in \mathcal{L}(TG^T)$. All the corresponding pairs (G^S, G^T) define the *model transformation relation* $MT_{FT,R} \subseteq \mathcal{L}(TG^S) \times \mathcal{L}(TG^T)$ based on TR_{FT}. The model transformation is *terminating* if there are no infinite TGT sequences via TR_{FT} starting with $G_0' = (Att^{\mathbf{F}}(G^S) \leftarrow \varnothing \rightarrow \varnothing)$ for some source graph G^S.   △

Example 7.35 (Model transformation via forward translation rules). Fig. 7.18 shows the resulting triple graph with translation attributes of a forward translation sequence. The execution starts by taking the source model G^S (see Fig. 7.13) and extending it with translation attributes according to Def. 7.34, i.e., $G_0' = (Att^{\mathbf{F}}(G^S) \leftarrow \varnothing \rightarrow \varnothing)$. We can execute the forward translation sequence shown in the bottom part of Fig. 7.18 with G_6' being the triple graph G' in Fig. 7.18. The triple graph G' is indeed completely translated, because all translation attributes are set to "**T**". No further forward translation rule is applicable and we derive the resulting target model G^T by restricting G' to its target component, i.e., $G^T = G'^T$. According to the equivalence of the model transformation concepts based on forward and forward translation rules in Fact 7.36 below, we can further conclude that G^T can be equivalently obtained via a source consistent forward transformation sequence based on forward rules without translation attributes.   △

By Fact 7.36 below we show that the model transformation sequences based on forward translation rules with NACs are in one-to-one correspondence with model transformation sequences based on forward rules with NACs, i.e., based on source consistent forward sequences. For this reason, we can equivalently use both concepts

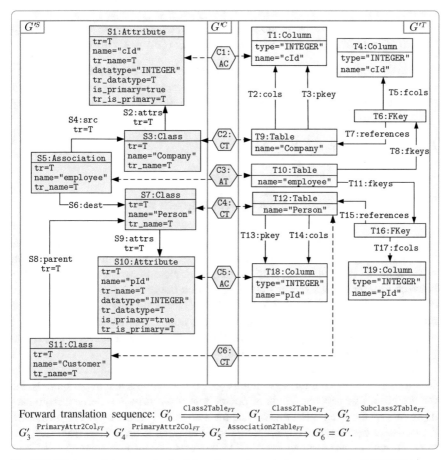

Forward translation sequence: $G'_0 \xrightarrow{\text{Class2Table}_{FT}} G'_1 \xrightarrow{\text{Class2Table}_{FT}} G'_2 \xrightarrow{\text{Subclass2Table}_{FT}}$
$G'_3 \xrightarrow{\text{PrimaryAttr2Col}_{FT}} G'_4 \xrightarrow{\text{PrimaryAttr2Col}_{FT}} G'_5 \xrightarrow{\text{Association2Table}_{FT}} G'_6 = G'.$

Fig. 7.18 Triple graph instance obtained from forward translation sequence for *CD2RDBM*

and choose one of them depending on the particular needs. While the concept based on source consistency shows advantages in formal proofs, the concept based on forward translation rules shows advantages concerning analysis and efficiency, as we will show in Sect. 8.2.1. It will be part of future work to extend the result to a corresponding result that generalises from the case with NACs to the case with general application conditions.

Fact 7.36 (Equivalence of forward transformation and forward translation sequences). *Given a source model $G^S \in \mathcal{L}(TG^S)$, the sets of forward rules TR_F and corresponding forward translation rules TR_{FT}, the following statements are equivalent for almost injective matches.*

1. There is a model transformation sequence $(G^S, G_0 \xrightarrow{tr_F^*} G_n, G^T)$ based on TR_F with $G_0 = (G^S \leftarrow \emptyset \rightarrow \emptyset)$ and $G_n = (G^S \leftarrow G^C \rightarrow G^T)$

2. *There is a model transformation sequence* $(G^S, G'_0 \overset{tr^*_{FT}}{\Longrightarrow} G'_n, G^T)$ *based on* TR_{FT}
   with $G'_0 = (Att^{\mathbf{F}}(G^S) \leftarrow \varnothing \rightarrow \varnothing)$ *and* $G'_n = (Att^{\mathbf{T}}(G^S) \leftarrow G^C \rightarrow G^T)$.

   Moreover, the model transformation relation $MT_{F,R}$ *for the model transformation based on forward rules coincides with the model transformation relation* $MT_{FT,R}$ *for the model transformation based on forward translation rules, i.e.,* $MT_{F,R} = MT_{FT,R}$.

   $\triangle$

Proof. See Appendix B.6.2. $\square$

7.5 Model Integration Based on Integration Rules

The main purpose of model integration is to establish a correspondence between various models, especially between source and target models. From the analysis point of view, model integration supports correctness checks of syntactical dependencies between different views and models. This section presents model integration based on triple graph grammars and shows the close relationship between model transformation and model integration. For each model transformation sequence there is a unique model integration sequence, and vice versa. The main concepts and results were first presented in [EEH08a, EEH08b].

The general problem of model integration is constructing an integrated model $G = (G^S \leftarrow G^C \rightarrow G^T)$ for a given pair (G^S, G^T) of source and target models. For this purpose, two separate kinds of operational triple rules are derived from each triple rule tr: the integration rule tr_I and the source–target rule tr_{ST}. These rules are the basis for defining and constructing model integration sequences from (G^S, G^T) to G. Of course, not each pair (G^S, G^T) allows us to construct such a model integration sequence. In Theorem 7.41, we characterise existence and construction of model integration sequences from (G^S, G^T) to G by model transformation sequences from G^S to G^T. This main result is based on the canonical decomposition result (see Theorem 7.21) and a similar decomposition result of triple transformation sequences into source–target and model integration sequences.

7.5.1 Model Integration Rules and Transformations

Given models $G^S \in \mathcal{L}(TG^S)$ and $G^T \in \mathcal{L}(TG^T)$, the aim of model integration is to construct an integrated model $G \in \mathcal{L}(TGG)$ such that G restricted to source and target is equal to G^S and G^T, respectively, i.e., $proj_S(G) = G^S$ and $proj_T(G) = G^T$. In analogy to model transformations, we use the operational rules derived from the given triple rules tr_i: the source–target rules $tr_{i,ST}$ and the integration rules $tr_{i,I}$.

Given a transformation sequence $G_0 \overset{tr^*_I}{\Longrightarrow} G_n$ via integration rules with $G_0 = (G^S \leftarrow \varnothing \rightarrow G^T)$, we want to make sure that the unrelated pair $(G^S, G^T) \in$

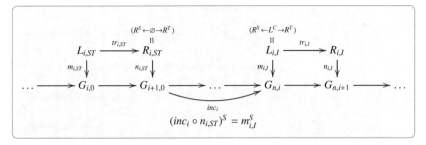

Fig. 7.19 Source–target consistency conditions

$\mathcal{L}(TG^S) \times \mathcal{L}(TG^T)$ is transformed into an integrated model $G = G_n$ with $proj_S(G) = G^S$ and $proj_T(G) = G^T$. Of course, this is not possible for all pairs $(G^S, G^T) \in \mathcal{L}(TG^S) \times \mathcal{L}(TG^T)$, but only for specific pairs. In order to be sure that $G_0 \overset{tr^*_I}{\Longrightarrow} G_n$ integrates all parts of G^S and G^T, we require that $\varnothing \overset{*}{\Longrightarrow} G_0$ be given by $\varnothing \overset{tr^*_{ST}}{\Longrightarrow} G_0$ based on the same triple rule sequence tr^* as $G_0 \overset{tr^*_I}{\Longrightarrow} G_n$. Moreover, the co-matches in $\varnothing \overset{tr^*_{ST}}{\Longrightarrow} G_0$ have to be compatible with the matches in $G_0 \overset{tr^*_I}{\Longrightarrow} G_n$. Finally, we need to ensure that a model integration can be performed for a given pair G^S, G^T if there is at least one integrated model $G = (G^S \leftarrow G^C \rightarrow G^T) \in \mathcal{L}(TGG)$ which contains both models. This leads to the formal condition of source–target consistency of transformation sequences via integration rules.

Definition 7.37 (Source–target consistency). Consider a sequence $(tr_i)_{i=1,\dots,n}$ of triple rules with *ST*-application conditions leading to corresponding sequences $(tr_{i,ST})_{i=1,\dots,n}$ and $(tr_{i,I})_{i=1,\dots,n}$ of source–target and integration rules (see Fig. 7.2 and Def. 7.12). A triple transformation sequence $G_{0,0} \overset{tr^*_{ST}}{\Longrightarrow} G_{n,0} \overset{tr^*_I}{\Longrightarrow} G_{n,n}$ via first $tr_{1,ST}, \dots, tr_{n,ST}$ and then $tr_{1,I}, \dots, tr_{n,I}$ with matches $m_{i,ST}$ and $m_{i,I}$ and co-matches $n_{i,ST}$ and $n_{i,I}$, respectively, is *match consistent* if the source and target components of the match $m_{i,I}$ are uniquely defined by the co-match $n_{i,ST}$ (see Fig. 7.19).

A triple transformation $G_{n,0} \overset{tr^*_I}{\Longrightarrow} G_{n,n}$ is called *source–target consistent* if there is a match consistent sequence $G_{0,0} \overset{tr^*_{ST}}{\Longrightarrow} G_{n,0} \overset{tr^*_I}{\Longrightarrow} G_{n,n}$. △

Definition 7.38 (Model integration based on integration rules). A *model integration sequence* $((G^S, G^T), G_0 \overset{tr^*_I}{\Longrightarrow} G_n, G)$ is given by a source graph G^S, a target graph G^T, a triple graph G, and a source–target-consistent transformation $G_0 \overset{tr^*_I}{\Longrightarrow} G_n$ with $G_0 = (G^S \overset{\varnothing}{\leftarrow} \varnothing \overset{\varnothing}{\rightarrow} G^T)$ and $G_n = G$.

A *model integration MI* $: \mathcal{L}(TGG)^S \times \mathcal{L}(TGG)^T \Rightarrow \mathcal{L}(TGG)$ is defined by all model integration sequences. △

Definition 7.39 (Decomposition and composition for model integration). A TGG satisfies the *decomposition and decomposition property for integration sequences* if the following holds:

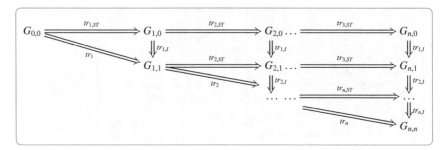

Fig. 7.20 Composition and decomposition of triple sequences for model integration

1. **Decomposition:** For each triple transformation sequence
   (1) $G_0 \xrightarrow{tr_1} G_1 \Rightarrow \ldots \xrightarrow{tr_n} G_n$ via rules in TR
   there is a corresponding match consistent triple transformation sequence
   (2) $G_0 = G_{0,0} \xrightarrow{tr_{1,ST}} G_{1,0} \Rightarrow \ldots \xrightarrow{tr_{n,ST}} G_{n,0} \xrightarrow{tr_{1,I}} G_{n,1} \Rightarrow \ldots \xrightarrow{tr_{n,I}} G_{n,n} = G_n$
   via rules in TR_{ST} and TR_I.
2. **Composition:** For each match consistent triple transformation sequence (2) as above there is a canonical triple transformation sequence (1) as above.
3. **Bijective Correspondence:** Composition and decomposition are inverse to each other.   △

Theorem 7.40 (Decomposition and composition for model integration). *For triple transformation sequences with S- and T-application conditions the decomposition and composition property for integration sequences holds.*   △

Proof. At first, we consider only triple rules without ACs (see Fig. 7.20).

1. *Decomposition:* Given a TGT sequence (1) $G_0 \xrightarrow{tr_1} G_1 \Rightarrow \ldots \xrightarrow{tr_n} G_n$ we first consider the case $n = 1$. The TGT step $G_0 \xrightarrow{tr_1} G_1$ can be decomposed uniquely into a match consistent TGT sequence $G_0 = G_{0,0} \xrightarrow{tr_{1,ST}} G_{1,0} \xrightarrow{tr_{1,I}} G_{1,1} = G_1$. By Fact 7.15 we have shown that tr_1 can be represented as E-concurrent rule $tr_1 = tr_{1,ST} *_E tr_{1,I}$. Using the Concurrency Theorem, the TGT step $G_0 \xrightarrow{tr_1} G_1$ can be decomposed uniquely into an E-related sequence as given above. In this special case, an E-relation is equivalent to the fact that the S- and T-components of the co-match of $G_{0,0} \xrightarrow{tr_{1,ST}} G_{1,0}$ and the match of $G_{1,0} \xrightarrow{tr_{1,I}} G_{1,1}$ coincide on the source and target components, which corresponds exactly to match consistency. Using this construction for $i = 1, \ldots, n$, the transformation sequence (1) can be decomposed canonically into an intermediate version between (1) and (2) called (1.5): $G_0 = G_{0,0} \xrightarrow{tr_{1,ST}} G_{1,0} \xrightarrow{tr_{1,I}} G_{1,1} \xrightarrow{tr_{2,ST}} G_{2,1} \xrightarrow{tr_{2,I}} G_{2,2} \Rightarrow \ldots \xrightarrow{tr_{n,ST}} G_{n,n1} \xrightarrow{tr_{n,I}} G_{n,n}$ where each subsequence $G_{i1,i1} \xrightarrow{tr_{i,ST}} G_{i,i1} \xrightarrow{tr_{i,I}} G_{i,i}$ is match consistent. Moreover, $G_{1,0} \xrightarrow{tr_{1,I}} G_{1,1} \xrightarrow{tr_{2,ST}} G_{2,1}$ is sequentially independent, because we have a morphism $d: L_{2,ST} \to G_{1,0}$, with $L_{2,ST} = (L_2^S \leftarrow \emptyset \to L_2^T)$

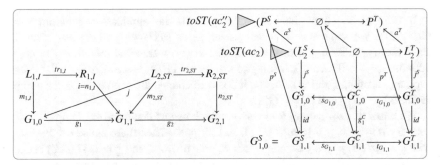

Fig. 7.21 Constructions for showing compatibility with ST-application conditions

and $d = (m_{2,ST}^S, \emptyset, m_{2,ST}^T)$. The morphism $m_{2,ST} : L_{2,ST} \rightarrow G_{1,1}$ is the match of $G_{1,1} \xrightarrow{tr_{2,ST}} G_{2,1}$, because the S- and T-components of $G_{1,0}$ and $G_{1,1}$ are equal according to the integration rule $tr_{1,I}$. Now, the Local Church–Rosser Theorem (Theorem 5.26) leads to an equivalent sequentially independent sequence $G_{1,0} \xrightarrow{tr_{2,ST}} G_{2,0} \xrightarrow{tr_{1,I}} G_{2,1}$ such that $G_{0,0} \xrightarrow{tr_{1,ST}} G_{1,0} \xrightarrow{tr_{2,ST}} G_{2,0} \xrightarrow{tr_{1,I}} G_{2,1} \xrightarrow{tr_{2,I}} G_{2,2}$ is match consistent. The iteration of this shift between $tr_{i,I}$ and $tr_{j,ST}$ leads to a shift-equivalent transformation sequence (2) $G_0 = G_{0,0} \xrightarrow{tr_{1,ST}} G_{1,0} \Rightarrow \ldots \xrightarrow{tr_{n,ST}} G_{n,0} \xrightarrow{tr_{1,I}} G_{n,1} \Rightarrow \ldots \xrightarrow{tr_{n,I}} G_{n,n} = G_n$, which is still match consistent.

2. *Composition:* Vice versa, each match consistent transformation sequence (2) leads to a canonical sequence (1.5) by inverse shift equivalence where each subsequence as above is match consistent. In fact, match consistency of (2) implies that the corresponding subsequences are sequentially independent in order to allow inverse shifts in an order opposite to that in Item 1 using again the Local Church–Rosser Theorem. Match consistent subsequences of (1.5) are E-related, as discussed in Item 1, which allows us to apply the Concurrency Theorem to obtain the TGT sequence (1).

3. *Bijective Correspondence:* The bijective correspondence of composition and decomposition is a direct consequence of the bijective correspondence in the Local Church–Rosser and the Concurrency Theorem, where the bijective correspondence for the Local Church–Rosser Theorem is not explicitly formulated in Theorem 5.26, but is a direct consequence of the proof in analogy to Theorem 5.30.

Now, we consider the case of triple rules with ACs. We use the facts that $tr_i = tr_{i,ST} *_{E_i} tr_{i,I}$, as shown in Prop. 7.16, and the transformations via $tr_{i,ST}$ and $tr_{j,I}$ are sequentially independent for $i > j$ as shown above for rules without application conditions. This result can be extended to triple rules with application conditions as shown in the following.

It suffices to show that the transformations $G_{1,0} \xrightarrow{tr_{1,I},m_1} G_{1,1} \xrightarrow{tr_{2,ST},m_2} G_{2,1}$ are sequentially independent. From the sequential independence without application conditions we obtain morphisms $i : R_{1,I} \rightarrow G_{1,1}$ with $i = n_1$ and $j : L_{2,ST} \rightarrow G_{1,0}$ with $g_1 \circ j = m_2$.

It remains to show the compatibility with the application conditions (see Fig. 7.21). We show that $j \models ac_{2,ST}$. $ac_{2,ST} = toST(ac_2)$, where ac_2 is an ST-application condition. For $ac_2 = $ true, also $ac_{2,ST} = $ true and therefore $j \models ac_{2,ST}$. Suppose $ac_2 = \exists\, (a, ac_2'')$, leading to $ac_{2,ST} = \exists\, ((a^S, id_\varnothing, a^T), toST(ac_2''))$. More-over, $tr_{1,I}$ is an integration rule, i.e., it does not change the source and target compo-nents: $G_{1,1}^S = G_{1,0}^S$ and $G_{1,1}^T = G_{1,0}^T$.

We know that $m_{2,ST} = g_1 \circ j \models ac_{2,S}$, which means that there exists $p : P \to G_{1,1}$ with $p \circ a = g_1 \circ j$, $p \models toST(ac_2'')$, and $p^C = \varnothing$. Then there exists $q : P \to G_{1,0}$ with $q = (p^S, \varnothing, p^T)$, $q \circ a = (p^S \circ a^S, \varnothing, p^T \circ a^T) = j$, and $q \models toST(ac_2'')$ because all objects occurring in $toST(ac_2'')$ have empty correspondence components. This means that $j \models ac_{2,S}$ for this case, and can be shown recursively for composed $ac_{2,S}$.   □

7.5.2 Model Integration as Model Transformation

From a general point of view, we want to analyse which pairs $(G^S, G^T) \in \mathcal{L}(TG^S) \times \mathcal{L}(TG^T)$ can be integrated. Intuitively, these are those which are related by the model transformation $MT : \mathcal{L}(TG^S) \Rightarrow \mathcal{L}(TG^T)$ (see also Theorem 8.4). In fact, model integration sequences can be characterised by unique model transformation sequences.

Theorem 7.41 (Characterisation of model integration sequences). *Each model integration sequence $((G^S, G^T), G_0 \overset{tr_I^*}{\Longrightarrow} G_n, G)$ corresponds uniquely to a model transformation sequence $(G^S, G_0' \overset{tr_F^*}{\Longrightarrow} G_n, G^T)$, where tr_I^* and tr_F^* are based on the same rule sequence tr^*.*   △

Proof. $((G^S, G^T), G_0 \overset{tr_I^*}{\Longrightarrow} G_n, G)$ is a model integration sequence

$\Leftrightarrow_{[def]}$ exists source–target consistent $G_0 \overset{tr_I^*}{\Longrightarrow} G_n$ with $G_0 = (G^S \leftarrow \varnothing \to G^T)$ and $G_n = (G^S \leftarrow G^C \to G^T) = G$

$\Leftrightarrow_{[def]}$ $\varnothing \overset{tr_{ST}^*}{\Longrightarrow} G_0 \overset{tr_I^*}{\Longrightarrow} G_n$ is ST-match consistent with with $G_0 = (G^S \leftarrow \varnothing \to G^T)$ and $G_n = G$

$\Leftrightarrow_{[Theorem\ 7.40]}$ exists $\varnothing \overset{tr^*}{\Longrightarrow} G_n$ with $G_n = (G^S \leftarrow G^C \to G^T)$

$\Leftrightarrow_{[Theorem\ 7.21]}$ exists $\varnothing \overset{tr_S^*}{\Longrightarrow} G_0' \overset{tr_F^*}{\Longrightarrow} G_n$ match consistent with $G_n = (G^S \leftarrow G^C \to G^T)$

$\Leftrightarrow_{[def]}$ exists source consistent $G_0 \overset{tr_F^*}{\Longrightarrow} G_n$ with $G_0 = (G^S \leftarrow \varnothing \to \varnothing)$ and $G_n = (G^S \leftarrow G^C \to G^T)$

$\Leftrightarrow_{[def]}$ $(G^S, G_0 \overset{tr_F^*}{\Longrightarrow} G_n, G^T)$ is a model transformation sequence.   □

Coming back to the example of a model transformation from class diagrams to database models, we describe the relevance and value of the given theorems from the more practical view.

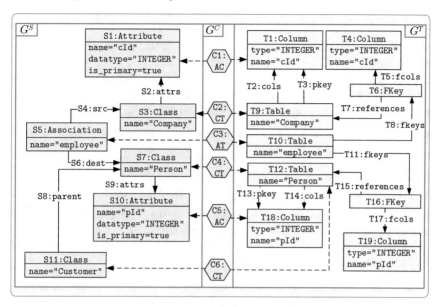

Fig. 7.22 Model integration for *CD2RDBM*: result graph of integration sequence

Integration sequence:
$$G_0 \xrightarrow{\text{Class2table}_I} G_1 \xrightarrow{\text{Class2table}_I} G_2 \xrightarrow{\text{Subclass2Table}_I} G_3 \xrightarrow{\text{PrimaryAttribute2Column}_I}$$
$$G_4 \xrightarrow{\text{PrimaryAttribute2Column}_I} G_5 \xrightarrow{\text{Association2Table}_I} G_6 = G_n.$$

Corresponding forward transformation sequence:
$$G_0 = H_0 \xrightarrow{\text{Class2table}_F} H_1 \xrightarrow{\text{Class2table}_F} H_2 \xrightarrow{\text{Subclass2Table}_F} H_3 \xrightarrow{\text{PrimaryAttribute2Column}_F}$$
$$H_4 \xrightarrow{\text{PrimaryAttribute2Column}_F} H_5 \xrightarrow{\text{Association2Table}_F} H_6 = G_6.$$

Step	Integration Sequence	
	Matched Elements	Created Elements
1	S3, T9	C2
2	S7, T12	C4
3	S7 − S8, S11, C4, T12	C6
4	S1 − S3, C2, T1 − T3, T9	C1
5	S7, S9 − S10, C4, T12 − T14, T18	C5
6	S3 − S7, C2, C4, T1, T3 − T13, T15 − T19	C3

Step	Matched Elements	Created Elements
1	S3	C2, T9
2	S7	C4, T12
3	S7 − S8, S11, C4, T12	C6
4	S1 − S3, C2, T9	C1, T1 − T3
5	S7, S9 − S10, C4, T12	C5, T13 − T14, T18
6	S3 − S7, C2, C4, T1, T3, T9, T12 − T13, T18	C3, T4 − T8, T10 − T11, T15 − T17, T19

Fig. 7.23 *CD2RDBM*: integration sequence (top) and corresponding forward sequence (bottom)

Example 7.42 (Model integration sequence). Fig. 7.22 shows the resulting triple graph G of a model integration sequence $((G^S, G^T), G_0 \xoverset{tr_I^*}{\Longrightarrow} G_n, G)$ via integration rules, where $G_0 = (G^S \leftarrow \varnothing \rightarrow G^T)$. The upper table in Fig. 7.23 shows the corresponding steps with numbers for matched and created elements. According to the numbers for the elements, the correspondence component is completely created during the model integration sequence. The source as well as the target elements of each match are created by the corresponding source–target rule application in $\varnothing \xrightarrow{tr_{ST}^*} G_0$. Therefore, $\varnothing \xrightarrow{tr_{ST}^*} G_0 \xrightarrow{tr_I^*} G_n$ is match consistent. According to Theorem 7.41, there is a corresponding model transformation sequence $(G^S, G_0 \xrightarrow{tr_F^*} G_n, G^T)$ via forward rules with $G_0 = (G^S \leftarrow \varnothing \rightarrow \varnothing)$. Thus, there is a match consistent transformation sequence $\varnothing \xrightarrow{tr_S^*} G_0' \xrightarrow{tr_F^*} G_n$. Numbers for the corresponding matched and created elements are provided in the lower table in Fig. 7.23, where co-matches of the source steps are given by the numbers for the source elements of the matches in the forward transformation sequence.   △

Remark 7.43 (Model integration with translation markers). The execution of model integrations can be performed using translation attributes in an analogous way to that presented for model transformations in the previous section.   △

7.5.3 Consistency Checking of Integrated Models

While model transformation and model integration aim to complete missing structures of triple graphs, consisteny checking is performed to validate that a given triple graph is consistent with respect to a given TGG. In order to perform a consistency check, we use a further kind of operational rule—the consistency creating rules for marking the currently consistent substructures. Technically, consistency creating rules are used to compute maximal subgraphs G_k of a given triple graph G typed over TG, such that $G_k \in \mathcal{L}(TGG)$. In the special case that $G \in \mathcal{L}(TGG)$, we know that $G_k \cong G$. Each consistency creating rule switches labels from **F** to **T** for those elements that would be created by the corresponding TGG rule in TR. This means that elements in the left hand side $L_{CC} = R$ are labelled with **T** if they are also contained in L, and they are labelled with **F** otherwise. Accordingly, all elements in the right hand side R_{CC} are labelled with **T**. We extend Def. 7.29 for forward translation rules to also define backward translation rules and consistency creating rules.

Definition 7.44 (Operational translation rules). Given a triple rule $tr = (L \rightarrow R)$ and its derived source rule $tr_S = (L_S \rightarrow R_S)$, target rule $tr_T = (L_T \rightarrow R_T)$, forward rule $tr_F = (L_F \rightarrow R_F)$, and backward rule $tr_B = (L_B \rightarrow R_B)$, the derived translation rules of tr are given by the *consistency creating rule* $tr_{CC} = (L_{CC} \xleftarrow{l_{CC}} K_{CC} \xrightarrow{r_{CC}} R_{CC})$, the *forward translation rule* $tr_{FT} = (L_{FT} \xleftarrow{l_{FT}} K_{FT} \xrightarrow{r_{FT}} R_{FT})$, and the *backward translation rule* $tr_{BT} = (L_{BT} \xleftarrow{l_{BT}} K_{BT} \xrightarrow{r_{BT}} R_{BT})$ defined in Fig. 7.24 using the notation based on translation attributes.

	main components			new AC for each ac of tr
tr_{CC}	$L_{CC} \xleftarrow{\quad l_{cc} \quad} K_{CC} \xhookrightarrow{\quad r_{cc} \quad} R_{CC}$ $\parallel$ $(R \oplus Att_L^T \oplus Att_{R\backslash L}^F)$	$\parallel$ $(R \oplus Att_L^T)$	$\parallel$ $(R \oplus Att_L^T \oplus Att_{R\backslash L}^T)$	$ac_{CC} = tExt(ac,$ $L_{CC},$ $\{S,C,T\})$
tr_{FT}	$L_{FT} \xleftarrow{\quad l_{FT} \quad} K_{FT} \xhookrightarrow{\quad r_{FT} \quad} R_{FT}$ $\parallel$ $(L_F \oplus Att_{L_S}^T \oplus Att_{R_S\backslash L_S}^F)$	$\parallel$ $(L_F \oplus Att_{L_S}^T)$	$\parallel$ $(R_F \oplus Att_{L_S}^T \oplus Att_{R_S\backslash L_S}^T)$	$ac_{FT} = tExt(ac,$ $L_{FT},$ $\{S\})$
tr_{BT}	$L_{BT} \xleftarrow{\quad l_{BT} \quad} K_{BT} \xhookrightarrow{\quad r_{BT} \quad} R_{BT}$ $\parallel$ $(L_B \oplus Att_{L_T}^T \oplus Att_{R_T\backslash L_T}^F)$	$\parallel$ $(L_B \oplus Att_{L_T}^T)$	$\parallel$ $(R_B \oplus Att_{L_T}^T \oplus Att_{R_T\backslash L_T}^T)$	$ac_{BT} = tExt(ac,$ $L_{BT},$ $\{T\})$

Fig. 7.24 Components of derived operational translation rules

Moreover, the application conditions are given by $ac_{CC} = tExt(ac, L_{CC}, \{S, C, T\})$, $ac_{FT} = tExt(ac, L_{FT}, \{S\})$, and $ac_{BT} = tExt(ac, L_{BT}, \{T\})$ (see Def. 7.28).

By TR_{CC}, TR_{FT}, TR_{BT} we denote the sets of all derived consistency creating, forward translation and backward translation rules, respectively.   △

Remark 7.45 (Construction of operational rules). Note that in Fig. 7.24 ($B +_A C$) is the union of B and C with shared A, as explained in Rem. 7.30. For instance, ($L_{FT} +_L P$) is the union of L_{FT} and P with shared L. Recall that $G \oplus Att_M^T$ denotes the addition of translation attributes for all the elements and attributes included in $M \subseteq G$ to the graph G. All these attributes are set to **T**.   △

As with the completeness of forward translation sequences in Def. 7.33, we define the execution via consistency creating sequences by Def. 7.46 below as an exhaustive application of the rules to the input graph and check whether the output graph contains any element marked with **F**. *Consistency creating sequences* are used for computing a maximal consistent part of a given triple graph. A consistency creating sequence starts at a triple graph $G_0' = Att^F(G)$, i.e., at a triple graph where all elements are marked with **F**. Each application of a consistency creating rule modifies some translation attributes of an intermediate triple graph G_i' from **F** to **T** and preserves the structural part G contained in G_i'. Therefore, the resulting triple graph G_n' extends G with translation attributes only, i.e., some are set to **T** and the remaining ones to **F**.

Definition 7.46 (Consistency creating sequence). Given a triple graph grammar $TGG = (TG, \varnothing, TR)$, let TR_{CC} be the set of all consistency creating rules of TR and let G be a triple graph G typed over TG. A *consistency creating sequence* $s = (G, G_0' \xRightarrow{tr_{CC}^*} G_n', G_n)$ is given by a TGT sequence $G_0' \xRightarrow{tr_{CC}^*} G_n'$ via TR_{CC} with $G_0' = Att^F(G)$ and $G_n' = G \oplus Att_{G_n}^T \oplus Att_{G\backslash G_n}^F$, where G_n is the subgraph of G derived from

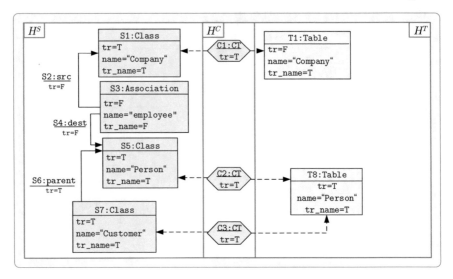

Fig. 7.25 Triple graph at the end of a consistency creating sequence

$G'_0 \xrightarrow{tr^*_{CC}} G'_n$ by restricting G'_n to all **T**-marked elements. The consistency creating sequence s is called *terminated* if there is no rule in TR_{CC} which is applicable to the result graph G'_n. In this case, the triple graph G'_n is called a maximal consistency marking of G. A triple graph G' is called *completely* **T**-*marked* if $G' = Att^{\mathbf{T}}(G)$ for a given triple graph G, i.e., all translation attributes in G' are set to "**T**".   △

Example 7.47 (Consistency creating sequence). Fig. 7.25 shows the resulting triple graph G'_n of a consistency creating sequence $s = (G, G'_0 \xrightarrow{tr^*_{CC}} G'_n, G_n)$ via consistency creating rules with $G'_0 = Att^{\mathbf{F}}(G)$. No further consistency creating rule is applicable. Some translation markers were not modified, such that the consistency creating sequence is not complete. In more detail, $G'_n = G \oplus Att^{\mathbf{T}}_{G_n} \oplus Att^{\mathbf{F}}_{G \setminus G_n}$, where G_n is the subgraph of G derived from $G'_0 \xrightarrow{tr^*_{CC}} G'_n$ by restricting G'_n to all **T**-marked elements. The **F**-marked elements are the `Association` node and its adjacent edges.   △

7.6 Flattening of Triple Graph Grammars

Triple graphs are a direct extension of single plain graphs, i.e., graphs consisting of one graph component instead of three in the case of triple graphs. As shown in the previous sections, this additional structural information provides the basis for an elegant and formal notion of model transformations and model integrations. The natural question arises whether this additional structure can be encoded within plain graphs. And in fact, many implementations [GHL12, SK08, LAVS12, HGN+14] for executing model transformations based on TGGs use plain graphs, where triple components are encoded by an additional pointer structure and mappings between the triple components are encoded as plain edges. This reduces the efforts for implementation, as existing graph transformation engines and development environments can be reused.

This section presents a general flattening construction from TGGs to plain graph grammars that is compatible with the encoding used in many tools. The main concepts and results were first presented in [EEH08c, EEH08d]. Since morphisms between the triple components are encoded by sets of edges, the class of suitable TGGs has to be restricted. Triple graphs may not contain edges in their correspondence components, i.e., correspondence graphs have to be discrete graphs. As the main result (see Theorem 7.57), we show that the encoding for model transformations based on the restricted class of TGGs is correct and complete, i.e., the underlying transformation sequences are in one-to-one correspondence.

Remark 7.48 (General assumption). The results for the flattening construction in this section are presented for TGGs without application conditions and where the triple graphs do not contain edges in the correspondence component. We are quite confident that the results can be extended to systems with application conditions, where we refer to [MEE13, MEE12] for the first general results in this direction based on the general results for $\mathcal{M}$-functors between $\mathcal{M}$-adhesive transformation systems.   △

Triple graphs can be interpreted as plain graphs consisting of three distinguishable subcomponents and edges of special type for interconnection between the components. This idea leads to the general flattening construction for triple graphs and triple graph morphisms. Since interconnections are encoded as plain edges, there is no possibility to encode interconnections between edges of different components. This means that edges in the correspondence component cannot be related to edges in the source and target components. For this reason, the correspondence component of a triple graph is required to contain only nodes to apply the flattening construction. This condition can be generally achieved by requiring that the type graph TG^C for the correspondence component be discrete, i.e., TG^C must not contain edges.

Definition 7.49 (Flattening construction). Given a triple graph $G = (G^S \xleftarrow{s_G} G^C \xrightarrow{t_G} G^T)$, the flattening $\mathcal{F}(G)$ of G is a plain graph defined by the disjoint union $\mathcal{F}(G) = G^S + G^C + G^T + Links_S(G) + Link_T(G)$ with additional edges (links) below:

- $Links_S(G) = \{(x, y) \mid x \in G_V^C, y \in G_V^S, s_G(x) = y\}$,
- $Link_T(G) = \{(x, y) \mid x \in G_V^C, y \in G_V^T, t_G(x) = y\}$,
- $s_{\mathcal{F}(G)}((x, y)) = x, (x, y) \in Links \cup Link_T$,
- $t_{\mathcal{F}(G)}((x, y)) = y, (x, y) \in Links \cup Link_T$.

Given a triple graph morphism $f = (f^S, f^C, f^T) : G \to G'$, the flattening $\mathcal{F}(f) : \mathcal{F}(G) \to \mathcal{F}(G')$ is defined by $\mathcal{F}(f) = f^S + f^C + f^T + f_{LS} + f_{LT}$ with $f_{LS} : Links_S(G) \to$

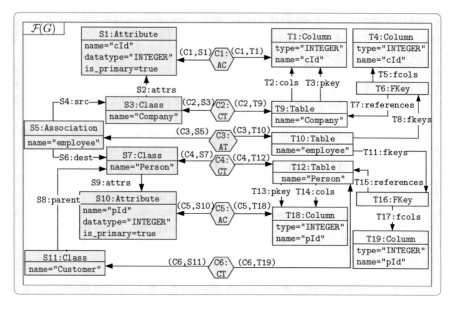

Fig. 7.26 Flattened triple graph $\mathcal{F}(G)$

$Link_S(G')$, $f_{LT} : Link_T(G) \to Link_T(G')$ defined by $f_{LS}((x,y)) = (f^C(x), f^S(y))$ and
$f_{LT}((x,y)) = (f^C(x), f^T(y))$.   △

Remark 7.50. Note that the flattening construction does not specify mappings of
edges in G_E^C. Therefore, we generally assume that $G_E^C = \emptyset$ by requiring that $TG_E^C = \emptyset$ for the type graph TG. Analogously, we require that TG^C does not contain any
attribute to ensure that G^C does not contain any attribute.   △

Example 7.51 (Flattening construction). The graph in Fig. 7.26 shows the plain
graph $\mathcal{F}(G)$ obtained by flattening the triple graph $G = (G^S \xleftarrow{s_G} G^C \xrightarrow{t_G} G^T)$ in
Fig. 7.13. The additional edges in $Link_S(G)$ and $Link_T(G)$ define the mappings s_G
and t_G from the correspondence component to the source and target components.
The flat graph consists of the following components:

- the subgraphs G^S, G^C and G^T,
- the edges in $Link_S(G)$ corresponding to the morphism $G^S \xleftarrow{s_G} G^C$, defined by
  $Link_S(G) = \{(C1, S1), (C2, S3), (C3, S5), (C4, S7), (C5, S10), (C6, S11)\}$ (where the
  numbers refer to the numbered nodes in Fig. 7.26), with $s_{\mathcal{F}(G)}((C1, S1)) =$
  $C1, t_{\mathcal{F}(G)}((C1, S1)) = S1$ (analogously for all other edges in $Link_S(G)$),
- and the edges in $Link_T(G)$ corresponding to the morphism $G^C \xrightarrow{t_G} G^T$, defined by
  $Link_T(G) = \{(C1, T1), (C2, T9), (C3, T10), (C4, T12), (C5, T18), (C6, T19)\}$, with
  $s_{\mathcal{F}(G)}((C1, T1)) = C1, t_{\mathcal{F}(G)}((C1, T1)) = T1$ (analogously for all other edges in
  $Link_T$).   △

The flattening construction induces a functor from the category of typed attributed triple graphs to typed attributed graphs. The functor ensures several important properties. The functor is compatible with typing, and preserves, creates and reflects pushouts and pullbacks.

Fact 7.52 (Properties of the flattening construction).

1. *The flattening construction defines a functor* $\mathcal{F}$: **ATrGraphs** $\rightarrow$ **AGraphs**, *which preserves pushouts.*
2. *Given an attributed triple type graph* $TG = (TG^S \leftarrow TG^C \rightarrow TG^T)$ *with* $TG^C_{E_G} = TG^C_{E_{NA}} = TG^C_{E_{EA}} = \emptyset$ *and flattening* $\mathcal{F}(TG)$, *the typed flattening construction is the functor* $\mathcal{F}_{TG}$: **ATrGraphs**$_{TG}$ $\rightarrow$ **AGraphs**$_{\mathcal{F}(TG)}$ *defined by* $\mathcal{F}_{TG}(G,t) = (\mathcal{F}(G), \mathcal{F}(t))$ *and* $\mathcal{F}_{TG}(f) = \mathcal{F}(f)$. *We sometimes write* $\mathcal{F}_{TG} = \mathcal{F}$ *for short.*
3. *The typed flattening* $\mathcal{F}_{TG}$ *is injective on objects.* $\mathcal{F}_{TG}$ *is injective on morphisms and creates morphisms, i.e., for all* m' : $\mathcal{F}_{TG}(L) \rightarrow \mathcal{F}_{TG}(G)$ *in* **AGraphs**$_{\mathcal{F}(TG)}$ *there is a unique morphism* m : $L \rightarrow G$ *with* $\mathcal{F}_{TG}(m) = m'$. *Especially we have* $\mathcal{F}_{TG}(A) \cong \mathcal{F}_{TG}(B)$ *iff* $A \cong B$.
4. $\mathcal{F}_{TG}$ *preserves and reflects pushouts, i.e.,* (1) *pushout in* **ATrGraphs**$_{TG}$ *iff* (2) *is pushout in* **AGraphs**$_{\mathcal{F}(TG)}$, *and* $\mathcal{F}_{TG}$ *creates pushouts, i.e., given* r : $L \rightarrow R$, m : $L \rightarrow G$ *in* **ATrGraphs**$_{TG}$ *and pushout* (3) *with* H, n', f' *in* **AGraphs**$_{\mathcal{F}(TG)}$, *there are unique* G', n, f *in* **ATrGraphs**$_{TG}$, *s.t.* (1) *is pushout in* **ATrGraphs**$_{TG}$ *with* $\mathcal{F}_{TG}(G') \cong H$, $\mathcal{F}_{TG}(n) = n'$, *and* $\mathcal{F}_{TG}(f) = f'$.

5. $\mathcal{F}_{TG}$ *preserves, reflects and creates pullbacks.* △

Proof. See Appendix B.6.3. □

Remark 7.53 (Flattening functor). The typed flattening construction $\mathcal{F}$: **ATrGraphs**$_{TG}$ $\rightarrow$ **AGraphs**$_{\mathcal{F}(TG)}$ is in general not surjective and hence defines no isomorphism or equivalence of categories **ATrGraphs**$_{TG}$ and **AGraphs**$_{\mathcal{F}(TG)}$. There are graphs $(H, type_H)$ in **AGraphs**$_{\mathcal{F}(TG)}$ which are not functional in the sense that for $TG = (TG^S \leftarrow TG^C \rightarrow TG^T)$ one node in $H^C = type_H^{-1}(TG^C)$ is connected in H with zero or more than one node in $H^S = type_H^{-1}(TG^S)$ or in $H^T = type_H^{-1}(TG^T)$. In this case, we do not obtain graph morphisms s_H : $H^C \rightarrow H^S$ or t_H : $H^C \rightarrow H^T$ and hence no triple graph $(H^S \leftarrow H^C \rightarrow H^T)$. Moreover, in the literature [GK08, KW07], triple graph applications exist where plain graphs are used which have multiple edges connecting the same correspondence node to various elements of the source and target language. This approach does not correspond to pure morphism-based triple graphs and hence is not covered by our translation construction. △

Using Fact 7.52 above, the flattening functor can be extended to translate triple graph grammars.

Definition 7.54 (Translation of triple graph grammars). Given a triple graph grammar $TGG = (TG, SG, TR)$ with triple type graph TG, start graph SG and triple rules $tr : L \to R$ in $\mathbf{TrGraphs}_{TG}$, the translation $\mathcal{F}(TGG)$ of TGG is the graph grammar $\mathcal{F}(TGG) = (\mathcal{F}(TG), \mathcal{F}(SG), \mathcal{F}(TR))$ with type graph $\mathcal{F}(TG)$, start graph $\mathcal{F}(SG)$, and rules $\mathcal{F}(TR) = \{\mathcal{F}(tr) : \mathcal{F}(L) \to \mathcal{F}(R) \,|\, (tr : L \to R) \in TR\}$.   △

Theorem 7.55 (Translation and creation of triple graph transformations). *Given a triple graph grammar $TGG = (TG, SG, TR)$ with translation $\mathcal{F}(TGG) = (\mathcal{F}(TG), \mathcal{F}(SG), \mathcal{F}(TR))$, the following hold.*

1. *Each TGT sequence trafo*: $SG \xRightarrow{tr_1,m_1} G_1 \Rightarrow \ldots \xRightarrow{tr_n,m_n} G_n$ *via TGG can be translated into a flattened graph transformation*
$$\mathcal{F}(trafo): \mathcal{F}(SG) \xRightarrow{\mathcal{F}(tr_1),\mathcal{F}(m_1)} \mathcal{F}(G_1) \Rightarrow \ldots \xRightarrow{\mathcal{F}(tr_n),\mathcal{F}(m_n)} \mathcal{F}(G_n) \text{ via } \mathcal{F}(TGG).$$
2. *Vice versa, each graph transformation sequence*
$$trafo' : \mathcal{F}(SG) \xRightarrow{\mathcal{F}(tr_1),m'_1} G'_1 \Rightarrow \ldots \xRightarrow{\mathcal{F}(tr_n),m'_n} G'_n \text{ via } \mathcal{F}(TGG)$$
creates a unique (up to isomorphism) TGT sequence
$$trafo: SG \xRightarrow{tr_1,m_1} G_1 \Rightarrow \ldots \xRightarrow{tr_n,m_n} G_n \text{ via } TGG$$
with $\mathcal{F}(trafo) = trafo'$, i.e., $\mathcal{F}(m_i) = m'_i$ and $\mathcal{F}(G_i) = G'_i$ for $i = 1 \ldots n$.   △

Proof. Using the general assumption that the TGG has no application conditions (see Rem. 7.48); this result follows from Fact 7.52.   □

Finally, we show that the flattening functor yields a one-to-one correspondence between the model transformation sequences via TGGs and their flattened versions for systems without application conditions (see Rem. 7.48 for this restriction).

Definition 7.56 (Translation of model transformation based on forward rules). Given a triple graph grammar $TGG = (TG, SG, TR)$ with model transformation $MT: \mathcal{L}(TG^S) \Rightarrow \mathcal{L}(TG^T)$, the following hold.

1. $MT = (\mathcal{L}(TG'^S), \mathcal{L}(TG'^T), TG, t_S, t_T, TR_F)$ is a model model transforamtion according to Def. 3.2 with $TG'^S = (TG^S \leftarrow \emptyset \to \emptyset)$, $TG'^T = (\emptyset \leftarrow \emptyset \to TG^T)$, inclusions $t_S = (inc_{TG^S}, \emptyset, \emptyset): (TG^S \leftarrow \emptyset \to \emptyset) \to TG$ and $t_T = (\emptyset, \emptyset, inc_{TG^T}): (\emptyset \leftarrow \emptyset \to TG^T) \to TG$.
2. The translated model transformation $\mathcal{F}(MT)$ is a plain model transformation defined by $\mathcal{F}(MT) = (\mathcal{F}(\mathcal{L}_S), \mathcal{F}(\mathcal{L}_T), \mathcal{F}(TG), \mathcal{F}(t_S), \mathcal{F}(t_T), \mathcal{F}(TR_F))$, where $\mathcal{F} : \mathbf{TrGraphs}_{TG} \to \mathbf{Graphs}_{\mathcal{F}(TG)}$ is the typed flattening functor (see Def. 7.54).
3. Each graph transformation sequence $trafo' : G'_0 \xRightarrow{\mathcal{F}(tr_{1,F}),m'_1} G'_1 \Rightarrow \ldots \xRightarrow{\mathcal{F}(tr_{n,F}),m'_n} G'_n$ satisfies the plain control condition if $G'_0 = \mathcal{F}(G_0)$ and the uniquely created triple graph transformation $trafo : G_0 \xRightarrow{tr_{1,F},m_1} G_1 \Rightarrow \ldots \xRightarrow{tr_{n,F},m_n} G_n$ (by Theorem 7.55) is source consistent.   △

Theorem 7.57 (Properties of translation). *Given a triple graph grammar $TGG = (TG, SG, TR)$ with model transformation $MT: \mathcal{L}(TG^S) \Rightarrow \mathcal{L}(TG^T)$ and the translated plain model transformation $\mathcal{F}(MT)$, the following hold.*

1. *There is a bijective correspondence (up to isomorphism) between model trans-*
   formation sequences of MT and $\mathcal{F}(MT)$, and
2. *$\mathcal{F}(MT)$ being functional is equivalent to MT being functional.*   △

Proof. Given an MT model transformation sequence $(G^S, G_1 \overset{tr_F^*}{\Longrightarrow} G, G^T)$, we ob-
tain by Def. 7.56 the $\mathcal{F}(MT)$ model transformation sequence $(\mathcal{F}(G^S), \mathcal{F}(G_1) \overset{\mathcal{F}(tr_F^*)}{\Longrightarrow}$
$\mathcal{F}(G_n), \mathcal{F}(G^T))$, because $\mathcal{F}(tr_F^*)$ satisfies the plain control condition, $t_S^<(G_S) = G_1$
implies $\mathcal{F}(t_S)^<(\mathcal{F}(G_S)) = \mathcal{F}(G_1)$, and $t_T^<(G_n) = G_T$ implies $\mathcal{F}(t_T)^<(\mathcal{F}(G_T)) =$
$\mathcal{F}(G_n)$ because $\mathcal{F}$ preserves pullbacks by Fact 7.52. Vice versa, each $\mathcal{F}(MT)$ model
transformation sequence creates a unique MT model transformation sequence using
again Def. 7.56 and the fact that $\mathcal{F}$ creates pushouts and pullbacks by Fact 7.52.
Injectivity of $\mathcal{F}$ by Fact 7.52 implies that we have a bijective correspondence (up to
isomorphism) between MT and $\mathcal{F}(MT)$ model transformation sequences. This implies that $\mathcal{F}(MT)$ being functional is equivalent to MT being functional.   □

Chapter 8
Analysis of Model Transformations

Model transformations based on TGGs as presented in Chap. 7 provide an excellent framework for analysing and verifying a major part of the properties that may have to be ensured in an application scenario with regard to the first dimension of challenges for model transformations—the functional dimension—presented in Sect. 3.1. The first two sections of this chapter (Sects. 8.1 and 8.2) present powerful analysis techniques that are based on the introduced model transformation concepts.

1. *Syntactical correctness and completeness:* Syntactical correctness of a transformation method means that if we can transform any source model G^S into a model G^T using the method, then the model G^T is a valid target model and, moreover, the pair (G^S, G^T) is consistent with respect to the specification of the model transformation provided by the triple graph grammar. Completeness, on the other hand, means that for any consistent pair (G^S, G^T) according to the specification our transformation method will be able to build G^T from G^S.
2. *Functional and strong functional behaviour:* Functional behaviour means that for each source model G^S each forward transformation starting with G^S leads to a unique valid target model G^T. Strong functional behaviour means, in addition, that also the forward transformation from G^S to G^T is essentially unique, i.e., unique up to switchings of independent transformation steps.
3. *Information and complete information preservation:* In case of bidirectional model transformations, information preservation means that for each forward transformation from G_S to G_T there is also a backward transformation from G_T to G_S. Complete information preservation means in addition that each backward transformation starting with G_T leads to the same G_S. △

It is the main aim of this chapter to analyse under which conditions the properties defined above can be guaranteed and how these conditions can be checked with suitable tool support. Additional important properties as listed in Sect. 3.1, like semantic correctness, are not considered in this chapter, but the interested reader is referred to [BHE09a, HHK10].

As the first main results, we show in Sect. 8.1 that the presented approaches for model transformations ensure syntactical correctness and completeness (see

Theorem 8.4, Cor. 8.5 and Theorem 8.9). In Sect. 8.2, we show as a second group of main results how functional behaviour of model transformations can be efficiently analysed (see Theorems 8.29 and 8.32) with automated tool support. Therefor, we provide a sufficient condition for termination (see Facts 8.13 and 8.21), which is often satisfied for practical applications, and if not, it can be usually achieved with minor efforts. Moreover, we present how model transformations based on TGGs are analysed with respect to information preservation (see Theorems 8.36 and 8.39) based on the techniques developed earlier. Information preservation is one aspect relevant for the bidirectional characteristics of model transformations, and thus already concerns the nonfunctional dimension of challenges. In Sect. 8.3, we study techniques for reducing nondeterminism. This chapter is based on [Her11, HEGO14, EEE+07, EEH08c, HHK10, GEH11].

Remark 8.1 (General assumption). The formal results in this chapter are presented for TGGs that ensure the composition and decomposition property for forward sequences (Def. 7.20) and for integration sequences (Def. 7.39). Chap. 7 presents sufficient conditions for these properties by Theorems 7.21 and 7.40. These conditions mainly require that the application conditions be compatible application condition schemata with almost injective morphisms and the execution be performed via almost injective matches.   △

Remark 8.2 (Validity of results for equivalent concepts). The formal results in this chapter are presented for model transformations based on forward rules. Using the equivalence results in Chap. 7, we automatically derive corresponding results for model transformations based on forward translation rules (Fact 7.36), flattened TGGs (Theorem 7.57), and model integrations (Theorem 7.41).   △

8.1 Syntactical Correctness and Completeness

The central challenges for model transformations are to ensure syntactical correctness and completeness. As one of the main advantages over other approaches for model transformation, we can generally ensure syntactical correctness and completeness for the presented approaches in Chap. 7 for model transformation (see Theorems 8.4 and 8.7 and Cor. 8.5) and for model integration (see Theorem 8.9). The main results of this section are based on [HEGO14]. Syntactical Correctness of a model transformation based on TGGs states that each successful execution of a model transformation starting with a valid source model G^S yields a target model G^T which exactly corresponds to G^S according to the language of integrated models $\mathcal{L}(TGG)$. Completeness means that all valid source models can be transformed. Moreover, we do not only show that our model transformations are left total with respect to source models, but they are also right total. This means that for each valid target model G^T there is a source model which can be transformed into G^T.

In [EEE+07, EEHP09] we have proven that source consistency ensures (syntactical) correctness and completeness of model transformations based on forward rules with respect to the language $\mathcal{L}(TGG)$ of integrated models. Syntactical correctness means that every model transformation sequence $(G^S, G_0 \overset{tr_F^*}{\Longrightarrow} G_n, G^T)$ via forward rules leads to an integrated model $G_n = (G^S \leftarrow G^C \rightarrow G^T)$ which is contained in $\mathcal{L}(TGG)$. In other words, source consistent forward transformations generate correct model transformations, according to the class of transformations specified by the given TGG. Completeness means that for any integrated model $G = (G^S \leftarrow G^C \rightarrow G^T) \in \mathcal{L}(TGG)$, there is a corresponding model transformation sequence $(G^S, G_0 \overset{tr_F^*}{\Longrightarrow} G, G^T)$. Intuitively, this means that any valid transformation specified by a TGG can be implemented by a source consistent forward transformation.

Note that the model transformation relation $MT_{F,R}$ is in general not a function from $\mathcal{L}(TG^S)$ to $\mathcal{L}(TG^T)$, but we study functional behaviour in Sect. 8.2.1.

Definition 8.3 (Syntactical correctness and completeness). A model transformation $MT : \mathcal{L}(TG^S) \Rightarrow \mathcal{L}(TG^T)$ based on forward rules is

- *syntactically correct* if for each model transformation sequence $(G^S, G_0 \overset{tr_F^*}{\Longrightarrow} G_n, G^T)$ there is $G \in \mathcal{L}(TGG)$ with $G = (G^S \leftarrow G^C \rightarrow G^T)$ implying further that $G^S \in \mathcal{L}(TGG)^S$ and $G^T \in \mathcal{L}(TGG)^T$, and it is
- *complete* if for each $G^S \in \mathcal{L}(TGG)^S$ there is $G = (G^S \leftarrow G^C \rightarrow G^T) \in \mathcal{L}(TGG)$ with a model transformation sequence $(G^S, G_0 \overset{tr_F^*}{\Longrightarrow} G_n, G^T)$ and $G_n = G$. Vice versa, for each $G^T \in \mathcal{L}(TGG)^T$ there is $G = (G^S \leftarrow G^C \rightarrow G^T) \in \mathcal{L}(TGG)$ with a model transformation sequence $(G^S, G_0 \overset{tr_F^*}{\Longrightarrow} G_n, G^T)$ and $G_n = G$.   △

Note that we define syntactical correctness and completeness concerning forward model transformations. If we consider Def. 8.3 for both directions of a bidirectional model transformation, i.e., for the forward and backward directions, we derive a more specific definition. In that case, the conditions for correctness and completeness are both required for all source and target models. The following result (based on [HEGO14, GEH11]) shows that model transformations based on forward rules are syntactically correct and complete.

Theorem 8.4 (Syntactical correctness and completeness). *Each model transformation $MT : \mathcal{L}(TG^S) \Rightarrow \mathcal{L}(TG^T)$ based on forward rules is syntactically correct and complete.*   △

Proof. 1. (Syntactical Correctness)

Given a model transformation sequence $(G^S, G_0 \overset{tr_F^*}{\Longrightarrow} G_n, G^T)$, the source consistency of $G_0 \overset{tr_F^*}{\Longrightarrow} G_n$ implies a match consistent sequence $\emptyset \overset{tr_S^*}{\Longrightarrow} G_0 \overset{tr_F^*}{\Longrightarrow} G_n$. Using the general assumption (see Rem. 8.1) we can apply the composition part of Def. 7.20 and have a corresponding TGT sequence $\emptyset \overset{tr^*}{\Longrightarrow} G_n$. This implies for $G = G_n$ that $G \in \mathcal{L}(TGG)$ with $G = (G^S \leftarrow G^C \rightarrow G^T)$, and hence also $G^S \in \mathcal{L}(TGG)^S$ and $G^T \in \mathcal{L}(TGG)^T$.

2. (Completeness)

   Given $G^S \in \mathcal{L}(TGG)^S$, we have by definition of $\mathcal{L}(TGG)^S$ some $G = (G^S \leftarrow G^C \rightarrow G^T) \in \mathcal{L}(TGG)$. This means that we have a TGT sequence $\varnothing \overset{tr^*}{\Longrightarrow} G$. Using the general assumption (see Rem. 8.1) we can apply the decomposition part of Def. 7.20 and have a match consistent sequence $\varnothing \overset{tr_S^*}{\Longrightarrow} G_0 \overset{tr_F^*}{\Longrightarrow} G$, which defines a model transformation sequence $(G^S, G_0 \overset{tr_F^*}{\Longrightarrow} G, G^T)$ using $G = (G^S \leftarrow G^C \rightarrow G^T)$. Vice versa (concerning $G^T \in \mathcal{L}(TGG)^T$), we use Rem. 7.22. $\square$

Based on the corresponding equivalence result in Chap. 7 (see Fact 7.36), we can directly conclude the following result (see also [Her11]), which shows that model transformations based on forward translation rules are syntactically correct and complete.

Corollary 8.5 (Syntactical correctness and completeness based on translation rules). *Each model transformation MT : $\mathcal{L}(TG^S) \Rightarrow \mathcal{L}(TG^T)$ based on forward translation rules is syntactically correct and complete.* $\triangle$

Proof. This follows direcly from Theorem 8.4 due to Fact 7.36.

Model transformations based on forward rules are source consistent. They define model transformations in the general notion of Def. 3.2 using source consistency as control condition.

Example 8.6 (Model transformation based on forward rules). Consider a triple graph grammar $TGG = (TG, \varnothing, TR)$ with source rules TR_S and forward rules TR_F defining the triple graph languages $\mathcal{L}_S = \mathcal{L}(TGG)^S$ and $\mathcal{L}_T = \mathcal{L}(TGG)^T$. Let TR be typed over $TG = (TG^S \leftarrow TG^C \rightarrow TG^T)$ with $t_S : (TG^S \leftarrow \varnothing \rightarrow \varnothing) \rightarrow TG$ and $t_T : (\varnothing \leftarrow \varnothing \rightarrow TG^T) \rightarrow TG$ being type graph embeddings and GTS $= (TR_F)$ with "source consistency" as control condition, i.e., $G_1 \overset{tr_F^*}{\Longrightarrow} G_n$ satisfies the control condition if it is source consistent. Then, the model transformation $MT : \mathcal{L}(TG^S) \Rightarrow \mathcal{L}(TG^T)$ based on forward rules is given by $MT = (\mathcal{L}(TG^S), \mathcal{L}(TG^T), TG, t_S, t_T, TR_F)$. $\triangle$

We show by Theorem 8.7 below that the general notions of correctness, totality, surjectivity and completeness in Def. 3.2 can be guaranteed for model transformations based on forward rules. This is possible, if the source language $\mathcal{L}_S$ coincides with the source language $\mathcal{L}(TGG)^S$ derived from the TGG and, vice versa, the target language $\mathcal{L}_T$ coincides with the target language $\mathcal{L}(TGG)^T$ derived from the TGG. Therefore, we require that $\mathcal{L}_S = \mathcal{L}(TGG)^S$ and $\mathcal{L}_T = \mathcal{L}(TGG)^T$. This allows us to apply the correctness and completeness results for TGGs (see Theorem 8.4) that ensure completeness (which implies totality and surjectivity by definition) and syntactical correctness concerning $\mathcal{L}_S = \mathcal{L}(TGG)^S$ and $\mathcal{L}_T = \mathcal{L}(TGG)^T$.

Theorem 8.7 (General properties of model transformation based on forward rules). *Let MT : $\mathcal{L}(TG^S) \Rightarrow \mathcal{L}(TG^T)$ be a model transformation in the sense of Def. 3.2 with $MT = (\mathcal{L}(TG^S), \mathcal{L}(TG^T), TG^{ST}, t_S, t_T, TR_F)$. Let forward rules TR_F be*

derived from the triple graph grammar TGG, source language $\mathcal{L}_S = \mathcal{L}(TGG)^S$,
and target language $\mathcal{L}_T = \mathcal{L}(TGG)^T$ *and let the consistency relation* MT_C *be given*
by $MT_C = \{(G^S, G^T) \mid \exists\, G = (G^S \leftarrow G^C \rightarrow G^T) \in \mathcal{L}(TGG)\}$. *Then*

- *each model transformation sequence* $(G^S, G_1 \xrightarrow{tr_F^*} G_n, G^T)$ *in the sense of*
  Def. 7.23 is a model transformation sequence in the sense of Def. 3.2 and vice
  versa.
- *Moreover, MT is syntactically correct, total, surjective and complete in the sense*
  of Def. 3.2. △

Proof. Each model transformation sequence in the sense of Def. 7.23 is also one in
the sense of Def. 3.2, and vice versa, because

- $\left[G_0 = t^{S>}(G^S) \right] \Leftrightarrow \left[G_0 \text{ typed over } (TG^S \leftarrow \varnothing \rightarrow \varnothing) \text{ and } proj_S(G_1) = G^S \right]$ and
- $\left[G^T = t^{T<}(G_n) \right] \Leftrightarrow \left[proj_T(G_n) = G^T \right]$.

MT is syntactically correct in the sense of Def. 3.2, because Theorem 8.4 ensures
that for each source consistent $G_0 \xrightarrow{tr_F^*} G_n$ with $G_0 = t^{S>}(G^S)$ we have $G = (G^S \leftarrow$
$G^C \rightarrow G^T) \in \mathcal{L}(TGG)$ with $G^S \in \mathcal{L}(TGG)^S \subseteq \mathcal{L}(TG^S)$ and $G^T = t^{T<}(G_n) \in$
$\mathcal{L}(TGG)^T \subseteq \mathcal{L}(TG^T)$, and $G_n \in \mathcal{L}(TGG)$ implies $(G^S, G^T) \in MT_C$.

MT is total, because for each $G^S \in \mathcal{L}(TGG)^S$ we have by definition $G \in \mathcal{L}(TGG)$
with $proj_S(G) = G^S$. $G \in \mathcal{L}(TGG)$ implies $\varnothing \xrightarrow{tr^*} G$, and hence, by Theorem 7.21, a match consistent sequence $\varnothing \xrightarrow{tr_S^*} G_0 \xrightarrow{tr_F^*} G$. This implies a model
transformation sequence $(G^S, G_0 \xrightarrow{tr_F^*} G, G^T)$ with $proj_S(G_0) = proj_S(G) = G^S$
and hence $(G^S, G^T) \in MT_R$, which implies that MT is total. Similarly we find, for
each $G^T \in \mathcal{L}(TGG)^T$, a triple graph $G \in \mathcal{L}(TGG)$ with $G^S = proj_S(G)$ such that
$(G^S, G^T) \in MT_R$. This shows that MT is surjective. □

Similarly to forward and backward model transformations based on TGGs, the
derived operation of model integration is syntactically correct and complete.

Definition 8.8 (Syntactical correctness and completeness of model integration).
A model integration $MI : \mathcal{L}(TG^S) \times \mathcal{L}(TG^S) \Rightarrow \mathcal{L}(TG)$ based on integration rules
is

- *syntactically correct if for each model integration sequence* $((G^S, G^T),$
  $G_0 \xrightarrow{tr_I^*} G_n, G)$ *we have a triple graph* $G \in \mathcal{L}(TGG)$ *with* $G = G_n = (G^S \leftarrow$
  $G^C \rightarrow G^T)$, *implying further that* $G^S \in \mathcal{L}(TGG)^S$ *and* $G^T \in \mathcal{L}(TGG)^T$, *and it is*
- *complete if for each* $G = (G^S \leftarrow G^C \rightarrow G^T) \in \mathcal{L}(TGG)$ *there is a model*
  integration sequence $((G^S, G^T), G_0 \xrightarrow{tr_I^*} G_n, G)$ *with* $G_n = G$. △

Theorem 8.9 (Syntactical correctness and completeness of model integration).
Each model integration $MI : \mathcal{L}(TG^S) \times \mathcal{L}(TG^T) \Rightarrow \mathcal{L}(TG)$ *based on integration*
rules is syntactically correct and complete. △

Proof. 1. Syntactical correctness: Given a model integration sequence $((G^S, G^T),$ $G_0 \overset{tr_I^*}{\Longrightarrow} G_n, G)$, the source–target consistency of $G_0 \overset{tr_I^*}{\Longrightarrow} G_n$ implies a match consistent sequence $\emptyset \overset{tr_{ST}^*}{\Longrightarrow} G_0 \overset{tr_I^*}{\Longrightarrow} G_n$. Using the general assumption (see Rem. 8.1) we can apply the composition part of Def. 7.39 and have a corresponding TGT sequence $\emptyset \overset{tr^*}{\Longrightarrow} G_n$. This implies for $G = G_n$ that $G \in \mathcal{L}(TGG)$ with $G = (G^S \leftarrow G^C \rightarrow G^T)$, and hence also $G^S \in \mathcal{L}(TGG)^S$ and $G^T \in \mathcal{L}(TGG)^T$.

2. Completeness: Given $G \in \mathcal{L}(TGG)$, we have a TGT sequence $\emptyset \overset{tr^*}{\Longrightarrow} G$, and using the general assumption (see Rem. 8.1) we can apply the decomposition part of Def. 7.39 and have a match consistent sequence $\emptyset \overset{tr_{ST}^*}{\Longrightarrow} G_0 \overset{tr_I^*}{\Longrightarrow} G$, which defines a model integration sequence $((G^S, G^T), G_0 \overset{tr_I^*}{\Longrightarrow} G_n, G)$ using $G = (G^S \leftarrow G^C \rightarrow G^T)$. $\square$

8.2 Functional Behaviour and Information Preservation

As shown in Sect. 8.1, we can ensure syntactical correctness and completeness for model transformations based on forward rules and equivalently for those based on forward translation rules using Fact 7.36. This section concentrates on the analysis of functional behaviour and information preservation. Several formal results are available concerning termination [EEHP09, GHL10], functional behaviour [HEOG10b, GHL10], and optimisation with respect to the efficiency of their execution [HEGO10, KLKS10, GHL10]. The main results of this section are based on [HEGO14, Her11, EEE+07].

8.2.1 Functional Behaviour and Efficient Execution

At first, we consider the general notion of functional behaviour that can be applied to arbitrary transformation systems, in particular to sets of operational rules of a TGG. Functional behaviour of a transformation system means that a transformation system yields unique results for the same input if the sequences are terminated. Termination of a transformation sequence means that the construction of this sequence ends at a graph to which no further forward translation rule is applicable.

Definition 8.10 (Functional behaviour of a transformation system). A transformation system $TS = (R)$ with transformation rules R has *functional behaviour* if for each two terminated transformation sequences $G \Rightarrow^* H_1$ and $G \Rightarrow^* H_2$ via TS and starting at G the resulting graphs are isomorphic, i.e., $H_1 \cong H_2$. $\triangle$

Functional behaviour of a model transformation means that each model of the source domain-specific language (DSL) $\mathcal{L}_S$ is transformed into a unique model of the target language, where we require $\mathcal{L}_S \subseteq \mathcal{L}(TGG)_S$ in order to ensure correctness and completeness by Theorem 8.4. The source DSL can form any subset of $\mathcal{L}(TGG)_S$ and it can be specified by the type graph TG^S together with additional well-formedness constraints. In many cases, model transformations should ensure the crucial property of functional behaviour. Moreover, in order to ensure efficient executions of model transformations, backtracking should be reduced or eliminated, respectively. Backtracking is necessary due to the possible choice of a suitable forward rule and match used for the translation of a particular source element. Therefore, backtracking is performed if a transformation sequence terminates and is not completed successfully, because some parts of the source model have not been translated. In the case of FT rules, this means that an execution of MT requires backtracking if there are terminating TGT sequences $(Att^F(G^S) \leftarrow \varnothing \rightarrow \varnothing) \xrightarrow{tr^*_{FT}} G'_n$ with $G'^S_n \ne Att^T(G^S)$. As we will show by Theorems 8.29 and 8.32, functional behaviour and elimination of backtracking are closely related topics (see [HEGO14]).

Definition 8.11 (Functional behaviour of model transformations). Given a source DSL $\mathcal{L}_S \subseteq \mathcal{L}(TGG)_S$, a model transformation MT based on forward translation rules has *functional behaviour* if each execution of MT starting at a source model $G^S \in \mathcal{L}_S$ leads to a unique (up to isomorphism) target model $G^T \in \mathcal{L}(TGG)_T$.   △

The standard way to analyse functional behaviour is to check whether the underlying transformation system is confluent, i.e., all diverging derivation paths starting at the same model finally meet again. According to Newman's Lemma [New42], confluence can be shown by proving local confluence and additionally ensuring termination. More precisely, local confluence means that whenever a graph K can be transformed in one step into two graphs P_1 and P_2, these graphs can be transformed into a graph K', as shown in the diagram on the right. Let us start with the analysis of termination.

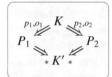

Definition 8.12 (Termination). A system of operational translation rules TR_X with $X \in \{CC, FT, BT\}$ is terminating if each transformation sequence via TR_X is terminating, i.e., the sequence ends at a graph to which no further translation rule (CC, FT, BT) is applicable.   △

For showing termination of a system of forward translation rules according to Def. 8.12, we have the following Fact 8.13, which is a direct extension of Thm. 1 in [HEGO10]. It provides a simple and sufficient condition for termination that can be checked statically.

Fact 8.13 (Termination). *Given a set of operational translation rules TR_X with $X \in \{CC, FT, BT\}$, TR_X is terminating if all input graphs are finite on the graph part (E-graph component) and each rule modifies at least one translation attribute from* **F** *to* **T**.   △

Proof. The input triple graphs are finite on the graph part, and thus contain finitely many translation attributes. Each translation rule modifies at least one translation attribute from **F** to **T**. According to Def. 7.44, none of the translation rules changes a translation attribute from **T** to **F**. Therefore, each transformation sequence stops after finitely many steps.   □

Local confluence can be shown by checking confluence of all *critical pairs* $(P_1 \Leftarrow K \Rightarrow P_2)$ (see Sects. 2.3.5 and 5.2.4), which represent the minimal objects where a confluence conflict may occur. A critical pair describes a minimal conflict, where minimality means that only overlappings of the rule components are considered for the graph K.

While termination of model transformations based on forward rules or forward translation rules can be ensured quite easily by checking that all TGGtriple rules are creating on the source component (see Fact 8.13), the conditions for local confluence are usually more restrictive. In fact, the system of forward translation rules of our case study *CD2RDBM* terminates but is not locally confluent. However, we can show in Ex. 8.34 that the model transformation has functional behaviour. Indeed, functional behaviour of a model transformation does not require general confluence of the underlying system of operational rules. Confluence only needs to be ensured for transformation paths which lead to completely translated models. More precisely, derivation paths leading to a point for backtracking do not influence the functional behaviour. For this reason, we introduce so-called *filter NACs* that extend the model transformation rules in order to avoid misleading paths that cause backtracking, such that the backtracking for the extended system is reduced substantially. By Fact 8.20 we ensure that the overall behaviour of the model transformation with respect to the model transformation relation is still preserved. As the first important result we show in Theorem 8.29 that functional behaviour of a model transformation is ensured by termination and strict confluence of all significant critical pairs of the system of forward translation rules enriched by filter NACs, where significant critical pairs are a subset of all critical pairs. Furthermore, we are able to characterise strong functional behaviour of a terminating model transformation based on forward translation rules with filter NACs in Theorem 8.32 by the condition that there is no significant critical pair at all. Compared with functional behaviour we additionally ensure by strong functional behaviour that the model transformation sequences are unique up to switch equivalence.

The addition of filter NACs therefore has two advantages. On the one hand, the analysis of functional behaviour is improved, because the possible conflicts between the transformation rules are reduced and we will show in this section that filter NACs allow us to verify functional behaviour for our case study *CD2RDBM*. On the other hand, filter NACs improve the efficiency of the execution by cutting off possible backtracking paths. Filter NACs are based on the following notion of misleading graphs, which can be seen as model fragments that are responsible for the backtracking of a model transformation.

Definition 8.14 (Translatable and misleading graphs). A triple graph with translation attributes G is *translatable* if there is a transformation sequence $G \overset{tr_{FT}^*}{\Longrightarrow} H$

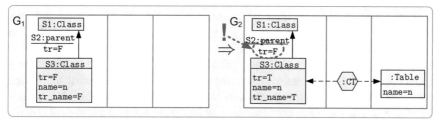

Fig. 8.1 Step $G_1 \xrightarrow{\textit{Class2Table}_{FT}} G_2$ with misleading graph G_2

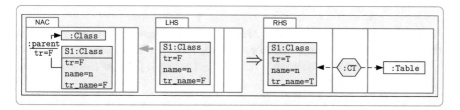

Fig. 8.2 A forward translation rule with filter NAC: $\textit{Class2Table}_{FN}$

via forward translation rules such that H is completely translated (see Def. 7.33). A triple graph with translation attributes G is *misleading* if for every triple graph G' with translation attributes that contains G ($G' \supseteq G$) we have that G' not translatable.   △

Example 8.15 (Misleading graph). Consider the transformation step shown in Fig. 8.1. The resulting graph G_2 is misleading according to Def. 8.14, because the edge S2 is labelled with a translation attribute set to **F**, but there is no rule which may change this attribute in any bigger context at any later stage of the transformation. The only rule which allows one to change the translation attribute of a parent-edge is $\textit{Subclass2Table}_{FT}$, but it requires that the source node S3 be labelled with a translation attribute set to **F**. However, forward translation rules do not modify translation attributes from **T** to **F**, and moreover they do not change the structure of the source component.   △

Definition 8.16 (Filter NAC). A filter NAC n for a forward translation rule tr_{FT} : $L_{FT} \leftarrow K_{FT} \rightarrow R_{FT}$ is given by a morphism $n : L_{FT} \rightarrow N$, such that there is a TGT step $N \xrightarrow{tr_{FT},n} M$ with M being misleading. The extension of tr_{FT} by some set of filter NACs is called forward translation rule tr_{FN} with filter NACs.   △

Example 8.17 (Forward translation rule with filter NACs). The rule $\textit{Class2Table}_{FT}$ is extended by a filter NAC in Fig. 8.2, which is obtained from the graph G_1 of the transformation step $G_1 \xrightarrow{\textit{Class2Table}_{FT}} G_2$ in Fig. 8.1, where G_2 is misleading according to Ex. 8.15.   △

A direct construction of filter NACs according to Def. 8.16 would be inefficient, because the size of the considered graphs to be checked is unbounded. For this reason we use efficient techniques as presented in [HEGO14, HEOG10b], which support the generation of filter NACs and allow us to bound the size without losing generality. At first we present an automated technique for a subset of filter NACs and thereafter an interactive generation technique leading to a much larger set of filter NACs. The first procedure in Fact 8.18 below is based on a sufficient criterion for checking the misleading property. Concerning our example, this automated generation leads to the filter NAC shown in Fig. 8.2 for the rule $Class2Table_{FT}$ for an incoming edge of type "*parent*".

Fact 8.18 (Automated generation of filter NACs). *Given a triple graph grammar, the following procedure applied to each triple rule* $tr \in TR$ *generates filter NACs for the derived forward translation rules* TR_{FT} *leading to forward translation rules* TR_{FN} *with filter NACs:*

- *Outgoing Edges: Check whether the following properties hold*

  - tr *creates a node* $(x : T_x)$ *in the source component and the type graph allows outgoing edges of type "* T_e *" for nodes of type "* T_x *", but* tr *does not create an edge* $(e : T_e)$ *with source node* x.
  - *Each rule in* TR *which creates an edge* $(e : T_e)$ *also creates its source node.*
  - *Extend* L_{FT} *to* N *by adding an outgoing edge* $(e : T_e)$ *at* x *together with a target node. Add a translation attribute for* e *with value* **F**. *The inclusion* $n : L_{FT} \to N$ *is a NAC-consistent match for* tr.

  For each node x *of* tr *fulfilling the above conditions, the filter NAC* $(n : L_{FT} \to N)$ *is generated for* tr_{FT}, *leading to* tr_{FN}.
- *Incoming Edges: Dual case, this time for an incoming edge* $(e : T_e)$.
- TR_{FN} *is the extension of* TR_{FT} *by all filter NACs constructed above.*   △

Proof. Consider a generated NAC $(n : L_{FT} \to N)$ for a node x in tr with an outgoing edge e in $N \setminus L$. A transformation step $N \xmapsto{tr_{FT},n} M$ exists according to Fact 7.32 and leads to a graph M, where the edge e is still labelled with a translation attribute set to "F", but x is labelled with "T", because it is matched by the rule. Now, consider a graph $H' \supseteq M$, such that H' is a graph with translation attributes over a graph without translation attributes H, i.e., $H' = H \oplus Att_{H_0}$ for $H_0 \subseteq H'$, meaning that H' has at most one translation attribute for each element in H without translation attributes.

In order to have M misleading (Def. 8.14), it remains to show that H' is not translatable. Forward translation rules only modify translation attributes from "F" to "T"; they do not increase the number of translation attributes of a graph and no structural element is deleted. Thus, each graph H_i in a TGT sequence $H' \xmapsto{tr_{FT}^*} \overline{H}_n$ will contain the edge e labelled with "F", because the rules, which modify the translation attribute of e, are not applicable due to x being labelled with "T" in each graph $\overline{H}_i$ in the sequence, and there is only one translation attribute for x in H'. Thus, each $\overline{H}_n$ is not completely translated, and therefore M is misleading. This means that $(n : L_{FT} \rightarrow N)$ is a filter NAC of tr_{FT}. Dualising the proof leads to the result for a generated NAC w.r.t. an incoming edge. □

The following interactive technique for deriving filter NACs was presented in [HEGO14, HEOG10b] and is based on the generation of critical pairs, which define conflicts of rule applications in a minimal context. By the completeness of critical pairs (Theorem 5.41) we know that for each pair of two parallel dependent transformation steps there is a critical pair which can be embedded. If a critical pair $P_1 \overset{tr_{1,FT}}{\Longleftarrow} K \overset{tr_{2,FT}}{\Longrightarrow} P_2$ contains a misleading graph P_1, we use the overlapping graph K as a filter NAC of the rule $tr_{1,FT}$. However, checking the misleading property needs manual interaction. But in some cases, these manual results of identified misleading graphs can be reused for more general static conditions. Indeed, the conditions used in Fact 8.18 were inspired by first applying the interactive method to our case study. Moreover, we are currently working on a technique that uses a sufficient criterion to check the misleading property automatically, and we are confident that this approach will provide a powerful generation technique.

Fact 8.19 (Interactive generation of filter NACs). *Given a set of forward translation rules, we generate the set of critical pairs $P_1 \overset{tr_{1,FT},m_1}{\Longleftarrow} K \overset{tr_{2,FT},m_2}{\Longrightarrow} P_2$. If P_1 (or similarly P_2) is misleading, we generate a new filter NAC $m_1 : L_{1,FT} \rightarrow K$ for $tr_{1,FT}$ leading to $tr_{1,FN}$, such that $K \overset{tr_{1,FN},m_1}{\Longrightarrow} P_1$ violates the filter NAC. Hence, the critical pair for $tr_{1,FT}$ and $tr_{2,FT}$ is no longer a critical pair for $tr_{1,FN}$ and $tr_{2,FT}$. But this construction may lead to new critical pairs for the forward translation rules with filter NACs. The procedure is repeated until no further filter NAC can be found or validated. This construction, starting with TR_{FT}, always terminates if the structural part of each graph of a rule is finite.* △

Proof. The constructed NACs are filter NACs, because the transformation step $K \overset{tr_{1,FT},m_1}{\Longrightarrow} P_1$ contains the misleading graph P_1. The procedure terminates, because the critical pairs are bounded by the number of possible pairwise overlappings of the left hand sides of the rules. The number of overlappings can be bounded by considering only constants and variables as possible attribute values. □

Based on the flattening construction presented in Sect. 7.6 we derive an equivalent plain graph transformation system from the system of forward translation rules. Since the system of forward translation rules ensures source consistency for complete transformation sequences by construction, the derived flattened grammar also ensures source consistency for complete transformation sequences. For this reason, we do not need to extend the analysis techniques for critical pairs and can use the critical pair analysis engine of AGG [AGG14].

Concerning our case study *CD2RDBM*, the interactive generation terminates after the second round, which is typical for practical applications, because the number of already translated elements in the new occurring critical pairs usually decreases. Furthermore, several NACs can be combined if they differ only on some translation attributes.

According to Fact 8.20 below, filter NACs do not change the behaviour of model transformations as presented in [HEGO14, Her11]. The only effect is that they filter out derivation paths which would lead to misleading graphs, i.e., to backtracking for the computation of the model transformation sequence. This means that the filter NACs filter out backtracking paths.

Fact 8.20 (Equivalence of transformations with filter NACs). *Given a triple graph grammar $TGG = (TG, \varnothing, TR)$ with forward translation rules TR_{FT} and filter NACs leading to TR_{FN}, let $G_0 = (G^S \leftarrow \varnothing \rightarrow \varnothing)$ be a triple graph typed over TG and $G'_0 = (Att^{\mathbf{F}}(G^S) \leftarrow \varnothing \rightarrow \varnothing)$; then the following statements are equivalent for almost injective matches:*

1. There is a complete TGT sequence $G'_0 \xRightarrow{tr^*_{FT}, m^*_{FT}} G'$ via TR_{FT}.

2. There is a complete TGT sequence $G'_0 \xRightarrow{tr^*_{FN}, m^*_{FT}} G'$ via TR_{FN}. $\quad\triangle$

Proof. Sequence 1 consists of the same transformation diagrams as Sequence 2. NAC consistency of sequence 2 implies NAC consistency of sequence 1, because each step in Sequence 2 involves a superset of the NACs for the corresponding step in Sequence 1. For the inverse direction, consider a step $G_{i-1} \xRightarrow{tr_{(i,FT)}, m_{(i,FT)}} G_i$, which leads to the step $G_{i-1} \xRightarrow{tr_{(i,FN)}, m_{(i,FT)}} G_i$ if NACs are not considered. Assume that m_{FT} does not satisfy some NAC of tr_{FN}. This implies that a filter NAC $(n : L_{i,FT} \rightarrow N)$ is not fulfilled, because all other NACs are fulfilled by NAC consistency of Sequence 1. Thus, there is a triple morphism $q : N \rightarrow G_{i-1}$ with $q \circ n = m_{i,FT}$. By Thm. 6.18 (Restriction Thm.) in [EEPT06] we have that the transformation step $G_{i-1} \xRightarrow{tr_{(i,FN)}, m_{(i,FT)}} G_i$ can be restricted to $N \xRightarrow{tr_{(i,FT)}, n} H$ with embedding $H \rightarrow G_i$. By Def. 8.16 of filter NACs we know that $N \xRightarrow{tr_{(i,FT)}, n} H$ and H is misleading, which implies by Def. 8.14 that G_i is not translatable. This is a contradiction to the completely translated graph G_n in sequence 1, and therefore the filter NAC is fulfilled, leading to NAC consistency of sequence 2. $\quad\square$

The equivalence above implies that we can check termination of a model transformation based on forward translation rules with filter NACs by checking that it is terminating without filter NACs.

Fact 8.21 (Termination with filter NACs). *Given TR_{FN} and TR_{FT} as in Fact 8.20, TR_{FN} is terminating if TR_{FT} is terminating.* $\quad\triangle$

Proof. Since TR_{FT} is terminating, we know that all transformation sequences via TR_{FT} terminate. Since $TR_{FN} \subseteq TR_{FT}$ by construction, we automatically derive that the transformation sequences via TR_{FN} are a subset of the ones via TR_{FT}. Hence, all transformation sequences via TR_{FN} terminate. $\quad\square$

In order to analyse functional behaviour we generate the critical pairs for the system of forward translation rules and show by Theorem 8.29 that strict confluence of "significant" critical pairs ensures functional behaviour. A critical pair is significant if it can be embedded into two transformation sequences via forward translation

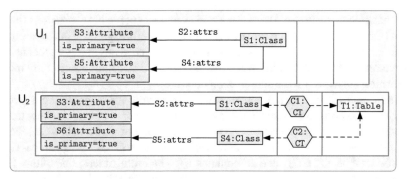

Fig. 8.3 Unreachable patterns U_1 and U_2

rules that start at the same source model G^S, which belongs to the source domain-specific language $\mathcal{L}_S$. This implies that a critical pair containing a misleading graph automatically is not significant. For this reason, some of the nonsignificant critical pairs can be eliminated with the presented automatic and interactive techniques for generating filter NACs in Facts 8.18 and 8.19.

Definition 8.22 (Significant critical pair). A critical pair $(P_1 \xLeftarrow{tr_{1,FN}} K \xRightarrow{tr_{2,FN}} P_2)$ for a set of forward translation rules with filter NACs TR_{FN} is called significant if it can be embedded into a parallel dependent pair $(G'_1 \xLeftarrow{tr_{1,FN}} G' \xRightarrow{tr_{2,FN}} G'_2)$ such that there is $G^S \in \mathcal{L}_S \subseteq \mathcal{L}(TGG)_S$ and $G'_0 \xRightarrow{tr^*_{FN}} G'$ with $G'_0 = (Att^F(G^S) \leftarrow \varnothing \rightarrow \varnothing)$.

$$G'_0 \xRightarrow{tr^*_{FN}} G' \underset{tr_{2,FN}}{\overset{tr_{1,FN}}{\rightleftarrows}} \begin{matrix} G'_1 \\ G'_2 \end{matrix} \qquad \triangle$$

The pragmatic solution for analysing critical pairs would be to start generating critical pairs and inspect overlapping graphs of some pairs. If we detect that an overlapping graph already contains an unreachable pattern, we can conclude that this critical pair is not significant, because it cannot be embedded in a forward translation sequence. Intuitively, an unreachable pattern cannot be reached by applying forward translation rules to a valid initial graph of a forward translation. Note that the notion of unreachable patterns is more restrictive than the notion of misleading graphs.

Definition 8.23 (Unreachable pattern). A graph U is called *unreachable pattern* if there is no forward translation sequence $G'_0 \Rightarrow^* G$ with $G'^S_0 = Att^F(G^S)$ and $G^S \in \mathcal{L}(TGG)^S$ such that there is an $\mathcal{M}$-morphism $u: U \rightarrow G$ (i.e., such that U is contained in G). $\triangle$

Example 8.24 (Unreachable patterns). The graph U_1 in Fig. 8.3 contains a Class node with two primary Attribute nodes. This pattern cannot occur in a valid source model $G^S \in \mathcal{L}(TGG)^S$, because the rule PrimaryAttr2Column is the only rule that creates `primary Attribute` nodes, and its source NAC prohibits the creation of a second one for the same `Class` node. Therefore, graph U_1 is unreachable: it cannot be embedded in a forward translation sequence starting with a consistent source model as required otherwise, because forward translation rules only modify translation attributes on the source component.

The graph U_2 in Fig. 8.3 contains two classes with each containing a primary attribute, and both `Class` nodes are connected to the same table. We show that U_2 is unreachable by contraposition. Assume that U_2 is not unreachable, i.e., there is a transformation sequence via forward translation rules $G' \Rightarrow^* H'$ such that $G'^S = Att^F(G^S)$, $G^S \in \mathcal{L}(TGG)^S$ and H' contains U_2. Therefore, at least one of the `Class` nodes was connected to the table via the rule `Subclass2Table`$_{FT}$. This means as well that both classes are within the same class hierarchy in G^S. Now, we explain why this leads to a contradiction. The condition $G^S \in \mathcal{L}(TGG)^S$ means that the source model G^S belongs to a consistent integrated triple graph G (i.e., $G = (G^S \leftarrow G^C \rightarrow G^T) \in \mathcal{L}(TGG)$). Thus, G is constructed by applying the triple rules of TGG. In order to create the two attribute nodes with `is_primary` = `true`, we have to apply the rule `PrimaryAttr2Column` twice using the same `Table` node, because the classes belong to the same hierarchy. This would mean that each of the two steps would create a primary key for that table. Thus, the second step would not be possible due to the target NAC of the rule. Therefore, the graph G cannot be constructed via the triple rules, and thus $G \notin \mathcal{L}(TGG)$. This is a contradiction, and therefore U_2 is an unreachable pattern.   △

Based on the notion of unreachable graphs, we define negative constraints, which we call filter constraints (Def. 8.25), because they are used to filter out nonsignificant critical pairs. They forbid the presence of unreachable patterns. As we will see for our example, filter constraints are only used for the generation of critical pairs, but can reduce the number of filter NACs required for showing functional behaviour. The reason for this is that the conditions for a graph to be an unreachable pattern and to be misleading partially overlap in a semantic way.

Definition 8.25 (Filter constraint). A *filter constraint* is given by a negative constraint $c = \neg \exists (p: \emptyset \rightarrow U, \text{true})$, where U is an unreachable pattern. We call c a *strict filter constraint* if for all transformation steps $G \xrightarrow{tr_{FT}} H$ via a forward translation rule tr_{FT}, where U is embedded into H via an $\mathcal{M}$-morphism $u: U \rightarrow H$, we can conclude that also G is unreachable.   △

Remark 8.26 (Relationship between filter constraints and filter NACs). Filter constraints are related to filter NACs, but none of them is a special case of the other. First of all, a filter NAC $\neg(\exists n: L \rightarrow N, \text{true})$ is a condition over L for a specific triple rule $tr = (L \rightarrow R)$, while a filter constraint $c = \neg \exists (p: \emptyset \rightarrow U, \text{true})$ is a condition over the initial object $\emptyset$. This means that the filter NAC concerns the applicability of a single rule, while a filter constraint is independent from the triple rules. Secondly, a filter constraint contains an unreachable pattern U (see Defs. 8.23 and 8.25), while a filter NAC contains a graph N that can be transformed into a misleading graph M (see Defs. 8.14 and 8.16). If we consider the system without

$$c_1 = \neg \exists (\varnothing: \varnothing \to U_1, \text{true})$$
$$c_2 = \neg \exists (\varnothing: \varnothing \to U_2, \text{true})$$

Fig. 8.4 Filter constraints based on graphs U_1 and U_2 from Fig. 8.3

filter NACs, we know that the unreachable patterns will not occur in any partial model translation sequence, but it might be that misleading graphs occur. Note that some misleading graphs can be unreachable patterns, e.g., one can combine a misleading graph with an unreachable pattern and obtain a new graph that is then both, unreachable and misleading. Vice versa, each unreachable pattern is misleading by definition, because there in no valid sequence in which it can be embedded.

Practically, a NAC with an unreachable pattern would have no effect as unreachable patterns never occur in valid sequences anyhow. On the other hand, a constraint with a misleading graph could be defined. However, it could be transformed into NACs, which usually improves efficiency as the condition checks are reduced. Thus, unreachable patterns are appropriate for constraints and misleading graphs are a suitable notion for obtaining relevant NACs. Concerning the analysis of critical pairs, a filter constraint generally reduces the number of critical overlapping graphs, while a filer NAC concerns only the overlapping graphs that are relevant for the specific rule.   △

Example 8.27 (Filter constraint). Consider the two constraints in Fig. 8.4. They are filter constraints, because they are based on the graphs U_1 and U_2, which are unreachable patters as shown in Ex. 8.24.   △

Using the concept of filter constraints, we show that they can be used as global constraints when generating critical pairs. They will ensure that all the nonsignificant critical pairs that contain the specified negative pattern will already be filtered out during the generation process.

Lemma 8.28 (Filtering of critical pairs). *Given a filter constraint $c = \neg \exists (p: \varnothing \to U, \text{true})$, all graphs (K, P_1, P_2) of significant critical pairs $(P_1 \Leftarrow K \Rightarrow P_2)$ satisfy c, i.e., $K \models c \wedge P_1 \models c \wedge P_2 \models c$.*   △

Proof. Let $c = \neg \exists (p: \varnothing \to U, \text{true})$ be a filter constraint; then U is unreachable. Let $c' = \exists (p: \varnothing \to U, \text{true})$, i.e., leaving out the negation, and let $(P_1 \Leftarrow K \Rightarrow P_2)$ be a critical pair. Assume that one of the graphs K, P_1 or P_2 does not satisfy c, i.e., $X \not\models c'$ with $X \in \{K, P_1, P_2\}$. This implies that there is an $\mathcal{M}$-morphism $x: U \to X$. By Def. 8.23 (unreachable pattern), we can conclude that there is no forward translation sequence $G'_0 \Rightarrow^* G$ with $G'_0{}^S = Att^{\mathsf{F}}(G^S)$ and $G^S \in \mathcal{L}(TGG)^S$. Therefore, we can conclude by Def. 8.22 that the critical pair is not significant. This is a contradiction, and we can conclude that the assumption is wrong. Therefore, all graphs of the critical pair satisfy c.   □

Using the notion of significant critical pairs, we can provide our first main result of this section on functional behaviour as presented in [HEGO14, HEOG10b]. It states that a model transformation has functional behaviour and does not require backtracking, if the significant critical pairs are strictly confluent.

Theorem 8.29 (Functional behaviour). *Let MT_{FT} be a model transformation based on forward translation rules TR_{FT} with model transformation relation $MT_{FT,R}$ and source DSL $\mathcal{L}_S$. Furthermore, let TR_{FN} extend TR_{FT} with filter NACs such that TR_{FN} is terminating and all significant critical pairs are strictly confluent. Then, MT_{FT} has functional behaviour. Moreover, the model transformation MT_{FN} based on TR_{FN} does not require backtracking and MT_{FN} defines the same model transformation relation, i.e., $MT_{FN,R} = MT_{FT,R}$.* △

Proof. For functional behaviour of the model transformation we have to show that each source model $G^S \in \mathcal{L}_S$ is transformed into a unique (up to isomorphism) completely translated target model G^T, which means that there is a completely translated triple model G' with $G'^T = G^T$, and furthermore $G^T \in \mathcal{L}(TGG)_T$.

For $G^S \in \mathcal{L}_S \subseteq \mathcal{L}(TGG)_S$ we have by definition of $\mathcal{L}(TGG)$ that there is a $G^T \in \mathcal{L}(TGG)_T$ and a TGT sequence $\varnothing \xrightarrow{tr^*} (G^S \leftarrow G^C \to G^T)$ via TR. Using the decomposition theorem with NACs, we obtain a match consistent TGT sequence $\varnothing \xrightarrow{tr_S^*} (G^S \leftarrow \varnothing \to \varnothing) \xrightarrow{tr_F^*} (G^S \leftarrow G^C \to G^T)$ by general assumption (Rem. 8.1), and by Fact 7.36 a complete TGT sequence $G'_0 = (Att^{\mathbf{F}}(G^S) \leftarrow \varnothing \to \varnothing) \xrightarrow{tr_{FT}^*} (Att^{\mathbf{T}}(G^S) \leftarrow G^C \to G^T) = G'$. This means that $(G^S, G'_0 \xrightarrow{tr_{FT}^*} G', G^T)$ is a model transformation sequence based on TR_{FT}. Assume that we also have a complete forward translation sequence $G'_0 = (Att^{\mathbf{F}}(G^S) \leftarrow \varnothing \to \varnothing) \xrightarrow{\overline{tr}_{FT}^*} (Att^{\mathbf{T}}(G^S) \leftarrow \overline{G^C} \to \overline{G^T}) = \overline{G}'$. By Fact 8.20 we also have the complete TGT sequences $G'_0 \xrightarrow{tr_{FN}^*} G'$ and $G'_0 \xrightarrow{\overline{tr}_{FN}^*} \overline{G}'$. Using the precondition that TR_{FN} is terminating and all significant critical pairs are strictly confluent, we show that all diverging transformation sequences can be merged again. Consider the possible transformation sequences starting at G'_0 (which form a graph of transformation steps) and two diverging steps $(G'_{i+1} \overset{p_1,m_1}{\Longleftarrow} G'_i \overset{p_2,m_2}{\Longrightarrow} G''_{i+1})$. If they are parallel independent, we can apply the local Church–Rosser theorem (LCR), Theorem 5.26, and derive the merging steps $(G'_{i+1} \overset{p_2,m'_2}{\Longrightarrow} H \overset{p_1,m'_1}{\Longleftarrow} G''_{i+1})$. If they are parallel dependent diverging steps, we know by completeness of critical pairs (Theorem 5.41) that there is a critical pair, and by Def. 8.22 we know that this pair is significant, because we consider transformations sequences starting at G'_0. This pair is strictly confluent by precondition. Therefore, these steps can be merged again. Now, any new diverging situation can be merged by either LCR for parallel independent steps or by strict confluence of critical pairs for parallel dependent steps. By precondition the system is terminating. In combination, this implies that $G' \cong \overline{G}'$, and hence $G^T \cong \overline{G}^T$.

Backtracking is not required, because the termination of TR_{FN} with strict confluence of significant critical pairs implies unique normal forms as shown above.

Therefore, any terminating TGT sequence $(Att^F(G^S) \leftarrow \varnothing \rightarrow \varnothing) \overset{tr^*_{FN}}{\Longrightarrow} G'_n$ leads to a unique G'_n up to isomorphism, and by correctness and completeness (Theorem 8.4 and Fact 7.36) we have that $G'^S_n = Att^T(G^S)$.

The model transformation relation is the same, because we have the equivalence of the model transformation sequences (Fact 8.20). □

If the set of generated critical pairs of a system of forward translation rules with filter NACs TR_{FN} is empty, we can directly conclude from Theorem 8.29 that the corresponding system TR_{FT} without filter NACs has functional behaviour. Moreover, from an efficiency point of view, the set of rules should be compact in order to minimise the effort for pattern matching. In the optimal case, the rule set ensures that each transformation sequence of the model transformation is itself unique up to switch equivalence, meaning that it is unique up to the order of sequentially independent steps. For this reason, we introduce the notion of strong functional behaviour with respect to a given source domain-specific language $\mathcal{L}_S$.

Definition 8.30 (Strong functional behaviour of model transformations). A model transformation based on forward translation rules TR_{FN} with filter NACs and the source DSL $\mathcal{L}_S \subseteq \mathcal{L}(TGG)_S$ has *strong functional behaviour* if for each $G^S \in \mathcal{L}_S$ there is a $G^T \in \mathcal{L}(TGG)_T$ and a model transformation sequence $(G^S, G'_0 \overset{tr^*_{FN}}{\Longrightarrow} G'_n, G^T)$ based on forward translation rules, and moreover,

- any partial TGT sequence $G'_0 \overset{tr^{i,*}_{FN}}{\Longrightarrow} G'_i$ terminates, i.e., there are finitely many extended sequences $G'_0 \overset{tr^{i,*}_{FN}}{\Longrightarrow} G'_i \overset{tr^{j,*}_{FN}}{\Longrightarrow} G'_j$, and
- each two TGT sequences $G'_0 \overset{tr^*_{FN}}{\Longrightarrow} G'_n$ and $G'_0 \overset{\overline{tr}_{FN}}{\Longrightarrow} \overline{G}'_m$ with completely translated graphs G'_n and $\overline{G}'_m$ are switch-equivalent up to isomorphism. △

Remark 8.31 (Strong functional behaviour).

1. The sequences being terminating means that no rule in TR_{FN} is applicable anymore. However, it is not required that the sequences be complete, i.e., that G'_n and $\overline{G}'_m$ are completely translated.
2. Strong functional behaviour implies functional behaviour, because G'_n and $\overline{G}'_m$ completely translated implies that $G'_0 \overset{tr^*_{FN}}{\Longrightarrow} G'_n$ and $G'_0 \overset{\overline{tr}^*_{FN}}{\Longrightarrow} \overline{G}'_m$ are terminating TGT sequences. △

The second main result of this section shows that strong functional behaviour of model transformations based on forward translation rules with filter NACs can be completely characterised by the absence of significant critical pairs, as presented in [HEGO14, HEOG10b].

Theorem 8.32 (Strong functional behaviour). *A model transformation based on terminating forward translation rules TR_{FN} with filter NACs has strong functional behaviour and does not require backtracking, leading to polynomial time complexity if and only if TR_{FN} has no significant critical pair.* △

Proof. **Direction "⇐":** Assume that TR_{FN} has no significant critical pair. As in the proof of Theorem 8.29 we obtain for each $G^S \in \mathcal{L}_S$ a $G^T \in \mathcal{L}(TGG)_T$ and a complete TGT sequence $G'_0 \overset{tr^*_{FT}}{\Longrightarrow} G'$ and a model transformation $(G^S, G'_0 \overset{tr^*_{FT}}{\Longrightarrow} G', G^T)$ based on TR_{FT} underlying TR_{FN}. By Fact 8.20 we also have a complete TGT sequence $G'_0 \overset{tr^*_{FN}}{\Longrightarrow} G'$, and hence also a model transformation $(G^S, G'_0 \overset{tr^*_{FT}}{\Longrightarrow} G', G^T)$ based on TR_{FT} underlying TR_{FN}. In order to show strong functional behaviour let $G'_0 \overset{tr^*_{FN}}{\Longrightarrow} G'_n$ and $G'_0 \overset{\overline{tr}^*_{FN}}{\Longrightarrow} \overline{G}'_m$ be two terminating TGT sequences with $m, n \geq 1$. We have to show that they are switch-equivalent up to isomorphism. We show by induction on the combined length $n + m$ that both sequences can be extended to switch-equivalent sequences.

For $n + m = 2$ we have $n = m = 1$ with $t1 : G'_0 \overset{tr_{FN}, m}{\Longrightarrow} G'_1$ and $\overline{t1} : G'_0 \overset{\overline{tr}_{FN}, \overline{m}}{\Longrightarrow} \overline{G}'_1$. If $tr_{FN} = \overline{tr}_{FN}$ and $m = \overline{m}$, then both are isomorphic with isomorphism $i : \overline{G}'_1 \overset{\sim}{\to} G'_1$, such that $t1 \approx i \circ \overline{t1}$. If not, then $t1$ and $\overline{t1}$ are parallel independent, because otherwise we would have a significant critical pair by completeness of critical pairs in Theorem 5.41. By the local Church–Rosser theorem, Theorem 5.26, we have $t2 : G'_1 \overset{\overline{tr}_{FN}}{\Longrightarrow} G'_2$ and $\overline{t2} : \overline{G}'_1 \overset{tr_{FN}}{\Longrightarrow} G'_2$, such that $t2 \circ t1 \approx \overline{t2} \circ \overline{t1} : G'_0 \Rightarrow^* G'_2$.

Now assume that for $t1 : G'_0 \Rightarrow^* G'_{n-1}$ and $\overline{t1} : G'_0 \Rightarrow^* \overline{G}'_m$ we have extensions $t2 : G'_{n-1} \Rightarrow^* H, \overline{t2} : \overline{G}'_m \Rightarrow^* H$, such that $t2 \circ t1 \approx \overline{t2} \circ \overline{t1}$.

$$
\begin{array}{ccccc}
G'_0 & \overset{t1}{\Longrightarrow}{}^* & G'_{n-1} & \overset{t}{\Longrightarrow} & G'_n \\
{\scriptstyle \overline{t1}} \big\Downarrow & & \big\Downarrow {\scriptstyle t2} & & \big\Downarrow {\scriptstyle t3} \\
\overline{G}'_m & \underset{\overline{t2}}{\Longrightarrow}{}^* & H & \underset{t3}{\Longrightarrow}{}^* & K
\end{array}
$$

For a step $t : G'_{n-1} \Rightarrow G'_n$ we have to show that $t \circ t1$ and $\overline{t1}$ can be extended to switch-equivalent sequences. By induction hypothesis and definition of significant critical pairs also t and $t2$ can be extended by $t3 : G'_n \Rightarrow^* K, \overline{t3} : H \Rightarrow^* K$, such that $t3 \circ t \approx \overline{t3} \circ t2$. Now, the composition closure of switch equivalence implies $t3 \circ t \circ t1 \approx \overline{t3} \circ \overline{t2} \circ \overline{t1} : G'_0 \Rightarrow^* K$. This completes the induction proof.

Now, we use that G'_n and $\overline{G}'_m$ are both terminal, which implies that $t3$ and $\overline{t3} \circ \overline{t2}$ must be isomorphisms. This shows that $G'_0 \overset{tr^*_{FN}}{\Longrightarrow} G'_n$ and $G'_0 \overset{\overline{tr}^*_{FN}}{\Longrightarrow} \overline{G}'_m$ are switch-equivalent up to isomorphism.

Direction "⇒": Assume now that TR_{FN} has strong functional behaviour and that TR_{FN} has a significant critical pair. We have to show a contradiction in this case.

Let $P_1 \overset{tr_{1,FN}}{\Longleftarrow} K \overset{tr_{2,FN}}{\Longrightarrow} P_2$ be the significant critical pair which can be embedded into a parallel dependent pair $G_1 \overset{tr_{1,FN}}{\Longleftarrow} G' \overset{tr_{2,FN}}{\Longrightarrow} G_2$, such that there is $G^S \in \mathcal{L}_S$ with $G'_0 \overset{tr^*_{FN}}{\Longrightarrow} G'$ and $G'_0 = (Att^\mathbf{F}(G^S) \leftarrow \varnothing \to \varnothing)$. Since TR_{FN} is terminating we have terminating sequences $G_1 \Rightarrow^* G_{1n}$ and $G_2 \Rightarrow^* G_{2m}$ via TR_{FN}. By composition we have the following terminating TGT sequences:

1. $G'_0 \overset{tr_{FN}}{\Longrightarrow} G' \overset{tr_{1,FN}}{\Longrightarrow} G_1 \Rightarrow^* G_{1n}$ and

2. $G'_0 \overset{tr_{FN}}{\Longrightarrow} G' \overset{tr_{2,FN}}{\Longrightarrow} G_2 \Rightarrow^* G_{2m}$.

Since TR_{FN} has strong functional behaviour both are switch-equivalent up to isomorphism. For simplicity assume $G_{1n} = G_{2m}$ instead of $G_{1n} \cong G_{2m}$. This implies $n = m$ and that $G' \xrightarrow{tr_{1,FN}} G_1 \Rightarrow^* G_{1n}$ is switch-equivalent to $G' \xrightarrow{tr_{2,FN}} G_2 \Rightarrow^* G_{1n}$. This means that $tr_{2,FN}$ occurs in $G_1 \Rightarrow^* G_{1n}$ and can be shifted in $G' \xrightarrow{tr_{1,FN}} G_1 \Rightarrow^*$ G_{1n}, such that we obtain $G' \xrightarrow{tr_{2,FN}} G_2 \Rightarrow^* G_{1n}$.

But this implies that in an intermediate step we can apply the parallel rule $tr_{1,FN}$ + $tr_{2,FN}$, leading to parallel independence of $G' \xrightarrow{tr_{1,FN}} G_1$ and $G' \xrightarrow{tr_{2,FN}} G_2$, which is a contradiction. Hence, TR_{FN} has no significant critical pair.

It remains to show that strong functional behaviour implies that backtracking is not required. This is a direct consequence of Theorem 8.29, since we do not have any significant critical pair, and therefore all of them are strictly confluent. □

Remark 8.33 (Analysis with AGG via flattening). We use the tool AGG to analyse critical pairs and dependencies of plain graph transformation rules. In order to analyse the operational rules of a TGG, we therefore apply the flattening construction (see Def. 7.49) and derive the plain graph transformation rules. By the equivalence result for the flattening construction (see Theorem 7.55), we know that there is a one-to-one correspondence between transformation sequences in each of the systems. Therefore, functional behaviour of one of the systems implies functional behaviour of the other. This means that the analysis in AGG based on the flattened rules is sound and complete with respect to the system of triple rules. △

Example 8.34 (Functional and strong functional behaviour). We analyse functional behaviour of the model transformation *CD2RDBM*. By Fact 8.21, *CD2RDBM* is terminating, because all TGGrules are creating in the source component. For analysing local confluence we use the tool AGG [AGG14] (version 2.07, see Rem. 8.33) for the generation of critical pairs. The set of derived forward translation rules from the rules *TR* in Fig. 3.8 is given by TR_{FT} = { Class2Table$_{FT}$, Subclass2Table$_{FT}$, Attr2Column$_{FT}$, PrimaryAttr2Column$_{FT}$, Association2Table$_{FT}$ }. We perform the following steps.

1. We obtain the initial table of critical pairs as shown in Fig. 8.5. In order to prevent a memory overflow, we set a limit for the the maximum number of generated critical pairs per rule pair and conflict kind to 200. This limit becomes effective for the pair $(5, 5)$, which shows that there are 200 or more critical pairs for this rule pair. In the next steps, we apply reduction techniques and do not reach this limit anymore.
2. Now, we use the concept of filter constraints to filer out nonsignificant critical pairs by setting the multiplicity (maximum values only) constraints depicted in Fig. 11.9. We can do this, because the multiplicity constraints are ensured by the triple rules as well as by the source rules, the forward rules and the forward translation rules. Formally, the multiplicity constraints correspond to filter constraints, which contain unreachable patterns that violate the maximum multiplicity constraints. As a result of this step, we obtain the table in Fig. 8.6.

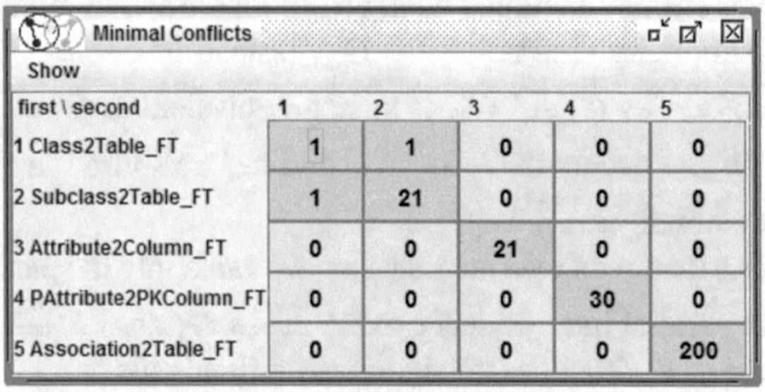
Minimal Conflicts
Show

first \ second	1	2	3	4	5
1 Class2Table_FT	1	1	0	0	0
2 Subclass2Table_FT	1	21	0	0	0
3 Attribute2Column_FT	0	0	21	0	0
4 PAttribute2PKColumn_FT	0	0	0	30	0
5 Association2Table_FT	0	0	0	0	200

Fig. 8.5 Table of critical pairs—initial table for the TGG *Class2Table*

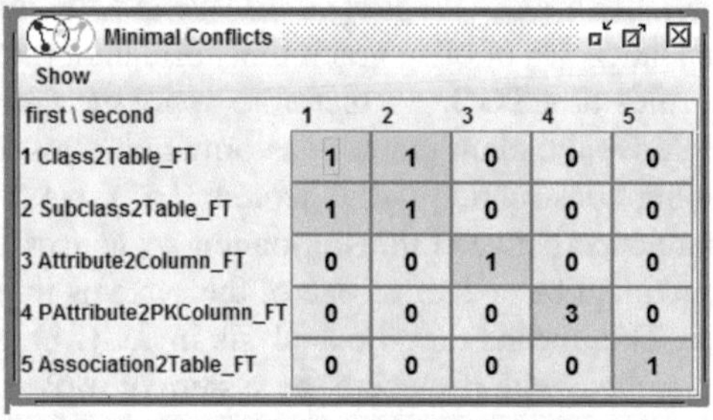
Minimal Conflicts
Show

first \ second	1	2	3	4	5
1 Class2Table_FT	1	1	0	0	0
2 Subclass2Table_FT	1	1	0	0	0
3 Attribute2Column_FT	0	0	1	0	0
4 PAttribute2PKColumn_FT	0	0	0	3	0
5 Association2Table_FT	0	0	0	0	1

Fig. 8.6 Table of critical pairs after setting the multiplicity constraints

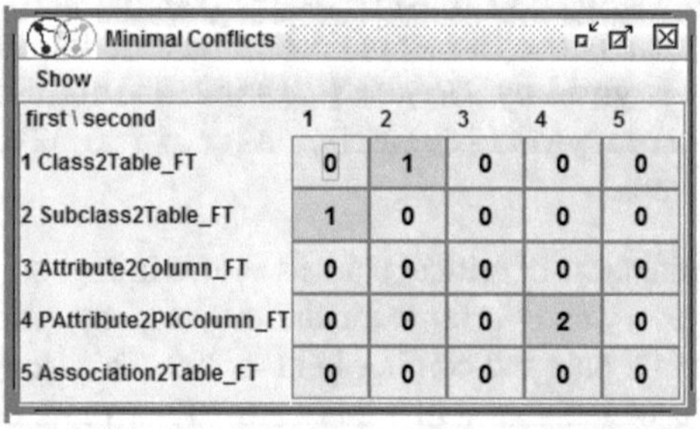
Minimal Conflicts
Show

first \ second	1	2	3	4	5
1 Class2Table_FT	0	1	0	0	0
2 Subclass2Table_FT	1	0	0	0	0
3 Attribute2Column_FT	0	0	0	0	0
4 PAttribute2PKColumn_FT	0	0	0	2	0
5 Association2Table_FT	0	0	0	0	0

Fig. 8.7 Table of critical pairs without pairs of identical rules and matches

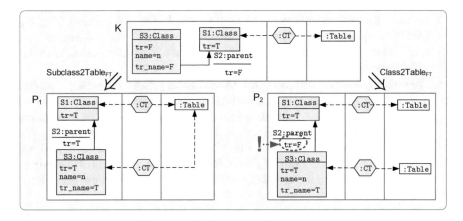

Fig. 8.8 Critical pair for the rules *Subclass2Table$_{FT}$* and *Class2Table$_{FT}$*

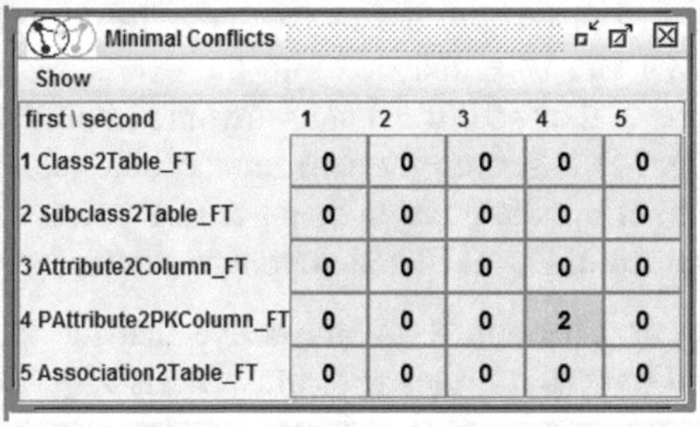

first \ second	1	2	3	4	5
1 Class2Table_FT	0	0	0	0	0
2 Subclass2Table_FT	0	0	0	0	0
3 Attribute2Column_FT	0	0	0	0	0
4 PAttribute2PKColumn_FT	0	0	0	2	0
5 Association2Table_FT	0	0	0	0	0

Minimal Conflicts — Show

Fig. 8.9 Table of critical pairs—after inserting the filter NAC for rule 2

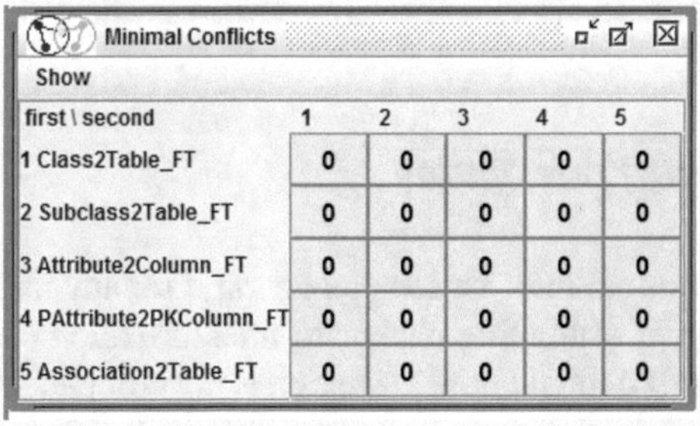

first \ second	1	2	3	4	5
1 Class2Table_FT	0	0	0	0	0
2 Subclass2Table_FT	0	0	0	0	0
3 Attribute2Column_FT	0	0	0	0	0
4 PAttribute2PKColumn_FT	0	0	0	0	0
5 Association2Table_FT	0	0	0	0	0

Minimal Conflicts — Show

Fig. 8.10 Table of critical pairs after inserting the filter constraints

3. Some of the critical pairs are pairs with identical rules and matches (diagonal line). These pairs are directly strictly confluent, because $P_1 \cong P_2$. We activate the corresponding AGG CPA option to omit the pairs of identical rules and matches and derive the table in Fig. 8.7.

4. The critical pair of the rule pair $(\texttt{SubClass2Table}_{FT}, \texttt{Class2Table}_{FT})$ is shown in Fig. 8.8. This critical pair describes the conflict that the rule $\texttt{Class2Table}_{FT}$ is about to translate a $\texttt{Class}$ node that has a $\texttt{parent}$ node. This conflict can by solved by the filter NAC discussed in Ex. 8.17 and shown in Fig. 8.2. We exchange the forward translation rule $\texttt{Class2Table}_{FT}$ with the extended rule with filter NACs $\texttt{Class2Table}_{FN}$ from Fig. 8.2, which we would also obtain by the automated generation according to Fact 8.18. This leads to the table shown in Fig. 8.9.

5. The remaining two critical pairs contain unreachable patterns in the overlapping graphs. They specify conflicts of the rule $\texttt{PrimaryAttr2Column}$ with itself. The corresponding overlapping graphs K of the critical pairs contain two primary attribute nodes, which belong in one case to one $\texttt{Class}$ and in the other case to two $\texttt{Classes}$ that are connected to the same $\texttt{Table}$. The two unreachable patterns in Fig. 8.3 can be embedded into the overlapping graphs. All of the overlapping graphs of the two critical pairs are unreachable. We use the two filter constraints in Fig. 8.4 based on the two unreachable patterns observed in the previous step to filter out the nonsignificant critical pairs. The resulting table of critical pairs is shown in Fig. 8.10 and no longer contains any critical pair.

Thus, we can apply Theorem 8.32 and derive that the model transformation based on the forward translation rules with filter NACs TR_{FN} has *strong functional behaviour* and does not require backtracking. Furthermore, by Theorem 8.29 we can conclude that the model transformation based on the forward translation rules TR_{FT} without filter NACs has *functional behaviour*. As an example, Fig. 7.18 shows the resulting triple graph of a model transformation starting with the class diagram G^S.   △

8.2.2 Information Preservation

Model transformations are information preserving if for any forward transformation sequence there is a corresponding backward transformation sequence yielding the initial source model. If a model transformation is not complete, this directly implies that it is not information preserving. This has a practical impact. In fact, several TGG tools do not support backtracking, such that they cannot ensure completeness. This implies that the execution of backward transformations may stop without creating a valid source model for some target models [GHL10, SK08, KLKS10]. This section provides results for analysing and ensuring information preservation for TGG model transformations according to Chap. 7 in general. In addition to that, we provide results for the stricter notion of complete information preservation. These results hold even if tools do not perform backtracking.

In this section, we analyse whether and how a source model can be reconstructed from the computed target model as presented in [HEGO14, EEE⁺07, Her11]. For this purpose, we distinguish between forward and backward model transformations. Interestingly, it turns out that complete information preservation is ensured by functional behaviour of the backward model transformation. We present the techniques for model transformations based on forward rules. According to the equivalence result in Fact 7.36, we also know that these techniques provide the same results for model transformations based on forward translation rules. Moreover, due to the symmetric definition of TGGs, the results can be applied dually for backward model transformations.

Definition 8.35 (Information preserving model transformation). A forward model transformation based on forward rules is *information preserving*, if for each forward model transformation sequence $(G^S, G_0 \overset{tr_F^*}{\Longrightarrow} G_n, G^T)$ there is a backward model transformation sequence $(G^T, G'_0 \overset{tr_B'^*}{\Longrightarrow} G'_m, G'^S)$ with $G^S = G'^S$, i.e., the source model G^S can be reconstructed from the resulting target model G^T via a target consistent backward transformation sequence.   △

By Theorem 8.36 we show that model transformations based on forward rules are information preserving as presented in [EEE⁺07, HEGO14].

Theorem 8.36 (Information preserving model transformation). *Each forward model transformation based on forward rules is information preserving.*   △

Proof. Given a set of triple rules TR with derived forward rules TR_F and backward rules TR_B. By Theorem 7.21 and Rem. 7.22 applied to the source consistent forward sequence $G_0 \overset{tr_F^*}{\Longrightarrow} G_n$ via TR_F we derive the target consistent backward transformation $G'_0 = (G^T \leftarrow \emptyset \rightarrow \emptyset) \overset{tr_B^*}{\Longrightarrow} G_n$ via TR_B with $G_n^S = G^S$. This means that we have a backward model transformation sequence $(G^T, G'_0 \overset{tr_B^*}{\Longrightarrow} G_n, G'^S)$ with $G^S = G'^S$.   □

Example 8.37 (Information preserving model transformation CD2RDBM). The model transformation *CD2RDBM* is information preserving, because it consists of model transformation sequences based on forward rules, which ensure source consistency of the forward sequences by definition. Therefore, the presented source model G^S of the triple graph in Fig. 7.18 can be reconstructed by a target consistent backward transformation sequence starting at the model $G'_0 = (\emptyset \leftarrow \emptyset \rightarrow G^T)$. But there are several possible target consistent backward transformation sequences starting at G'_0. The reason is that the rule *Subclass2Table$_B$* can be applied arbitrarily often without having an influence concerning the target consistency, because the rule is identical on the target component. This means that the inheritance information within a class diagram has no explicit counterpart within a relational database model.

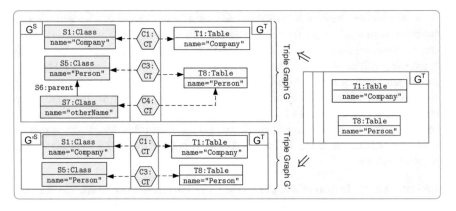

Fig. 8.11 Two possible target consistent backward transformations

There are many possible target consistent backward transformation sequences for the same derived target model G^T where two of them are presented in Fig. 8.11. The source model G^S can be transformed into $G = (G^S \leftarrow G^C \rightarrow G^T)$. But starting with G^T, both depicted backward transformation sequences are possible and target consistent. The resulting source graphs G^S and G'^S, however, differ with respect to the class node $S7$ and the edge $S6$ in G^S. Hence, some information of G^S cannot be reconstructed uniquely and therefore, is partially lost in the target model G^T.   △

According to Theorem 8.36 each model transformation based on forward rules is information preserving. But the reconstruction of a corresponding source model from a derived target model is in general not unique. In order to ensure uniqueness of the reconstruction we now present the notion of complete information preservation. This stronger notion ensures that all information contained in a source model of a source domain-specific language (DSL) can be reconstructed from the derived target model itself. More precisely, starting with the target model, each backward model transformation sequence will produce the original source model. This ensures that only one backward model transformation sequence has to be constructed. Intuitively, this means that the model transformation is invertible.

Definition 8.38 (Complete information preservation). A forward model transformation with source DSL $\mathcal{L}_S$ is *completely information preserving* if it is information preserving, and furthermore, given a source model $G^S \in \mathcal{L}_S$ and the resulting target model G^T of a forward model transformation sequence, each partial backward transformation sequence starting with G^T terminates and produces the given source model G^S as result.   △

We can verify complete information preservation by showing functional behaviour of the corresponding backward model transformation with respect to the derived target models $\mathcal{L}'_T \subseteq MT(\mathcal{L}_S) \subseteq \mathcal{L}(TGG)_T$ as presented in [Her11, HEGO14].

Theorem 8.39 (Completely information preserving model transformation).
Given a forward model transformation MT, it is completely information preserv-
ing if the corresponding backward model transformation according to Rem. 7.22
has functional behaviour with respect to the target language $\mathcal{L}'_T = MT(\mathcal{L}_S)$. △

Proof. By Theorem 8.36 we know that *MT* is information preserving. For a model
transformation sequence $(G^S, G_0 \xRightarrow{tr_F^*} G_n, G^T)$, we additionally know that $G^T \in$
$\mathcal{L}(TGG)_T$ by Theorem 8.4, and furthermore, that $G^T \in \mathcal{L}'_T = MT(\mathcal{L}_S)$. Using
the functional behaviour of the corresponding backward model transformation ac-
cording to Def. 8.11 for the language $\mathcal{L}'_T$ we know that for each model H^T the
backward model transformation yields a unique $H^S \in \mathcal{L}(TGG)_S$. Therefore, each
backward model transformation sequence $(G^T, G'_0 \xRightarrow{tr_B^*} G'_n, G'^S)$ leads to a unique
$G'^S \in \mathcal{L}(TGG)_S$. Furthermore, there is a backward model transformation sequence
$(G^T, G''_0 \xRightarrow{tr_B^*} G''_n, G^S)$ by Theorem 8.36, implying $G^S \cong G'^S$, i.e., the model transformation is completely information preserving.   □

Example 8.40 (Complete information preservation). The model transformation
$MT_1 = CD2RDBM$ is not completely information preserving. Consider, e.g., the
source model G^S in Fig. 8.11 of Ex. 8.37, where two backward model transformation sequences are possible starting with the same derived target model G^T. This
means that the backward model transformation has no functional behaviour with
respect to $MT_1(\mathcal{L}_S) = MT(\mathcal{L}(TGG)_S) = \mathcal{L}(TGG)_T = \mathcal{L}_T$.

However, we can also consider the inverse model transformation, i.e., swap-
ping the forward and backward direction, leading to the model transformation
$MT_2 = RDBM2CD$ from relational database models to class diagrams. In this case,
the model transformation is completely information preserving, meaning that each
relational database model M_{DB} can be transformed into a class diagram M_{CD}, and
each database model M_{DB} can be completely and uniquely reconstructed from its derived class diagram M_{CD}. In other words, each class diagram resulting from a model
transformation sequence of *RDBM2CD* contains all information that was present in
the given database model. According to Ex. 8.34 we know that the model transformation *CD2RDBM* has functional behaviour, and hence the backward model transformation of *RDBM2CD* has functional behaviour with respect to $\mathcal{L}(TGG)_T$ being
equal to the source language $\mathcal{L}(TGG)_S$ of *CD2RDBM*. For this reason, we can apply Theorem 8.39, and have that *RDBM2CD* is completely information preserving.
In particular, foreign keys are completely represented by associations, and primary
keys by primary attributes. There is no structure within the database model which is
not explicitly represented within the class diagram.   △

8.3 Reduction of Nondeterminism

Transformation systems in general cannot ensure deterministic behaviour. Nondeterminism is caused by the choice of the transformation rule and its match at each

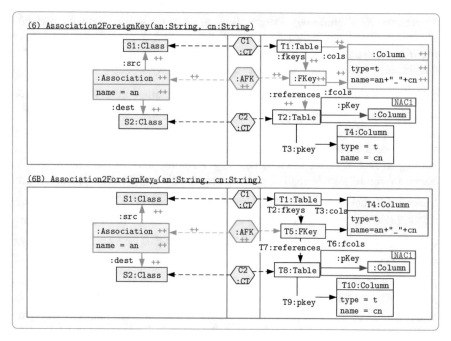

Fig. 8.12 Additional TGG rule for showing the effect of conservative policies

step during a transformation. The general concept for reducing nondeterminism is to analyse functional behaviour based on critical pairs as presented in Sect. 8.2 before and refine the rule set using filter NACs. This section presents two further practical concepts for reducing nondeterminism while ensuring completeness. The first one is using policies for transformation rules and the second one is restricting the operational rules to an effective subset called kernel translation rules.

Example 8.41 (Additional rule with nondeterminism). Consider the triple rule (6) Association2ForeignKey in Fig. 8.12, which we can use as an additional rule for the TGG *CD2RDBM* to handle $1 - n$ associations in the class diagram via foreign keys in the source table in the database model. The rule contains an attribute computation in the target component: the value of name is derived by combining the name of the association an and the name of the primary key of the destination Column. We now consider the corresponding backward rule (6B) Association2ForeignKey$_B$. In order to apply the rule, the matching process of a transformation engine has to find assignments for all variables, i.e., for an and cn. The match for the nodes on the left hand side of the rule (black part without ++) provides enough information to assign the variable cn, but for the variable an we need to find a value solving the constraint that (an+ "_"+cn) is equal to the name of node T4. In general, there can be several solutions if the character "_" occurs several times in the string expression. An efficient approach would be to assign an to the substring from the beginning until the first occurrence of "_". But this would neglect further potential choices. However, this is enough if we are only interested in obtaining one solution, as long as the backward transformation for the possible target models can be completed.   △

In order to reduce nondeterminism for attribute assignments, we present the concept of policies. The main idea of a policy for an operational rule is to restrict the matches using additional attribute conditions in order to eliminate ambiguous results. Attribute conditions are given by equations over attribute values, i.e., they require that some expressions be evaluated equally. In our case study, we use one attribute condition (see Ex. 8.44).

A policy can be arbitrarily restrictive in general. However, if a policy is too restrictive, the model transformation may no longer be complete. Thus, we need to ensure that the model transformation can still be executed successfully for all valid inputs. For this reason, we introduce the notion of a *conservative* policy. In the case of forward transformations, a policy for the set of forward translation rules is conservative if all valid source models can be translated. This ensures that the model transformation is still complete.

Definition 8.42 (Policy for operational translation rules). An *attribute condition* *attCon* for a (triple) rule $tr : L \rightarrow R$ is a set of equations for attribute values. A match $m : L \rightarrow G$ satisfies *attCon*—written $m \models attCon$—if the evaluation of attribute values satisfies each equation. Given a TGG, let TR_{FT} be the derived set of forward translation rules. A *policy pol* : $TR_{FT} \rightarrow TR'_{FT}$ for restricting the applications of the rules in TR_{FT} maps each rule $tr_{FT} \in TR_{FT}$ to an extended rule $tr'_{FT} \in TR'_{FT}$, where tr'_{FT} is given by tr_{FT} extended by a set of additional attribute conditions $AttC_{pol}(tr_{FT})$. The policy *pol* is called *conservative* if the derived model transformation relation $MT'_{FT,R} \subseteq \mathcal{L}(TGG)_S \times \mathcal{L}(TGG)_T$ based on TR'_{FT} is left total and is contained in the model transformation relation $MT_{FT,R}$ derived from TR_{FT}, i.e., $MT'_{FT,R} \subseteq MT_{FT,R}$.

A policy for backward translation rules TR_{BT} is defined analogously by replacing *FT* with *BT* and it is conservative if the derived model transformation relation $MT'_{BT,R} \subseteq \mathcal{L}(TGG)_T \times \mathcal{L}(TGG)_S$ is left total and contained in $MT_{BT,R}$.   △

In order to automatically check that a policy is conservative we provide a sufficient condition by Fact 8.43 below based on the analysis of dependencies between rules [EEPT06]. Intuitively, two transformation steps $G_0 \xrightarrow{p_1,m_1} G_1 \xrightarrow{p_2,m_2} G_2$ are sequentially independent if (1) there is no use–delete dependency (the first step uses (creates or reads) an element (node, edge, or attribute) that is deleted by p_2 in the second step) and (2) there is no forbid–produce dependency. A produce–forbid dependency occurs if the first step forbids a pattern by a negative application condition of p_1 and the second step produces some elements of it, such that applying the second step first will disable the execution of the first step thereafter.

A policy restricts the applicability of rules. The main challenge is to ensure that the restrictions are not too strict. In more detail, for each valid input model of an operational transformation sequence we have to ensure that there is an equivalent transformation sequence respecting all restrictions of the policy. The key idea is to check for each restriction of a rule p whether there are rules that could depend on the execution of p. If we can show that there is no dependency on all possible subsequent steps in an operational transformation sequence, we can conclude that all steps via p can be shifted to the end of the sequence. This allows us to focus on p itself. As stated by Fact 8.43 below, it is then sufficient to show that for each match of p there is an equivalent match satisfying the conservative policy.

Fact 8.43 (Conservative policy). *Let* $pol : TR_{FT} \rightarrow TR'_{FT}$ *be a policy, such that for each rule* $tr'_{FT} = pol(tr_{FT})$ *in* TR'_{FT} *with* $tr : L \rightarrow R$ *the following conditions hold.*

1. *Given a match* $m : L \rightarrow G$ *for* tr_{FT}, *there is also a match* $m' : L \rightarrow G$ *for* tr'_{FT} *satisfying* $AttC_{pol}(tr_{FT})$.
2. *If* $AttC_{pol}(tr_{FT}) \neq \emptyset$, *then for each rule* $tr_2 \in TR_{FT}$ *with* $tr_{FT} \neq tr_2$ *the pair* (tr_{FT}, tr_2) *is sequentially independent.*

Then, the policy pol *is conservative (cf. Def. 8.42). A similar fact holds for a policy* $pol : TR_{BT} \rightarrow TR'_{BT}$ *concerning backward translation rules.*   △

Proof (Idea). According to Def. 8.42, the policy pol is conservative if the derived model transformation relation $MT'_{FT,R}$ is left total. The model transformation relation MT_R based on TR_{FT} is left total due to the completeness result for TGG model transformations based on forward translation rules (cf. Theorem 8.4). Thus, given a source model $G^S \in \mathcal{L}(TGG)_S$, there is a complete forward translation sequence s_{FT} via TR_{FT}. We have to show that there is also a complete forward translation sequence s'_{FT} via TR'_{FT}. First of all, $MT'_{FT,R} \subseteq MT_{FT,R}$, because the additional attribute conditions only restrict the possible transformation sequences and no additional ones are possible. Item (1) in Fact 8.43 ensures that for each step $s_{i,FT}$ in s_{FT} via TR_{FT}, there is a step $s'_{i,FT}$ via TR'_{FT}, but this step may differ on the resulting triple graph. However, item (2) ensures that there is no subsequent step in s_{FT} via a different rule that is sequentially dependent on neither $s_{i,FT}$ nor $s'_{i,FT}$. Therefore, we can iteratively exchange the original steps with corresponding ones via TR'_{FT}, shift them to the end of the the sequence, and continue with the next step that is not via TR'_{FT}. Finally, we derive a complete forward translation sequence s'_{FT} via TR'_{FT}. For the full proof see Fact 7 in [HEO$^+$11b].   □

Example 8.44 (Nonconservative policy). Fig. 8.13 shows the backward translation rule (6BT) and its extension (6BT′) with a policy. The policy is an attribute condition concerning the variable an. It ensures that the match for the nodes will fully determine the values for all variables. For each match of rule (6BT), there is a match for rule 6BT′, because the condition requires that an be equal to a term. This term uses string functions substr and pos, which have to be left total relations in order to be algebra operations. Therefore, the policy satisfies the first condition of Def. 8.42. However, it does not satisfy the second condition. Since the rule translates nodes of type Column, there are possible dependencies on rules that use a node of type Column as context node. This is the case for the rule Association2Table$_{BT}$.

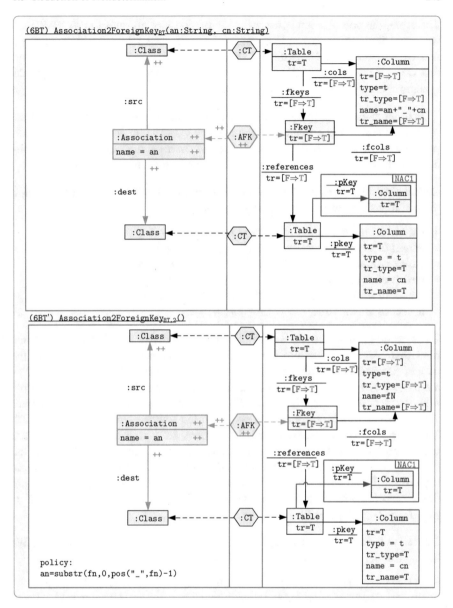

Fig. 8.13 Backward translation rule (*6BT*) without and rule (*6BT'*) with policy

Therefore, the policy is not conservative if we consider all rules of *CD2RDBM*. If we would drop the rule $\texttt{Association2Table}_{BT}$, then the policy would be conservative, because the second condition of Def. 8.42 would be satisfied in that case. Note that we will use a conservative policy explicitly for optimising a TGG for model synchronisation in Ex. 9.21 in Chap. 9. $\triangle$

In order to ensure termination of the sets of operational rules using Fact 8.13, we restrict the sets to those that modify at least one translation attribute. For this purpose, we distinguish between several subsets of the triple rules of a TGG depending on their effects concerning the creation of elements in the triple components.

Definition 8.45 (Creating and identic triple rules). Let *TR* be a set of triple rules. We distinguish between the following subsets:

- the set of *creating* rules $TR^+ = \{tr \in TR \mid tr \neq id\}$,
- the set of *source creating* rules $TR^{+s} = \{tr \in TR \mid tr^S \neq id\}$,
- the set of *source identic* rules $TR^{1s} = \{tr \in TR \mid tr^S = id\}$,
- the set of *target creating* rules $TR^{+t} = \{tr \in TR \mid tr^T \neq id\}$,
- the set of *target identic* rules $TR^{1t} = \{tr \in TR \mid tr^T = id\}$, and
- the set of *identic* rules $TR^1 = \{tr \in TR \mid tr = id\}$. $\triangle$

Based on the different kinds of creating rules, we derive the effective operational rules that ensure termination. We call these rules kernel translation rules. In the case of forward translation rules, the kernel forward translation rules $TR_{FT}^{+s} \subseteq TR_{FT}$ are those forward translation rules that are derived from the source creating triple rules $TR^{+s} \subseteq TR$ of the triple rules *TR*. The remaining forward translation rules $TR_{FT}^{1s} = TR_{FT} \setminus TR_{FT}^{+s}$ are those derived from the source identic triple rules TR^{1s}. Vice versa, the kernel backward translation rules $TR_{BT}^{+t} \subseteq TR_{BT}$ are the backward translation rules that are derived from the target creating triple rules $TR^{+t} \subseteq TR$, and TR_{BT}^{1t} are the remaining backward translation rules derived from the target identic triple rules. Finally, the kernel consistency creating triple rules $TR_{CC}^{+} \subseteq TR_{CC}$ are those consistency creating rules that are derived from the creating triple rules $TR^{+} = \{(tr\colon L \to R) \in TR \mid L \neq R\}$.

Definition 8.46 (Kernel translation rules). Let *TR* be a set of triple rules. We distinguish between the following sets of rules:

- the set of *kernel consistency creating* rules $TR_{CC}^{+} = \{tr_{CC} \in TR_{CC} \mid tr \in TR^{+}\}$,
- the set of *kernel forward translation* rules $TR_{FT}^{+s} = \{tr_{FT} \in TR_{FT} \mid tr \in TR^{+s}\}$, and
- the set of *kernel backward translation* rules $TR_{BT}^{+t} = \{tr_{BT} \in TR_{BT} \mid tr \in TR^{+t}\}$. $\triangle$

The notion of kernel translation rules automatically ensures termination according to Lem. 8.47 below. We generally assume that the input models are finite on the structure part, i.e., the carrier sets of the data values can be infinite, but the graph nodes and all sets of edges are finite.

Lemma 8.47 (Termination of rules with conservative policies). *Let* $TGG =$ $(TG, \emptyset, TR)$ *be a triple graph grammar. Let further* $TR_{CC}^+, TR_{FT}^{+s},$ *and* TR_{BT}^{+t} *be the derived sets of operational translation rules for consistency creating, forward translation, and backward translation, respectively, according to Def. 7.44 and possibly extended by some policies. Then, the transformation systems* $TR_{CC}^+, TR_{FT}^{+s},$ *and* TR_{BT}^{+t} *are terminating for any input triple graph that is finite on the graph part.* △

Proof. This is a direct consequence of Fact 8.13, because each rule of the sets $TR_{CC}^+, TR_{FT}^{+s},$ and TR_{BT}^{+t} changes at least one translation attribute. □

The restriction of the set of operational rules to those that change the marking can cause the model transformation not to be complete anymore. Thus, it cannot be ensured anymore that for an arbitrary valid input model there is a valid operational transformation sequence via forward or backward translation rules, respectively. However, we can use the same idea as for conservative policies and check that the remaining rules do not depend on the omitted ones (TR_{FT}^{1s} and TR_{BT}^{1t}), as stated by Rem. 8.48 below. The main idea is the following. If we can show that none of the remaining triple rules depends on the source identic triple rules, we can actually omit the source identic ones. The reason is that for each forward transformation sequence, we can shift the steps along source identic rules to the end and obtain an equivalent sequence. Since all steps along source identic triple rules do not change the marking of the source model, we further derive that these steps can be removed, yielding still a complete forward translation sequence. This ensures that the rules that do not change any translation attribute can be omitted while still all valid input models can be processed successfully.

Remark 8.48 (Shifting of independent steps). Consider two sets P_1 and P_2 of rules such that each pair $(p_1, p_2) \in P_1 \times P_2$ is sequentially independent. Then, there is a transformation sequence $(G \xRightarrow{r^*} H)$ via $(P_1 \cup P_2)$ if and only if there are transformation sequences $s_1 = (G \xRightarrow{p^*} G_1)$ via P_2 and $s_2 = (G_1 \xRightarrow{q^*} H)$ via P_1 with the same G_1. This result is shown by Fact 3 in App. A.2 in [HEO$^+$11b]. △

Based on the result on shifting independent steps, we introduce the notion of kernel-grounded operational translation rules and show thereafter that this property allows us to restrict the sets of rules appropriately, such that termination and completeness are ensured.

Definition 8.49 (Kernel-grounded and deterministic sets of operational translation rules). Let $TGG = (TG, \emptyset, TR)$ be a triple graph grammar from which we obtain the operational translation rules $TR_{CC}, TR_{FT},$ and TR_{BT}. They are called *kernel-grounded* if the pairs $(TR_{FT}^{1s}, TR_{FT}^{+s})$ and $(TR_{BT}^{1t}, TR_{BT}^{+t})$ are sequentially independent. This means that there is no pair (p_1, p_2) of sequentially dependent rules with either $(p_1, p_2) \in (TR_{FT}^{1s} \times TR_{FT}^{+s})$ or $(p_1, p_2) \in (TR_{BT}^{1t} \times TR_{BT}^{+t})$.

The sets of operational translation rules $TR_{CC}, TR_{FT},$ and TR_{BT} (possibly extended by conservative policies) are called *deterministic* if they have functional behaviour and do not require backtracking. △

Example 8.50 ((Non-)kernel-grounded operational rules and determinism). The operational rules of the TGG *CD2RDBM* are not kernel-grounded, because some backward translation rules depend on the target-identic backward translation rule (2BT) SubClass2Table$_{BT}$. For instance, rule (3BT) Attr2Column$_{BT}$ depends on it: (2BT) SubClass2Table$_{BT}$ creates a C2T node and (3BT) Attr2Column$_{BT}$ uses a C2T node as context node. If we consider a very restricted TGG *CD2RDBM$_2$*, which uses the rule set *TR* = { Class2Table, Attr2Column }, then we obtain sets of operational translation rules that are kernel grounded and deterministic. By Fact 9.24 in Chap. 9, we show that the operational rules for the TGG of our case study on model synchronisation are indeed kernel-grounded and deterministic as well.   △

The tool AGG [AGG14] supports the automated analysis of dependencies between rules. We apply this analysis engine to check whether a policy is conservative and that the reduced sets of operational rules are sufficient to ensure completeness of the propagation operations.

Remark 8.51 (Analysis of operational rules). In order to check that the sets of operational translation rules are kernel-grounded and deterministic, we describe how the preconditions of Def. 8.49 are checked using the tool AGG. The condition that they have functional behaviour and do not require backtracking can be checked via Theorem 8.29.

1. Sequential independence of the pairs $(TR^{1s}_{FT}, TR^{+s}_{FT})$ and $(TR^{1t}_{BT}, TR^{+t}_{BT})$: we can use the tool AGG for the analysis of rule dependencies based on the generation of critical pairs according to Fact 2 in [HEO$^+$11b].
2. Applied policies are conservative: According to Fact 8.43, this requires that the additional application conditions according to the policy restrict the evaluation of attribute values only, i.e., the assignment of variables. We have to show that the existence of matches is preserved for each rule and that other rules are not sequentially dependent. For the latter, we can again use the tool AGG and validate that the corresponding table entries show the value 0. The preservation of the existence of matches can be ensured by checking that the affected variables are free in the unmodified rule (tr_{FT} or tr_{BT}), i.e., they are not part of a term that is connected to a node in the LHS (L_{FT} or L_{BT}).   △

Moreover, we can apply the presented results for showing that the derived model transformation relations are left total. In particular, left totality of the relations is required in Chap. 9 for ensuring completeness of model synchronisations via TGGs.

Remark 8.52 (Left totality). If the sets of operational translation rules of a TGG are kernel-grounded, we can conclude that the forward model transformation relations $MT_{F,R}: \mathcal{L}(TG^S) \Rightarrow \mathcal{L}(TG^T)$ based on TR^{+s}_{FT} and the backward model transformation relation $MT_{B,R}: \mathcal{L}(TG^T) \Rightarrow \mathcal{L}(TG^S)$ based on TR^{+s}_{BT} specify left total relations as shown by Fact 5 in [HEO$^+$11b]. This means that the model transformations can be performed on reduced sets of operational translation rules. Source identic triple rules TR^{1s}_{FT} are not used for forward translations and target identic triple rules TR^{1t}_{BT} are not used for backward translations. According to Def. 8.42, we can specify conservative policies in order to reduce the number of possible transformation sequences and derive left total model transformation relations $MT'_{FT,R}$ and $MT'_{BT,R}$ that use these policies. △

Chapter 9
Model Synchronisation

Bidirectional model transformations are a key concept for model generation and synchronisation within model-driven engineering (MDE, see [Ste10, QVT15, CFH+09]). Triple graph grammars (TGGs) have been successfully applied in several case studies for bidirectional model transformation, model integration and synchronisation [KW07, SK08, GW09, GH09], and in the implementation of QVT [GK10]. This chapter provides a TGG framework for model synchronisation that ensures correctness and completeness based on the theory of TGGs. It is inspired by work on incremental synchronisation by Giese et al. [GW09, GH09], and the model synchronisation framework by Diskin [Dis11]. The chapter is based on [HEO+11a, HEEO12, HEO+13]. The main ideas and results are the following:

1. Models are synchronised by propagating changes from a source model to a corresponding target model using forward and backward propagation operations. The operations are specified by a TGG model framework, inspired by symmetric replica synchronisers [Dis11] and realised by model transformations based on TGGs [EEHP09] (see Chap. 7). The specified TGG also defines consistency of source and target models.
2. Since TGGs define, in general, nondeterministic model transformations, the derived synchronisation operations are, in general, nondeterministic. But we are able to provide sufficient static conditions based on TGGs to ensure that the operations are deterministic.
3. The first main result shows that a TGG synchronisation framework with *deterministic* synchronisation operations is correct, i.e., consistency preserving, and complete (see Theorems 9.25 and 9.29). We also give sufficient static conditions for invertibility and weak invertibility of the framework, where "weak" restricts invertibility to a subclass of inputs.
4. The second main result shows that a TGG synchronisation framework for *concurrent* model synchronisation based on *deterministic* propagation operations is correct and complete (see Theorem 9.41). Concurrent model synchronisation means that updates may occur on both domains simultaneously, which requires addi-

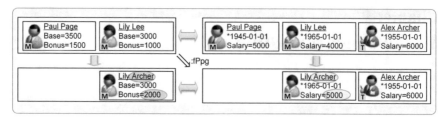

Fig. 9.1 Forward propagation

tional conflict resolution and may cause nondeterminism due to user interaction during conflict resolution.

Deriving a synchronisation framework from a TGG has the following practical benefits. Consistency of the related domains is defined declaratively and in a pattern-based style, using the rules of a TGG. Consistency of source and target models is always ensured after executing the synchronisation operations (correctness) and the synchronisation can be performed for all valid inputs (completeness). The required static conditions for deterministic behaviour and the additional conditions for invertibility can be checked automatically using the tool support of AGG [AGG14]. The extension of this approach to the general case of nondeterministic synchronisation operations based on *nondeterministic* TGGs is described in [GHN+13].

Remark 9.1 (General assumption). As in Chap. 8, the formal results in this chapter are presented for TGGs that ensure the composition and decomposition property for forward sequences (Def. 7.20) and for integration sequences (Def. 7.39). Chap. 7 presents sufficient conditions for these properties by Theorems 7.21 and 7.40. These conditions mainly require that the application conditions be compatible application condition schemata with almost injective morphisms and the execution be performed via almost injective matches. Moreover, the formal results in this chapter are presented for triple graph grammars with negative application conditions (NACs). An extension to general nested application conditions is future work. Several case studies show that NACs are usually sufficient to restrict the applicability of triple rules in the context of model synchronisation.   △

Throughout this chapter, we use a simple running example, which is adapted from [DXC11a, HEO+13]. The example considers the synchronisation of two organisational diagrams as shown in Fig. 9.1. Diagrams in the first domain—depicted left—provide a view on employees of the marketing department of a company, while diagrams in the second domain—depicted right—show all employees. Furthermore, both domains differ on the type of information they specify. Diagrams on the left show the base and bonus salary values of each person, while diagrams in the second domain show only the total salary for each person, but additionally, they provide the birth dates (marked by "*"). Therefore, both domains contain exclusive information and none of them can be interpreted as a view—defined by a query—of the other.

Both diagrams together with some correspondence structure build up an integrated model, where we refer to the first diagram as the source model and to the second diagram as the target model. An integrated model is called *consistent* if:

- corresponding persons coincide on names,
- salary values are equal to the sums of corresponding base and bonus values, and
- persons in the source domain are exactly those who are marked with M in the target domain.

Example 9.2 (Update propagation). The first row of Fig. 9.1 shows a consistent integrated model *M* in a visual notation. The source model of *M* consists of two persons belonging to the marketing department (depicted as persons with label M and without pencils) and the target model additionally contains the person Alex Archer belonging to the technical department (depicted as a person with label T and with pencil). The first column shows an update of the source model, where the person Paul Page is removed and some attribute values of the person Lily Lee are modified. This change is propagated to the target domain, leading to a target update (right column) and a new integrated model (bottom row).   △

The synchronisation problem is to propagate a model update in such a way that the resulting integrated model is consistent. Looking at Fig. 9.1, this requires that the source model update of removing the person Paul Page and changing the attributes LastName and Bonus of the person Lily Lee is propagated in an appropriate way to the target domain. In this example, this means that the executed forward propagation (fPpg) shall remove the person Paul Page and update the attribute values of Lily Lee in the target model, such that the unchanged birth date value and consistency is preserved.

Remark 9.3 (Choice of the example). This chapter uses a rather simple example for model synchronisation to keep the figures and constructions compact. It is sufficient to illustrate the relevant aspects and can serve as a reference to design more complex ones. In fact, it is closely related to the even simpler benchmark example for bidirectional model transformation and synchronisation presented in [ACG$^+$14]. Note that the example *CD2RDBM* used throughout Chapters 7 and 8 does not ensure all conditions for our general results in this chapter concerning correctness and completeness and we discuss the corresponding problems of them in Rem. 9.27.   △

Synchronisation scenarios like the one in our example are present in many domains. Consider for example synchronisations between different kinds of visual models for software development, models for software analysis, and even source code. Synchronisations between these domains often need to provide mechanisms that do not require that one model be completely obtainable from the other. In other words, none of the models is just a view of the other. In this chapter, we show how this flexibility in the synchronisation process is possible based on the formal notion of TGGs. Stepwise, we develop the required formal techniques and illustrate them on the running example in Fig. 9.1, whose intermediate steps are presented in Fig. 9.16 of Sect. 9.2.

Sect. 9.1 presents the general concept for change propagation and Sect. 9.2 introduces the model synchronisation framework based on TGGs for the basic case, i.e., where updates are propagated from one side to the other. Sect. 9.2.3 shows that the derived synchronisation framework is correct and complete under sufficient static conditions. Moreover, it provides sufficient conditions that ensure compatibility of the propagation operations, namely invertibility. Sect. 9.3 extends the framework to the concurrent case, where updates may occur on both domains simultanously. The propagation of both updates additionally requires us to resolve occurring conflicts with optional user input.

9.1 General Concept for Change Propagation

This section describes the basic framework for model synchronisation, where triple graphs describe pairs of interrelated models and *triple graph grammars* (TGGs) are used as a tool to specify classes of consistent interrelated models (see Chap. 7). The framework is a simplified version of symmetric delta lenses proposed by Diskin et al. [Dis11]. Based on the notion of a TGG model framework, we define *synchronisation problems* and the *propagation operations* which are used to solve them. We present and discuss explicit properties concerning correctness, completeness and invertibility.

In general, model synchronisation aims to achieve consistency among interrelated models. We consider a model as a kind of graph. Moreover, we assume that a pair of interrelated models (M^S, M^T), called *source and target models*, are represented by a triple graph $G = (G^S \leftarrow G^C \rightarrow G^T)$, which we also call an *integrated model*. The source graph G^S represents M^S and the target graph G^T represents M^T. The two graph morphisms $s_G : G^C \rightarrow G^S$ and $t_G : G^C \rightarrow G^T$ specify a *correspondence* $r : G^S \leftrightarrow G^T$, which relates the elements of G^S with their corresponding elements of G^T and vice versa. For simplicity, we use double arrows ($\leftrightarrow$) as an equivalent shorter notation for triple graphs whenever the explicit correspondence graph can be omitted.

Example 9.4 (Type graph). The triple type graph TG of our example is shown in Fig. 9.2. It specifies that models of the source domain contain persons, including their detailed salary information (bonus and base salary) and their names. Models of the target domain additionally contain the department to which a person is assigned, his or her birth date, and a single value for his or her complete salary, while the details about bonus and base salary are not provided.   △

Example 9.5 (Triple rules). The triple rules of the TGG are depicted in a compact notation in Fig. 9.3. Left and right hand sides of a rule are depicted in one triple graph, where the elements to be created have the label "++". They exist in the right-hand side of the triple rule only. The first rule (Person2FirstMarketingP) inserts a new department with name Marketing and the NAC ensures that none of the existing departments is named equally. The rule creates a person of the new department

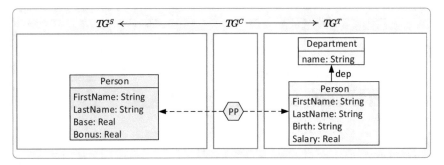

Fig. 9.2 Triple type graph *TG*

in the target model as well as a corresponding person in the source model. Note that the left hand side of this rule is empty, i.e., it does not require existing structures. The rule Person2NextMarketingP is used to extend both models with further persons in the marketing department. The left hand side of this rule contains the department node with name Marketing. Note that the attributes of the created persons are not set with these rules. This is possible in our formal framework of attributed graph transformation based on the notion of E-graphs (see Chap. 2, [EEPT06]). The main advantage is that we can propagate changes of attribute values without the need for deleting and recreating the owning structural nodes. This is important from the efficiency and application point of view. Thus, rules 3–6 concern the creation of attribute values only. Rules 3 (FName2FName) and 4 (LName2LName) create new corresponding values for first and last names, respectively. The next rule (Empty2Birth) assigns the birth date of a person in the target component and does not change the source component. Finally, rule 6 (DetailedSalary2Salary) assigns the detailed salary values (bonus and base) in the source component and the sum of them in the target component. Rule 7 (Empty2OtherDepartment) creates a new department that is not named Marketing, but does not change the source model. The negative application condition (NAC) ensures that the used attribute value is different from Marketing. The last rule Empty2OtherP of the TGG creates a new person of a department that is different from the marketing department. Therefore, there are no correspondences to the source model and the rule directly creates the person, including all attribute values.   △

A *TGG model framework* specifies the possible correspondences between models and updates of models for a given TGG according to Def. 9.6 below. More precisely, a model framework is defined as consisting of the classes of well-typed source and target models, the class of correspondences between source and target models (i.e., the class of well-typed triple graphs), the subset of consistent correspondences (i.e., the class of triple graphs defined by the given TGG) and the classes of source and target updates. In particular, a model update $\delta : G \rightarrow G'$ is specified as a *graph modification* consisting of two inclusions, $\delta : G \hookleftarrow I \hookrightarrow G'$. This notion is inspired by the derived spans of graph transformation sequences (see Def. 5.31). The intuition of a graph modification is that the inclusion $I \hookrightarrow G$ specifies the elements that

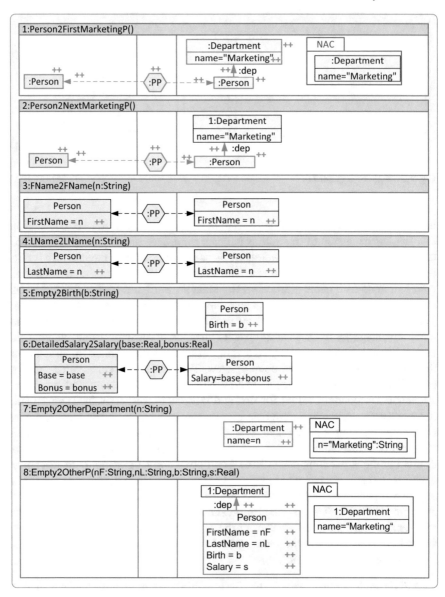

Fig. 9.3 Triple rules

are deleted from G (all the elements that are not in I) and $I \hookrightarrow G'$ specifies all the elements that are added by δ (all the elements in G' that are not in I). Therefore, the elements in I are the elements that remain invariant after the modification. Intuitively, one can also interpret a graph modification as a DPO rule that describes the complete update as one step and contains the complete graphs and not just some parts of them. Finally, it may be noted that graph modifications look like triple graphs; however, their role is different: triple graphs are used to make explicit the interrelations between two integrated models, while graph modifications are used to describe updates on a given model.

Given a TGG with type graph $TG = (TG^S \leftarrow TG^C \rightarrow TG^T)$, we refer by $\mathcal{L}(TGG)$ to the language of consistent integrated models, and by $\mathcal{L}(TG^S)$, $\mathcal{L}(TG^T)$ to the languages of source and target models typed over TG^S and TG^T, respectively.

Definition 9.6 (TGG model framework). Let $TGG = (TG, \emptyset, TR)$ be a triple graph grammar with empty start graph $\emptyset$ and triple type graph TG containing source and target components TG^S and TG^T, and a set TR of triple rules. The derived TGG *model framework* $MF(TGG) = (\mathcal{L}(TG^S), \mathcal{L}(TG^T), R, C, \Delta_S, \Delta_T)$ consists of source domain $\mathcal{L}(TG^S)$, target domain $\mathcal{L}(TG^T)$, the set R of correspondence relations given by $R = \mathcal{L}(TG)$, the set C of consistent correspondence relations $C \subseteq R$ given by $C = \mathcal{L}(TGG)$, (i.e., R contains all integrated models and C all consistent integrated ones), and sets Δ_S, Δ_T of graph modifications for the source and target domains, given by $\Delta_S = \{a : G^S \rightarrow G'^S \mid G^S, G'^S \in \mathcal{L}(TG^S)$, and a is a graph modification$\}$ and $\Delta_T = \{b : G^T \rightarrow G'^T \mid G^T, G'^T \in \mathcal{L}(TG^T)$, and b is a graph modification$\}$, respectively.   △

Given a TGG model framework, *the synchronisation problem* is to provide suitable *total* and *deterministic* forward and backward operations fPpg and bPpg that propagate updates on one model (G^S or G^T) to the other model. The propagation operations are executed via graph transformations. The operations are deterministic if they ensure unique results for any input. They are total if they provide results for all inputs. In other words, the propagation operations have to be proper functions which is not satisfied by arbitrary graph transformation systems. Note that one can consider also scenarios where update propagation is not necessarily deterministic, i.e., the propagation of a source update would provide different possible target updates. For full details of this extended case and the corresponding results for the nondeterministic scenario we refer you to [GHN⁺13].

The conceptual idea of forward propagation is the following. Given an integrated model (a correspondence relation) $G^S \leftrightarrow G^T$ and an update $a : G^S \rightarrow G'^S$, the operation fPpg must propagate the update a to G^T, returning as results an update $b : G^T \rightarrow G'^T$ and a correspondence relation $G'^S \leftrightarrow G'^T$. Similarly, bPpg is the dual operation that propagates updates on target models to updates on source models. The effect of these operations is depicted schematically in the diagrams in Fig. 9.4, which we call *synchronisation tiles*, where we use solid lines for the inputs and dashed lines for the outputs [Dis11]. Note that, in a common tool environment, the required input for these operations is either available directly or can be obtained. For example, the graph modification of a model update can be derived

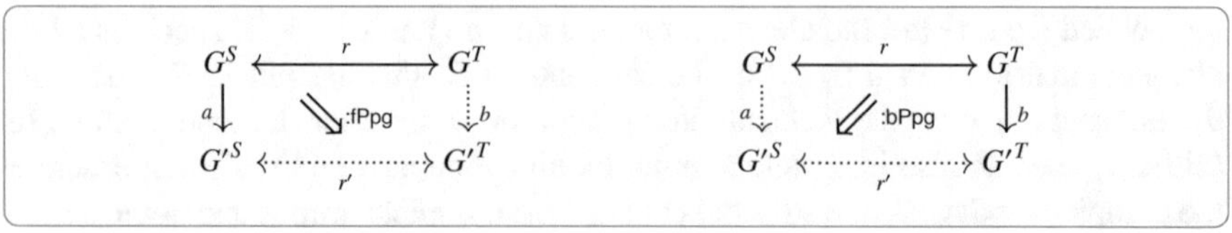

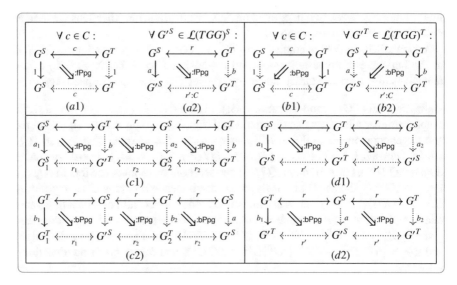

Fig. 9.4 Synchronisation operations

Fig. 9.5 Laws for correct and (weak) invertible synchronisation frameworks

via standard difference computation and the initial correspondence can be computed based on TGG integration concepts (see Chap. 7, [EEH08a, KW07]). Note also that determinism of fPpg means that the resulting correspondence $G'^S \leftrightarrow G'^T$ and the update $b : G^T \rightarrow G'^T$ are uniquely determined up to isomorphism. The propagation operations are *correct* if they additionally preserve consistency as specified by laws $(a1)$–$(b2)$ in Fig. 9.5. Law $(a2)$ means that fPpg always produces consistent correspondences from consistent updated source models G'^S. Law $(a1)$ means that if the given update is the identity and the given correspondence is consistent, then fPpg changes nothing. Laws $(b1)$ and $(b2)$ are the dual versions concerning bPpg. Moreover, the sets $\mathcal{L}(TGG)^S$ and $\mathcal{L}(TGG)^T$ specify the *consistent source and target models*, which are given by the source and target components of the integrated models in $C = \mathcal{L}(TGG)$.

Definition 9.7 (Synchronisation problem and framework). Let $MF = (\mathcal{L}(TG^S), \mathcal{L}(TG^T), R, C, \Delta_S, \Delta_T)$ be a TGG model framework. The *forward synchronisation problem* is to construct a total and deterministic operation fPpg : $R \otimes \Delta_S \rightarrow R \times \Delta_T$ leading to the left diagram in Fig. 9.4, where $R \otimes \Delta_S = \{(r, a) \in R \times \Delta_S \mid r: G^S \leftrightarrow G^T, a: G^S \rightarrow G'^S\}$, i.e., a and r coincide on G^S. The pair

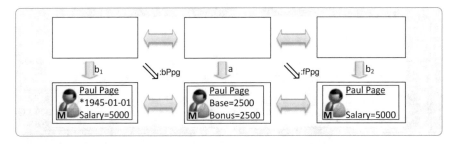

Fig. 9.6 Counterexample for invertibility

$(r, a) \in R \otimes \Delta_S$ is called premise and $(r', b) \in R \times \Delta_T$ is called solution of the forward synchronisation problem, written $\mathsf{fPpg}(r, a) = (r', b)$. The *backward synchronisation problem* is to construct a total and deterministic operation bPpg leading to the right diagram in Fig. 9.4. The operation fPpg is called *correct* with respect to C if axioms $(a1)$ and $(a2)$ in Fig. 9.5 are satisfied and, symmetrically, bPpg is called *correct* with respect to C if axioms $(b1)$ and $(b2)$ are satisfied.

Given total and deterministic propagation operations fPpg and bPpg derived from TGG, the derived *synchronisation framework Synch(TGG)* is given by $Synch(TGG) = (MF(TGG), \mathsf{fPpg}, \mathsf{bPpg})$. It is called *correct* if fPpg and bPpg are correct; it is *weakly invertible* if axioms $(c1)$ and $(c2)$ in Fig. 9.5 are satisfied; and it is *invertible* if additionally axioms $(d1)$ and $(d2)$ in Fig. 9.5 are satisfied. $\triangle$

Invertibility (laws $(d1)$ and $(d2)$) means that the propagation operations are essentially inverse of each other. For instance, axiom $(d1)$ states that if we propagate an update $a_1 : G^S \rightarrow G_1^S$ to G^T, obtaining as result an update b, and now we propagate update b to G^S, we obtain the same result G_1^S. However, notice that we do not require that the resulting update $a_2 \colon G^S \rightarrow G_1^S$ coincide with a_1. In particular, it may be possible that the set of elements of G^S that are not modified by a_1 may not coincide with the set of elements that are not modified by a_2, even if they produce the same result G_1^S (see Ex. 9.8 below). However, as we show in Sect. 9.2.3, we are able to ensure the more flexible notion of weak invertibility (laws (c_1) and (c_2)) for our example. More precisely, weak invertibility expresses that the two operations are the inverse of each other, up to certain information that may be lost when applying the operations. For instance, in axiom (c_1) the intuition is that update b, the result of propagation of update a_1, may ignore part of the information added by a_1, because this kind of information may not be relevant for target models. As a consequence, when propagating b to G^S this information would be lost. However, weak invertibility also states that no information added by update b would be ignored when propagating it back to G^S. Thus, update b is recovered in the last propagation step. The reason is that all that information was, in some sense, included in update a_1, so it must be relevant for source models.

Example 9.8 (Invertibility and weak invertibility). Consider a model update b_1 of a given target model, as depicted in Fig. 9.6, where a new person (Paul Page) is

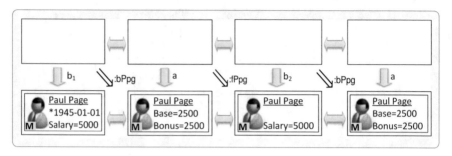

Fig. 9.7 Example for weak invertibility

added together with his birth date, leading to a target model G'^T. The propagation via bPpg yields an update a, whose resulting source model G'^S includes that person without his birth date. Now, the propagation of a via fPpg yields an update b_2 whose resulting target model G''^T does not contain any information about the birth date. Therefore, $G'^T \neq G''^T$, meaning that $Synch(TGG)$ is not invertible, since law ($d2$) does not hold. However, if we continue the diagram and perform an additional backward propagation as in Fig. 9.7, we derive a source update that coincides again with a, i.e., the diagrams satisfy law ($c2$) of weak invertibility.   △

9.2 Basic Model Synchronisation

This section shows how to construct the synchronisation operations for the basic case, where updates from one domain are propagated to the other domain to achieve a consistent state. The more general case of simultaneous updates on both domains is presented in Sect. 9.3.

The synchronisation operation fPpg of a TGG synchronisation framework (see Def. 9.7) is derived as a composition of auxiliary operations, which are executed based on the sets of operational rules of the TGG.

9.2.1 Derived Operational Rules for Synchronisation

The auxiliary operations for the propagation operations fPpg and bPpg are based on the sets of operational rules of the specified TGG. The used sets are the derived consistency creating rules, the forward translation rules and the backward translation rules (see Def. 7.44). The consistency creating rules are used to mark the still consistent parts of the current state of the integrated model while the forward and backward translation rules are used to propagate the update from one domain to the other.

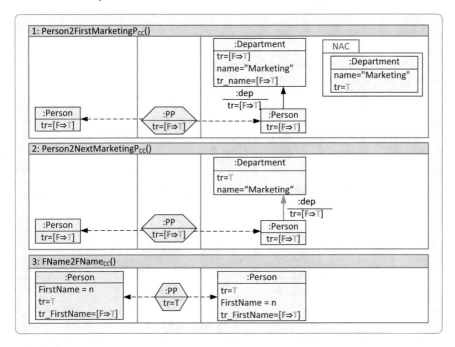

Fig. 9.8 Derived operational triple rules: TR_{CC} (part 1)

In general, each intermediate phase during the execution of a propagation operation prepares the application of the operational rules in the subsequent phase. Hence, the sets of operational rules have to be compatible up to a certain extent.

Remark 9.9 (Interdependencies between operational rules). The consistency creating rules (TR_{CC}) are used for marking the already consistent parts of a given integrated model in the second sub-phase of the synchronisation. The forward and backward translation rules are used for the third sub-phase. This third sub-phase can be interpreted as a completion of the computed sequence of the second sub-phase. We show in Sect. 9.2.3 that this continuation is always possible if the sets of operational rules are deterministic (Theorem 9.25), for which we also provide an automated check and analysis. If a TGG does not ensure deterministic sets of operational rules, the computed maximal subgraph via TR_{CC} may be too large to find a corresponding completion via forward (backward) translation rules. In this case, a possible solution would be to perform backtracking for sub-phases 2 and 3 of the synchronisation, as discussed in Sect. 9.3.   △

Example 9.10 (Derived sets of consistency creating rules). Figs. 9.8 and 9.9 show the set of the consistency creating rules derived from the triple rules in Ex. 9.5 according to Def. 7.44. They do not modify the structure of a triple graph, but only the translation attributes. They are used for marking consistent substructures of a

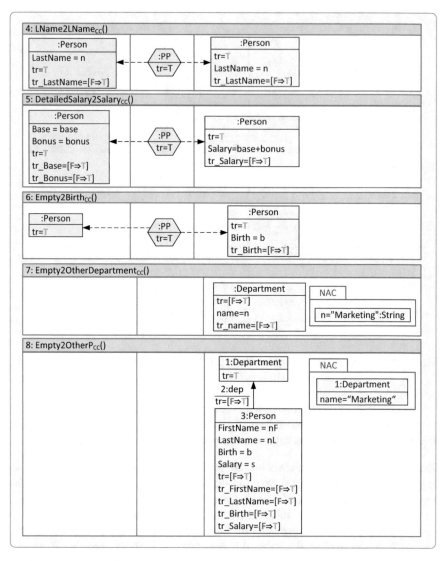

Fig. 9.9 Derived operational triple rules: TR_{CC} (part 2)

given triple graph, i.e., of a given integrated model. For that purpose, we apply
all derived consistency creating rules as long as possible to a given triple G with all
translation attributes set to "**F**". Thus, we compute a maximal consistent triple graph
that is contained in G. Intuitively, for each element $x \in R$ (node, edge, or attribute) of
a triple rule $tr = (L \rightarrow R)$, a separate translation attribute ($\mathtt{tr}$ or $\mathtt{tr\_x}$) is added for
the consistency creating rule tr_{CC}. If an element $x \in R$ is preserved by the triple rule
tr ($x \in L$), then the consistency creating rule preserves it as well and the translation

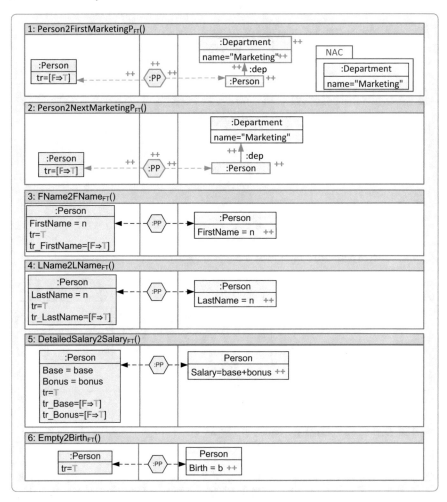

Fig. 9.10 Derived operational triple rules: TR_{FT} (part 1)

attribute has value **T**. Otherwise, if $x \in R$ is created by tr ($x \in R \setminus L$), then it becomes a preserved element in the consistency creating rule tr_{CC} and the corresponding translation attribute is changed from **F** to **T**. In visual notation, this means that all plus signs are replaced by additional translation attributes whose values are changed from **F** to **T**, and we denote such a modification by $[\mathbf{F} \Rightarrow \mathbf{T}]$. △

Example 9.11 (Derived sets of forward translation rules). Figs. 9.10 and 9.11 show the set of the forward translation rules derived from the triple rules in Ex. 9.5 according to Def. 7.44. These rules are used for translating a source model into its corresponding target model. For this reason, the rules are only modifying the translation attributes on the source component. Intuitively, for each element x in the source

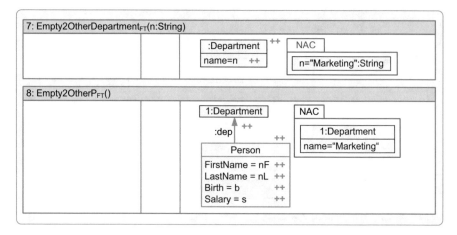

Fig. 9.11 Derived operational triple rules: TR_{FT} (part 2)

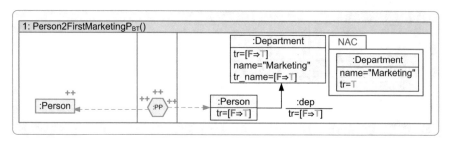

Fig. 9.12 Derived operational triple rules: TR_{BT} (part 1)

component R^S (node, edge, or attribute) of a triple rule $tr = (L \rightarrow R)$ a separate translation attribute (tr or tr_x) is added for the forward translation rule tr_{FT}. If an element $x \in R^S$ is preserved by the triple rule tr, then the forward translation rule preserves it as well and the translation attribute has value **T**. Otherwise, if $x \in R^S$ is created by tr, then it becomes a preserved element in the forward translation rule tr_{FT} and the corresponding translation attribute is changed from **F** to **T**. In visual notation, this means that each plus sign in the source component of a triple rule is replaced by an additional translation attribute whose value changes from **F** to **T**.

Note that the rules 6–8 are contained in TR_{FT}^{ls}, i.e., they are identities on the source component and, according to Def. 9.7, they are not used for fPpg, which is based on TR_{FT}^{+s}. This is important to ensure termination (see Lem. 8.47) and we show by Fact 9.24 that the derived sets of operational rules are kernel-grounded (see Def. 8.49). This is a sufficient condition to guarantee that the reduced set still ensures completeness according to Rem. 8.52 and Theorem 9.25.   △

Example 9.12 (Derived sets of backward translation rules). Figs. 9.12 and 9.13 show the set of the backward translation rules derived from the triple rules in Ex. 9.5

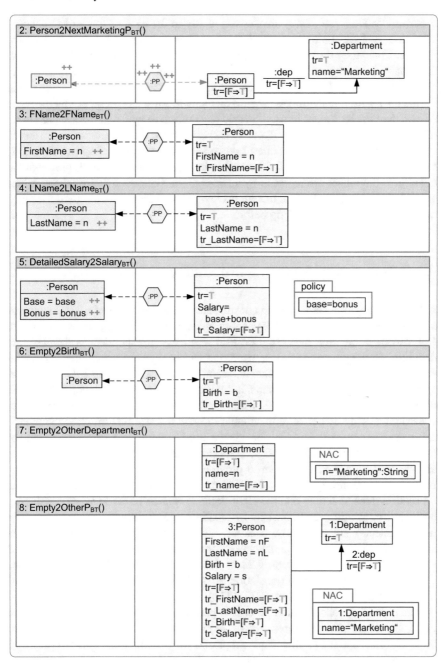

Fig. 9.13 Derived operational triple rules: TR_{BT} (part 2)

according to Def. 7.44. They are derived dually to the case of forward translation rules and used for the translation of target models into their corresponding source models. Thus, they do only modify translation attributes on the target component. Intuitively, for each element x in the target component R^T (node, edge, or attribute) of a triple rule $tr = (L \to R)$, a separate translation attribute ($\mathtt{tr}$ or $\mathtt{tr\_x}$) is added for the backward translation rule tr_{BT}. If an element $x \in R^T$ is preserved by the triple rule tr, then the backward translation rule preserves it as well and the translation attribute has value $\mathbf{T}$. Otherwise, if $x \in R^T$ is created by tr, then it becomes a preserved element in the backward translation rule tr_{BT} and the corresponding translation attribute is changed from $\mathbf{F}$ to $\mathbf{T}$. In visual notation, this means that all plus signs in the target component are replaced by additional translation attributes whose values are changed from $\mathbf{F}$ to $\mathbf{T}$. Note that all backward translation rules are used for bPpg in contrast to the operation fPpg before.   △

9.2.2 Execution of Basic Synchronisation

In the following, we show how to construct the operation fPpg of a TGG synchronisation framework (see Def. 9.7) as a composition of auxiliary operations ⟨fAln, Del, fAdd⟩. Intuitively, the operation fAln removes correspondences that become dangling via the given update, the operation Del computes the maximal consistent sub triple graph of the current state and removes inconsistent elements on the correspondence and target components, but not on the source component. Finally, the operation fAdd propagates the elements on the source component that are not yet consistent already by performing suitable forward transformation steps. Symmetrically, the operations ⟨bAln, Del, bAdd⟩ are used to define the operation bPpg for the backward direction. By Def. 9.7, the propagation operations have to be *total* and *deterministic*, i.e., they have to provide unique results for all inputs. Therefore, we will require that the given TGG provide *deterministic* sets of operational translation rules, meaning that the algorithmic execution of the forward translation, backward translation, and consistency creating rules ensure functional behaviour (unique results) and not require backtracking. For this purpose, additional policies can be defined that restrict the matches of operational translation rules as presented in Sect. 8.3 by Fact 8.43. Rem. 9.22 in Sect. 9.2.3 provides sufficient conditions for deterministic operational translation rules. Additional static conditions and automated checks are provided in [HEO$^+$11b].

The general synchronisation process is performed as follows (see Def. 9.13 and Fig. 9.14, where we use double arrows ($\leftrightarrow$) for correspondence in the signature of the operations, and the explicit triple graphs for the construction details). Given two corresponding models G^S and G^T and an update of G^S via the graph modification $a = (G^S \xleftarrow{a_1} D^S \xrightarrow{a_2} G'^S)$ with $G'^S \in \mathcal{L}(TGG)^S$, the forward propagation fPpg of the model update a is performed in three steps via the auxiliary operations fAln, Del, and fAdd. At first, the deletion performed in a is reflected in the correspondence relation between G^S and G^T by calculating the forward alignment remainder via

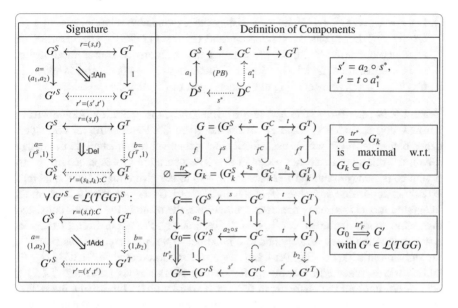

Fig. 9.14 Auxiliary operations fAln, Del and fAdd

the operation fAln. This step deletes all correspondence elements whose elements in G^S have been deleted. In the second step, performed via the operation Del, the two maximal subgraphs $G_k^S \subseteq G^S$ and $G_k^T \subseteq G^T$ are computed such that they form a consistent integrated model in $\mathcal{L}(TGG)$ according to the TGG. All elements that are in G^T but not in G_k^T are deleted, i.e., the new target model is given by G_k^T. Finally, in the last step (operation fAdd), the elements in G'^S that extend G_k^S are transformed to corresponding structures in G'^T, i.e., G_k^T is extended by these new structures. The result of fAdd, and hence also fPpg, is an integrated model $G' = (G'^S \leftrightarrow G'^T)$. Since graph transformation is nondeterministic in general, we require that the sets of operational translation rules be deterministic (see Def. 8.49) in order to ensure unique results for both the second and the third step of the propagation operation fPpg.

Definition 9.13 (Auxiliary TGG operations). Let $TGG = (TG, \varnothing, TR)$ be a TGG with deterministic sets TR_{CC}, TR_{FT}^{+s}, and TR_{BT}^{+s} of operational translation rules and let further $MF(TGG)$ be the derived TGG model framework.

1. The auxiliary operation fAln computing the forward alignment remainder is given by $\text{fAln}(r, a) = r'$, as specified in the upper part of Fig. 9.14. The square marked by (PB) is a pullback (see Def. A.22, [EEPT06]), meaning that D^C is the intersection of D^S and G^C.
2. Let $r = (s, t): G^S \leftrightarrow G^T$ be a correspondence relation; then the result of the auxiliary operation Del is the maximal consistent subgraph $G_k^S \leftrightarrow G_k^T$ of r, given by $\text{Del}(r) = (a, r', b)$, which is specified in the middle part of Fig. 9.14.

3. Let $r = (s, t): G^S \leftrightarrow G^T$ be a consistent correspondence relation, $a = (1, a_2):$ $G^S \rightarrow G'^S$ be a source modification and $G'^S \in \mathcal{L}(TGG)^S$. The result of the auxiliary operation fAdd, for propagating the additions of source modification a, is a consistent model $G'^S \leftrightarrow G'^T$ extending $G^S \leftrightarrow G^T$, and is given by fAdd$(r, a) = (r', b)$, according to the lower part of Fig. 9.14.   △

Remark 9.14 (Auxiliary TGG operations). Intuitively, the operation fAln constructs the new correspondence graph D^C from the given G^C by deleting all correspondence elements in G^C whose associated elements in G^S are deleted via the update a and, for this reason, do not occur in D^S. The operation Del is executed by applying consistency creating rules (see Def. 7.44) to the given integrated model until no rule is applicable anymore. If, at the end, $G^S \leftrightarrow G^T$ is completely marked, the integrated model is already consistent; otherwise, the result is the largest consistent integrated model included in $G^S \leftrightarrow G^T$. Technically, the application of the consistency creating rules corresponds to a maximal triple rule sequence, as shown in the right middle part of Fig. 9.14 and discussed in more detail in [HEO+11a]. Finally, fAdd is executed by applying forward translation rules (see Sect. 7.4.2) to $G'^S \leftrightarrow G^T$ until all the elements in G'^S are marked with **T**. Intuitively, these TGT steps form a model transformation of G'^S extending G^T. Technically, the application of the forward translation rules corresponds to a source-consistent forward sequence from G_0 to G', as shown in the right lower part of Fig. 9.14. By correctness of model transformations (see Theorem 8.7, [EEHP09, HEGO14]), the sequence implies consistency of G' as stated above. The constructions for these auxiliary operations are provided in full detail in [HEO+11b]. Note that the constructions for Del and fAdd yield unique results due to the requirement that the operational translation rules be deterministic (see Def. 9.13).   △

The auxiliary operation Del is based on the execution of consistency creating rules. The computed resulting triple graph G_k is required to be consistent ($G_k \in \mathcal{L}(TGG)$). This result is ensured by the equivalence of maximal triple and complete extended consistency creating sequences according to Fact 9.15 below and shown by Fact 11 in [HEO+11b].

Fact 9.15 (Equivalence of maximal triple and complete extended consistency creating sequences). *Given a set of nonidentic consistency creating rules TR_{CC} and $G \in \mathcal{L}(TG)$, the following statements are equivalent for almost injective matches:*

1. *There is a TGT sequence $s = (\emptyset \overset{tr^*}{\Longrightarrow} G_k)$ via TR with injective embedding $f : G_k \rightarrow G$, such that s is f-maximal, i.e., any extension of s via TR is not compatible with f.*

2. *There is a terminated consistency creating sequence $s' = (G'_0 \overset{tr^*_{CC}}{\Longrightarrow} G'_k)$ via TR_{CC} with $G'_0 = Att^F(G)$, i.e., all translation attributes are set to **F**.*

Moreover, the sequences correspond via $G'_k = G \oplus Att^T_{G_k} \oplus Att^F_{G \setminus G_k}$.   △

Proof. For the full proof, see [HEO+11b].   □

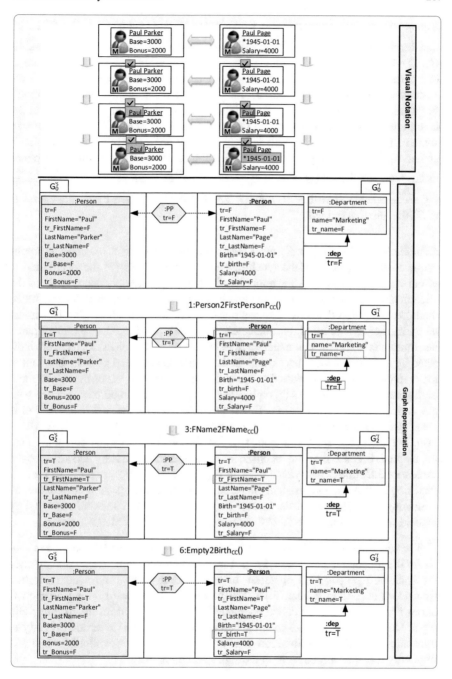

Fig. 9.15 Marking sequence: visual notation and graph representation

Example 9.16 (Marking sequence). Consider the marking sequence in Fig. 9.15 with transformation steps $G'_0 \xRightarrow{\text{1:Person2FirstPersonP}_{CC}} G'_1 \xRightarrow{\text{3:FName2FName}_{CC}} G'_2 \xRightarrow{\text{6:Empty2Birth}_{CC}} G'_3$. The upper part of the figure depicts the steps in visual notation, where consistent parts are indicated by gray boxes with checkmarks. The lower part shows the abstract syntax including the modification of the translation attributes. Each modification is highlighted via a box around the changed translation attribute. All translation attributes of the initial graph $G'_0 = G \oplus Att_G^{\mathbf{F}}$ are set to $\mathbf{F}$, and in each step some markers are set to $\mathbf{T}$. The graph G'_3 still contains some markers with value $\mathbf{F}$ and no further rule is applicable. Thus, the sequence is terminated and corresponds to an f-maximal triple sequence $\emptyset = G_0 \xRightarrow{\text{1:Person2FirstPersonP}} G_1 \xRightarrow{\text{3:FName2FName}} G_2 \xRightarrow{\text{6:Empty2Birth}} G_3$ with $f: G_3 \to G$. The graph G_3 is given by all the elements in G'_3 that are marked with $\mathbf{T}$.   △

Example 9.17 (Forward propagation via operation fPpg*).* Fig. 9.16 shows the application of the three steps of the synchronisation operation fPpg to the visual models of our running example. After removing the dangling correspondence node of the alignment in the first step (fAln), the maximal consistent subgraph of the integrated model is computed (Del) by stepwise marking the consistent parts. Explicit translation markers are omitted, but indicated by visually marking the consistent parts, i.e., those elements whose translation markers are set to $\mathbf{T}$. Consistent parts are indicated by gray boxes with checkmarks in the visual notation and by bold face font in the graph representation. Note that the node Alex Archer is part of the target graph in this maximal consistent subgraph, even though it is not in correspondence with any element of the source graph. This is possible, because the node Alex Archer is connected to a different department (see rule 8:Empty2OtherP in Fig. 9.3). Moreover, the attributes Base and Bonus of Lily Archer in the source component are not marked, because they are inconsistent with the attribute Salary according to the triple rule 5:DetailedSalary2Salary$_{CC}$ in Fig. 9.9 (Base + Bonus $\neq$ Salary). In the final step (fAdd), the inconsistent elements in the target model are removed and the remaining new elements of the update are propagated towards the target model by model transformation, such that all elements are finally marked as consistent.   △

The constructions for the auxiliary operations fAln, Del, and fAdd provide the basis for the propagation operation fPpg. Together with its symmetric version, namely the backward propagation operation bPpg, we derive the TGG synchronisation framework according to Def. 9.18. The forward and backward propagation operations fPpg and bPpg are called *complete* if they yield valid results for any valid input. Completeness of the synchronisation operations is an important property in the context of TGGs, and therefore it is worth emphasising it explicitly, while it is implicitly included already within the signature in Fig. 9.17.

Definition 9.18 (Derived TGG synchronisation framework). Let $TGG = (TG, \emptyset, TR)$ be a TGG with deterministic sets TR_{CC}, TR_{FT}^{+s}, and TR_{BT}^{+s} of derived operational translation rules (consistency creating, source creating forward translation, and target creating backward translation rules) and with derived model framework

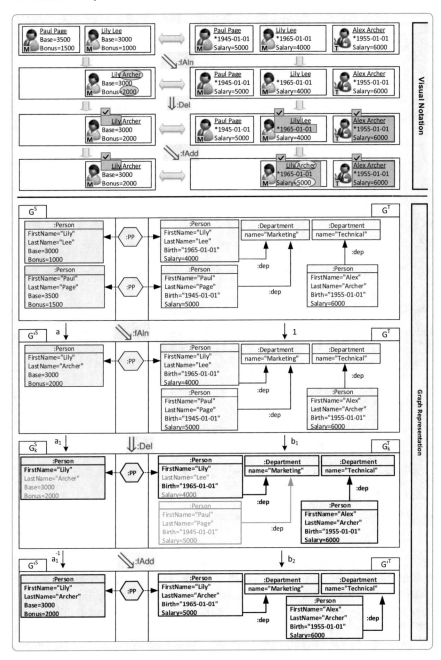

Fig. 9.16 Forward propagation in detail: visual notation and graph representation

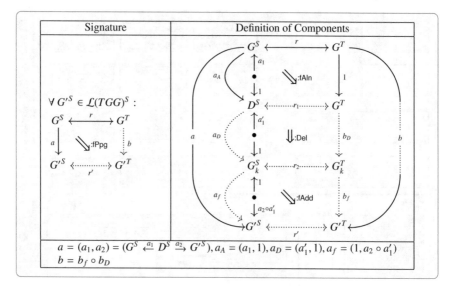

Fig. 9.17 Synchronisation operation fPpg—formal definition

```
/* == alignment remainder == */
forall(correpondence nodes without image in the source model){
   delete these elements }
/* ==== delete === */
while(there is a triple rule tr:L->R with R\L is unmarked in G){
   apply the consistency creating rule of tr to G }
forall(unmarked nodes and edges from the target model){
   delete these elements }
/* ===== add ===== */
while(there is a forward translation rule applicable to G){
   apply to G the forward translation rule }
```

Fig. 9.18 Synchronisation operation fPpg—algorithm

$MF(TGG)$; then the operation fPpg of the *derived TGG synchronisation framework Synch(TGG)* is given by the composition of the auxiliary operations for the forward direction (fAln, Del, fAdd), as described in Rem. 9.19 according to Fig. 9.17. Symmetrically—not shown explicitly—we obtain bPpg as the composition of the auxiliary operations for the backward direction (bAln, Del, bAdd). *Synch(TGG)* is called *complete* if its propagation operations are complete, i.e., they always yield a result for any valid input.   △

Remark 9.19 (Construction of fPpg *according to Fig. 9.17).* Consider a not necessarily consistent integrated model $r: G^S \leftrightarrow G^T$ and a source model update $a: G^S \to G'^S$. If $G'^S \in \mathcal{L}(TGG)^S$, we compute fPpg$(r, a)$ as follows. First, fAln computes the correspondence $(D^S \leftrightarrow G^T)$, where D^S is the part of G^S that is preserved by update a. Then, Del computes its maximal consistent integrated submodel $(G_k^S \leftrightarrow G_k^T)$. Finally, fAdd composes the embedding $G_k^S \rightarrow G'^S$ with correspondence $(G_k^S \leftrightarrow G_k^T)$ leading to $(G'^S \leftrightarrow G_k^T)$, which is then extended into the integrated model $(G'^S \leftrightarrow G'^T)$ via forward transformation. Note that this execution is only possible if $G'^S \in \mathcal{L}(TGG)^S$. If $G'^S \notin \mathcal{L}(TGG)^S$, the above execution fails and the result is given by $b = (1, 1)\colon G^T \rightarrow G^T$, together with the correspondence relation $r' = (\varnothing, \varnothing)$, and additionally an error message is provided. Fig. 9.18 describes this construction algorithmically in pseudocode, leaving out the error handling.   △

Fact 9.20 (Case study: termination of synchronisation operations). *The derived synchronisation operations* fPpg *and* bPpg *for our example TGG terminate.*   △

Proof. According to Def. 9.18, the synchronisation operations are based on the sets TR_{CC}^+, TR_{FT}^{+s}, and TR_{BT}^{+s} of operational translation rules. Hence, we can apply Lem. 8.47 and derive that the synchronisation operations are terminating.   □

9.2.3 Correctness and Invertibility of Model Synchronisation

In this section, we present our main results for unidirectional model synchronisation concerning the properties correctness, completeness and invertibility of the synchronisation framework. According to Def. 9.7, correctness requires that the synchronisation operations ensure laws $(a1)$–$(b2)$ and are deterministic (see Def. 8.49), i.e., they have functional behaviour (see Def. 8.10) and do not require backtracking. Concerning determinism, Theorem 9.23 below provides a sufficient condition based on the notion of critical pairs (see Def. 2.39 and Def. 5.40, based on [EEPT06]). In order to ensure this condition, Sect. 8.3 presents the concept of additional propagation policies that eliminate nondeterminism. They can be seen as application conditions for the rules, and are called *conservative* if they preserve the completeness result. Fact 8.43 provides a sufficient static condition for checking this property and we perform the automated analysis of this condition for our example TGG using the tool AGG [AGG14] as described below. Note again that we generally require almost injective matching (see Def. 7.3 in Sect. 7.4.2).

Example 9.21 (Conservative policy). In Fig. 9.19, the backward translation rule 5:DetailedSalary2Salary$_{BT}$ from Ex. 9.12 is extended to the rule 5': DetailedSalary2Salary$_{BT,2}$ by a policy in the form of an additional application condition in order to ensure determinism. Since the left hand side of this rule specifies only the sum of the salary of a person, the values of the base and bonus components are not fixed via a match. The application condition (see Def. 2.9 and Def. 5.12 based on [EEPT06]) requires that both values be set to half the amount of the salary sum. Now, this is possible for each number, such that we can conclude that the policy is conservative (Fact 8.43), which is important for ensuring completeness of the propagation operation bPpg (see Theorem 9.25).   △

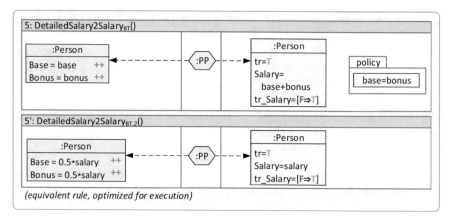

Fig. 9.19 Backward translation rule without (5) and with (5′) conservative policy

Now we investigate the most important property that has to be checked for the operational translation rules in order to ensure correct propagation operations—*deterministic behaviour*. First of all, this means that their execution has *functional behaviour*, i.e., ensures unique results (see Sect. 8.2.1 and Def. 8.10). In addition to that, their execution does not require *backtracking*. This means that once an operational translation rule is applied, we do not have to undo the step during the synchronisation process.

A system of operational translation rules has functional behaviour and does not require backtracking if all significant critical pairs are strictly confluent, as shown by Fact 9 in [HEO⁺11b], based on the corresponding result for forward translation rules (see Theorem 8.29).

Remark 9.22 (Analysis of functional behaviour and backtracking). The tool AGG [AGG14] provides an analysis engine for generating the complete set of critical pairs. On this basis, Rem. 8.51 provides sufficient conditions for deterministic operational translation rules and we provide the analysis results for our example TGG in Fact 9.24.   △

Theorem 9.23 below shows that termination and strict confluence of the set of significant critical pairs ensures the required conditions for deterministic behaviour.

Theorem 9.23 (Deterministic synchronisation operations). *Let TGG be a triple graph grammar and let TR_{CC}^+, TR_{FT}^{+s}, and TR_{BT}^{+t} be the derived sets of kernel translation rules. If the significant critical pairs of the sets of operational translation rules are strictly confluent and the systems of rules are terminating, then the sets of operational translation rules are deterministic (see Def. 8.49), which implies that the derived synchronisation operations fPpg and bPpg are deterministic as well.*   △

Proof (Idea). The operations fAln and bAln are given by pullback constructions, which are unique up to isomorphism by definition. Therefore, they are deterministic.

first \ second	1: Per...	2: Per...	3: FNa...	4: LNa...	5: Det...	6: Em...	7: Em...	8: Em...
1: Person2FirstMarketingP_FT	0	1	1	1	0	1	0	1
2: Person2NextMarketingP_FT	0	0	1	1	0	1	0	0
3: FName2FName_FT	0	0	0	0	0	0	0	0
4: LName2LName_FT	0	0	0	0	0	0	0	0
5: DetailedSalary2Salary_FT	0	0	0	0	0	0	0	0
6: Empty2Birth_FT	0	0	0	0	0	0	0	0
7: Empty2OtherDepartment_FT	0	0	0	0	0	0	0	1
8: Empty2OtherPerson_FT	0	0	0	0	0	1	0	0

Fig. 9.20 Dependency analysis with AGG for TR_{FT}—fields with "1" contain dependencies

Termination of Del, fAdd, and bAdd is ensured according to Lem. 8.47, because the operational translation rules are given by TR_{CC}^+, TR_{FT}^{+s}, and TR_{BT}^{+t}. By Theorem 8.29 we know that functional behaviour of the transformation systems is ensured and backtracking is not required if all significant critical pairs are strictly confluent and the system is terminating. This ensures that the operations Del, fAdd, and bAdd are deterministic. Thus, also the operations fPpg and bPpg are deterministic. For the full proof, see Fact 1 in [HEO+11b]. □

By Fact 9.20, we know that the synchronisation operations in the running example are terminating. Fact 9.24 below shows that the derived sets of operational rules are deterministic and kernel-grounded (see Def. 8.49) using the fact that they are terminating (Theorem 9.23) and using the sets of critical pairs generated with the tool AGG.

Fact 9.24 (Case study: determinism). *The derived sets of operational rules for* fPpg *and* bPpg *of our example TGG are deterministic and kernel-grounded.* △

Proof. We use the critical pair analysis engine of the tool AGG to show that the sets are kernel-grounded and deterministic (see Def. 8.49). First, we show that they are kernel grounded, i.e., the marking changing forward translation rules TR_{FT}^{+s} (backward translation rules TR_{BT}^{+t}) do not depend on the marking preserving rules TR_{FT}^{1s} (TR_{BT}^{1t}).

Concerning the set TR_{FT}, we used AGG to derive the dependency table depicted in Fig. 9.20. The source identic rules are the rules with numbers 6 to 8. There is no dependency (entry > 0) for any pair (p, q) with $p \geq 6$ and $q \leq 5$. Moreover, there are no target identic backward translation rules, because all triple rules are

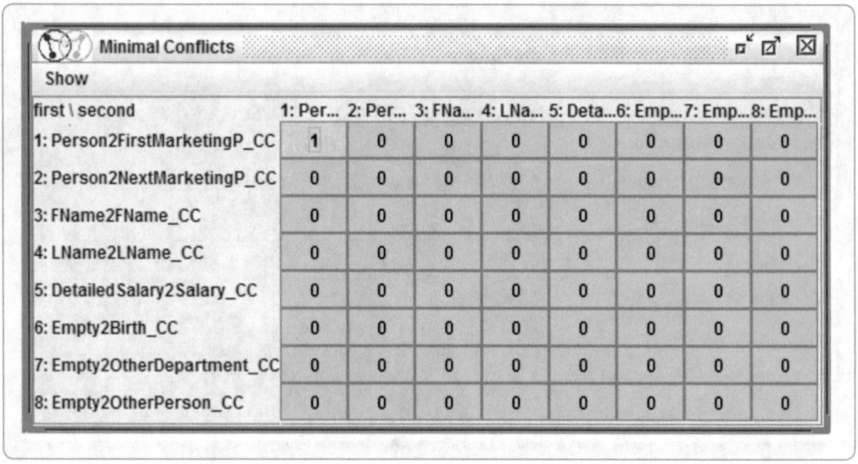

Fig. 9.21 Critical pair analysis with AGG for TR_{CC}—fields with "1" contain conflicts

Fig. 9.22 Critical pair analysis with AGG for TR_{FT}—fields with "1" contain conflicts

creating on the target component. Therefore, the sets of operational translation rules are kernel-grounded.

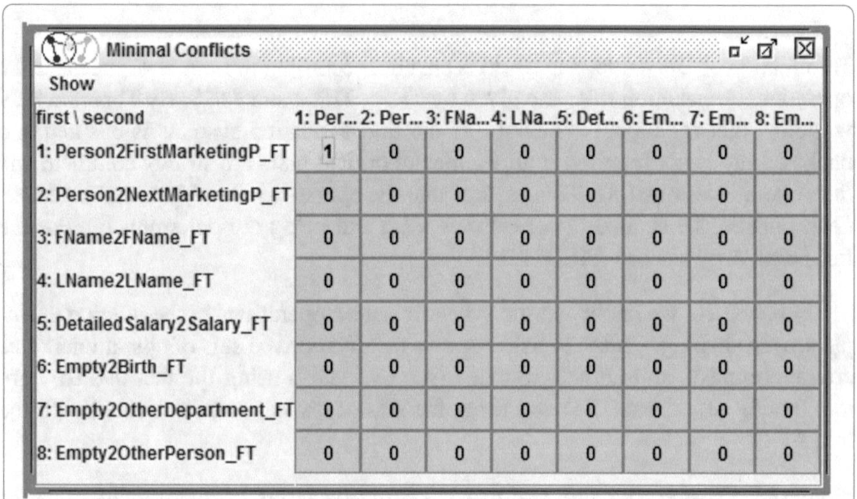

By Fact 9.20, we know that the transformation systems based on the operational translation rules are terminating. Concerning the set TR_{CC}, we derive the resulting table of critical pairs via AGG as depicted in Fig. 9.21. The only generated critical pair is (p_1, p_1) for $p_1 = Person2FirstMarketingP_{CC}$ and it is strictly confluent by

Minimal Dependencies

Show

first \ second	1: Per...	2: Per...	3: FNa...	4: LNa...	5: Det...	6: Em...	7: Em...	8: Em...
1: Person2FirstMarketingP_BT	0	1	1	1	1	1	0	0
2: Person2NextMarketingP_BT	0	0	1	1	1	1	0	0
3: FName2FName_BT	0	0	0	0	0	0	0	0
4: LName2LName_BT	0	0	0	0	0	0	0	0
5: DetailedSalary2Salary_BT	0	0	0	0	0	0	0	0
6: Empty2Birth_BT	0	0	0	0	0	0	0	0
7: Empty2OtherDepartment_BT	0	0	0	0	0	0	0	1
8: Empty2OtherPerson_BT	0	0	0	0	0	0	0	0

Fig. 9.23 Dependency analysis with AGG for TR_{BT}—fields with "1" contain dependencies

Minimal Conflicts

Show

first \ second	1: Per...	2: Per...	3: FNa...	4: LNa...	5: Det...	6: Emp...	7: Emp...	8: Emp...
1: Person2FirstMarketingP_BT	1	0	0	0	0	0	0	0
2: Person2NextMarketingP_BT	0	0	0	0	0	0	0	0
3: FName2FName_BT	0	0	0	0	0	0	0	0
4: LName2LName_BT	0	0	0	0	0	0	0	0
5: DetailedSalary2Salary_BT	0	0	0	0	0	0	0	0
6: Empty2Birth_BT	0	0	0	0	0	0	0	0
7: Empty2OtherDepartment_BT	0	0	0	0	0	0	0	0
8: Empty2OtherPerson_BT	0	0	0	0	0	0	0	0

Fig. 9.24 Critical pair analysis with AGG for TR_{BT}—fields with "1" contain conflicts

applying rule $p_2 = Person2NextMarketingP_{CC}$ to the remaining structure, and since p_2 does not contain any NAC we automatically have strict confluence.

Concerning the set TR_{FT}, we derived the resulting table depicted in Fig. 9.22, where we used the constraint that there are no two departments with name Marketing. This is always ensured for the language $\mathcal{L}(TGG)$ due to the NACs of the first two rules (see Def. 8.25 and Lem. 8.28).

The only significant critical pair is strictly confluent via one transformation step using the rule $p_2 = \text{Person2NextMarketingP}_{FT}$, where no NAC is involved.

The set TR_{BT} is not functional, because there is a choice of how to split the salary into base and bonus. We can restrict the choice for the rule DetailedSalary2Salary to $base = bonus = 1/2 \cdot salary$ as a policy, which is shown by the additional positive application condition in Fig. 9.12. We apply Fact 8.43 and derive that the policy is conservative. First of all, no other rule depends on this rule, which we have verified by the generated dependency table by AGG in Fig. 9.23. Moreover, any match for the original rule implies that there is a match for the restricted rule, because the restricted values are real numbers. We derive the table of generated critical pairs depicted in Fig. 9.24, where the only significant critical pair is again strictly confluent via one transformation step using rule $p_2 = $ Person2NextMarketingP$_{BT}$, where no NAC is involved.

Summing up, the sets of operational translation rules are kernel-grounded and all significant critical pairs are strictly confluent, such that we can apply Theorem 9.23 and derive that the derived sets of operational rules are deterministic.   □

We now analyse correctness and completeness. A correct synchronisation framework has to satisfy laws $(a1)$–$(b2)$ in Def. 9.7. Intuitively, the propagation operations have to preserve consistent inputs. First of all, if the given integrated model is already consistent and the given update does not change anything, then the resulting integrated model has to be the given one and the resulting update on the opposite domain has to be the identity (laws $(a1)$ and $(b1)$). Most importantly, given an arbitrary integrated model together with a source update $d^S : G^S \rightarrow G'^S$ with consistent new source model $G'^S \in \mathcal{L}(TGG)^S$, the forward propagation via fPpg has to provide a new consistent integrated model $G'^S \leftrightarrow G'^T \in \mathcal{L}(TGG)$. *Completeness* of a synchronisation framework *Synch(TGG)* requires that the operations fPpg and bPpg can be successfully applied to all consistent source models $G'^S \in \mathcal{L}(TGG)^S$ and target models $G'^T \in \mathcal{L}(TGG)^T$, respectively. This property is of general importance in the context of TGGs, and therefore we explicitly show it together with correctness in Theorem 9.25 below. Both results are ensured if the sets of the operational rules are deterministic as in Theorem 9.23 and, additionally, if they are kernel-grounded (see Def. 8.49), i.e., the effective forward and backward translation rules do not depend on any source or target identic translation rule, respectively. This second condition is important for laws $(a1)$–$(b2)$, because it ensures that the computed transformation sequences via auxiliary operations Del, fAdd, and bAdd can be composed in a consistent way.

Theorem 9.25 (Correctness and completeness). *Let Synch(TGG) be a derived TGG synchronisation framework such that the sets of operational translation rules derived from TGG are kernel-grounded and deterministic (see Def. 8.49). Then Synch(TGG) is correct and complete.*   △

Proof (Idea). By Theorem 9.23, the provided constructions of operations fPpg and bPpg based on the operational translation rules have functional behaviour, i.e., for each input the computation yields a unique output. Thus, the derived synchronisation framework is complete.

In order to show correctness, we have to show laws $(a1)$ and $(a2)$ of Def. 9.7. The precondition $G \in \mathcal{L}(TGG)$ of law $(a1)$ implies that there is a triple sequence

$\emptyset \overset{tr^*}{\Longrightarrow} G$ via TR, and by Fact 9.15 there is a corresponding complete consistency creating sequence. Moreover, there is a corresponding forward translation sequence via TR_{FT} by Thm. 1 in [HEGO10]. Using the precondition that the operational translation rules are kernel-grounded, we can conclude that all steps via TR_{FT}^{1s} can be shifted to the end. Thus, no further forward translation rule in TR_{FT}^{+s} is applicable. The functional behaviour of the operation fPpg and the given identical source update $d_s = id_{G^S}$ ensure the requested result, i.e., we derive the target update $d^T = id_{G^T}$ and the integrated model $G' = G$. In order to show law $(a2)$, we can use precondition $G'^S \in \mathcal{L}(TGG)^S$, which implies that there is a source consistent forward sequence s_F starting at G'^S and a corresponding complete forward translation sequence. Since the operational rules are kernel-grounded we can conclude by Rem. 8.52 that there is a complete forward translation sequence s_{FT}^{+s} via TR_{FT}^{+s}. Due to functional behaviour of the operation Del we derive a consistency creating sequence that corresponds to the first part of s_F, and therefore to a sequence s_{FT} via forward translation rules. Since the sets of operational rules are kernel-grounded, we can conclude that the steps via TR_{FT}^{+s} do not depend on TR_{FT}^{1s}. This allows us to complete s_{FT} using TR_{FT}^{+s}, where we can shift the source identic steps via TR_{FT}^{1s} to the end. Thus, we derive a complete forward translation sequence, where we can omit the steps via TR_{FT}^{1s} at the end. Functional behaviour of TR_{FT}^{+s} implies that this sequence corresponds to the complete forward translation sequence s_{FT}^{+s}, and therefore to a source consistent forward sequence s_F^{+s} leading to G'. Thus, $G' \in \mathcal{L}(TGG)$ by Theorem 8.7. For the full proof see Lemma 3 in [HEO$^+$11b]. □

The initially derived set of backward transformation rules for our running example is not completely deterministic because of the nondeterministic choice of base and bonus values for propagating the change of a salary value. Therefore, we have defined a conservative policy for the responsible backward triple rule by fixing the propagated values of modified salary values to $bonus = base = 0.5 \cdot salary$. By Fact 8.43 in Sect. 8.3, we have provided a sufficient static condition for checking that a policy is conservative; we have validated our example and have shown that the derived sets of operational rules for fPpg and bPpg are deterministic and kernel-grounded (see Fact 9.24 in Sect. 9.2.2). For this reason, we can apply Theorem 9.25 and conclude that the derived TGG synchronisation framework is correct and complete (see Fact 9.26 below).

Fact 9.26 (Case study: correctness and completeness). *The derived synchronisation framework for our example TGG is correct and complete.* △

Proof. By Fact 9.24, we know that the sets of operational rules of our example TGG are deterministic and kernel-grounded. This allows us to apply Theorem 9.25 and we derive that the derived synchronisation framework is correct and complete. □

Remark 9.27 (Model synchronisation for CD2RDBM). The TGG *CD2RDBM* of our example for Chapters 7 and 8 does not satisfy the conditions we require in Theorem 9.25 to ensure a correct and complete model synchronisation framework. In fact, the operational translation rules are not deterministic as required by Theorem 9.25 for the following reason (see also Ex. 8.50). The backward translation rules are not terminating, because the rule SubClass2Table$_{BT}$ is identic on the target domain and it cannot be omitted, because several other rules depend on it, e.g., an application of rule Attr2Column$_{BT}$ (see Fig. 3.8 for the TGG rules) may depend on the correspondence node that is created by SubClass2Table$_{BT}$. However, from an application point of view, one can argue that the inheritance information of a class diagram gets lost via the forward model transformation, and thus it is just natural to omit the creation of new inheritance links via the backward transformation. Indeed, leaving out the rule SubClass2Table$_{BT}$ seems to be the practical choice to obtain a model synchronisation framework that is possibly correct and complete. Future work may provide alternative conditions that also handle TGGs like *CD2RDBM*.   △

Now we present techniques and results for analysing invertibility of a model synchronisation framework. Intuitively, invertibility means that the propagation operations are inverse to each other (see Def. 9.7). Weak invertibility requires this property for a restricted set of inputs, namely those where the given update on one domain can be interpreted as the result of a propagation of an update from the corresponding opposite domain. In addition to the conditions for ensuring a correct synchronisation framework (Theorem 9.25), the notions of pure and tight TGGs allow us to ensure these properties in Theorem 9.29 below. If the source identic triple rules are empty rules on the source and correspondence components, and analogously for the target-identic triple rules, then we say that the TGG is *pure*. This condition is used to ensure weak invertibility according to Theorem 9.29 below. In the more specific case that all triple rules of a TGG are creating on the source and target components ($TR = TR^{+s} = TR^{+t}$), the TGG is called *tight*, because the derived forward and backward rules are strongly related. Effectively, a tight TGG ensures for the operational forward and backward translation rules that each of them changes at least one translation attribute. In other words, for each triple rule *tr* there is a derived forward translation rule $tr_{FT} \in TR_{FT}^{+s}$ and a derived backward translation rule $tr_{BT} \in TR_{BT}^{+t}$. This additional property ensures invertibility according to Theorem 9.29 below.

Definition 9.28 (Pure and tight TGG). A TGG is called *pure* if $TR^{1s} \subseteq TR_T$ and $TR^{1t} \subseteq TR_S$. It is called *tight* if the sets of source and target creating rules TR^{+s} and TR^{+t} coincide with the set of triple rules TR, i.e., $TR = TR^{+s} = TR^{+t}$.   △

Theorem 9.29 (Invertibility and weak invertibility). *Let Synch(TGG) be a derived TGG synchronisation framework such that the sets of operational translation rules of TGG are kernel-grounded and deterministic (see Def. 8.49), TGG is pure and at most one set of operational translation rules was extended by a conservative policy; then Synch(TGG) is weakly invertible. If, moreover, TGG is tight and there was no policy applied at all, then Synch(TGG) is also invertible.*   △

Proof (Idea). To prove the weak invertibility law (c1) in Fig. 9.5, we can first show that the intermediate triple graphs after applying (bAln, Del) and (fAln, Del) according to Figs. 9.14 and 9.17 are the same for the steps in the last two diagrams of (c1). We compute all three diagrams of (c1) and obtain consistency creating sequences via Del for each diagram using the precondition that the operational rules are deterministic (which subsumes termination). Moreover, we derive that the second and the third diagrams contain the same intermediate triple graph G_I. Afterwards, the auxiliary operations fAdd and bAdd for all three diagrams can be executed. We can use the composition and decomposition result for TGGs and the requirements that the TGG be pure, deterministic and preserve functional behaviour. If at most one set of operational translation rules is extended by a conservative policy, the proof shows that backward transformation sequences are not eliminated by the policy. This allows us to obtain the resulting diagrams according to law (c1). The proof for axiom (c2) follows out of the symmetry of the definitions. To prove invertibility (laws (d1) and (d2)), we use the preconditions that no policy is applied and that the TGG is tight, i.e., all rules are source and target creating. This ensures that for each forward translation sequence there is a corresponding backward translation sequence. For the full proof see Thm. 1 in [HEO$^+$11b], where sets of operational rules are called deterministic if they are kernel-grounded and deterministic using the notions of this chapter.   □

In our example TGG, the sets of operational translation rules are kernel-grounded and deterministic according to Fact 9.24 in Sect. 9.2.2. Moreover, the TGG is pure and we have used the conservative policy for the backward direction only. Thus, Theorem 9.29 ensures that $Synch(TGG)$ is weakly invertible (see Fact 9.30 below).

Fact 9.30 (Case study: weak invertibility). *The derived synchronisation framework for our example TGG is weakly invertible.*   △

Proof. In order to apply Theorem 9.29 concerning weak invertibility, we have to show that the TGG is pure (see Def. 9.28) and at most one set of operational rules was restricted by a conservative policy (see Def. 8.49). The used policy for the set of backward translation rules is conservative, which we have shown already in Fact 9.26. No further policy is applied and the TGG is pure, because each rule is either creating on the source and target component, or it is creating either on the source or the target component and empty on the other components. Therefore, we can apply Theorem 9.29 and derive weak invertibility.   □

An intuitive example for weak invertibility is shown in Ex. 9.8 in Sect. 9.1, where we also show by counterexample that the derived synchronisation framework for our example TGG is not invertible in the general sense. The reason is that information about birth dates is stored in one domain only. The automated validation for our example TGG with eight rules was performed in 25 seconds on a standard consumer notebook via the analysis engine of the tool AGG [AGG14]. We are confident that the scalability of this approach can be significantly improved with additional optimisations.

Remark 9.31 (Applicability of the approach). We have provided sufficient conditions ensuring correctness and completeness (Theorem 9.25) which can be checked statically. In the following, we discuss these restrictions with respect to relevant application scenarios.

1. *Determinism:* Most importantly, we require that the derived sets of operational rules be deterministic, i.e., the forward and backward propagation operations ensure unique results. In several application domains, this property is already a requirement by the domain experts, i.e., has to be ensured anyhow. For example, unique results are often required for the synchronisation between visual models and implementation code, i.e., for code generation and reverse engineering. Note as well that one can modify existing triple rules to enforce determinism based on the discussed critical pair analysis of a TGG using the tool AGG. For example, the designer may insert additional correspondence nodes (trace links) to enforce determinism and avoid conflicts between rules. The condition for determinism does not seem to confine the expressiveness of TGG rules. In a large-scale industrial project, we have used a TGG for the fully automated translation of satellite control software [HGN⁺14], where the used TGG contains more than 200 rules. The forward translation has functional behaviour as required by the industrial partner. As a general recommendation based on the experiences from this project, we can state that a designer of a TGG should divide the rules in small groups, such that there are no cyclic dependencies between the groups.

2. *Kernel-grounded sets of operational rules:* Intuitively, the restriction to kernel-grounded rules concerns the possibility that one domain may contain information that is not present in the corresponding opposite domain. When translating from one domain to another, we apply only those rules that are changing at least one translation attribute (TR_{FT}^{+s} and TR_{BT}^{+t}). Thus, we require that the structures that concern only one domain be handled separately by triple rules that are the identity on the corresponding opposite domain (TR_{FT}^{1s} and TR_{BT}^{1t}). In addition to that, these sets of rules do not create structures that may be needed by the first group of rules. This means that the restriction to kernel-grounded sets of operational rules restricts the freedom mainly of how to design the TGG and usually not of the problem and application domain.

The result on invertibility (Theorem 9.29) requires additional properties. Weak invertibility is ensured if the TGG is pure and at most one of the sets of operational rules is extended by a conservative policy. While this condition is not very restrictive in the experience of the authors, the stronger condition for invertibility requiring a tight TGG practically means that all pieces of information in one domain are also reflected in the corresponding opposite domain. This result is consistent with Diskin et al.'s analysis of strong invertibility [DXC⁺11b]. △

In the case that the specified TGG does not ensure deterministic synchronisation operations, there are still two options for performing synchronisation that ensure correctness and completeness. On the one hand, the triple rules can be modified in a suitable way, such that the TGG can be verified to be deterministic. For this purpose, the critical pair analysis engine of the tool AGG [AGG14] can be used to analyse conflicts between the generated operational translation rules. Moreover, backtracking can be reduced or even eliminated by generating additional application conditions for the operational translation rules using the automatic generation of filter NACs (see Fact 8.18 based on [HEGO10, HEGO14]). On the other hand, the

TGG can be used directly, leading to nondeterministic synchronisation operations, which may provide several possible synchronisation results [GHN⁺13].

9.3 Concurrent Model Synchronisation

Based on the basic framework for model synchronisation in the previous sections, we now provide a correct TGG framework for *concurrent* model synchronisation, where concurrent model updates in different domains have to be merged into a consistent solution. In this case, we have the additional problem of detecting and solving conflicts between given updates. Such conflicts may be hard to detect, since they may be caused by concurrent updates on apparently unrelated elements of the given models. Furthermore, there may be apparently contradictory updates on related elements of the given domains which may not be real conflicts.

This section is based on [HEEO12]. The main idea and results for the approach for concurrent model synchronisation based on TGGs are as follows:

1. Model synchronisation is performed by propagating the changes from one model of one domain to a corresponding model in another domain using forward and backward propagation operations. The propagated changes are compared with the given local update. Possible conflicts are resolved in a semiautomated way.
2. The operations are realised by model transformations based on TGGs [HEO⁺11a] and tentative merge constructions solving conflicts [EET11]. The specified TGG also defines consistency of source and target models.
3. In general, the operation of model synchronisation is nondeterministic, since there may be several conflict resolutions. The different possible solutions can be visualised to the modellers, who then decide which modifications to accept or discard.
4. The main result shows that the concurrent TGG synchronisation framework is correct and compatible with the basic synchronisation framework (see Sect. 9.2), where only single updates are considered at the same time.

9.3.1 Concurrent Synchronisation Problem

Concurrent model synchronisation aims to provide a consistent merging solution for a pair of concurrent updates that are performed on two interrelated models. This section provides a formal specification of the concurrent synchronisation problem and the corresponding notion of correctness. At first, we motivate the general problem with a compact example.

Example 9.32 (Concurrent model synchronisation problem). Fig. 9.25 shows two models in correspondence. Two model updates have to be synchronised concurrently: on the source side (model update d_1^S), the node Paul Page is deleted and

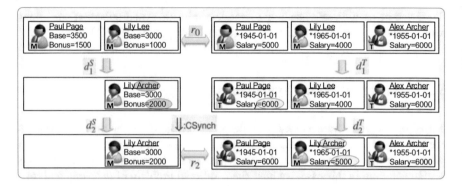

Fig. 9.25 Concurrent model synchronisation: compact example

the family name of Lilly Lee changes due to her marriage; moreover, since she is married, her bonus is raised from 1,000 to 2,000. On the target side (model update d_1^T), Paul Page is switching from the marketing to the technical department (in the visualisation in Fig. 9.25 this is indicated by a different role icon and the label M is replaced by label T). The department change is combined with a salary raise from 5,000 to 6,000. After performing the updates d_2^S and d_2^T, a "consistently integrated model" is derived that reflects as many changes as possible from the original updates in both domains and resolves inconsistencies, e.g., by computing the new Salary of Lily Lee in the target domain as the sum of the updated source attributes Base and Bonus. Note that Paul Page is not deleted in the target domain by the concurrent model synchronisation because in this case the changes required by d_1^T could not have been realised. This conflict can be considered an apparent one. If a person leaves the marketing department, but not the company, its node should remain in the target model. Thus, a concurrent model synchronisation technique has to include an adequate conflict resolution strategy.   △

The concurrent model synchronisation problem is visualised in Fig. 9.26, where we use solid lines for the inputs and dashed lines for the outputs. Given an integrated model $G_0 = (G_0^S \leftrightarrow G_0^T)$ and two model updates $d_1^S = (G_0^S \rightarrow G_1^S)$ and $d_1^T = (G_0^T \rightarrow G_1^T)$, the required result consists of updates $d_2^S = (G_1^S \rightarrow G_2^S)$ and $d_2^T = (G_1^T \rightarrow G_2^T)$ and a consistently integrated model $G_2 = (G_2^S \leftrightarrow G_2^T)$. The solution for this problem is a concurrent synchronisation operation CSynch, which is left total but in general nondeterministic, which we indicate by a wiggly arrow "$\rightsquigarrow$" in Def. 9.33 below. The set of inputs is given by $(Rel \otimes \Delta_S \otimes \Delta_T) = \{(r, d^S, d^T) \in Rel \times \Delta_S \times \Delta_T \mid r: G_0^S \leftrightarrow G_0^T, d^S: G_0^S \rightarrow G_2{}^S, d^T: G_0^T \rightarrow G_2{}^T\}$, i.e., r coincides with d^S on G_0^S and with d^T on G_0^T.

Definition 9.33 (Concurrent model synchronisation problem and framework).
Given a triple graph grammar TGG, the *concurrent model synchronisation problem* is to construct a left total and nondeterministic operation CSynch : $(Rel \otimes \Delta_S \otimes \Delta_T) \rightsquigarrow (Rel \times \Delta_S \times \Delta_T)$ leading to the signature diagram in Fig. 9.26, called concurrent syn-

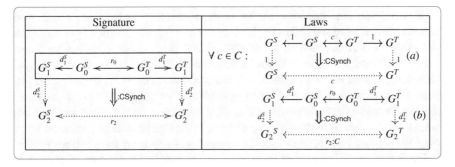

Fig. 9.26 Signature and laws for correct concurrent model synchronisation frameworks

chronisation tile with concurrent synchronisation operation CSynch. Given a pair $(prem, sol) \in$ CSynch, the triple $prem = (r_0, d_1^S, d_1^T) \in Rel \otimes \Delta_S \otimes \Delta_T$ is called premise and $sol = (r_2, d_2^S, d_2^T) \in Rel \times \Delta_S \times \Delta_T$ is called a solution of the synchronisation problem, written $sol \in$ CSynch$(prem)$. The operation CSynch is called *correct* with respect to the consistency relation C if laws (a) and (b) in Fig. 9.26 are satisfied for all solutions. Given a concurrent synchronisation operation CSynch, the *concurrent synchronisation framework CSynch* is given by $CSynch = (TGG, \text{CSynch})$. It is called *correct* if the operation CSynch is correct.   △

Correctness of a concurrent synchronisation operation CSynch according to Fig. 9.26 ensures that any resulting integrated model $G_2 = (G_2^S \leftrightarrow G_2^T)$ is consistent (law (b)) and the synchronisation of an unchanged and already consistently integrated model always yields the identity of the input as output (law (a)).

9.3.2 Concurrent Model Synchronisation with Conflict Resolution

In addition to the propagation operations used for the basic model synchronisation framework in Sect. 9.2, the concurrent case requires additional steps. The most important one is conflict resolution, and we use the constructions and results for conflict resolution in a single domain according to [EET11]. Note that we apply conflict resolution either to two conflicting target model updates (one of them induced by a forward propagation operation fPpg) or to two conflicting source model updates (one of them induced by backward propagation). Hence, we here consider updates over *standard graphs* and not over triple graphs. Moreover, we use additional TGG-specific operations that restrict the constructed intermediate models to those that are consistent with the given TGG.

Two graph modifications $(G \leftarrow D_i \rightarrow H_i)$, $(i = 1, 2)$ are called *conflict-free* if they do not interfere with each other, i.e., if one modification does not delete a graph element, the other one needs to perform its changes. Conflict-free graph modifications can be merged to one graph modification $(G \leftarrow D \rightarrow H)$ that realises both original graph modifications simultaneously.

If two graph modifications are not conflict-free, then at least one conflict occurs which can be of the following kinds: (1) *delete–delete conflict*: both modifications delete the same graph element, or (2) *delete–insert conflict*: m_1 deletes a node which shall be the source or target of a new edge inserted by m_2 (or vice versa). Of course, several of such conflicts may occur simultaneously. In [EET11], we propose a *merge construction* that resolves conflicts by giving *insertion* priority over *deletion* in case of delete–insert conflicts. The result is a merged graph modification where the changes of both original graph modifications are realised as far as possible.[1] We call this construction *tentative* merge because usually the modeller is asked to finish the conflict resolution manually, e.g., by opting for deletion instead of insertion of certain conflicting elements. The resolution strategy to prioritise insertion over deletion preserves all model elements that are parts of conflicts and allows us to highlight these elements to support manual conflict resolution. We summarise the main effects of the conflict resolution strategy by Fact 9.34 below (see also Thm. 3 in [EET11] for the construction).

Fact 9.34 (Conflict resolution by tentative merge construction). *Given two conflicting graph modifications* $m_i = G \overset{D_i}{\Longrightarrow} H_i$ $(i = 1, 2)$ *(i.e., they are not conflict-free). The tentative merge construction yields the merged graph modification* $m = (G \leftarrow \overline{D} \to H)$ *and resolves conflicts as follows:*

1. *If* (m_1, m_2) *are in* delete–delete *conflict, with both* m_1 *and* m_2 *deleting* $x \in G$, *then* x *is deleted by* m.
2. *If* (m_1, m_2) *are in* delete–insert *conflict, there is an edge* e_2 *created by* m_2 *with* $x = s(e_2)$ *or* $x = t(e_2)$ *preserved by* m_2, *but deleted by* m_1. *Then* x *is preserved by* m *(and vice versa for* (m_2, m_1) *in delete–insert conflict).*   △

Note that attributed nodes, which shall be deleted on the one hand and change their values on the other hand, would cause delete/insert–conflicts and therefore would not be deleted by the tentative merge construction. Attributes which are differently changed by both modifications would lead (tentatively) to attributes with two values. In many cases, the domain languages require single-valued attributes, which means that the user has to restore the conflict and choose the value for the attribute.

Throughout the paper, we depict conflict resolution based on the tentative merge construction and manual modifications as shown to the right, where m_1 and m_2 are conflicting graph modifications, and H is their merge after conflict resolution. The dashed lines correspond to derived graph modifications $(G_1 \leftarrow D_3 \to H)$ and $(G_2 \leftarrow D_4 \to H)$ with interfaces D_3 and D_4.

Example 9.35 (Conflict resolution by tentative merge construction). Consider the conflict resolution square 3:Res in the upper right part of Fig. 9.29. The first modification $d_{1,F}^T$ deletes the node for Paul Page and updates the attribute values for

---

[1] Note that the conflict-free case is a special case of the tentative merge construction.

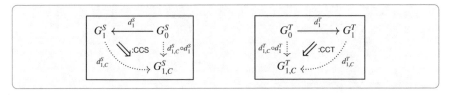

Fig. 9.27 Consistency creating operations

Surname and Salary of Lily Lee. The second modification d_1^T relinks the node of Paul Page from the marketing department to the technical department and updates his Salary attribute. The result of the tentative merge construction keeps the Paul Page node, due to the policy that nodes that are needed as source or target for newly inserted edges or attributes will be preserved. Technically, the attribute values are not preserved automatically. This means that the tentative merge construction only yields the structure node of Paul Page (and the updated attribute), and the modeller should confirm that the remaining attribute values should be preserved (this is necessary for the attribute values for FirstName, LastName and Birth of the node for Paul Page).

Variant: As a slight variant to the above example, let us consider the case that the modification d_1^T also modifies the surname of Lily from "Lee" to "Smith". Since the same attribute is updated differently by both modifications, we now have two tentative attribute values for this attribute (we would indicate this by <Archer|Smith> as attribute value for the Surname attribute of Lily). This can be solved by the user as well, who should select the proper attribute value.   △

The merge construction cannot be applied directly to detect and solve conflicts in concurrent model synchronisation. The problem is that source and target updates occur in different graphs and not the same one. To solve this problem we use the forward and backward propagation operations (Sect. 9.2), allowing us to see the effects of each source or target update on the other domain, so that we can apply the merge construction. In addition, we use two further operations, CCS and CCT, to reduce a given domain model to a maximal consistent submodel according to the TGG.

Given a source update $d_1^S : G_0^S \rightarrow G_1^S$, the consistency creating operation CCS (left part of Fig. 9.27) computes a maximal consistent subgraph $G_{1,C}^S \in \mathcal{L}(TGG)^S$ of the given source model G_1^S. The resulting update from G_0^S to G_1^S is derived by update composition $d_{1,C}^S \circ d_1^S$. The dual operation CCT (right part of Fig. 9.27) works analogously on the target component.

Remark 9.36 (Execution of consistency creating operation CCS). Given a source model G_1^S, the consistency creating operation CCS is executed by computing terminated forward sequences $(H_0 \xRightarrow{tr_F^*} H_n)$ with $H_0 = (G_1^S \leftarrow \varnothing \rightarrow \varnothing)$. If the sets of operational rules of the TGG are deterministic (see Def. 8.49 and Rem. 8.51), then backtracking is not necessary. If G_1^S is already consistent, then $G_{1,C}^S = G_1^S$,

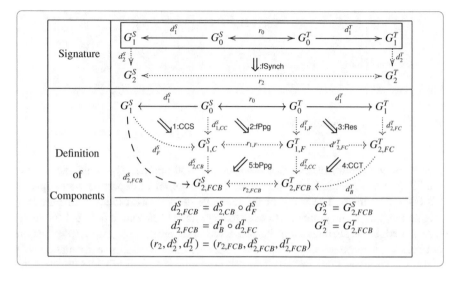

Fig. 9.28 Concurrent model synchronisation with conflict resolution (forward case: fSynch)

which can be checked via the operation CCS. Otherwise, the operation CCS creates a maximal consistent subgraph $G^S_{1,C}$ of G^S_1. $G^S_{1,C}$ is maximal in the sense that there is no larger consistent submodel H^S of G^S_1, i.e., with $G^S_{1,C} \subseteq H^S \subseteq G^S_1$ and $H^S \in \mathcal{L}(TGG)^S$. From the practical point of view, the operation CCS is performed using forward translation rules (see Sect. 7.4), which mark in each step the elements of a given source model that have been translated so far. This construction is well defined due to the equivalence with the corresponding triple sequence $(\varnothing \xRightarrow{tr^*} H_n)$ via the triple rules TR of the TGG (see Fact 7.36 and App. B in [HEEO11]).   △

The concurrent model synchronisation operation CSynch derived from the given TGG is executed in five steps. Moreover, it combines the operations fSynch and bSynch depending on the order in which the steps are performed. The used propagation operations fPpg, bPpg are required to be correct and we can take the derived propagation operations according to Sect. 9.2. The steps of the operation fSynch are depicted in Fig. 9.28 and Construction 9.37 describes the steps for both operations.

Construction 9.37 (Operations fSynch *and* bSynch*).* In the first step (operation CCS), a maximal consistent subgraph $G^S_{1,C} \in \mathcal{L}(TGG)^S$ of G^S_1 is computed (see Rem. 9.36). In Step 2, the update $d^S_{1,CC}$ is forward propagated to the target domain via the operation fPpg. This leads to the pair $(r_{1,F}, d^T_{1,F})$, and thus to the pair $(d^T_{1,F}, d^T_1)$ of target updates, which may show conflicts. Step 3 applies the conflict resolution operation Res including optional manual modifications. In order to ensure consistency of the resulting target model $G^T_{2,FC}$, we apply the consistency creating operation CCT (see Rem. 9.36) for the target domain and derive the target model $G^T_{2,FCB} \in \mathcal{L}(TGG)^T$ in Step 4. Finally, the derived target update $d^T_{2,CC}$ is

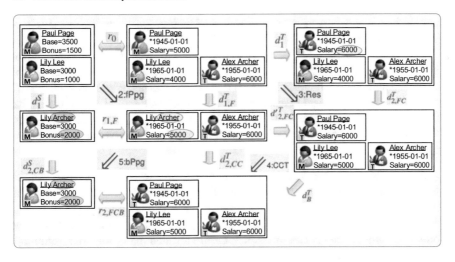

Fig. 9.29 Concurrent model synchronisation with conflict resolution applied to the organisational model example

backward propagated to the source domain via the operation bPpg, leading to the source model $G_{2,FCB}^S$ and the source update $d_{2,CB}^S$. Altogether, we have constructed a nondeterministic solution (r_2, d_2^S, d_2^T) of the operation fSynch for the premise (r_0, d_1^S, d_1^T) with $(r_2, d_2^S, d_2^T) = (r_{2,FCB}, d_{2,FCB}^S, d_{2,FCB}^T)$ (see Fig. 9.28). The concurrent synchronisation operation bSynch is executed analogously via the dual constructions. Starting with CCT in Step 1, it continues via bPpg in Step 2, Res in Step 3, and CCS in Step 4, and finishes with fPpg in Step 5. The nondeterministic operation CSynch = (fSynch ∪ bSynch) is obtained by joining the two concurrent synchronisation operations fSynch and bSynch. △

Example 9.38 (Concurrent model synchronisation with conflict resolution). The steps in Fig. 9.29 specify the execution of the concurrent synchronisation in Ex. 9.32. Since the given model G_0^S is consistent, Step 1 (1:CCS) can be omitted, i.e., $G_{1,C}^S = G_1^S$ and $d_{1,CC}^S = d_1^S$. Step 2:fPpg propagates the source update to the target domain: The attributes of the node for Lilly Lee are updated and the node representing Paul Page is deleted. The resolution 3:Res resolves the conflict between the target model update d_1^T and the propagated source model update on the target side $d_{1,F}^T$ (see Ex. 9.35). We assume that the user selected the old attribute value for the birthday of Paul Page. Step 4:CCT does not change anything, since the model is consistent already. Finally, all elements that were introduced during the conflict resolution and concern the source domain are propagated to the source model via (5:bPpg). This concerns only Paul Page, who now is assigned to the technical department. According to the TGG, such persons are not reflected in the source model, such that the backward propagation does not change anything in the source model. The result of the concurrent model synchronisation with conflict resolution is $r_{2,FCB}$, where as many as possible of both proposed update changes have been kept and insertion got priority over deletion.

Variant: Let us consider the case that both modifications d_1^T $d_{1,F}^T$ insert additionally an edge of type married between the nodes of Lilly Lee and Alex Archer. The conflict resolution operation 3:Res would yield two married edges between the two nodes. But the subsequent consistency creating operation 4:CCT would detect that this is an inconsistent state and would delete one of the two married edges. Note that the user can already detect this conflict in Step 3 and resolve it by deleting one of the edges.   △

Remark 9.39 (Execution and termination of concurrent model synchronisation). Note that the efficiency of the execution of the concurrent synchronisation operations can be significantly improved by reusing parts of previously computed transformation sequences as described in App. B in [HEEO11]. In [HEO⁺11a], we have provided sufficient static conditions that ensure termination of the propagation operations and they can be applied similarly for the consistency creating operations. Update cycles cannot occur, because the second propagation step does not lead to a new conflict.   △

Note that the operation CSynch is nondeterministic for several reasons: the choice between fSynch and bSynch, the reduction of domain models to maximal consistent sub graphs, and the semi-automated conflict resolution strategy.

Definition 9.40 (Derived concurrent TGG synchronisation framework). Let fPpg and bPpg be correct basic synchronisation operations for a triple graph grammar *TGG* and let the operation CSynch be derived from fPpg and bPpg according to Construction 9.37. Then the *derived concurrent TGG synchronisation framework* is given by $CSynch = (TGG, \text{CSynch})$.   △

9.3.3 Correctness and Compatibility

Our main results show correctness of the derived concurrent TGG synchronisation framework (Def. 9.40) and its compatibility with the derived basic TGG synchronisation framework (Sect. 9.2). Correctness of a concurrent model synchronisation framework requires that the nondeterministic synchronisation operation CSynch ensures laws (*a*) and (*b*) in Def. 9.33. In other words, CSynch guarantees consistency of the resulting integrated model and, moreover, the synchronisation of an unchanged and already consistently integrated model always yields the identity of the input as output (law (*a*)).

According to Theorem 9.41 below, correctness of given forward and backward propagation operations ensures correctness of the concurrent model synchronisation framework.

Theorem 9.41 (Correctness of concurrent model synchronisation). *Let* fPpg *and* bPpg *be correct basic synchronisation operations for a triple graph gram-*

mar TGG. Then the derived concurrent TGG synchronisation framework CSynch =
(TGG, CSynch) (see Def. 9.40) is correct (see Def. 9.33). △

Proof.

Law *(a)* **in Fig. 9.26:** Let $(r, d_1^S, d_1^T) \in$ $(R \otimes \Delta_S \otimes \Delta_T)$ with $r = c \in C =$ $\mathcal{L}(TGG)$ and identities $d_1^S = id^S :$ $G_0^S \to G_0^S$, $d_1^T = id^T : G_0^T \to G_0^T$, such that $G = (G^S \leftrightarrow G^T) =$ $(G_0^S \leftrightarrow G_0^T) = G_0$.

$$G^S \xleftarrow{1} G^S \xleftrightarrow{c} G^T \xrightarrow{1} G^T$$
$$\forall c \in C: \quad 1 \Big\downarrow \qquad \Downarrow:\text{CSynch} \qquad \Big\downarrow 1 \quad (a)$$
$$G^S \dashleftarrow{\cdots\cdots\cdots\cdots\cdots}_{c} \dashrightarrow G^T$$

We have to show that the operation CSynch yields (c, id^S, id^T) as result, i.e., no further result is possible.

We apply the operation fSynch according to Fig. 9.28 and Construction 9.37. Since $r = G_0 \in \mathcal{L}(TGG)$ and $G_1^S = G_0^S \in \mathcal{L}(TGG)^S$ we know that there is a model transformation sequence $s_F = (G_1^S, G_0' \xRightarrow{tr_F^*} G_n', G_1^T)$ based on forward rules using the completeness result for model transformations based on forward rules (see Theorem 8.4). Therefore, the operation CCS yields the maximal consistent subgraph $G_{1,C}^S = G_0^S$ and update $d_{1,CC}^S = id^S$. By correctness of the operation fPpg (law $(a1)$ in Fig. 9.5) we derive that the second step yields the target model $G_{1,F}^T = G_0^T$, the correspondence $r_{1,F} = r$ and the target update $d_{1,F}^T = id^T$. Therefore, the resolution in Step 3 concerns the updates $d_1^T = id^T$ and $d_{1,F}^T = id^T$, which are parallel independent by definition, leading to the merging result $G_{2,FC}^T = G_0^T$ and updates $d_{2,FC}^T = id^T$ and $d'_{2,FC}^T = id^T$. This means that manual modification is not necessary and therefore not executed. Again, since $G_0^T \in \mathcal{L}(TGG)^T$ we know that there is a model transformation sequence $s_B = (G_0^T, H_0' \xRightarrow{tr_B^*} H_n', G_1^S)$ based on backward translation rules using the dual version of the completeness result for model transformations based on forward rules (see Theorem 8.4). Therefore, the operation CCT yields the maximal consistent subgraph $G_{2,FCB}^T = G_0^T$ and update $d_B^T = id^T$. By correctness of the operation bPpg (law $(b1)$ in Fig. 9.5) we derive that Step 5 yields the source model $G_{2,FCB}^S = G_0^S$, the correspondence $r_{2,FCB} = r_{1,F} = r = c$ and the source update $d_{2,CB}^S = id^S$. Altogether, the operation fSynch yields the result (c, id^S, id^T) as required by law (a). The same result holds for the operation bSynch using the symmetry of the precondition and the symmetric definition of TGGs and the derived operations. Therefore, the result holds for the concurrent synchronisation operation CSynch.

Law *(b)* **in Fig. 9.26:** Let $(r_0, d_1^S, d_1^T) \in (R \otimes \Delta_S \otimes \Delta_T)$ as depicted on the right and in Fig. 9.26. We have to show that the concurrent synchronisation operation yields a consistent correspondence $r_2 \in \mathcal{L}(TGG)$.

$$G_1^S \xleftarrow{d_1^S} G_0^S \xleftrightarrow{r_0} G_0^T \xrightarrow{d_1^T} G_1^T$$
$$d_2^S \Big\downarrow \qquad \Downarrow:\text{CSynch} \qquad \Big\downarrow d_2^T \quad (b)$$
$$G_2^S \dashleftarrow{\cdots\cdots\cdots\cdots\cdots}_{r_2:C} \dashrightarrow G_2^T$$

We apply the concurrent synchronisation operation fSynch according to Fig. 9.28 and Construction 9.37. All steps are well

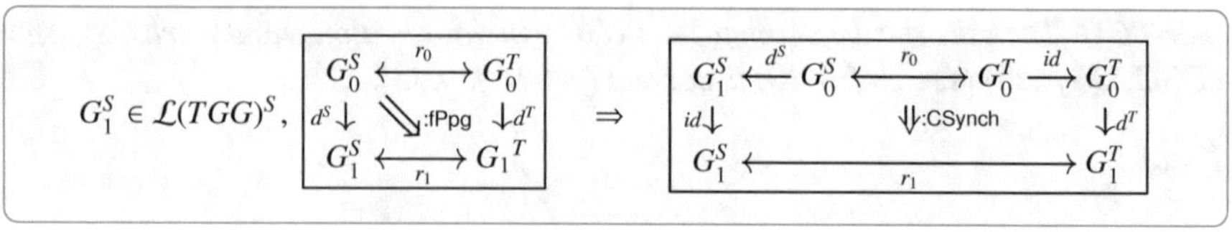

$$G_1^S \in \mathcal{L}(TGG)^S, \quad \begin{array}{ccc} G_0^S & \xrightarrow{r_0} & G_0^T \\ d^S \downarrow & \searrow_{:fPpg} & \downarrow d^T \\ G_1^S & \xrightarrow{r_1} & G_1^T \end{array} \quad \Rightarrow \quad \begin{array}{ccccc} G_1^S & \xleftarrow{d^S} & G_0^S & \xrightarrow{r_0} & G_0^T & \xrightarrow{id} & G_0^T \\ id \downarrow & & & \Downarrow_{:CSynch} & & & \downarrow d^T \\ G_1^S & & \xleftarrow{\hspace{5cm}} & & & & G_1^T \\ & & & r_1 & & & \end{array}$$

Fig. 9.30 Compatibility with synchronisation of single updates (forward case)

defined according to Rem. 9.39. Step 4 (operation CCT) provides a maximal consistent subgraph $G_{2,FCB}^T \in \mathcal{L}(TGG)^T$. Therefore, we can apply law (b2) in Fig. 9.5 and derive that the operation bPpg in Step 5 yields a consistent correspondence r_2 (see Theorem 9.25). The proof for the concurrent synchronisation operation bSynch is analogous using the symmetric definition of TGGs and the dual definitions of the steps according to Rem. 9.39. Therefore, the concurrent synchronisation operation CSynch always yields a consistent correspondence $r_2 \in \mathcal{L}(TGG)$.   □

Example 9.42 (Correctness and compatibility). In Sect. 9.2, we have presented a suitable realisation of correct propagation operations derived from the given TGG. This allows us to apply the following main results in Theorem 9.41 and Theorem 9.44 to our case study and we derive a concurrent model synchronisation framework that is correct and compatible with the basic model synchronisation framework.   △

In addition to correctness, we show that the concurrent TGG synchronisation framework is compatible with the basic synchronisation framework. This means that the propagation operations (fPpg, bPpg) (see Sect. 9.2) provide the same result as the concurrent synchronisation operation CSynch if one update of one domain is the identity. Fig. 9.30 visualises the case for the forward propagation operation fPpg. Given a forward propagation (depicted left) with solution (r_1, d^T), a specific solution of the corresponding concurrent synchronisation problem (depicted right) is given by $sol = (r_1, id, d^T)$, i.e., the resulting integrated model and the resulting updates are the same. Due to the symmetric definition of TGGs, we can show the same result concerning the backward propagation operation, leading to the general result of compatibility in Theorem 9.44.

Definition 9.43 (Compatibility of concurrent with basic model synchronisation). Let fPpg, bPpg be basic TGG synchronisation operations and let CSynch be a concurrent TGG synchronisation operation for a given TGG. The nondeterministic synchronisation operation CSynch is *compatible* with the propagation operations fPpg and bPpg if the following condition holds for the forward case (see Fig. 9.30) and a similar one for the backward case:

$$\forall (d^S, r_0) \in \Delta_S \otimes Rel, \text{ with } (d^S : G_0^S \to G_1^S) \wedge (G_1^S \in \mathcal{L}(TGG)^S):$$
$$(id, fPpg(d^S, r_0)) \in CSynch(d^S, r_0, id)$$

                                                                              △

Theorem 9.44 (Compatibility of concurrent with basic model synchronisation).
Let fPpg *and* bPpg *be correct basic synchronisation operations for a given TGG and let the operation* CSynch *be derived from* fPpg *and* bPpg *according to Construction 9.37. Then, the* derived concurrent TGG synchronisation operation CSynch *is compatible with propagation operations* fPpg, bPpg.   △

Proof. Let CSynch be obtained from the derived forward and backward synchronisation operations fSynch and bSynch, i.e., CSynch = (fSynch ∪ bSynch). According to Def. 9.43, we have to show for the forward case that $\forall$ $(d^S, r_0) \in \Delta_S \otimes R$ with $d^S : G_0^S \to G_1^S \wedge G_1^S \in \mathcal{L}(TGG)^S$ it holds that $(id, \text{fPpg}(d^S, r_0)) \in \text{CSynch}(d^S, r_0, id)$. The result for the backward case holds by dualisation due to the symmetric definition of TGGs and the derived operations.

Let $r_0 = (G_0^S \leftrightarrow G_0^T)$ with $G_0^S \in \mathcal{L}(TGG)^S$ and let $d^S : G_0^S \to G_1^S$ be a source model update. Let further (r_1, d^T) be the result of applying fPpg to (d^S, r_0) with $r_1 = (G_1^S \leftrightarrow G_1^T)$ and $d^T : G_0^T \to G_1^T$. We show that $(id, r_1, d^T) \in \text{fSynch}(d^S, r_0, id)$ according to Fig. 9.28 and Construction 9.37. Since $G_1^S \in \mathcal{L}(TGG)^S$, we have that Step 1 (CCS) yields the maximal consistent subgraph $G_{1,C}^S = G_1^S$ (see Rem. 9.36) and source update $d_F^S = id: G_1^S \to G_1^S$. Thus, Step 2 applies the operation fPpg to the same input as in the precondition, such that we derive the correspondence $r_{1,F} = r_1$ and target model update $d_{1,F}^T = d^T$. By correctness of fPpg (law (a2) in Fig. 9.5) we know that $r_1 = r_{1,F} \in \mathcal{L}(TGG)$, and therefore $G_1^T \in \mathcal{L}(TGG)^T$. In Step 3 (operation Res), the merge construction is applied to d^T and the target update id, which does not delete or create anything (the corresponding minimal rule is the empty rule). This means that the updates are conflict-free and the merge construction yields the target updates $d'^T_{2,FC} = id$ and $d^T_{2,FC} = d^T$. Since $G_1^T \in \mathcal{L}(TGG)^T$ (see Step 2 above), Step 4 yields the maximal consistent subgraph $G_{2,FCB}^T = G_1^T$ (see Rem. 9.36) and the target update $d_B^T = id: G_1^T \to G_1^T$. Therefore, the target update $d_{2,CC}^T = d_B^T \circ d'^T_{2,FC}$ is given by $d_{2,CC}^T = id$. By $r_1 = r_{1,F} \in \mathcal{L}(TGG)$ (see Step 2 above) and correctness of operation bPpg (law (b1) in Fig. 9.5) we know that Step 5 yields the source model update $d_{2,CB}^S = id$ and correspondence $r_{2,FCB} = r_1 \in \mathcal{L}(TGG)$. This leads to the source model update $d_{2,FCB}^S = d_{2,CB}^S \circ d_F^S = id$. All together, we have that $s = (id, r_1, d^T) \in \text{CSynch}(d^S, r_0, id)$, i.e., s is a valid solution for the concurrent synchronisation problem of $\text{CSynch}(d^S, r_0, id)$.   □

9.4 Related and Future Work

The presented approach to (concurrent) model synchronisation is based on the formal results and the constructions for performing model transformations via TGGs in Chapters 7 and 8. It is inspired by Schürr et al. [Sch94, SK08] and Giese et al. [GH09, GW09], respectively. The constructions formalise the main ideas of model synchronisation based on TGGs in order to show correctness and completeness of the approach based on the results known for TGG model transformations. Moreover, we have extended the approach to the case of concurrent model synchronisation, where updates can occur concurrently on both domains, including the resolution of possible merging conflicts.

Given an integrated model $G^S \leftrightarrow G^T$ and an update on one domain, either G^S or G^T, the basic synchronisation problem is to propagate the given changes to the other domain. This problem has been studied at a formal level by several authors (see, for instance, [FGM$^+$07, HMT08, Ste10, BCF$^+$10, XSHT11, HPW11, DXC11a, DXC$^+$11b, HEO$^+$11a]). Many of these approaches [FGM$^+$07, HMT08, Ste10, XSHT11] are state-based, meaning that they consider that the synchronisation operations take as parameter the states of the models before and after the modification and yield new states of models. However, in [BCF$^+$10, DXC11a] it is shown that state-based approaches are not adequate in general for solving the problem. Instead, our approach in this chapter as well as a number of other approaches (see, for instance, [BCF$^+$10, HPW11, DXC$^+$11b]) are delta-based, meaning that the synchronisation operations take modifications as parameters and return modifications as results. These results can be seen as an instantiation, in terms of TGGs, of the abstract algebraic approach presented in [DXC$^+$11b]. Let us look into the related approaches and concepts in more detail.

Egyed et. al [EDG$^+$11] discuss challenges and opportunities for change propagation in multiple view systems based on model transformations concerning consistency (correctness and completeness), partiality, and the need for bidirectional change propagation and user interaction. Our presented approach based on TGGs reflects these issues. In particular, TGGs automatically ensure consistency for those consistency constraints that can be specified with a triple rule. This means that the effort for consistency checking with respect to domain language constraints is substantially reduced.

Stevens developed an abstract state-based view on symmetric model synchronisation based on the concept of constraint maintainers [Ste10], and Diskin described a more general delta-based view within the *tile algebra* framework [Dis11, DXC$^+$11b]. These tile operations inspired the constructions for the basic synchronisation operations (Sects. 9.2 and 9.3). Concurrent updates are a central challenge in multi-domain modelling, as discussed in [XSHT11], where the general idea of combining propagation operations with conflict resolution is used as well. However, the paper does not focus on concrete propagation and resolution operations and requires that model updates be computed as model differences. The latter can lead to unintended results by hiding the insertion of new model elements that are similar to deleted ones.

Merging of model modifications usually means that nonconflicting parts are merged automatically, while conflicts have to be resolved manually. A survey on model versioning approaches and on (semiautomatic) conflict resolution strategies is given in [ASW09]. A category-theoretical approach formalising model versioning is given in [RRLW09]. As in our approach, modifications are considered as spans of morphisms for describing a partial mapping of models, and merging of model changes is based on pushout constructions. In contrast to [RRLW09], we consider an automatic conflict resolution strategy according to [EET11] that is formally defined.

Bidirectional transformation frameworks originate from the lens framework proposed by Foster et al. [FGM⁺07]. Lenses consider the asymmetric synchronisation, where one model is a view of the other, and define a state-based framework for asymmetric synchronisation. "State-based" means that the synchroniser takes the states of models before and after update as input, and produces new states of models as output. Inspired by the lens framework, several researchers propose state-based frameworks for symmetric synchronisation [Ste10, HPW11, Dis08]. As a more general case, symmetric synchronisation allows neither of the models to be a view of the other. However, as Diskin et al. [DXC11a] point out, state-based bidirectional transformations actually mix two different operations, namely delta discovery (correspondence relations between models or between different versions of a model) and delta propagation, leading to several semantic problems. To fix these problems, several researchers [BCF⁺10, DXC11a, Dis11, DXC⁺11b, HPW12] propose delta-based frameworks, where deltas are taken as input and output. Typical delta-based frameworks include delta lens [DXC11a] for the asymmetric cases, and symmetric delta lens [DXC⁺11b] and edit lens [HPW12] for the symmetric cases.

The model synchronisation framework used in this chapter is a simplified version of the symmetric delta lens (sd-lens) framework proposed by Diskin et al. [DXC⁺11b]. The difference is that we do not consider the weak undoability laws (fUndo) and (bUndo) defined there. In addition, Diskin et al. [DXC⁺11b] also refine an sd-lens as an alignment framework and a consistency maintainer. Our approach is consistent with this refinement as well. The alignment framework corresponds to fAln and bAln operations. The consistency maintainer is implemented by Del, fAdd, and bAdd operations, which first mark the consistent parts of the integrated model, then propagate the changes, and finally delete the remaining inconsistent parts. As a result, the presented TGG approach serves as a proof of concept for the theory of symmetric delta lenses.

The BiG system proposed by Hidaka et al. [HHI⁺10] is a bidirectional graph synchronisation system. Different from our work based on symmetric TGG specification, the BiG system is based on an unidirectional graph transformation language, UnQL [BFS00], and thus is asymmetric by nature. Accordingly, the BiG system adopts an asymmetric synchronisation framework (a variant of the basic lens framework [FGM⁺07]), while our work adopts a simplified version of the symmetric delta lens [DXC⁺11b]. In an asymmetric framework, one model has to be a view of the other, and it is not possible to synchronise two models each containing information not presented in the other.

Giese et al. [GW09] introduced incremental synchronisation techniques based on TGGs in order to preserve consistent structures of the given models by revoking previously performed forward propagation steps and their dependent ones. This idea is generalised by the auxiliary operation Del in the present framework, which ensures the preservation of maximal consistent substructures and extends the application of synchronisation to TGGs that are not tight or contain rules with negative application conditions. Giese et al. [GH09] and Greenyer et al. [GPR11] proposed extending the preservation of substructures by allowing for the reuse of any *partial* substructure of a rule causing nondeterministic behaviour. However, a partial reuse can cause unintended results. Consider, e.g., the deletion of a person A in the source domain and the addition of a new person with the same name; then the old birth date of person A could be reused.

In order to improve efficiency, Giese et al. [GW09, GH09] proposed avoiding the computation of already consistent substructures by encoding the matches and dependencies of rule applications within the correspondences. In the present framework, the operation Del can be extended conservatively by storing the matches and dependency information separately, such that the provided correctness and completeness results can be preserved as presented in Sect. 9.2.3.

Becker et al. presented a generally nondeterministic synchronisation approach based on TGGs [BNW08] using the PROGRES approach [SWZ99] with the focus on integration, i.e., construction of missing correspondence links. The algorithm requires user interaction at each rule application, where some integration rules are in conflict for partial matches. For general TGGs, such integrations may require backtracking to achieve a resulting model that is fully integrated. In principle, it might be possible to adapt this algorithm in order to apply the main results in this chapter on correctness and completeness, since the actual steps are performed via the operational rules of a TGG.

In future work, we plan to develop extended characterisations of the correctness and maximality criteria of a concurrent synchronisation procedure. In Sect. 9.3, correctness is defined explicitly in terms of the two laws formulated in Sect. 9.2 and, implicitly, in terms of the properties of compatibility with basic model synchronisation proven in Theorem 9.44. We think that this can be strengthened by relating correctness of a synchronisation procedure with the total or partial *realisation* of the given source and target updates, for a suitable notion of realisation. At a different level, we also believe that studying in detail, from both theoretical and practical viewpoints, the combination of fSynch and bSynch operations should also be a relevant matter. The main parts of the basic model synchronisation framework have been implemented in the tool HenshinTGG (Sect. 12.4) and we plan to extend the implementation to the concurrent case.

Part IV
Application Domains, Case Studies and Tool Support

This fourth part of this book treats different application domains and case studies according to different parts of the theory given in Parts II and III, respectively. Moreover we give an overview of different tools, which support modeling and analysis of systems using graph transformation techniques presented in this book. In Chap. 10, we introduce self-adaptive systems and show how they can be modelled and analysed using graph transformation systems in Chap. 2, including a case study concerning business processes. The application domain of enterprise modelling is considered in Chap. 11, based on Chapters 3, 7 and 8, together with a case study on model transformation between business and IT service models. Chap. 12 includes a discussion of the following tools:

1. The Attributed Graph Grammar system AGG 2.0,
2. ActiGra: Checking consistency between control flow and functional behaviour,
3. Controlled EMF model transformation with EMF Henshin,
4. Bidirectional EMF model transformation with Henshin$_{TGG}$.

Chapter 10
Modelling and Static Analysis of Self-adaptive Systems by Graph Transformation

Software systems nowadays require continuous operation despite changes both in user needs and in their operational environments. Self-adaptive systems are typically instrumented with tools to autonomously perform adaptation to these changes while maintaining some desired properties. In this chapter, we model and analyse self-adaptive systems by means of typed, attributed graph grammars. The interplay of different grammars representing the application and the adaptation logic is realised by an adaptation manager. Within this formal framework we define consistency and operational properties that are maintained despite adaptations, and we give static conditions for their verification. The overall approach is supported by AGG 2.0 (see Sect. 12.1). A case study modelling a business process that adapts to changing environment conditions is used to demonstrate and validate the formal framework [BKM+12]. The modelling framework described in this chapter is based on joint work of the authors with Antonio Bucchiarone, Patrizio Pellicione and Olga Runge [EER+10, BEE+13, BEE+15].

The chapter is organised as follows. In Sect. 10.1, we show how self-adaptive systems can be modelled in our approach based on graph transformation. Section 10.2 describes how to formally verify desirable consistency and operational properties of self-adaptive systems. Section 10.3 discusses aspects related to the automation of the approach. In Sect. 10.4, we compare our approach with related work, and we conclude in Sect. 10.5. For full details of our case study, the reader is referred to our technical report [BEE+13].

10.1 Modelling Self-adaptive Systems

The high degree of variability that characterises modern systems requires us to design them with runtime evolution in mind. Self-adaptive systems are a variant of fault-tolerant systems that autonomously decide how to adapt the system at runtime to the internal reconfiguration and optimisation requirements or to environment changes and threats [BMSG+09]. A classification of modelling dimensions for self-adaptive systems can be found in [ALMW09], where the authors distinguish between *goals* (what the system is supposed to do), *changes* (causes for adaptation), *mechanisms* (system reactions to changes) and *effects* (the impact of adaptation upon the system).

The initial four self-* properties of self-adaptive systems are self-configuration, self-healing,[1] self-optimisation, and self-protection [KC03]. Self-configuration comprises component installation and configuration based on some high-level policies. Self-healing deals with automatic discovery of system failures, and with techniques to recover from them. Typically, the runtime behaviour of the system is monitored to determine whether a change is needed. Self-optimisation monitors the system status and adjusts parameters to increase performance when possible. Finally, self-protection aims to detect external threats and mitigate their effects [WHW+06].

In [BPVR09], the authors modelled and verified dynamic software architectures and self-healing systems (called self-repairing systems) by means of hypergraphs and graph grammars. The work in [EER+10] shows how to formally model self-healing systems by using algebraic graph transformations [EEPT06] and to prove consistency and operational properties. In this chapter, we extend the work in [EER+10] by formally modelling and analyzing self-adaptive systems based on the framework of algebraic graph transformation. Our modelling and validation framework is supposed to be used offline to evaluate and evolve a self-adaptive system: the framework helps the developer to decide which adaptation solutions used and logged in the past have desired properties and should become part of the final system model. Since we aim at modelling in a general way the concepts of self-awareness, context-awareness, self-monitoring and self-adaptation, our modelling framework is in principle applicable to systems with different kinds of self-* properties. The aim of our analysis is to show operational properties of self-adaptive systems concerning overall conflicts and dependencies of normal system behaviour and adaptations.

Self-adaptive systems are modelled in our approach as a set of typed attributed graph grammars where three kinds of system rules are distinguished: normal, context, and adaptation rules. *Normal rules* define the normal and ideal behaviour of the system. *Context rules* define context flags (adaptation hooks) that trigger adaptation rules. *Adaptation rules* in different adaptation grammars define the adaptation logic.

10.1.1 Running Example: A Car Logistics System (CLS)

The *Car Logistics System (CLS)* scenario will be used throughout the chapter to explain our approach to modelling self-adaptive systems. At the automobile terminal of the Bremerhaven sea port [BPSR09], nearly two million new vehicles are handled each year; the business goal is *to deliver them from the manufacturer to the dealer*. To achieve that goal, several intermediate business activities are involved. These in-

---

[1] Following [RGSS09], we consider self-healing and self-repair as synonymous.

clude unloading and storing cars from a ship, applying to them treatments to meet the customer's requirements and distributing them to the retailers. The company "Logistics IT Solutions" wants to develop a service-based application (the CLS) to support the delivery of vehicles from the ship to the retailers. The CLS must implement the business process depicted in Fig. 10.1 by invoking and orchestrating the set of available services in a proper way. Each business activity of the process is executed invoking a set of available services (i.e., Car Check Service, Unloading Service, etc.) that can be atomic or composite (i.e., Store Car Service). Additional services, i.e., services that are not directly attached to the business process, are defined and they can be used during the application execution. For example, the Wait For Treatment Service may be invoked when a vehicle that needs a treatment has to wait some time because of a long queue in the treatment station.

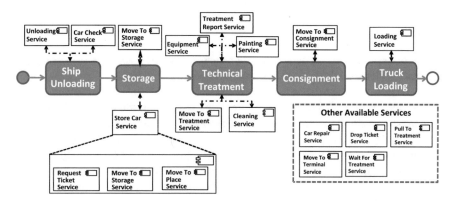

Fig. 10.1 Business process and services of the Car Logistics Scenario

The CLS executes the business process presented before for each vehicle under the following assumptions: (i) each business activity is executed in the defined order; (ii) the context in which the business process is executed can evolve in time.

Assume now the following two cases that may happen at runtime:

- *A vehicle is severely damaged during its movement from the ship to the storage area:* The vehicle has been unloaded from the ship and has requested a ticket (using the Request Ticket Service) to park in the storage area. It receives a precise ticket and starts to move to the storage (using the Move To Storage Service). While moving, the vehicle is severely damaged and then it stops. In this case, the business process does not know how to proceed and the booked ticket cannot be used.
- *A vehicle arrives at a service point in the treatment area, but the required service is busy and cannot be executed immediately:* The vehicle is ready to undergo a service as ordered by the customer (concerning, e.g., painting or equipment), but there are already a number of vehicles waiting for this service. In this case, the corresponding business service cannot proceed in an expected way (such as treating the vehicles right after their arrival).

In each of these cases, the business process should not proceed as planned. A system adaptation is required. In the next section we present a framework for rule-based dynamic adaptation to model and analyse systems that exhibit the aforementioned characteristics and problems.

10.1.2 Framework for Rule-Based Dynamic Adaptation

The framework manages the dynamic adaptation by specifying when and how adaptation is triggered, how the choice among the possible adaptations is performed, and, finally, how the nature of adaptations can be characterised.

Requirements

To be able to execute system behaviour also in case of unexpected situations, an adaptation framework needs to address the following problems:

Context-awareness: To relate the application execution to the context, the application must be context-aware, i.e., during the execution, information on the underlying environment can be obtained (e.g., relevant information on entities involved, status of the business process execution, human activities, etc.). To be adaptable, an application should provide adaptation hooks, i.e., context information on parts of the application's structure and behaviour. The adaptation hooks should be used to select the most suitable adaptation strategy.

Separation of concerns: The adaptation logic should be developed separately from the application logic by some adaptation manager. The adaptation logic can be created and/or changed after the application has been deployed without modifying the running application. At runtime, the adaptation manager should check the context (adaptation hooks) to control whether any adaptations are required, and reconfigure the system in the best suitable way.

Components

Our adaptation framework (AF) is composed of three fundamental components as illustrated by Fig. 10.2: the *Application Logic* describes how the application evolves (by *application rules*); the *Context Monitor* watches properties of the application operational environment and how they evolve (by *adaptation hooks* and *context rules*); the *Adaptation Manager* specifies how a system is adapted in the case of adaptation needs (using *adaptation rules*).

According to the scenario in Sect. 10.1.1, the considered self-adaptive system is the Car Logistics application. Its application logic describes what the different

Fig. 10.2 Adaptation framework

activities are that can be executed (i.e., Ship Unloading, Storage, Technical Treatment, Consignment, and Truck Loading), the set of available services that can be used to realise such activities (i.e., Store Car Service, Move To Treatment Service, Cleaning Service, etc.) and the assumed behaviour of the overall application. The behaviour describes the order of the activities a car must execute plus a set of business policies (in terms of activity preconditions).

The *Context Monitor* continuously monitors the context at fixed intervals. It is defined as a set of rules, called *context rules*, which once applied add *adaptation hooks* to the system to trigger the adaptation process. The system is monitored at regular time intervals, and an *adaptation problem* is sent to the *Adaptation Manager* if one or more adaptation hooks are found. In response, the *Adaptation Manager* returns an *adaptation solution* to the application logic that aims to do its best to "recover", so that the blocked activity can be executed and the main process can continue.

Formalisation

The formal model of a self-adaptive system is a set of graph grammars typed over the same type graph. A main system grammar consists of *system rules* modelling normal behaviour (the *Application Logic*) and *context rules* modelling changes that require adaptation by generating adaptation hooks. *Context Monitoring* is modelled by context constraints in the main system grammar that are violated in the presence of adaptation hooks and trigger the (semiautomatic) selection of a corresponding adaptation grammar (the *Adaptation Logic*). An *adaptation grammar* contains *adaptation rules* modelling reactions to the detection of context changes. The interplay of the different grammars representing the adaptation logic is realised by the *Adaptation Manager*.

Ordering adaptations

Different adaptations may be applicable during the system execution. The choice of which adaptation to apply may influence the time required for performing the adaptation, or even the final result. In our framework, adaptations are selected by the adaptation manager.

Nature of adaptations

We consider two classes of adaptations that can be applied and treated in different ways [CHK+01, LBM10], in particular:

Corrective Adaptations take care of adapting the application when the current implementation instance cannot proceed with the execution in the current context (i.e., a car is damaged). The main objective is to recover the application and hence focus on the *self-healing* property. The adaptation starts from the actual context state and performs the necessary changes to bring the application and its context to the expected state where it can be executed again. In our framework, an adaptation is *corrective* if each adaptation state can be *repaired*, i.e., the normal state before the adaptation became necessary is reestablished.

Enhancing Adaptations augment existing services of the application; this may for instance change the nonfunctional properties of the service, or provide new services with the same or expanded functionalities. In our framework, an adaptation is *enhancing* if each adaptation state can become a normal state, possibly by adding new functionalities and services. The adaptation result is *not necessarily* identical with the normal state before the adaptation became necessary.

10.1.3 Modelling SA Systems by Graph Transformation

In this section, we show how to model self-adaptive systems in the formal framework of algebraic graph transformation. Specifically, typed attributed graphs, introduced in Def. 2.5 in Chap. 2, are used to model the static part of the system. Typed attributed graphs are enriched with constraints that self-adaptive systems have to satisfy even during adaptation. Moreover, we model the behaviour and the adaptation of self-adaptive systems by means of graph grammars, introduced in Def. 2.20. We use the *Car Logistics System* running example and the AGG tool (see Sect. 12.1) to show how practitioners can model and simulate self-adaptive systems.

Example 10.1 (CLS type graph and initial state). Figure 10.3 depicts the type graph for the Car Logistics System, which contains types used for modelling the "normal" aspects of the car logistics scenario, as well as the context types used for adaptation, e.g., the hooks (context flags) that trigger the adaptation rules.

   In the integrated type graph in Fig. 10.1, we have the following types for normal behaviour:

- Start, End and BusinessActivity are the main business activities, which are ordered, i.e., linked by directed arcs of type next.
- Service is a service station belonging (linked) to a BusinessActivity. A service may be a composite service. Then it contains other services which are ordered (linked

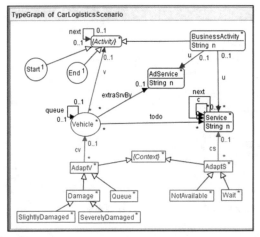

Fig. 10.3 CLS type graph

by next arcs). Containment of sub-services in a composite service is modelled by c edges from the sub-services to its composite service.

- Vehicle is a car running through the business process. At the beginning it will be linked (by a v link) to the Start activity and is ready to enter a service.
- A todo link between a vehicle and each service of each BusinessActivity is generated when a Vehicle starts the business process. The successful processing of a service leads to the deletion of the corresponding todo link. When all services belonging to the business process have been processed (all todo links are removed), the Vehicle arrives at the End activity as a completed product with a precise treatment executed and ready to be delivered to a retailer.

For adaptation handling we have the following types:

- Context is the super-type for all possible context signals, including adaptation hooks. These hooks are used for triggering the adaptation grammars. We specify two main context types, AdaptV and AdaptS.
- AdaptV with refinements Damage and Queue denotes that a car is damaged and needs to be repaired (SlightlyDamaged) or disposed (SeverelyDamaged), or a car is in a queue.
- AdaptS with refinements NotAvailable and Wait denotes that a service is not available or there is a queue at a service, respectively. The refinement Wait denotes that cars in the current business activity should queue up and wait to be processed.
- AdService is an adaptation service not directly attached to a BusinessActivity. Such additional services are used according to an adaptation scenario.
- An edge of type extraSrvBy connects a Vehicle to an adaptation service.
- Edges of type queue connect the Vehicles in a queue at a service.

Figure 10.4 displays the initial state graph of a scenario with two vehicles.   △

In order to model global consistency and adaptation constraints of a self-adaptive system, we use (*TG*-typed) graph conditions and constraints. A *graph condition* is

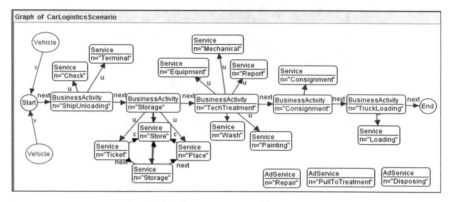

Fig. 10.4 Initial state graph of the Car Logistics case study

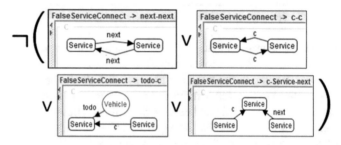

Fig. 10.5 Graph constraint noFalseServiceConnect

given by a graph morphism $c : P \rightarrow C$ (where P is called *premise* and C *conclusion*). The condition $c : P \rightarrow C$ is *satisfied by a graph* G, written $G \models c$, if the existence of an injective graph morphism $p : P \rightarrow G$ implies the existence of an injective graph morphism $q : C \rightarrow G$, such that $q \circ c = p$. Graph conditions can be negated or combined by logical connectors (e.g., $\neg c$). See Defs. 2.7 and 5.1 and Ex. 2.16 for some examples of graph conditions.

Technically, *graph constraints* describe global requirements for objects.. They are conditions over the initial object in the category of typed, attributed graphs. This means that, for instance, a constraint $\exists (i_C, \text{true})$ with the initial morphism i_C into C is valid for a graph G if there exists a morphism $c : C \rightarrow G$. This constraint expresses that the existence of C as a part of G is required (see Def. 5.10 in Chap. 5).

Example 10.2 (Graph constraints for the Car Logistics system). The set $C_{consist} = \{noFalseServiceConnect, sameBAforComp, noEqualContextFlags\}$ contains consistency constraints that have to be satisfied throughout all states of the car logistics model. Constraint *noFalseServiceConnect*, depicted in Fig. 10.5, means that there must be no next or containment loops, a vehicle must not be served (todo edge) by a service which is a container of other services, and a sub-service must not be a composite service.

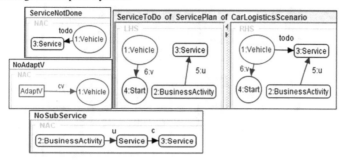

Fig. 10.6 The normal behaviour rule ServiceToDo

Constraint *sameBAforComp* (not depicted) means that all sub-services belonging to the same composite service are linked to the same BusinessActivity and to the same composite service node. Constraint *noEqualContextFlags* (not depicted) requires that the same element (vehicle or service) be not marked by more than one adaptation hook of the same type.

Moreover, we have two sets of adaptation constraints, $C_{adapt}^1 = \{Damage\}$ and $C_{adapt}^2 = \{Wait\}$, describing adaptation hooks that are required to hold for certain adaptations to occur, i.e., constraint *Wait* requires the existence of a Wait flag at a service, and constraint *Damage* requires a Damage flag at a vehicle. △

We now model the behaviour of the main scenario and the adaptations by different graph grammars (according to Def. 2.20 in Chap. 2). Whenever an adaptation becomes necessary, the respective adaptation rules are added to the main grammar.

In the main grammar *CarLogisticsScenario* (Ex. 10.3), normal behaviour is modelled. Moreover, context flags (e.g., adaptation hooks) can be generated.

Example 10.3 (Rules modelling the Car Logistics scenario). Here, we describe the rules for the normal behaviour of Vehicles running through BusinessActivities smoothly.[2] To save space, we depict only the left- (*LHS*) and right-hand sides (*RHS*) for a rule. The interface consists of those nodes and edges that are present both in *LHS* and in *RHS* and mapped to each other by equal numbers.

The first step for each Vehicle is to enter the business process. Then, all services of the Vehicle's BusinessActivities are marked as *to do* by creating todo edges between the Vehicle and each service not yet marked (by applying rule ServiceToDo in Fig. 10.6 as long as possible). Three negative application conditions (NACs) ensure that a todo edge is created only if there is not already a todo edge between the vehicle and the service (NAC ServiceNotDone), the vehicle is not in an adaptation state (NAC NoAdaptV) and it is not linked to a composite service (NAC NoSubService).

Next, rule EnterBP moves the Vehicle to the first BusinessActivity. When a service is processed, the corresponding todo edge is removed by rule DoService. For processing a composite service consisting of ordered sub-services, rule DoSubService processes a sub-service only if its previous services in the queue have already been processed. A Vehicle can move to the next BusinessActivity when all services

---

[2] The complete set of rules with all details is given in [BEE+13]

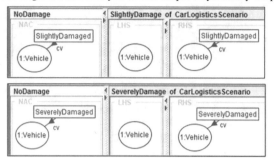

Fig. 10.7 The context rules SlightlyDamage and SeverelyDamage

of the previous one are done (rule NextBA). Finally, when there are no more services to do, the business process for the Vehicle is finished (rule FinishBP).

In addition to the *normal behaviour* rules, the main grammar contains *context rules* that are applicable at any time and mark a Vehicle or Service with an adaptation hook, i.e., they create a node of one of the context node types AdaptV or AdaptS, respectively. In case a car is damaged, rules SlightlyDamage or SeverelyDamage mark it by an adaptation hook of kind either "SlightlyDamaged" or "SeverelyDamaged" as illustrated in Fig. 10.7.

Another context rule Queue (not depicted) marks a service of the technical treatment area and a vehicle with context nodes of type Wait and Queue, respectively, if a service station in the technical treatment area is busy and the treatment cannot be executed immediately.   △

All above introduced adaptation hooks guide the adaptation manager to select a suitable adaptation grammar realising the adaptation. We may have different adaptation grammars that are suitable for the same adaptation hook. For instance, if a damaged car is in the midst of a composite service, first a rollback adaptation has to be performed, and then a repair adaptation. The adaptation manager coordinates the different adaptation grammars such that the rules of the most suitable adaptation grammars are imported into the main grammar. These rules perform the necessary adaptations at the host graph so that the "normal behaviour rules" can proceed and maybe new adaptation hooks are set. Note that in AGG the modeller plays the role of an adaptation manager and imports the necessary adaptation rules into the main grammar. After adaptation, the adaptation rules are removed from the main grammar, and the application of the "normal behaviour rules" continues.

Example 10.4 (Rollback adaptation). A rollback adaptation is needed when a damaged vehicle is in the midst of a composite service, as depicted in Fig. 10.8. In this case, the already finished sub-services are "rolled back" by applying rule RollBack as long as possible before the vehicle is moved to the treatment area to be repaired.   △

Example 10.5 (Repair adaptation). A repair adaptation becomes necessary when a vehicle is marked by a context flag as being slightly damaged. We require that the vehicle to be repaired be not in the midst of a composite service anymore. (If it is, the

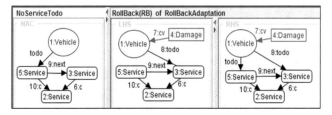

Fig. 10.8 The rollback adaptation rule RollBack

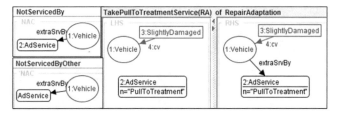

Fig. 10.9 The repair adaptation rule TakePullToTreatmentService

rollback adaptation has to be selected by the adaptation manager to be performed before the repair adaptation.) When repair is needed, two additional services are evoked, i.e., the Vehicle is linked to them, one after the other. Rule TakePullToTreatmentService in Fig. 10.9 uses an extra service to pull up the damaged Vehicle to the treatment area.

Rule TakeRepairService (not depicted) allocates an extra repair service for the slightly damaged Vehicle. A slightly damaged Vehicle is repaired (i.e., the adaptation hook is removed) by rule RepairVehicle.After repair, the Vehicle should continue its normal behaviour at the point where the adaptation hook was set. The composite service it left before (which has rolled back by applying the rollback adaptation) may now start again from the beginning. Note that the Vehicle does not forget which services have been done already and which are still to be done.   △

Example 10.6 (Dispose adaptation). A severely damaged vehicle that cannot be repaired is disposed of by the dispose adaptation. This adaptation grammar also contains the rules TakePullToTreatmentService and TakeRepairService that are analogous to the corresponding rules in the repair adaptation in Ex. 10.5, with the slight difference that the context flag is now always of kind SeverelyDamaged.

Using rule TakeDisposingService in Fig. 10.10, a severely damaged Vehicle is picked up by the disposing service.

Before the vehicle can be disposed of, its todo links and all flags of kind SlightlyDamaged or SeverelyDamaged are removed by applying the rules RemoveToDo and RemoveDamageFlag as long as possible. Finally, the vehicle is disposed of by rule DisposeVehicle. In our model, a disposed vehicle remains in the graph without any links to other objects.   △

Example 10.7 (Wait-at-queue adaptation). The wait-at-queue adaptation becomes necessary when a service station in the technical treatment area is busy and the

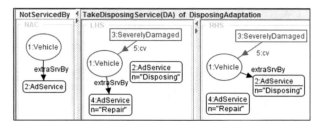

Fig. 10.10 The dispose adaptation rule TakeDisposingService

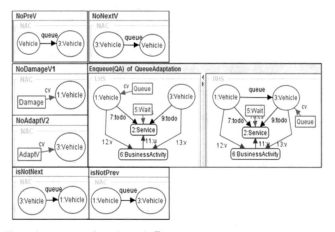

Fig. 10.11 The wait-at-queue adaptation rule Enqueue

treatment cannot be executed immediately. When a car arrives at the treatment area and discovers that there is a queue it should queue up and wait. Rule Enqueue in Fig. 10.11 enqueues all Vehicles waiting for this service. In doing so the Queue flag is shifted along the queue link.

Vehicles in a queue are served in the order of their arrival by applying rule DoService. Finally, the Wait flag at the busy service and the Queue flag at the last Vehicle are removed by rule RemoveWait after the whole queue has been processed.   △

An SA system is defined in Def. 10.8 by a typed graph grammar where the system rules can be partitioned into *normal*, *context* and *adaptation* rules. Moreover, we have two kinds of constraints, namely *consistency* and *adaptation* constraints.

Definition 10.8 (Self-adaptive system in AGT framework). A self-adaptive system (SA system) is given by $SAS = (GG, C_{sys})$, where:

- $GG = (TG, G_{init}, R_{sys})$ is a typed graph grammar with type graph TG, a TG-typed initial graph G_{init}, and a set of TG-typed rules R_{sys} (*system rules*), defined by $R_{sys} = R_{norm} \cup R_{cont} \cup R_{adapt}$, where R_{norm} (*normal rules*), R_{cont} (*context rules*) and R_{adapt} (*adaptation rules*) are pairwise disjoint.

- C_{sys} is a set of TG-typed graph constraints, called *system constraints*, with $C_{sys} = C_{consist} \cup C_{adapt}$, where $C_{consist}$ (called *consistency constraints*) and C_{adapt} (called *adaptation constraints*) are pairwise disjoint.

We distinguish *reachable*, *adaptation* and *normal* states, where reachable states are partitioned into normal and adaptation states.

- $Reach(SAS) = \{G \mid G_{init} \stackrel{*}{\Longrightarrow} G$ via $R_{sys}\}$, i.e., all states reachable via system rules,
- $Adapt(SAS) = \{G \mid G \in Reach(SAS) \wedge \exists C \in C_{adapt} : G \vDash C\}$, the *adaptation states*, i.e., all reachable states satisfying some adaptation constraints,
- $Norm(SAS) = \{G \mid G \in Reach(SAS) \wedge \forall C \in C_{adapt} : G \nvDash C\}$, the *normal states*, i.e., reachable states not satisfying any adaptation constraints.

For SA systems *SAS*, we require that

1. each pair of a context and a normal rule $(p, r) \in R_{cont} \times R_{norm}$ be sequentially independent (see Def. 5.21); this means that context rules can be applied independently of the normal system behaviour occurring in different parts of the system),
2. *SAS* is system consistent: all reachable states are consistent, i.e., they fulfill the consistency constraints: $\forall G \in Reach(SAS), \forall C \in C_{consist} : G \vDash C$,
3. *SAS* is normal-state consistent, i.e., normal rules must not create adaptation hooks: the initial state is normal and all normal rules preserve and reflect normal states: $G_{init} \in Norm(SAS)$ and $\forall G_0 \stackrel{r}{\Longrightarrow} G_1$ via $r \in R_{norm}$ $[G_0 \in Norm(SAS) \Leftrightarrow G_1 \in Norm(SAS)]$,
4. the set of adaptation rules R_{adapt} is confluent and terminating, i.e., adaptation results are unique and do not depend on the order or location of the adaptation rule applications. △

The requirements of SA systems can be concluded in a static way by inspecting the corresponding rules. This means, e.g., that we do not need to check the consistency of all states reachable via system rules; instead, we only check G_{init} for consistency and then check whether the system rules preserve consistent states. In particular, we can check statically that different adaptations do not interfere with each other, i.e., they are confluent and terminating (see Requirement 4). This property is interesting if more than one set of adaptation rules has to be used to adapt a given state, which is a highly relevant practical problem.

Example 10.9 (Car Logistics system as SA system). We define the Car Logistics SA system $CLS = (GG, C_{sys})$ by the type graph TG in Fig. 10.3, the initial state G_{init} in Fig. 10.4, and the following sets of rules and constraints:

- $R_{norm} = \{$ServiceToDo, EnterBP,DoService, DoSubService,NextBA, FinishBP$\}$,
- $R_{cont} = \{$SlightlyDamage, SeverelyDamage, Queue$\}$,
- $R_{adapt} = R_{adapt}^1 \cup R_{adapt}^2 \cup R_{adapt}^3 \cup R_{adapt}^4$ with $R_{adapt}^1 = \{$Rollback$\}$,
- $R_{adapt}^2 = \{$TakePullToTreatmentService, TakeRepairService, RepairVehicle$\}$,

- R^3_{adapt} = {TakePullToTreatmentService, TakeRepairService, TakeDisposingService, RemoveTodo, RemoveDamageFlag, DisposeVehicle},
- R^4_{adapt} = {Enqueue, DoService, RemoveWait},
- $C_{consist}$ = {$noFalseServiceConnect$, $sameBAforComp$, $noEqualContextFlags$},
- $C_{adapt} = C^1_{adapt} \cup C^2_{adapt}$ with C^1_{adapt} = {$Damage$} and C^2_{adapt} = {$Wait$}.

The normal rules in R_{norm} and the context rules in R_{cont} have been explained in Ex. 10.3. The adaptation rules in R^1_{adapt} to R^4_{adapt} have been introduced in Ex. 10.4 to Ex. 10.7, respectively. The consistency constraints in $C_{consist}$ have been explained in Ex. 10.2 and model the desired structural properties. The adaptation constraints in C_{adapt} model properties that have to be valid only if the corresponding adaptation is running. The adaptation constraints *Damage* and *Wait* require the existence of the corresponding adaptation hook.

Checking the requirements for SA systems in Def. 10.8, we find that:

1. We have sequential independence for each pair of context and normal rules $(p, r) \in R_{cont} \times R_{norm}$, which we have checked using the automatic dependency analysis of AGG. Each of the pairs (SlightlyDamage,r), (SeverelyDamage,r) and (Queue,r) with r being a normal rule is sequentially independent (i.e., there are no dependencies for each rule pair).

2. *CLS* is system consistent, because for all $C \in C_{consist}$, $G_{init} \models C$ and for all $G_0 \overset{r}{\Longrightarrow} G_1$ via $r \in R_{sys}$ and $G_0 \in Consist(SAS)$ we also have $G_1 \in Consist(SAS)$. This can be concluded since no system rules manipulate the structure of services.

3. *CLS* is normal-state consistent, because $G_{init} \in Norm(SAS)$ and for all $G_0 \overset{r}{\Longrightarrow} G_1$ via $r \in R_{norm}$ and for all $C \in C_{Adapt}$ we have that $[G_0 \not\models C \Leftrightarrow G_1 \not\models C]$. This can be concluded since no normal rule manipulates (inserts or deletes) any adaptation hooks (subtype of type Context), which is required by the adaptation constraints to hold.

4. With regard to confluence, we have used AGG to check the sets of adaptation rules for critical pairs (i.e., minimal conflicts; see Def. 5.40) and found that there are no critical pairs, and hence no conflicts for any rule pairs within the same adaptation rule set. AGG has computed some conflicts between different adaptation rule sets: there is, e.g., a conflict when the adaptation rule Rollback would be applied after the adaptation rule RepairVehicle. Note that this conflict can be disregarded since the adaptation manager has to make sure to apply the rollback adaptation (if necessary) before evoking the repair adaptation, and not afterwards. Similarly, conflicts between rules for repairing and rules for disposing vehicles can be ignored since we expect that in presence of severely damaged cars, the adaptation manager selects the *Dispose* adaptation first, and applies the *Repair* adaptation afterwards, when all severely damaged cars have been disposed of. Under these restrictions on the application order of adaptation rule sets, the union R_{adapt} of all adaptation rule sets is confluent.

   With regard to termination, we argue as follows: The rollback adaptation R^1_{adapt} terminates as there are only a finite number of services to roll back within a composite service, and a todo edge may be inserted only once between a vehicle and a service. The repair adaptation R^2_{adapt} terminates due to a finite number of slightly damaged vehicles, and due to the NACs of the repair rules ensuring that each rule is applicable only once for each damaged vehicle. For the dispose adaptation R^3_{adapt}, we have one rule, RemoveDamageFlag, that might be applicable twice, if a car has two Damage flags. No car has more than two flags, due to NACs of the corresponding context rules. For the wait-at-queue adaptation R^4_{adapt}, NACs ensure that each vehicle is enqueued only once by the rule Enqueue. The remaining rules only delete todo edges and are terminating due to the finite number of vehicles in the system. Hence, all adaptation rule sets are confluent and terminating. $\triangle$

10.2 Static Analysis of Self-adaptive Systems

In this section, we define desirable operational properties of SA systems and propose static analysis techniques to verify them. One of the main ideas of SA systems is that they are monitored in regular time intervals and it is checked whether the current system state is an adaptation state. In this case one or more adaptation hooks have been created in the last time interval by context rules. With the *enhancing*-adaptation property below, we require that a system in an adaptation state be eventually adapted, i.e., transformed again to a normal state, possibly by adding new functionalities to the system and using new services. Moreover, *corrective* self-adaptation means that the state will be recovered, i.e., the normal state after adaptation is the same as if no adaptation had occurred. In the following, we use the notation $G \Rightarrow^! G'$ to denote a transformation where the rules are applied as long as possible; we write $G \Rightarrow^* G'$ to denote a transformation where the rules are applied arbitrarily often, and the transformation $G \Rightarrow^+ G'$ consists of at least one rule application.

Definition 10.10 (Self-adaptation classes). An SA system SAS is called

1. *enhancing* if each adaptation state is adapted to become a normal state (unique up to isomorphism), possibly by adding new functionalities and services. In more detail:
$\forall\, G_{init} \Rightarrow^* G$ via $(R_{norm} \cup R_{cont})$ with $G \in Adapt(SAS)\ \exists\ G \Rightarrow^! G'$ via R_{adapt} with $G' \in Norm(SAS)$.
2. *corrective* if each adaptation state is adapted in a corrective way (repaired). In more detail:
$\forall\, G_{init} \Rightarrow^* G$ via $(q_1 \ldots q_n) \in (R_{norm} \cup R_{cont})^*$ with $G \in Adapt(SAS)\ \exists\ G \Rightarrow^! G'$ via R_{adapt} with $G' \in Norm(SAS)$ and $\exists\ G_{init} \Rightarrow^* G'$ via $(r_1 \ldots r_m) \in R^*_{norm}$, where $(r_1 \ldots r_m)$ is the subsequence of all normal rules in $(q_1 \ldots q_n)$. $\triangle$

Remark 10.11. By definition, each *corrective* SAS is also *enhancing*, but not vice versa. The additional requirement for corrective self-adaptation means that the system state G' obtained after adaptation is not only normal, but can also be generated by all normal rules in the given mixed sequence $(q_1 \ldots q_n)$ of normal and context rules, as if no context rule had been applied. We will see that our SA system CLS is corrective, considering only the *repair* adaptation for slightly damaged vehicles, but CLS together with the *wait-at-queue* adaptation is enhancing only, but not corrective.   △

In Def. 10.12, we define adaptation properties, which imply that the SA system is corrective/enhancing under suitable conditions, stated in Theorem 10.14. We want to ensure that for each context rule that adds an adaptation hook, there is a suitable adaptation grammar containing one or more adaptation rules leading again to a state without this adaptation hook, even if they are not applied immediately after its occurrence but later, when the context monitor reveals that the adaptation must be invoked. This means that other normal and context rules may have been applied before the occurrence of the adaptation hook is monitored.

Definition 10.12 (Self-adaptation (SA) properties). Let G_0 be a reachable state in the SA system *SAS*. *SAS* has the

1. *direct adaptation property* if the adaptation can be performed directly, i.e., $\forall\, G_0 \stackrel{p}{\Longrightarrow} G_1$ via $p \in R_{cont}\ \exists\ G_1 \Rightarrow^* G_0$ via R_{adapt},
2. *normal adaptation property* if the necessary adaptation can be performed up to normal transformations, leading to a possibly different normal state that is reachable from the state before the adaptation hook was set, i.e., $\forall\, G_0 \stackrel{p}{\Longrightarrow} G_1$ via $p \in R_{cont}\ \exists\ G_1 \Rightarrow^+ G_2$ via R_{adapt} s.t. $\exists\ G_0 \Rightarrow^* G_2$ via R_{norm},
3. *rollback adaptation property* if the necessary adaptation can be performed up to normal transformations, leading to a possibly different normal state from which the state before the adaptation hook was set is reachable, i.e., $\forall\, G_0 \stackrel{p}{\Longrightarrow} G_1$ via $p \in R_{cont}\ \exists\ G_1 \Rightarrow^+ G_2$ via R_{adapt} s.t. $\exists\ G_2 \Rightarrow^* G_0$ via R_{norm}.   △

Remark 10.13. Note that the normal and rollback adaptation property only differ in the direction of $G_0 \Rightarrow^* G_2$ and $G_2 \Rightarrow^* G_0$ via R_{norm}. For the normal and the rollback adaptation properties, it is required that the adapted state G_2 be related to the old state G_0 by a normal transformation. The direct adaptation property implies both the normal and the rollback property using $G_2 = G_0$.   △

Theorem 10.14 (Self-adaptation classes and their SA properties).
An SA system SAS is

I. *corrective, if we have property 1 below*
II. *enhancing, if we have*
    a) property 2 or b) properties 3 and 4 below.

1. *SAS has the direct adaptation property.*
2. *SAS has the normal adaptation property.*
3. *SAS has the rollback adaptation property.*
4. *Each pair $(r, q) \in R_{norm} \times R_{adapt}$ is sequentially independent.*   △

Proof.

I. Given $G_{init} \Rightarrow^* G$ via $(q_1, \ldots q_n) \in (R_{norm} \cup R_{cont})^*$ with $G \in Adapt(SAS)$, we have $n \geq 1$, because $G_{init} \in Norm(SAS)$ since SAS is normal-state consistent. By sequential independence we can switch the order of $(q_1, \ldots q_n)$, s.t. first all normal rules $r_i \in R_{norm}$ and then all context rules $p_i \in R_{cont}$ are applied. For example let us consider $G_{init} \Rightarrow^+ G$ via $(r_1, p_1, r_2, p_2, r_3)$ with $r_i \in R_{norm}$ and $p_j \in R_{cont}$. Then sequential independence leads by the Local Church–Rosser theorem to equivalent sequences in subdiagrams (1), (2), (3) respectively.

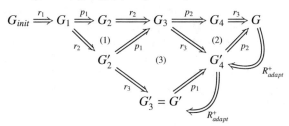

By the direct adaptation property 1, we have that $G'_4 \Rightarrow^* G'_3$ and $G \Rightarrow^* G'_4$ via R^*_{adapt}. With $G' = G'_3$ we have that $G \Rightarrow^+ G'$ via R^*_{adapt} and $G_{init} \Rightarrow^* G'$ via $(r_1, r_2, r_3) \in R^*_{norm}$, where (r_1, r_2, r_3) is the subsequence $(r_1, p_1, r_2, p_2, r_3)$, which consists of only normal rules, and normal-state consistency implies $G_{init}, G' \in Norm(SAS)$. Note that in the adaptation sequence $G \Rightarrow^+ G'$, the (possible) adaptations due to the adaptation hooks caused by $p_1, p_2 \in R_{cont}$ are performed in opposite order. In general, the sequence $(q_1, \ldots q_n)$ $(n \geq 1)$ contains at least one rule in R_{cont}, because otherwise $G \notin Adapt(SAS)$ (due to normal-state consistency), which is a contradiction to the assumption $G \in Adapt(SAS)$. This implies that we have an adaptation sequence $G \Rightarrow^+ G'$ via R_{adapt}. Since adaptation rules are confluent and terminating, all possible adaptation transformations $G \Rightarrow^+ \overline{G}$ lead to the same result $\overline{G} \cong G'$. Hence SAS is corrective.

II a) We can proceed as above up to the point, where first all normal and then all context rules are applied. As shown in our example, the normal adaptation property 2 leads first to an adaptation transformation $G \Rightarrow^+ G_5$ via R_{adapt} with $G'_4 \Rightarrow^* G_5$ via R_{norm}, and then we can switch rules in (4) according to the sequential independence of context and normal rules. Finally the normal adaptation property 2 leads to $G_5 \Rightarrow^+ G_6$ via R_{adapt}, with $G'_5 \Rightarrow^* G_6$ via R_{norm}.

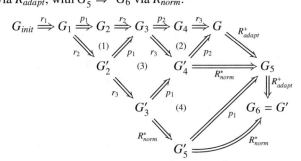

Altogether, we obtain for $G' = G_6$ an adaptation transformation $G \Rightarrow^+ G_5 \Rightarrow^+ G_6 = G'$ and a normal rule transformation $G_{init} \Rightarrow^* G_3' \Rightarrow^* G_5' \Rightarrow^* G_6 = G'$, which implies $G' \in Norm(SAS)$ by normal-state consistency. In general, $G \in Adapt(SAS)$ implies that we have at least one context rule in the given sequence and hence an adaptation transformation $G \Rightarrow^+ G'$ of length $n \geq 1$. Since adaptation rules are confluent and terminating, all possible adaptation transformations $G \Rightarrow^+ \overline{G}$ lead to the same result, $\overline{G} \cong G'$. Hence, SAS is enhancing.

II b) Again, we proceed as above up to the point, where first all normal and then all context rules are applied. Due to property 3, we have the rollback adaptation property which leads in our example to a normal transformation sequence $G_5 \Rightarrow^* G_4'$ via R_{norm} after an adaptation transformation $G \Rightarrow^+ G_5$ via R_{adapt} and to a normal transformation sequence $G_5' \Rightarrow^* G_3'$ via R_{norm} after an adaptation transformation $G_4' \Rightarrow^+ G_5'$ via R_{adapt}. Due to property 4, we can switch rules in square (4), leading to $G \Rightarrow^+ G_6$ via R_{adapt} and to $G_6 \Rightarrow^* G_3'$ via R_{norm}.

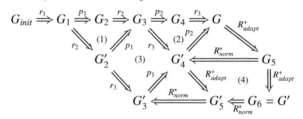

Altogether, we obtain for $G' = G_6$ an adaptation transformation $G \Rightarrow^+ G_5 \Rightarrow^+ G_6 = G'$ and normal rule transformations $G_{init} \Rightarrow^* G_3' \Leftarrow^* G_5' \Leftarrow^* G_6 = G'$. Finally, since normal rules preserve and reflect also adaptation states due to normal state consistency, we can conclude that $G_6 = G' \in Norm(SAS)$: if G' was an adaptation state, then also G_3' would be an adaptation state since normal rules preserve adaptation states. Since adaptation rules are confluent and terminating, all possible adaptation transformations $G \Rightarrow^+ \overline{G}$ lead to the same result $\overline{G} \cong G'$. Hence, SAS is enhancing.   □

Remark 10.15. Note that our sufficient conditions for Theorem 10.14 are also necessary in case that the context rules are sequentially independent. It is advisable to model the set of context rules in this way because usually the need for adaptation may arise in any possible state from independent sources of disturbances issued by the environment. In our example, the context rules are all independent, i.e., if they are applicable in a sequence, their order can be swapped.   △

In Theorem 10.18 we define static conditions for the direct, normal and rollback adaptation properties. By Theorem 10.14, these static conditions are also sufficient conditions for the nature of our self-adaptive system. We then make use of the static conditions to verify the properties of the different adaptations of our Car Logistics Systems case study.

In part 1 of Theorem 10.18 we require that for each context rule p the *inverse rule* p^{-1} be *SAS-equivalent* to the *concurrent rule* q^* constructed from an adaptation rule sequence $(q_1, \ldots, q_n) \in R_{adapt}$. We will explain shortly the notions *SAS-equivalent rules*, *inverse rule* and *concurrent rule* before stating Theorem 10.18.

SAS-equivalent rules model the same possible system changes:

Definition 10.16 (*SAS*-equivalent rules). Let *SAS* be an SA system and $r_1, r_2 \in R_{sys}$ be two system rules of *SAS*. Two rules r_1 and r_2 are called *SAS-equivalent* (written $r_1 \simeq r_2$) if $(\exists\, G \xRightarrow{r_1} G') \Longleftrightarrow (\exists\, G \xRightarrow{r_2} G')$ with $G \in Reach(SAS)$.   △

Remark 10.17 (Sufficient conditions for SAS-equivalent rules). An obvious sufficient condition for checking *SAS*-equivalence of two rules is that the rules are isomorphic (two rules are isomorphic if they are componentwise isomorphic). Sometimes, this condition is too strong (as we will see later for our example). A weaker sufficient condition is that the two rules are "isomorphic up to fixed objects", i.e., one rule may contain more elements than the other rule under the condition that these additional elements are 1) preserved by the rule and 2) available in all states reachable by rules of the corresponding grammar. Moreover, 3) all NACs that are not isomorphic for both rules have to hold in all reachable states. Obviously, in this case the rules are applicable at the same matches and result in the same transformation. Note that conditions 1) to 3) can be checked statically by inspecting the initial state and the rules.   △

For $p = (L \leftarrow I \rightarrow R)$ with negative application condition $nac : L \rightarrow N$ it is possible to construct the inverse rule $p^{-1} = (R \leftarrow I \rightarrow L)$ with equivalent $nac' : R \rightarrow N'$, such that $G \xRightarrow{p} G'$ implies $G' \xRightarrow{p^{-1}} G$, and vice versa (see Def. 5.15).

A concurrent rule summarises a given rule sequence in one equivalent rule (see Def. 5.27). In a nutshell, a concurrent rule $p *_E q$ is constructed from two rules, $p = (L_p \leftarrow K_p \rightarrow R_p)$ and $q = (L_q \leftarrow K_q \rightarrow R_q)$, that may be sequentially dependent via an overlapping graph E by modelling all deletions and creations of elements that are modelled either in p or in q. The application of the concurrent rule then has the same effect as applying p first, and applying q subsequently, where the co-match of p and the match of q overlap, as defined in their overlapping graph E (see Theorem 5.30).

The concurrent adaptation rule $(p_1 * \ldots * p_n)_E$ with $E = (E_1, \ldots, E_{n-1})$ for a longer sequence is constructed in an iterated way by $(p_1 * \ldots * p_n)_E = p_1 *_{E_1} p_2 *_{E_2} \cdots *_{E_{n-1}} p_n$.

In Theorem 10.18 we require as weaker conditions that each context rule p have a corresponding adaptation sequence $(q_1 * \ldots * q_n) \in R_{adapt}$, which is not necessarily inverse to p. We require that q *be applicable after p has been applied*, which is not a real static condition, but it can be argued from the context whether it is fulfilled. In part 2 of Theorem 10.18 it is sufficient to require for the normal adaptation property that we be able to construct a concurrent rule $p *_{E_0} (q_1, \ldots, q_n)_E$ which is *SAS*-equivalent to a concurrent rule r constructed from a normal rule sequence $(r_1, \ldots, r_m) \in R_{norm}$. Analogously, for the rollback adaptation property we require that $p *_{E_0} (q_1, \ldots, q_n)_E$ be *SAS*-equivalent to an inverse concurrent normal rule r^{-1}.

Theorem 10.18 (Verification of self-adaptation properties).
Let SAS be an SA system and G a reachable system state. SAS has

1. *the* direct adaptation property *if for each context rule p there is an adaptation rule sequence that directly reverses the effect of the context rule, i.e.,* $\forall\ p \in R_{cont}$ $\exists\ q = (q_1 * \ldots * q_n)_E$ *via* $E = (E_1, \ldots, E_{n-1})$ *and* $n \geq 1, q_i \in R_{adapt}$ *with* $q \simeq p^{-1}$.

2. *the* normal (resp. rollback) adaptation property *if for each context rule p there is an adaptation rule sequence that reverses the effect of the context rule up to normal rule applications, i.e.,* $\forall\ p \in R_{cont}$ *we have*

   (a) $\exists\ q = (q_1 * \ldots * q_n)_E$ *via* $E = (E_1, \ldots, E_{n-1})$ *and* $n \geq 1, q_i \in R_{adapt}$, *and q is applicable after p has been applied,*

   (b) $\forall$ *overlappings* E_0 *of p and q leading to a concurrent rule* $p *_{E_0} q$ $\exists\ r = (r_1 * \ldots * r_m)_{E'}$ *with* $m \geq 1$ *via* $E' = (E'_1, \ldots, E'_{m-1})$ *with* $r_i \in R_{norm}$ *such that* $p *_{E_0} q \simeq r$ *(resp.* $p *_{E_0} q \simeq r^{-1}$ *in case of rollback).*   △

Proof.

1. Given the context rule $p = (L \xleftarrow{l} K \xrightarrow{r} R)$ with NACs $nac_{i,L} : L \rightarrow N_{i,L} (i \in I)$, the inverse rule is given by $p^{-1} = (R \xleftarrow{r} K \xrightarrow{l} L)$ with corresponding NACs $nac_{i,R} : R \rightarrow N_{i,R}$. Now, given $p \in R_{cont}$ and $G_0 \in Reach(SAS)$ with $G_0 \xRightarrow{p} G_1$, we have by assumption $(q_1, \ldots, q_n) \in R^*_{adapt}$ with $(q_1 * \ldots * q_n)_E \simeq p^{-1}$, and by construction of p^{-1} also $G_1 \xRightarrow{p^{-1}} G_0$ and hence also $G_1 \xRightarrow{(q_1*\ldots*q_n)_E} G_0$ by SAS equivalence, which implies by the Concurrency Theorem that $G_1 \xRightarrow{*} G_0$ via R_{adapt}.

2. Given $G_0 \in Reach(SAS)$ and $G_0 \xRightarrow{p} G_1$ with $p \in R_{cont}$, by (a) $\exists\ q = (q_1 * \ldots * q_n)_E$ with $n \geq 1$ and $q_i \in R_{adapt}$, and q is applicable after p has been applied. Since $G_0 \xRightarrow{p} G_1$ we also have $G_1 \xRightarrow{q} G_2$ for some G_2 and hence $G_0 \xRightarrow{p} G_1 \xRightarrow{q} G_2$. According to Fact 5.29, an overlapping E_0 of p and q exists, such that $G_0 \xRightarrow{p} G_1 \xRightarrow{q} G_2$ are E_0-related and hence $G_0 \xRightarrow{p*_{E_0}q} G_2$. By (b), we have $\exists\ r = (r_1 * \ldots * r_m)_{E'}$ with $m \geq 1$ and $r_i \in R_{norm}$ such that $p *_{E_0} q \simeq r$ in the case of the normal adaptation property (resp. $p *_{E_0} q \simeq r^{-1}$ in the case of the rollback adaptation property). Hence, in case of the normal adaptation property, $G_0 \xRightarrow{p*_{E_0}q} G_2$ implies $G_0 \xRightarrow{r} G_2$. Now, $G_0 \xRightarrow{r} G_2$ and $G_1 \xRightarrow{q} G_2$ via concurrent rules r and q imply by Concurrency Theorem sequences $G_0 \xRightarrow{+} G_2$ via R_{norm} and $G_1 \xRightarrow{+} G_2$ via R_{adapt}, leading to the normal adaptation property. Similarly, in the case of the rollback adaptation property, we have that $G_0 \xRightarrow{p*_{E_0}q} G_2$ implies $G_0 \xRightarrow{r^{-1}} G_2$, such that $G_2 \xRightarrow{r} G_0$. Now, $G_2 \xRightarrow{r} G_0$ and $G_1 \xRightarrow{q} G_2$ via concurrent rules r and q with $n, m \geq 1$ imply sequences $G_2 \xRightarrow{+} G_0$ via R_{norm} and $G_1 \xRightarrow{+} G_2$ via R_{adapt}, leading to the rollback adaptation property.   □

In the following Examples 10.19 and 10.20, we verify the self-adaptation properties for different variants of our Car Logistics System case study, *CLS*, considering different adaptations. Two more examples are elaborated on in [BEE⁺13], including a counterexample, where the self-adaptation properties do not hold. In all

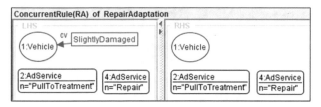

Fig. 10.12 Concurrent adaptation rule q constructed from sequence s in CLS_{Repair}

examples we have the same normal rules, R_{norm} = {ServiceToDo, EnterBP, DoService, DoSubService, NextBA, FinishBP}, and a subset of the context rules, R_{cont} = {SlightlyDamage,SeverelyDamage, Queue}, of CLS. Hence, we have sequential independence of context rules and normal rules, and, since CLS is normal-state consistent, also $CLS_{Repair}, CLS_{RollbackAndRepair}, CLS_{Queue}$, and $CLS_{Dispose}$ are normal-state consistent.

Example 10.19 (SA system CLS_{Repair} is corrective). In CLS_{Repair}, we analyse the *repair* adaptation of CLS. This means that we have one context rule R_{cont} = {SlightlyDamage} (Fig. 10.7), and the set of adaptation rules $R_{adapt} = R_{adapt}^2$ = {TakePullToTreatmentService, TakeRepairService, RepairVehicle}. Moreover, C_{adapt} = {*Damage*} is the set of adaptation constraints.

According to Theorem 10.14, we have to show that CLS_{Repair} has the direct adaptation property (property 1). According to Theorem 10.18, CLS_{Repair} has the direct adaptation property if for p = SlightlyDamage we have $(q_1, \ldots, q_n) \in R_{adapt}$ with $q = (q_1 * \ldots * q_n)_E \simeq p^{-1}$, where q is the concurrent rule of the adaptation rule sequence $(q_1, \ldots, q_n)$. This means that we have to find an adaptation rule sequence that results in a concurrent rule q which is SAS-equivalent to the inverse context rule p = SlightlyDamage (i.e., it removes the SlightlyDamaged flag).

We consider the adaptation rule sequence s = {TakePullToTreatmentService, TakeRepairService, RepairVehicle} together with suitable dependencies (overlappings) of the right-hand side of q_i and the left-hand side of q_{i+1}, and construct a concurrent rule from this sequence in an iterated way. In AGG, the construction of concurrent rules from rule sequences can be computed automatically. For our sequence, we get the concurrent adaptation rule depicted in Fig. 10.12.

The concurrent adaptation rule q is SAS-equivalent to the inverse context rule p = SlightlyDamage due to the following argumentation: The additional elements in rule q w.r.t. rule p (the PullToTreatment and Repair nodes) are preserved by q, and are always there in all possible states, since no system rule ever adds or deletes PullToTreatment and Repair nodes.

Hence, CLS_{Repair} has the direct adaptation property, and due to Theorem 10.14, we can conclude that CLS_{Repair} is corrective.   △

Example 10.20 (SA system CLS_{Queue} is enhancing). In CLS_{Queue}, we consider the *wait-at-queue* adaptation. This means that we have one context rule R_{cont} = {Queue} and the following set of adaptation rules: R_{adapt} = {Enqueue, DoService, RemoveWait}. Moreover, C_{adapt} = {*Wait*} is the set of adaptation constraints.

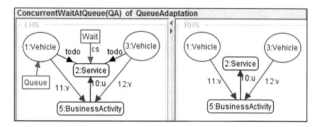

Fig. 10.13 Concurrent adaptation rule q constructed from sequence s in CLS_{Queue}

According to Theorem 10.14, we show that CLS_{Queue} has the normal adaptation property (property 2). According to Theorem 10.18, CLS_{Queue} has the normal adaptation property if for $p =$ Queue we have that

(a) $\exists\, q = (q_1 * \ldots * q_n)_E$ via $E = (E_1, \ldots, E_{n-1})$ and $n \geq 1, q_i \in R_{adapt}$ and q is applicable after p has been applied,

(b) $\forall$ overlappings E_0 of p and q, leading to a concurrent rule $p *_{E_0} q$ $\exists\, r = (r_1 * \ldots * r_m)_{E'}$ via $E' = (E'_1, \ldots, E'_{m-1})$ with $r_i \in R_{norm}$ such that $p *_E q \simeq r$.

For the context rule $p =$ Queue, we have a sequence of adaptation rules $s =$ {Enqueue, DoService, RemoveWait} that results in the concurrent rule q shown in Fig. 10.13. Since the right-hand side of p equals the left-hand side of q, we can conclude that q is applicable after p has been applied, as required in (a).

We then construct according to (b) the concurrent rule Queue $*_E q$ which equals rule q in Fig. 10.13 but does not contain the context nodes Wait and Queue and their adjacent edges. Obviously, this rule Queue $*_E q$ is isomorphic to a concurrent rule r constructed by the sequence of normal rules (DoService $*_{E'}$ DoService), which removes two todo edges from two different vehicles to the same service in one step. Hence, Queue $*_E q$ and r are *SAS*-equivalent.

Thus, CLS_{Queue} has the normal adaptation property, and due to Theorem 10.14, we can conclude that CLS_{Queue} is enhancing.   △

We give two further explicit examples in [BEE$^+$13], using our static conditions to show that $CLS_{RollbackAndRepair}$ *is enhancing* and that $CLS_{Dispose}$ *does not satisfy the sufficient conditions for SA properties.*

10.3 Automating the Approach by AGG

AGG^3 (see Sect. 12.1) is a well-established tool environment for typed attributed graph transformation. Graphs in AGG are defined by a type graph with node type inheritance and may be attributed by any kind of Java object. Graph transformations

---

^{3} AGG (Attributed Graph Grammars): http://www.tfs.tu-berlin.de/agg

can be equipped with arbitrary computations on Java objects described by Java expressions. Sect. 12.1 presents an extensive description of the AGG environment, consisting of several visual editors, an interpreter, and a set of validation tools.

As shown in the previous sections, our framework for modelling and analysing SA systems is supported by the AGG tool for modelling both the initial configuration of the system and also the possible configurations that the system can reach in the case of adaptations. A simulation of adaptations is performed by applying adaptation rules within AGG so that practitioners can get confidence on the system and its evolutions.

From the modelling point of view, referring to the example, within AGG the model engineer can perform the system design in a visual way: the business process and services shown in Fig. 10.1 can be directly mapped to elements of the initial state graph depicted in Fig. 10.4. Furthermore, graph constraints can be graphically represented, as illustrated in Sect. 10.1.3. The behaviour and the evolution of the system are also graphically represented within AGG in a rule-based, intuitive way.

From the analysis point of view, dependencies between rules and conflicts between rules in a minimal context (critical pairs) can be computed fully automatically. The results support our argumentation, showing that the sufficient conditions for our two theorems are satisfied in our examples. Moreover, the construction of concurrent rules from rule sequences is also fully automatic. It is only required that the modeller define a suitable object flow between the rules of the sequence to define the overlapping graphs of rules and to get a unique resulting concurrent rule.

In the following, we discuss some aspects to assess the applicability of the proposed analysis techniques.

Practical relevance of assumptions: For the Car Logistics case study, we found the assumptions for SA systems very helpful for structuring the model (by distinguishing different sets of rules for normal behaviour, context changes and different adaptations). The sufficient conditions we checked for applying our results did not prove to be too strong. Instead, whenever our model did not satisfy one of the sufficient conditions, the changes we implemented in the model did not only result in the satisfaction of the conditions but also in a more systematic and concise model.

Achieved degree of automation: All our analysis techniques are static, i.e., they check rule properties only. Yet, some of the techniques require manual effort, i.e., reasoning about rule properties that are not supported by AGG in a fully automatic way. For instance, although AGG implements checking sufficient conditions for termination of rule sets, for our example these sufficient conditions turned out to be too strong. So we had to argue about termination by inspecting the rules "manually".

Similarly, since up to now there is no automatic check for *SAS* equivalence of rules implemented, we had to perform these checks by hand by inspecting the rules ourselves. Critical pair analysis is a powerful instrument assisting with checking confluence of rule sets. A sufficient condition for the confluence of a system is (termination and) the absence of critical pairs. But if critical pairs are found, indicating potential conflicts, manual effort is needed to show whether these critical pairs could really lead to conflicts in a specific system or not. Usually, these hints are very helpful for the modeller and show where problems lie and the model should be adapted.

Currently, the adaptation manager selects the most suitable adaptation grammar manually. In future, this decision could be supported by defining priorities for adaptation grammars depending on the severity of the disturbance ("treat worst case first"). This is currently not supported by AGG. Finding suitable adaptation sequences using an adaptation grammar is realised in a semiautomatic way by simulation: by applying the adaptation rules nondeterministically, the adaptation is performed automatically, and by keeping track manually of the used rules and their matches, the rule sequence and its object flow are determined. The object flow of the used rule sequence is then the input to AGG's automatic construction of a concurrent adaptation rule. It would be desirable if for larger grammars AGG could automatically record the rule application order and their matches when applying rules from the grammar. The recorded sequence and its object flow derived from the match overlappings could then be used to construct a concurrent rule fully automatically from a simulation run.

Time and memory consumption: The time consumption for the fully automatic computation of dependency and conflict tables in our example[4] depends on the number of rules in the corresponding grammar and the number of objects (nodes and edges) in the rules. Time increases exponentially for large rules. For analysing conflicts of the wait-and-queue adaptation with the largest number of objects in rules, AGG took 115 seconds; all other conflict and dependency tables were computed in less than 4 seconds.[5] It is hence advisable to use more but smaller rules instead of describing the system by fewer and larger rules. For the fully automatic computation of concurrent rules from rule sequences that were used in our example, AGG used less than one second.

Scalability: Obviously, time and memory consumption grows when the modelled system becomes larger. However, when the system becomes more complex w.r.t. the number of rules and the size of the system graphs, we find that the size of rules (modelling only local effects) remains nearly stable. It is in the hands of the modeller to formulate a rule set that can remain small enough to be analysed properly. The size of the system graphs does not influence the performance of our static analyses. Up to now, AGG can analyse rules with up to $20 - 30$ elements in reasonable time.

Addition of rules at runtime: Since our framework for modelling SA systems is modular, i.e., based on different rule sets, we can simulate adding new rule sets (adaptation grammars) to the system at runtime. To add a new adaptation hook to the system, the type graph must be updated by adding the new adaptation hook type. Ideally, this type is a subtype of a more general adaptation hook type. In this case,

---

[4] Measured on a standard notebook with an Intel dual-core processor and 2 GB of memory.

[5] Explicit benchmark values for our case study can be found in [BEE+13].

the existing rules do not have to be updated at all, only a new context rule and a new adaptation grammar specific to the new adaptation hook have to be added to the system. The adaptation manager realises the selection of available adaptation grammars and this may include new grammars when needed.

10.4 Related Work

In this section, we compare our work with related approaches to adaptive system modelling and analysis, focusing on *context-awareness*, *self-adaptiveness*, *formalisation* and *tool support*, which we see as main aspects of SA system modelling.

Own previous work: In [BPVR09], we proposed for the first time an approach to modelling self-repairing system architectures as typed (hyper-)graph grammars and verified them w.r.t. dependencies and conflicts of rules modelling the environment, the normal system and repair actions. This approach was extended in [EER$^+$10] to modelling self-healing (SH) systems using algebraic graph transformation. In this book, we have built on the preliminary sufficient conditions formulated in [EER$^+$10] and generalised the approach to the class of adaptive systems that is identified in Sect. 10.1.1. This approach includes *enhancing* adaptive systems, i.e., systems that enhance existing services of the application and can be adapted to become normal again, but up to new functionalities or services. For ensuring that a system is indeed enhancing, we now check statically the *normal* (or *rollback*) *adaptation property*, requiring that there be adaptation *rule sequences* leading back to normal states (instead of looking only for inverse adaptation rules as in [EER$^+$10]). We also now allow for more general static conditions to ensure corrective behaviour; for the *direct adaptation property*, we require the existence of a *sequence* of adaptation rules that reverses the effect of the context rule (up to normal rules) instead of a single inverse rule. [BCG$^+$12] presents a conceptual framework, where adaptation is defined as the runtime modification of the control data. Our approach is compatible with this framework, in the sense that the control data can be identified by the dividing of the set of rules into rules that correspond to normal system behaviour and rules that implement adaptation mechanisms.

Software Architectures: There is a wealth of Architecture Description Languages (ADLs) and architectural notations which provide support for dynamic software architectures analysis [BCDW04].

[BB09] proposed an approach called *Genie* that offers management of structural variability of adaptive systems. Genie can be considered as an ADL with generative capabilities for reconfiguring from one system structure to another according to changes in the environment and for deciding what kind of structural reconfiguration has to be performed. The main limit of this approach is the absence of a way to guarantee desired properties of the systems after each adaptation execution; the language is not supported by any formal framework. From the modelling point of view, the approach is specifically architectural, whereas we propose a general approach that can be used at different levels of abstraction. For instance, our case study presents the business process of a service-oriented scenario.

[GCH⁺04] introduced an SA-based self-adaptation framework, called Rainbow, using external mechanisms and an SA model to monitor a managed system, detect problems, determine a course of action, and carry out the adaptation actions. Rainbow, by making use of architectural styles, provides infrastructures with explicit customisation points. The definition of these customisation points limits the dynamicity of the approach; in particular, the context is not considered as a part of the system model that can evolve during the system lifecycle. In our approach, we do not rely on predefined customisation points to manage the adaptation, but we monitor properties of the context to understand where and how to adapt the system.

[HIM00] presented an approach specifying software architecture styles using hyperedge replacement systems. The authors use graph rewriting combined with constraint solving to specify how components evolve and communicate. With respect to our approach, the limits are: (i) the way they check the system correctness, and (ii) the tool support. Regarding system correctness, they need to inspect all the reachable system states for each property to be verified. In our approach, we define operational properties (corrective, enhancing, direct (normal) adaptation, etc.) that we check in a static way by inspecting only the related rules without producing all reachable states explicitly. Moreover, their formal framework is not supported by a tool, whereas our formal framework is supported by AGG to model and analyse self-adaptive systems.

[BHTV05] presented an approach checking whether an architecture is a refinement of another one by defining relationships between abstract and concrete styles. Refinement involves a reachability analysis for a target configuration from a given initial configuration. Reachability is analysed by model checking and simulation, which requires us to a priori restrict the systems to finite state systems by restricting the number of dynamic model elements that can be created by the transformation rules. Static analysis techniques, as applied by us, do not have this limitation. Moreover, we do not only consider refinements but more general system adaptations.

[BG08] presented a graph transformation-based approach to modelling correct self-adaptive systems on a high level of abstraction. The approach considers different levels of abstraction according to the reference architecture by [KM07]. The correctness of the modelled self-adaptive systems is checked by using simulation and invariant checking techniques. Invariant checking is mainly used to verify that a given set of graph transformations will never reach a forbidden state. The verification is efficient and the complexity is linear in the number of rules and properties to be checked. The limitation of this approach is that a unique model is used for application and adaptation logics. This means that when a new adaptation case is added, the overall model must be refined. In our approach, the adaptation logic is developed separately from the application logic in terms of adaptation rules. The adaptation logic can be modified without requiring us to change the application.

Service-Oriented Computing: In the community of Service Oriented Computing, various approaches supporting self-healing have been defined, e.g.: for triggering re-

pair strategies as a consequence of a requirement violation [SZK05]; for optimising
QoS of service-based applications [CPEV05]; for satisfying some application constraints [VGS+05]. Repair strategies usually are specified by means of policies for
managing the dynamics of the execution environment [BGP07, CNM06]. The goals
of the strategies proposed by the aforementioned approaches range from service
selection to rebinding and application reconfiguration [PLS08]. Some techniques
enable the definition of various adaptation strategies but they all lack a coherent design approach for supporting designers in this complex task.

Summarising, our approach abstracts from particular languages and notations
and can be applied at different levels of granularity. We provide a coherent design
approach that allows software engineers to model and analyse self-adaptive systems
within the same framework. Once a suitable level of abstraction has been identified,
the system can be modelled together with adaptation strategies and mechanisms.
The system specification is then used to formally verify operational properties.

10.5 Summary

In this chapter, we have modelled and analysed self-adaptive systems using algebraic graph transformation and graph constraints. We have defined consistency
properties that include system consistency, normal state consistency, and adaptation
state consistency. Furthermore, we have defined operational properties that include
self-adaptation, corrective self-adaptation and enhancing self-adaptation; we also
have defined direct, normal, and rollback adaptation properties concerning the behaviour of adaptations w.r.t. their influence on the normal system behaviour. Our
analysis detects in which class of self-adaptive systems a given system belongs (enhancing, corrective), and which properties we have with respect to the kind of adaptation (direct, normal, rollback). The classification helps us to reason about system
behaviour, where systems with the rollback adaptation property may be in more dan-
ger of repeated failures than systems with normal adaptation property, since states
that preceded failures are reached again after the adaptation. Note that the operational properties concern *all* reachable system states, whereas they are checked
in a static way by inspecting only the rules without producing all reachable states
explicitly.

The main results concerning operational properties of SA systems are summarised in Fig. 10.14, where most of the static conditions in Theorems 10.14
and 10.18 can be automatically checked by the AGG tool. We have needed manual effort to show the termination of properties and *SAS* equivalence of rules, but
it was always possible to perform the analysis statically. Although static conditions
lead to overapproximation of systems, we have found that the conditions to check
were reasonable enough to be expected to hold in SA systems and did not restrict
our intuitive notion of SA system properties. Exemplarily, the different properties

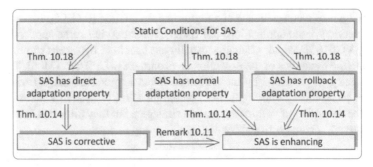

Fig. 10.14 Static SAS analysis

are verified for different adaptations of our car logistics system in a seaport terminal.

In our approach, the selection of an adaptation grammar is done by a human adaptation manager who selects the most suitable adaptation. In an extended approach, a distinguished kind of *control attribute* (like, e.g., failure counter or timeout parameter) might be used to select automatically between different variants of adaptations. Note that states with additional control attributes (typed over control types) would still be *normal states* in our framework (not violating adaptation constraints). Our results hence can be extended in a straightforward way by reformulating the more flexible operational properties of enhancing and corrective systems to hold "up to changed control attributes". We refrained from this extension here to keep proofs clear and simple.

Future work is needed to further automate the checks currently needing manual effort with AGG. To enhance the practical usability of static analysis, a continuous optimisation of the performance of the critical pair analysis AGG is in progress. Finally, the implementation of a logging feature for recording the order and matches of applied rules in AGG would be very helpful for automating the selection of rule sequences in our adaptation framework.

Chapter 11
Enterprise Modelling and Model Integration

The aim of enterprise modelling is to support and improve the design, documentation, analysis and administration of business objects and operations based on adequate modelling techniques [FG98, SAB98]. For this purpose, domain-specific enterprise models shall provide the basis for communication between people with different professional backgrounds [Fra02]. This chapter presents how model transformation and integration techniques presented in Part III can be applied to automate and improve the modelling tasks within a distributed enterprise modelling framework. Sect. 11.1 describes the main aspects of enterprise modelling and presents the used enterprise modelling framework. Sect. 11.2 illustrates how the alignment of different domains within the framework can be specified by triple graph grammars. Sects. 11.3 and 11.4 demonstrate the application of model transformation and integration techniques to concrete domain models. Finally, Sect. 11.5 discusses the achievements, their relevance and related work. This chapter is based on the results of a research collaboration between Technische Universität Berlin, the University of Luxembourg and Credit Suisse, which were published in [BHE09c, BHEE10, BH10, Her11, Bra13].

11.1 Enterprise Modelling

Enterprise models provide representations of the structures, processes, resources, involved actors, executed functions, goals, and constraints relevant for the modelled enterprise. For this reason, enterprise modelling has to provide an agile modelling process, which is integrated across the different business functions [FG98]. An agile modelling process additionally reduces the required time frames for adapting the models according to change requests, which can occur quite frequently during the lifetime of an enterprise. Moreover, adequate modelling techniques should support the propagation of changes from one domain to others. This way, the knowledge and expertise of enterprise modellers can be focussed on their main domain, which is an

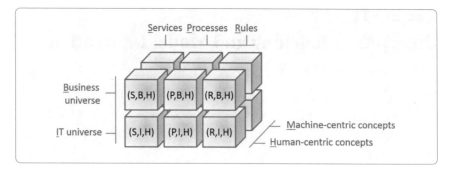

Fig. 11.1 Enterprise model framework

important requirement for decentralised and distributed models occurring especially in large multinational enterprises.

In order to master the high complexity of an enterprise in its whole, visual modelling techniques have been successfully applied. They provide intuitive notations and high abstraction capabilities. Clearly, visual models cannot replace all textual models in all domains. But still, where suitable, they often show high benefits. Furthermore, enterprise models are usually not only used for the design and documentation of enterprises, but also for the analysis and management of operations. In particular, process analysis concerns, e.g., the question whether certain business processes can be performed in a different but more suitable way, such that some goals can be achieved in an optimised way. In combination with formal abstract syntax definitions, visual modelling techniques can enable verified automated analyses, which can support error detection and thus quality assurance of the models.

In order to satisfy the requirements in enterprise modelling, we apply formal techniques based on graph and model transformation of Parts II and III, which provide powerful and efficient techniques for model transformation and integration. Their formal foundation ensures correct analysis results and the automated tool support provides efficient checks. The techniques provide the basis for an agile and decentralised modelling process, including distribution of models and efficient change propagation.

The integration of different enterprise models requires, on the one hand, the application of techniques that ensure certain quality and consistency requirements and, on the other hand, the application of techniques for setting up and maintaining a common understanding of the enterprise by the modellers. While this chapter applies suitable techniques for the first requirement based on the concepts in Part III, the common understanding is supposed to be set up and maintained based on sophisticated techniques in the area of ontology engineering [FG98].

The main domains of enterprise modelling can be captured by the enterprise modelling framework in Fig. 11.1, introduced by Brandt et al. [BHE09c, BH10, Bra13]. It shows different coordinates (X, Y, Z) and each of them represents a container for several domain-specific models. For instance, coordinate (S, B, M) represents all service models for the business universe using a machine-centric modelling language. We denote the set of the models in one coordinate by $\mathcal{M}_{X,Y,Z}$ and refer to a specific model by the notation $M_{X,Y,Z}$ for a model $M_{X,Y,Z} \in \mathcal{M}_{X,Y,Z}$. Clearly, there are several further perspectives and aspects within an enterprise which are also relevant for enterprise modelling, but they go beyond the scope of this chapter. However, the general techniques provided in this book show good potential that they can also improve the modelling process in further domains, as discussed and evaluated in [BH10, Bra13].

The separation of models into business and IT domain models is a common standard (see, e.g., [CIO99]). The main idea is to separate the application and task oriented view of services, processes and rules in the business universe from specific realisations via implemented components in the IT domain to provide automation for these applications and tasks. The alignment between both domains, however, is usually not complete. Parts of a business domain model can be automated via a corresponding IT model. Other parts of the business domain model may stay without IT support. Vice versa, parts of an IT model may automate a corresponding business model. However, the IT may have their own specific services, processes and rules for which there is no correspondence in the business domain models. Therefore, neither a top-down nor a bottom-up relationship is appropriate in the general case.

Besides differentiating between business and IT model domains, Fig. 11.1 categorises their content into service, process, and rule models. Moreover, it separates according to the primarily used concepts, depending on whether they are machine-centric concepts or human-centric concepts, i.e., whether they are meant for documentation and human interaction (human-centric domain) or primarily for execution and automated analysis (machine-centric domain). The separation into different model spaces leads to a separation of concerns while allowing for an overlapping on common aspects. This reduces the complexity for domain modelling, but requires a sound approach for model transformation and integration to handle the inter-model dependencies.

In this chapter we will use ABT-Reo diagrams [Arb04, Arb05, BHEE10], which are based on Abstract Behaviour Types and Reo connectors. ABTs specify system components and are represented by blocks having ports. Reo connectors specify the different types of links between ports of ABTs and such structures can be composed to form a single new ABT.

Remark 11.1 (Scope of the chapter within the model framework). The enterprise model framework as shown in Fig. 11.1 encompasses the development of many different aspects. This chapter presents suitable intra- and and inter-modelling techniques focussed on machine-centric business and IT service models given by ABT-Reo diagrams, which can be used as a single modelling language for both types of service models. These ABT-Reo models are intended to be aligned with their corresponding human-centric service models. The scope of this chapter concerns the model transformation and integration of ABT-Reo service models as illustrated in Fig. 11.2. The techniques are general with respect to different domain-specific languages, because they are based on the underlying abstract syntax graphs of the

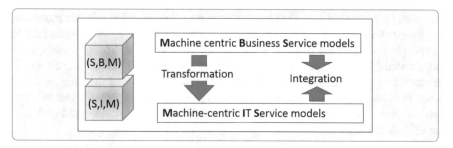

Fig. 11.2 Scope of this chapter

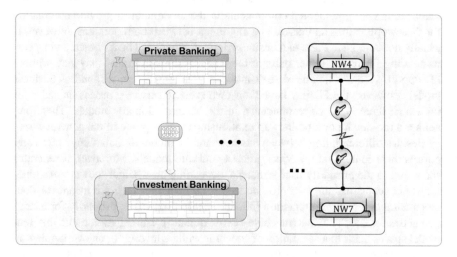

Fig. 11.3 Human-centric service models (left: business domain, right: IT domain)

models. Thus, there is a good potential that they can be applied for several other dimensions in the enterprise model framework, too.   △

Human-centric service models are diagram-like language artefacts that sketch service instances as well as their connections, and we present two examples for human-centric service models in the business and IT domains.

Example 11.2 (Human-centric service models for business and IT domains). The model on the left of Fig. 11.3 is a fragment of a human-centric business service model. The departments "Investment Banking" and "Private Banking" are departments at Credit Suisse. Information exchanged between the two parties must comply with the Chinese Wall Policy [BN89]. The policy defines what information is allowed to be exchanged between the two departments. To guarantee that the policy is respected a filter will suppress illegal messages between the two service instances in the diagram. Each service instance represents a department. Therefore, the policy is realised as a filter in a service model. This business view completely abstracts

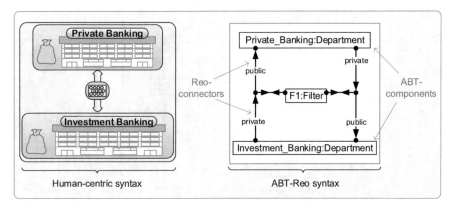

Fig. 11.4 ABT-Reo instance in the business universe

away IT details. The fragment of a human-centric IT service model on the right
of Fig. 11.3 corresponds to the business model and shows interconnected network
zones "NW4" and "NW7", which are connected via a secured connection.   △

For a concrete alignment between the business and IT model fragments in
Fig. 11.3, the private banking department can be mapped to the network zone
"NW4" and the investment banking department can be mapped to the network zone
"NW7". The connector between the investment and the banking department in the
business universe is then related to the connections between networks "NW4" and
"NW7".

Human-centric service models as they have just been introduced are only syntax
artefacts. A corresponding semantics can be assigned by the help of an alignment
with machine-centric models, as presented in [BHE09c, BHEE10, BH10]. The corresponding types of machine-centric models that we use in this case study are abstract behaviour types and Reo connectors [Arb05, KB06, Arb04, AR02]. The reason why we use abstract behaviour types and Reo connectors to specify services and
service landscapes is because of their support for exogenous coordination. In addition to that, abstract behaviour types focus on incoming and outgoing messages and,
therefore, abstract away implementation details of services. This frees an ABT-Reo
model from implementation aspects and reduces the overall complexity of models.

Example 11.3 (Machine-centric business model). Fig. 11.4 shows the human-centric
business model from Fig. 11.3 and a corresponding machine-centric business service model. Here, two abstract behaviour types are used to represent the investment
and the private banking department. Messages running between these two abstract
behaviour types have to pass through different Reo connectors. While messages via
private connectors are not visible to the outside, those along public connectors are
visible. The filter in the middle of the diagram listens to messages and will suppress
private messages targeting a public connector. This filter can be interpreted as a formal specification of an organisational security policy, like the Chinese Wall Policy.

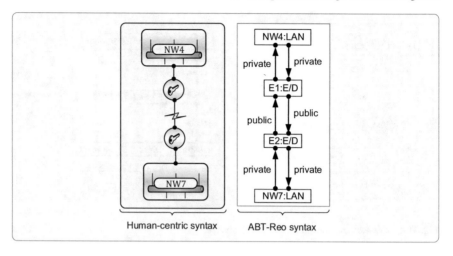

Fig. 11.5 ABT-Reo instance M_1 of the IT universe

For example, the filter may ensure that the communicated data does not contain files which contain both address and balance information. Using model checking it is possible to prove that no private message will finally pass by a public connector.   △

Example 11.4 (Machine-centric IT model). Fig. 11.5 shows the human-centric IT model from Fig. 11.3 and a corresponding ABT-Reo diagram M_1 specifying the structure of a part of a network composed of local area networks (LANs). It contains four ABT elements that are connected via Reo connectors. The two outer ABT elements represent the LANs "NW4" and "NW7" while the two inner ones denote encryption/decryption nodes, i.e., the communication between both LANs is encrypted.   △

    While the concrete (DSL) syntax of ABT-Reo models is more compact and intuitive, a precise and detailed specification and analysis is based on the abstract syntax, which enables us, e.g., to explicitly specify properties of ports that are implicit only in the concrete notion.

Example 11.5 (Abstract syntax of machine-centric model). Fig. 11.6 shows fragments of the models from Fig. 11.5 on the left and the corresponding abstract syntax graph on the right. Bold bullets in ABT-Reo notation correspond to nodes of type "Point" and arrows (Reo connectors) correspond to nodes of type "Reo" in the abstract syntax graph. They are attached respectively to external input and external output ports according to the direction of the Reo connectors. Each point glues one input to one output port, e.g., the left Reo connector in the ABT-Reo diagram corresponds to the left Reo node in the abstract syntax graph and the communication data enters the connector via the input port at the bottom and exits the connector via the output port at the top.   △

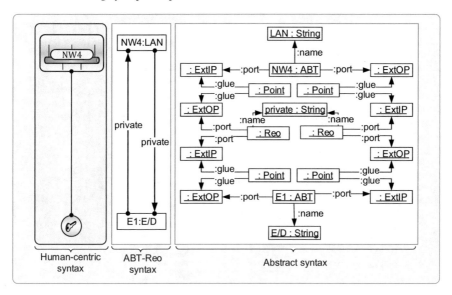

Fig. 11.6 Part of M_1 in Fig. 11.5 in visual notation and abstract syntax

Similarly to the definition of a meta model of a visual language according to the OMG MOF approach [Obj14a], a type graph specifies the general structure of abstract syntax graphs. A graph of the language is typed over its type graph via a graph morphism that maps each element to its type element in the type graph, i.e., each node to a node and each edge to an edge in the type graph (see Def. 2.2).

Example 11.6 (Type graph). The structure of ABT-Reo diagrams in abstract syntax is given by the type graph $TG_{ABT-Reo}$ in Fig. 11.7 containing the main types "ABT" for abstract behaviour type nodes, "Reo" for Reo connectors, "Port" for ports and "Point" for points that glue together input and output ports of both ABT nodes and Reo connectors. Ports of elementary ABT nodes and Reo connectors are external, i.e., they are used for external communication with other elements. ABT nodes can also be composite, i.e., they may contain a further specified internal structure involving other ABT nodes and Reo connectors. In this case their external ports are connected to complementary internal ports, such that the communication is transferred through the borders of the composite ABT nodes.   △

11.2 Inter-modelling by Triple Graph Grammars

In the following, we use the concept of triple graphs, an extension of plain graphs dividing elements into source, target and correspondence sections which are connected by graph morphisms as presented Chap. 7. This extension improves the def-

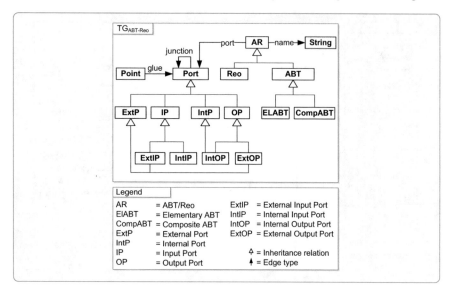

Fig. 11.7 Type graph $TG_{ABT-Reo}$ for ABT-Reo models

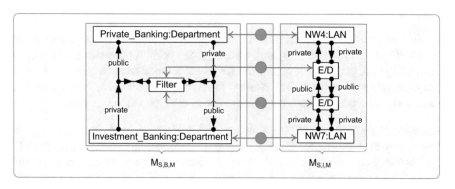

Fig. 11.8 Triple graph with models $M_{S,B,M}$ and $M_{S,I,M}$

inition of model transformations, where models of a source language are translated to models of a target language and correspondence elements can be used to guide the creation of the sequence of transformation steps. Similarly, triple graph transformations are suitable for model integration, which takes a source and a target model and sets up the missing correspondences between both models—if possible.

Example 11.7 (Triple graph). The triple graph in Fig. 11.8 shows an integrated model consisting of a business service model in the source component (left) and an IT service model in the target component (right). These models are the ABT-Reo diagrams from Figs. 11.4 and 11.5. The corresponding elements of both models are related by graph morphisms (indicated in grey) from the correspondence graph

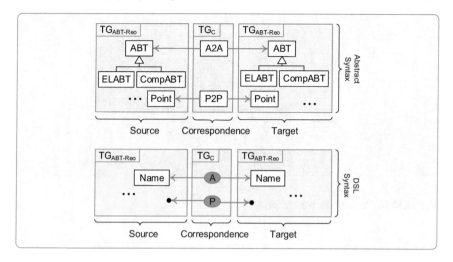

Fig. 11.9 Triple type graph TG_{B2IT}

(middle) to source and target, respectively. In detail, the departments in the business model correspond to the LAN nodes in the IT model as the local area networks "NW4" and "NW7" are used in the private banking and investment banking departments. The filter in the business model corresponds to the composed structure of two encryption nodes in the IT model. The node type "E/D" denotes encryption and decryption capabilities.   △

Model transformation as well as model integration do not require deletion during the transformation. Technically, both techniques compute graph transformation sequences, yielding a consistent triple graph (see Chap. 7 for the formal details). In the case of model integration, the result is exactly the computed triple graph of the transformation sequence and in the case of model transformations the result is obtained by restricting the resulting triple graph to its target component. For this reason, it is sufficient to consider triple rules that are nondeleting. This implies that the first step in the DPO graph transformation approach (see Chap. 2) can be omitted, because the creation of elements is performed in the second step.

Example 11.8 (Triple graph grammar). The triple graph grammar $TGG_{B2IT} = (TG_{B2IT}, S_{B2IT}, TR_{B2IT})$ specifies how business service models and IT service models given as ABT-Reo diagrams are related and its type graph is shown in Fig. 11.9 in abstract syntax and DSL syntax (syntax of the domain-specific languages). The language of ABT-Reo diagrams is used for both, the source and the target language, and the type graph shows a correspondence between the relevant types, which are "ABT" for abstract behaviour type elements and "Point" for the gluing points between input and output ports of ABT elements and Reo connectors. The start graph S_{B2IT} is empty and Figs. 11.10 and 11.11 show some of the triple rules of TR_{B2IT}.

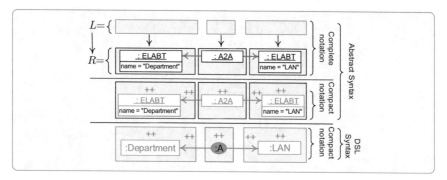

Fig. 11.10 Triple rule "DepartmentToLAN"

Each rule specifies a pattern that describes how particular fragments of business and IT models shall be related.

The first rule "DepartmentToLAN" synchronously creates two ABT elements and a correspondence node that relates them. This reflects the general correspondence between departments in the business view and the installed local area networks in the IT view. The rule is presented in complete and in compact notation as well as using the DSL syntax. The complete notion shows that a triple rule consists of a triple graph for the left-hand side (the upper one in the figure), a triple graph for the right-hand side (the lower one in the figure) and the relating morphism in between. Recall that triple rules are nondeleting and the intermediate graph K as it appears for plain graph transformation rules in Chap. 2 is not needed.

The compact notation for triple rules combines the left- and the right-hand sides of a rule, i.e., the rule is shown by a single triple graph with special annotations. All elements that are created by the rule, i.e., which appear in the right-hand side only, are marked by double plus signs. The DSL syntax of the rule shows the ABT diagrams in visual notation, where the attribute "name" is used as the label of the visual elements.

The rules in Fig. 11.11 are presented in DSL syntax using the compact notation. Similarly to the first rule, rule "PublicToPublic" has also an empty left-hand side. It synchronously creates two Reo connectors on both sides and they are related by the points at their input and output ports. The rule "FilterToED" is slightly more complex and shows how filters in business models correspond to encrypted connections in the related IT model. This reflects the abstract business requirement of hiding confidential information and its possible implementation by encryption in the IT domain. Note that the left-hand side of this rule corresponds to the right-hand side of the rule "PublicToPublic".

Private connections leading to related and secured public connections are related by the rule "PrivateInToPrivateIn". The last rule, "FilteredOutToPrivateOut" in Fig. 11.11, specifies how outgoing communication from a secured connection is handled. The private outgoing connection in an IT model corresponds to the gluing of the filtered public connection to the target ABT element, which is defined by the

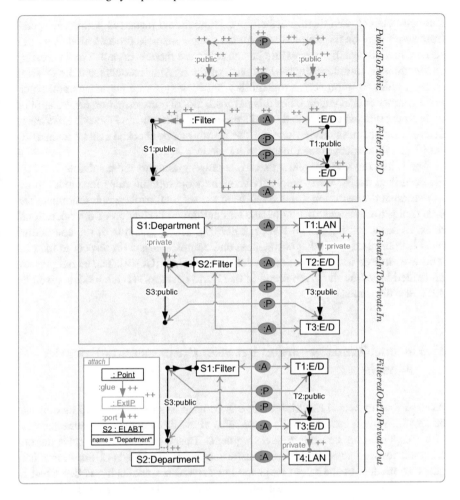

Fig. 11.11 Further triple rules of TGG_{B2IT}

box with the label "attach". This box specifies the explicit creation of elements in
the source component based on the underlying abstract syntax, where an input port
for the ABT node "S2" and the linking edges to the existing nodes are created.

The presented rules suffice for the examples in this chapter. However, the TGG
can be extended by further rules to cover the complete language of ABT-Reo models [BHEE10]. For example, symmetric communication via public channels as depicted in Fig. 11.4 would require an additional triple rule to extend public connections by corresponding ones in opposite direction.   △

Considering the model framework in Fig. 11.1, one may experience the following scenarios during the development of the models. First of all, two models that
should be integrated may be unrelated; thus performing model integration may de-

tect conflicts between them. Furthermore, some model instances within the model framework may be missing at the beginning, e.g., business process models and IT service models usually exist while business service models do not. The interesting challenge is to automatically retrieve parts of the missing models from the existing ones in order to improve interoperability between system components and enterprise components in general. For this purpose, model transformation can be applied on business process models to derive IT process models, and on IT service models to derive basic business service models. The results can be checked against integration conflicts with respect to the other existing models.

The following sections show that triple graph grammars are a suitable basis for the described needs. We exemplarily show how operational rules for model transformation and integration are derived from the original triple graph grammar. The underlying formal construction enables an automatic derivation of the operational rules, as described in [Sch94, KW07, EEE$^+$07]. The application of the operational rules is controlled, such that correctness and completeness with respect to the patterns are ensured for the resulting models [EEHP09, HEGO14]. The techniques are illustrated based on the given scenario of IT and business service models given by ABT-Reo diagrams.

11.3 Model Transformation Between Business and IT Service Models

As described in Sect. 11.2, triple rules can be used to specify how two models can be created simultaneously. Thus, triple rules allow the modeller to define patterns of correspondences between model fragments. Based on these triple rules the operational forward rules for model transformations from models of the source language to models of the target language are derived automatically as described in Chap. 7. Since triple rules have a symmetric character, the backward rules for backward model transformations from models of the target to models of the source language are also derived automatically. In this section we present both directions and show the application in our scenario for the forward case.

The operational rules for forward model transformations are source and forward rules. Both kinds are derived from the triple rules that specify pattern by pattern how integrated models are created, i.e., how source and target models are developed synchronously. Given a triple rule, its source rule is given by removing all elements in the correspondence and target component. Its forward rule is derived by replacing the source component on the left-hand side by the source component on the right-hand side. The forward rule requires a complete fragment in the source component of an integrated model and completes the missing parts for the correspondence and target components. Intuitively, the source rules specify how a source model can be constructed and the forward rules specify how source models can be completed to integrated models. Given a source model, the forward rules are used for the actual

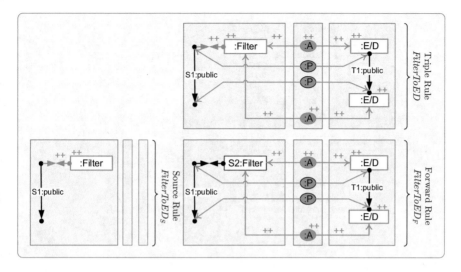

Fig. 11.12 Derived source and forward rules

transformation steps leading to an integrated model and the source rules are used for controlling the application of the forward rules [EEHP09, HEGO14].

Example 11.9 (Operational rules). Fig. 11.12 shows the triple rule "*FilterToED*", its derived source rule "*FilterToED_S*" and its forward rule "*FilterToED_F*". Note that the source component is now identical in the left- and right-hand side of the forward rule by construction. The forward rule is used to transform a filter into two ABT nodes of the type "E/D", which implement the encryption and decryption of communication data. As described in Sect. 11.1, the underlying idea is that confidential communication in a business universe is filtered out while in the IT universe the data is encrypted for public channels. Since the left-hand side of the forward rule already contains all source elements of the right-hand side of the triple rule, the items "S1" and "S2" appear on both the left- and the right-hand side.   △

A forward model transformation is performed by taking a given source model and first extending it to a triple graph G_0 with an empty correspondence and an empty target component. The transformation starts with this triple graph. Each step of the transformation starts with the computation of the possible matches from the left-hand sides of the forward rules to the current triple graph. A valid match according to the on-the-fly construction in [EEHP09] is chosen and the forward step is performed. The matching can be performed equivalently using special translation markers as presented in [HEGO14]. The resulting target model is obtained by restricting the final triple graph to its target component. This construction leads to a source consistent forward sequence that defines the model transformation sequence in the sense of Part III.

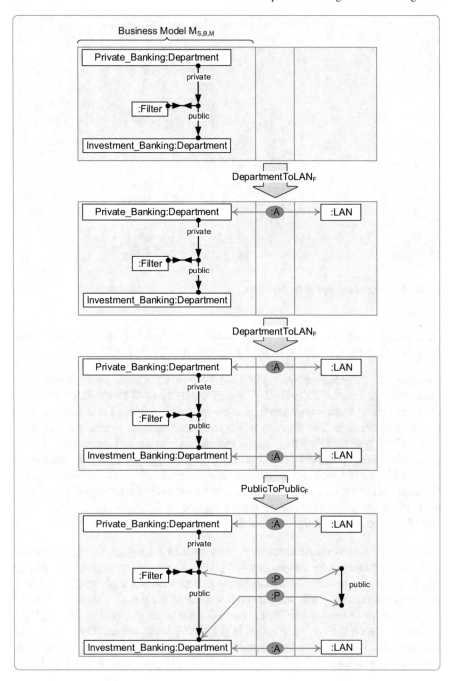

Fig. 11.13 Model transformation sequence, part 1

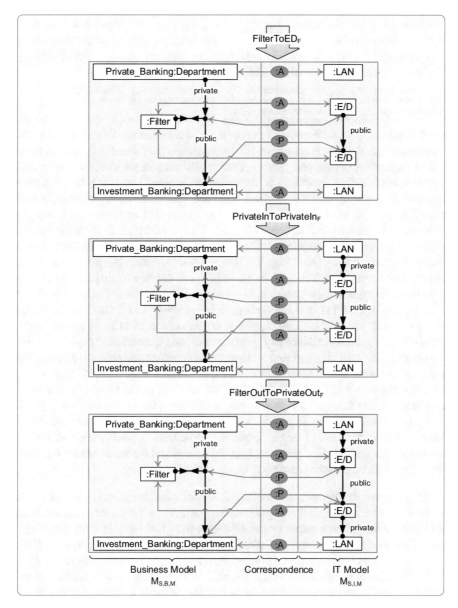

Fig. 11.14 Model transformation sequence, part 2

Example 11.10 (Model transformation). Using the presented triple rules and their derived forward rules, the business service model $M_{S,B,M}$ is transformed to the IT service model $M_{S,I,M}$. The model transformation sequence consists of the following six forward transformation steps and is shown in Figs. 11.13 and 11.14:

$$G_0 \xrightarrow{\textit{DepartmentToLAN}_F} G_1 \xrightarrow{\textit{DepartmentToLAN}_F} G_2 \xrightarrow{\textit{PublicToPublic}_F} G_3 \xrightarrow{\textit{FilterToED}_F} G_4$$
$$\xrightarrow{\textit{PrivateInToPrivateIn}_F} G_5 \xrightarrow{\textit{FilteredOutToPrivateOut}_F} G_6, \text{ where } G_0 = (M_{S,B,M} \leftarrow \emptyset \rightarrow \emptyset).$$

Each step completes a fragment of the source model by the missing elements in the correspondence and target component. The matches of the forward rules do not overlap on their effective elements. Those effective elements are the ones that are created by the corresponding triple rule in its source component ($R^S \setminus L^S$), i.e., elements that are visualised by plus signs in the source components of the triple rule. Furthermore, the full transformation sequence has to ensure that each element in $M_{S,B,M}$ is matched by exactly one effective element of a forward rule. Both properties are ensured by the formal condition called "source consistency", which controls the forward transformation (see Chap. 7). Therefore, each source element is translated exactly once and the resulting triple graph is a well-formed integrated model. The resulting triple graph G_6 is shown at the bottom of Fig. 11.14 and coincides almost with the triple graph in Fig. 11.8. Only the labels "NW4" and "NW7" for the nodes of type "LAN" are left blank. Such labels (explicit object identifiers) cannot be determined by the information of any source model, and therefore the triple rules create blank labels. But note that real attributes are processed and computed during the model transformation, e.g., the connector type "public" is specified as an attribute of a Reo connector in the abstract syntax. The result of of the forward transformation is given by the target component $M_{S,I,M}$ of $G_6 = (M_{S,B,M} \leftarrow G_{6,C} \rightarrow M_{S,I,M})$. Fig. 11.15 shows the model transformation in one step from its source model as input to its target model as output. Note that some transformation steps in this sequence are sequentially independent, e.g., the second and the third step are independent. They can be switched leading to an equivalent sequence.   △

As presented in Theorem 8.4 in Chap. 8, model transformations based on TGGs ensure correctness in the following sense. Each model transformation sequence from a source model G^S to a target model G^T guarantees that there is a corresponding consistent integrated model $G \in \mathcal{L}(TGG)$, i. e., $G = (G^S \leftarrow G^C \rightarrow G^T) \in \mathcal{L}(TGG)$. This ensures that the descriptive triple patterns as specified by the triple rules of the TGG are adhered to by the model transformation. Moreover, model transformations based on TGGs are complete in the sense that they always yield a target model G^T for a given source model G^S of the source language $\mathcal{L}(TGG)^S$, which is the projection of the language of consistent integrated models to the source component. The result in Theorem 8.4 goes beyond the available results for other graph transformation approaches, which only ensure that the resulting target models are correctly typed according to the target meta model. In addition to the correctness and completeness result, the triple graph transformation approach benefits from its intuitive triple patterns of corresponding fragments, which substantially increases usability and maintainability.

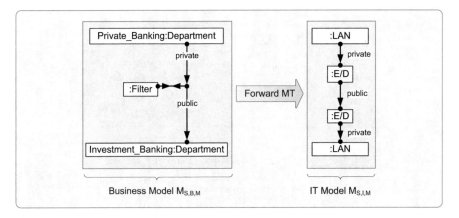

Fig. 11.15 Model transformation of service models

11.4 Model Integration Between Business and IT Service Models

The purposes of model integration are interoperability in general and the analysis of consistency within the overall enterprise model in particular. Conceptually, the challenge of model integration is different from model transformation. However, both techniques can be based on triple graph transformation, and thus they are strongly related in our case. The used TGG is the same—the only differences occur in the derivation of operational rules and the definition of the control conditions for the execution. This section presents the constructions and the application to the case study.

Analogously to forward rules, integration rules are derived from the set of triple rules, which describe the patterns of the relations between two models. An integration rule is obtained from a triple rule by replacing the source and the target components of the left-hand side by the source and target components of the right-hand side, such that only the correspondence part remains different. This way, a match of the rule requires that all fragments of the triple pattern in the source and target component be present before applying the rule, such that only the correspondences are added to complete the triple pattern. Similarly to source rules for the forward model transformation, source–target rules are used to control the execution of model integrations. They are obtained be removing the correspondence part from the original triple rule.

Example 11.11 (Source–target and integration rules). Fig. 11.16 shows the triple rule *"FilterToED"*, its derived source–target rule *"FilterToED$_{ST}$"* and its integration rule *"FilterToED$_I$"*, according to Def. 7.11. The triple rule synchronously extends a public connector by a filter in the business model and its corresponding two de/encryption elements in the IT model. Thus, an application of the integration rule completes the correspondence structure of existing and already related public connectors that have adjacent filter and de/encryption elements. △

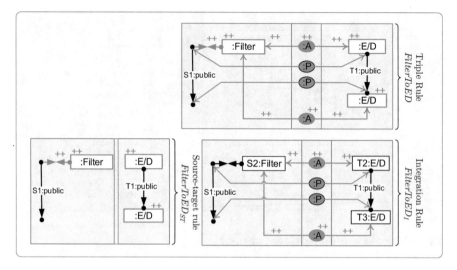

Fig. 11.16 Derived integration rule *FilterToED$_I$*

Model integration based on triple graph transformation is defined by integration sequences, in which the derived model integration rules are applied. Given a source and a target model, their integration is performed by completing the correspondence structure that relates both models. The consistency of a model integration is ensured by a formal condition, called source–target $(S–T-)$consistency (see Def. 7.37). This condition ensures that the given source and target models are completely parsed using the inverted triple rules restricted to the source and target component, respectively. Thus, each fragment of the models is processed and integrated exactly once.

Example 11.12 (Model integration). Figs. 11.17 to 11.19 show the integration of the models $M_{S,B,M}$ and $M_{S,I,M}$, i.e., the creation of the correspondence relation between them via a correspondence graph. The model integration is based on an $S–T$-consistent integration sequence consisting of the following six steps using the derived model integration rules:

$$G_0 \xrightarrow{\;DepartmentToLAN_I\;} G_1 \xrightarrow{\;DepartmentToLAN_I\;} G_2 \xrightarrow{\;PublicToPublic_I\;} G_3 \xrightarrow{\;FilterToED_I\;} G_4$$
$$\xrightarrow{\;PrivateInToPrivateIn_I\;} G_5 \xrightarrow{\;FilteredOutToPrivateOut_I\;} G_6, \text{ with } G_0 = (M_{S,B,M} \leftarrow \emptyset \rightarrow M_{S,I,M}).$$

In the first four steps, some fragments of the source and target components are completed by the missing elements in the correspondence component. But note that in the two last steps, nothing changes on the correspondence component. The reason is the following. The triple rules *"PrivateInToPrivateIn"* and *"FilteredOutToPrivateOut"* extend integrated fragments in the source and target model by connecting Reo elements and they do not create any correspondence node. Therefore, the derived integration rules do not create correspondences either. But these integration rules are necessary to ensure correct integrations in the way that the positions of the corresponding private Reo connectors are checked.

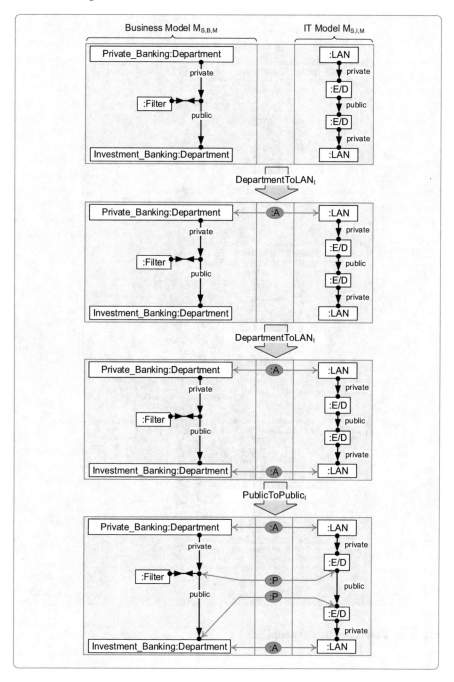

Fig. 11.17 Model integration sequence, part 1

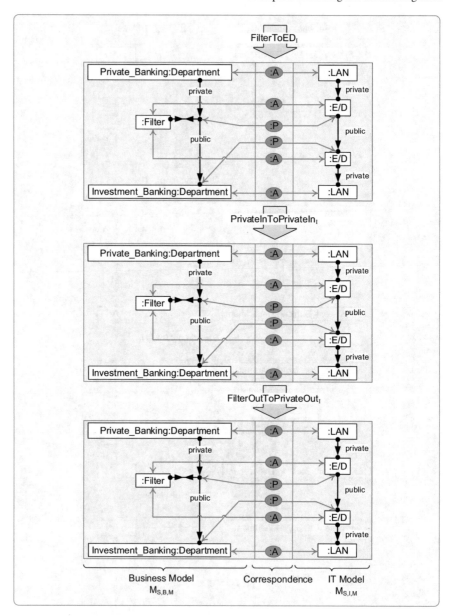

Fig. 11.18 Model integration sequence, part 2

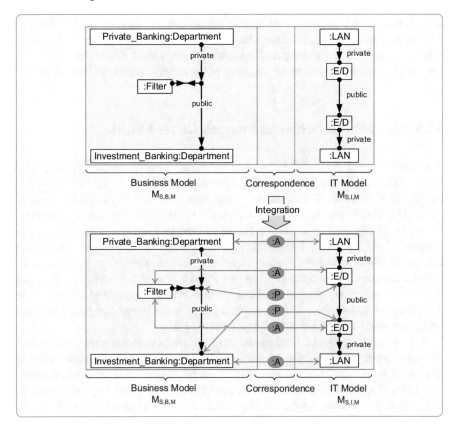

Fig. 11.19 Model integration for service models

The matches of the integration rules do not overlap on their effective elements, which are those that are created by the triple rule in the source and target components $(R^S \cup R^T \setminus L^S \cup L^T)$. Furthermore, each element in $M_{S,B,M}$ and $M_{S,I,M}$ is matched by an effective element of an integration rule. Both properties are ensured formally by the S–T consistency condition (see Def. 7.37), which controls the integration sequence. This way, each source element and each target element is integrated exactly once and the resulting triple graph is a well-formed integrated model. The resulting triple graph G_6 is shown at the bottom of Fig. 11.18 and coincides with the triple graph in Fig. 11.8.

The result of the model integration is the completely integrated triple graph of the sequence and it is shown in the bottom part of Fig. 11.19, where the model integration is presented in a single step from the pair of the source and target model as input to the integrated model as output. △

Integration rules are used to establish or update the correspondences between two models. The model framework in Fig. 11.1 shows many coordinates, where models should be integrated. Two models are consistent with each other if they can be completely integrated; otherwise they show conflicts. Consistency between the models is ensured by checking $S-T$ consistency (see Def. 7.37), which means that the integration has to conform to a parsing of the existing source and target model.

11.5 Summary of Achievements and Related Work

The described and illustrated techniques for model transformation and integration improve the interoperability between the business and the IT service models given by ABT-Reo diagrams. The techniques are general, such that they should be applicable to other visual languages and coordinates in the model framework for enterprise modelling in Fig. 11.1 as well. The benefits of the techniques can be described as follows. Model transformation enables the construction of model stubs that can be refined by the experts for the specific domain. Model integration establishes the correspondences between the existing models, e.g., for data interchange, and, furthermore, conflicts between models can be detected and highlighted. According to Theorems 8.4 and 8.9, model transformation and model integration based on triple graph grammars are syntactically correct and complete.

Both techniques can be used as the basis for more complex scenarios, including those where conflicts exist between existing models in different dimensions of the model framework. For instance, the IT model may contain some communication paths that are not modelled in the business model. Those fragments have to be synchronised and can be detected as remainders when computing the integration sequences with maximal coverage. The synchronisation can be supported by the application of the derived forward and backward triple rules in the way that additional fragments in one model are translated to new ones in the other model (see Chap. 9 for the formal details on model synchronisation based on TGGs). Conflicts are detected and presented to the modeller, who is responsible for solving them. Moreover, we further studied in [BHEE10, EHSB11, EHSB13] how to propagate constraints between different domains and to analyse and ensure their validity. Such constraints can specify functional and nonfunctional properties that can be relevant for different domains in the model framework.

There is a wide range of literature spanning various aspects of enterprise modelling. We like to discuss some of the well-known available approaches on a general basis. The concept of multi-perspective enterprise modelling (MEMO) is a comprehensive framework presented in [Fra02]. The MEMO method provides semiformal domain-specific modelling languages that are focussed on the concrete modelling domains of enterprises. The ISO standard 19439:2006 [ISO06] specifies a general framework for enterprise modelling and enterprise integration based on the GERAM framework (generalised enterprise reference architecture and methodology, [BN96]), where several aspects and dimensions are distinguished in a partly similar way to that in the MEMO method.

Compared with enterprise modelling based on general modelling languages like UML, as presented in [Mar00], domain-specific modelling techniques provide specific modelling concepts and notations which are appropriate for the modelling tasks instead of providing generic ones from which the specific ones have to be manually constructed and derived. For this reason, the unified enterprise modelling language [Ver02, ABV07] does not aim to provide one universal enterprise modelling language, but rather provides techniques for integrating models of existing domain-specific languages based on merging their ontologies. We have also followed this line in this chapter, which illustrates model transformation and integration techniques on one particular DSL: ABT-Reo (abstract behaviour types (ABT) and Reo connectors). The new contribution is the application of the formal techniques based on graph transformation to ensure several important properties, e.g., syntactical correctness and completeness.

Chapter 12
Tool Support

The more graph transformations are applied in various application domains, the more tools supporting modelling, simulation and analysis of graph transformation system become crucial for the promotion of graph transformation in industry. In this chapter, we present four related modelling environments that have been developed at Technische Universität Berlin and that support the specification, simulation and analysis of behavioural models and model transformations based on algebraic graph transformation.

We start in Sect. 12.1 with the tool environment AGG. For over 20 years, the strength of AGG has been its consequent implementation of analysis techniques for algebraic graph transformation systems [LB93, TER99, Tae04, EEPT06]. Recently, AGG has been extended to better support new concepts for rule application control (like nested application conditions) and rule synthesis (like amalgamation of rules), as presented in Part II of this volume. This has made AGG better suited for defining and executing complex model transformation systems [RET12]. Sect. 12.2 presents ActiGra, a visual modelling language combining activity models to define the control flow and graph transformation rules specifying the activity semantics. Based on AGG's conflict and dependency analysis techniques, favourable and critical signs concerning model consistency are visualised directly in the activity model. In Sect. 12.3, we introduce EMF Henshin, an Eclipse plugin supporting modelling and execution of EMF model transformations, i.e., transformations of models conforming to a meta model given in the EMF Ecore format. In EMF Henshin, we have lifted implicit and explicit control structures from graph transformation to EMF model transformation. The visual EMF Henshin environment provides intuitive visual editors for rule application conditions and explicit control structures on rules. Recently, EMF Henshin has been extended to support model transformations based on triple graph grammars (TGGs). In Sect. 12.4, we present the visual TGG modelling and analysis environment Henshin$_{TGG}$, building on EMF Henshin and AGG. Related tools to our four modelling environments are discussed in Sect. 12.5.

12.1 The Attributed Graph Grammar System AGG 2.0

AGG [AGG14, Tae04] (the *Attributed Graph Grammar* system) is a well-established integrated development environment for algebraic graph transformation systems, developed and extended over the past 20 years at Technische Universität Berlin. The environment supports the specification of algebraic graph transformation systems based on typed attributed graphs with node type inheritance, graph rules with application conditions, and graph constraints. It offers several analysis techniques for graph transformation systems, including graph parsing, consistency checking of graphs as well as conflict and dependency detection in transformations by critical pair analysis of graph rules, an important instrument to support the confluence check of graph transformation systems.

Model transformations have recently been identified as a key subject in model-driven development (MDD). Graph transformations offer useful concepts for MDD, while the software engineering community can generate interesting challenges for the graph transformation community. In the past six years, those new challenges have led to the augmentation of AGG by new tool features. In this section, we describe some of those challenges, the formal approaches developed to solve them, and the impact they have had on recent developments of new features of AGG, leading to AGG 2.0 [RET12]. In particular, the modelling scope and validation support in AGG has been extended by features for rule application control and rule synthesis. AGG 2.0 now supports the specification of complex control structures for rule application comprising the definition of control and object flow for rule sequences and nested application conditions.

Structure of this section: The basic features and the tool environment of AGG in 2006 have been introduced in [EEPT06]. We summarise these features in Sect. 12.1.1 along a running example on semantics definition for visual modelling languages (VLs). In Sect. 12.1.2, we introduce the new features for rule application control. Furthermore, new possibilities for rule synthesis by constructing rules from existing ones (e.g., inverse, minimal, amalgamated, and concurrent rules) have been realised. We introduce the new rule synthesis support in Sect. 12.1.3. Last but not least, AGG 2.0 provides more flexible usability of the critical pair analysis. We review the basic concepts of critical pair analysis in AGG in Sect. 12.1.4 and demonstrate the new features when analyzing graph transformation systems for conflicts and dependencies.

12.1.1 Editing and Simulating Graph Transformation Systems

In AGG, *graphs* consist of two disjoint sets of *nodes* and *edges*. In correspondence to the theory (see Def. 2.1), we may have multiple edges between a single pair of nodes. Each graph object (node or edge) is associated with exactly one *type* from a given type set. The type set consists of the set of node types and the set of edge types. As in the theory, node types and edge types may be represented as nodes

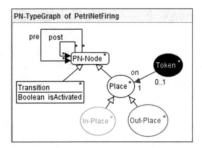

Fig. 12.1 Attributed type graph with node type inheritance in AGG 2.0

and edges of a *type graph* (see Def. 2.2). The type graph may contain *multiplicity constraints* on edge types that constrain how many objects may be connected by instances of a certain edge type. A multiplicity constraint on a node type restricts the number of instances of this node type.

Another important principle for handling complex graph structures stems from object-oriented programming, where class *inheritance* extends typing by allowing the definition of more abstract types, and more concrete types, inheriting from the abstract ones. In graph transformation, we speak of *node type inheritance* [BEdLT04, EEPT05, EEPT06]. In AGG 2.0, a type graph with node type inheritance may be defined by inserting *generalisation* edges between concrete and abstract node types (parent types). Inheritance allows a much denser form of graph transformation systems, since similar rules can be abstracted into one rule: a node in a rule's left-hand side that is typed over an abstract node type may be matched to any instance graph node that is in the *clan* of the rule node, i.e., the instance node's type node in the type graph is connected by a path of generalisation edges to the abstract node's type node.

In AGG, *attributes* may be defined for graph objects. An attribute is declared for a graph object type in the type graph by specifying a *name* and a *type* (a Java type) for it. As in the theory, all graph objects of the same type share their attribute declarations (see Def. 2.5). The values of attributes may be chosen individually for graph objects in instance graphs. AGG requires that each attribute occurring in a graph object in an instance graph be bound to exactly one value. In rules, attributes may be bound to variables (in a rule's LHS) or can be equipped with arbitrary computations on Java objects described by Java expressions (in a rule's RHS).

Figure 12.1 depicts an attributed type graph with node type inheritance, modelling the language of elementary Petri nets. An elementary Petri net consists of PN-Nodes which can be either Transitions or Places. PN-Nodes are connected via pre-arcs and post-arcs. We distinguish between two special kinds of Places, namely In-Places and Out-Places, where In-Places are meant to be connected to Transitions via pre-arcs only, and Out-Places should be connected by post-arcs only. The clan of the PN-Node consists of the node types Transition, Place, In-Place, OutPlace and PN-Node itself. All Places can carry Tokens, which is modelled by the on-edge from type Token to type Place, where the multiplicity "1" at its end means that each Token instance must be connected via on-edge to exactly one Place instance. On the other hand, multiplicity "0..1" at the beginning of the on-edge means that a Place instance may carry at most one Token. The multiplicity "*" at pre- and post-arcs means that in principle a PN-Node may be connected to any number of PN-Nodes. Transitions are attributed by a Boolean attribute called isActivated. In instance graphs (concrete Petri nets), the value of this attribute is meant to be true if the transition is enabled to fire; else it will be set to false.

Basically, a *graph transformation rule* according to Def. 5.12 is represented in AGG by a left-hand side L, a right-hand side R and mappings from L to R for all graph objects that are preserved by the rule.[1] The mappings are visualised by equal numbers in the rule sides L and R. Hence, a rule in AGG corresponds to the span $p = (L \xleftarrow{l} K \xrightarrow{r} R)$ defining a DPO rule, where the gluing K consists of those graph objects that are preserved (i.e., marked by numbers in AGG). Alternatively, AGG offers an option for the current graph grammar to be interpreted according to the SPO approach (by disabling the dangling and identification condition). In this case, dangling edges will be deleted, and identification conflicts are handled by automatic deletion as well. Attributes can be defined also for rules, where the left-hand side may contain constants or variables as attribute values, but no Java expressions. The right-hand side may contain Java expressions in addition, but no unbound variables (that do not occur already in the left-hand side). The scope of a variable is its rule.

A rule is applied to an instance graph G by finding a *match* (graph morphism) from its left-hand side L into G. If no match can be found, or if the gluing condition according to the DPO approach is not satisfied (see Def. 5.13), the rule is not applicable. Otherwise, graph objects in the image of the match that are not mapped to R by the rule are deleted, and graph objects in R that have not been mapped from L by the rule are created. A graph G is transformed *in-place* to the new instance graph that results from the rule application. In the case of multiple matches from L to G, AGG may choose a match randomly, or the user may choose a match manually.

The AGG environment consists of a graphical user interface comprising several visual editors, an interpreter, and a set of validation tools. The basic interpreter supports simulation of graph transformation systems by stepwise transformation of graphs as well as by applying rules in random order at nondeterministic matches as long as possible.

Figure 12.2 displays the basic graphical user interface of AGG. The *tree view* [1] shows the elements of the current graph grammar (a type graph, one or more instance graphs, an arbitrary number of rules, and, optionally, graph constraints defining required or forbidden graph patterns in instance graphs; see Sect. 12.1.2). The *rule editor* [2] shows a rule for firing an elementary Petri net transition (according to the type graph of elementary Petri nets in Fig. 12.1). Below, the *instance graph editor* [3] contains an elementary Petri net. The current rule is applicable to this instance graph, and the match is indicated by corresponding mapping numbers in the rule's left-hand side and the graph. Some Place nodes from the rule's left-hand side

---

[1] We consider rules with application conditions in Sect. 12.1.2.

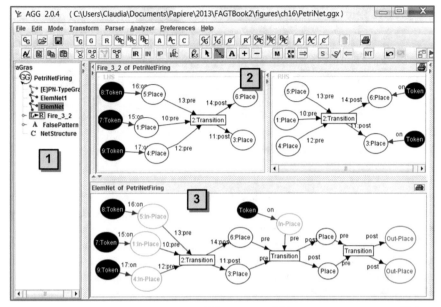

Fig. 12.2 AGG user interface with tree view $\boxed{1}$, rule editor $\boxed{2}$ and instance graph editor $\boxed{3}$

are mapped to nodes of type In-Place, which is type-compatible due to the node type inheritance relation defined in the type graph.

Note that this rule works only for firing a transition with exactly three input and two output places. We would run into problems with the Petri net semantics if we tried to apply this rule to a transition with four marked input places: the rule would remove only three of the input tokens. We will see in the next section how rule application conditions can help to solve such problems.

12.1.2 Rule Application Control

To further enhance the expressiveness of graph transformations, rules may contain application conditions (see Defs. 5.1 and 5.12, which are also called *nested application conditions* in [HP09] and *general application conditions* in AGG 2.0). Application conditions are as powerful as first-order logic on graphs and provide a mechanism to control the rule application (see [GBEE11] for an extensive case study). In addition to application conditions, AGG supports the definition of *attribute conditions*, which can be defined in a text editor. A rule may have a set of attribute conditions (Boolean Java expressions over a rule's set of variables). All attribute conditions have to be evaluated to true for the rule to be applicable.

AGG also supports *graph constraints* according to Def. 5.10, i.e., graph conditions that are required to hold globally (for every graph in a transformation) and

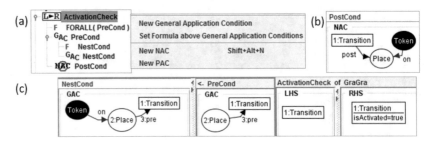

Fig. 12.3 Nested application conditions in AGG

are not coupled with rules. The concept of graph constraints allows for consistency checking. AGG can be configured in a way to stop the running transformation when a graph constraint is violated by the current instance graph.

Furthermore, *control structures* defined on rules, such as layers or sequences, restrict the choice of rules to be applied. Often, rule sequences are supposed to be applied to the same subset of graph objects. Here, AGG supports the definition of *object flow* between rules in a sequence, restricting the matches of subsequent rules.

Application Conditions

Let us recall that a rule with application condition in AGG corresponds to the span $p = (L \leftarrow K \rightarrow R, ac)$ (see Def. 5.12). Application conditions in AGG basically consist of *positive application conditions (PACs)*, *negative application conditions (NACs)*, and *general application conditions (GACs)*, which may be defined separately in any number for each rule. PACs and NACs require or forbid the presence of certain structures in the graph for the rule to be applied. Note that PACs and NACs are special GACs, defined by a morphism from L to the PAC (or NAC) graph. A rule with a PAC is applicable if there exists a morphism from the PAC graph to the instance graph that extends the match injectively. Vice versa, a rule with a NAC is applicable only if such a morphism from the NAC graph to the instance graph does *not* exist. According to the theory, GACs are arbitrary conditions over L (see Def. 5.1). The conjunction of all of its PACs, NACs and GACs comprises a rule's application condition ac.

An example is displayed in Fig. 12.3, where the activation of an elementary Petri net transition is checked: The rule ActivationCheck sets the transition attribute isActivated to true if two conditions, called PreCond and PostCond, are satisfied: PostCond is the NAC shown in Fig. 12.3 (b) which forbids the existence of a marked place in the transition's post-domain, and PreCond is a nested application condition shown in Fig. 12.3 (c) which requires that on each place in the transition's pre-domain, there must be one token. Note that this condition cannot be expressed by using simple NACs or PACs. Figure 12.3 (a) depicts the context menu entries for generating general application conditions (GACs) in AGG 2.0.

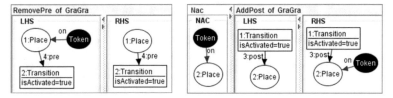

Fig. 12.4 Rules RemovePre and AddPost for Petri net firing

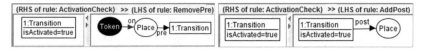

Fig. 12.5 Object flow definition for rule sequences in AGG

Object Flow for Rule Sequences

Object flow between rules has been defined in [JLM⁺09] as partial rule dependencies relating nodes of the RHS of one rule to (type-compatible) nodes of the LHS of a (not necessarily direct) subsequent rule in a given rule sequence. Object flow thus enhances the expressiveness of graph transformation systems and reduces the match finding effort.

In AGG 2.0, object flow can be defined between subsequent rules in a rule sequence, and the rule sequence can be applied to a given graph respecting the object flow. An example is the definition of a Petri net transition firing step by the rule sequence (ActivationCheck, RemovePre*, AddPost*, DeActivate) with object flow. The sequence defines that the rule ActivationCheck (see Fig. 12.3) is applied once, followed by the rules RemovePre (removing a token from a pre-domain place), AddPost (adding a token to a post-domain place), which are shown in Fig. 12.4, and DeActivate (setting the transition attribute isActivated back to false).

In the sequence, the rules RemovePre and AddPost are applied as long as possible (denoted by "*"), and the rule DeActivate is applied once. To restrict the application of the rule sequence to exactly one transition, we need to express that the transition in the matches of all rules is the same. This is done by defining the object flow, e.g., by mapping the transition from the RHS of rule ActivationCheck to the LHSs of rules RemovePre and AddPost, as depicted in Fig. 12.5.

An example of a firing step by applying the rule sequence with object flow is shown in Fig. 12.6, where the left transition in the net was selected, found activated and fired.

12.1.3 Rule Synthesis

AGG 2.0 supports various ways to automatic construction of new rules from existing ones (rule synthesis). Obviously, the definition of a rule sequence with object flow

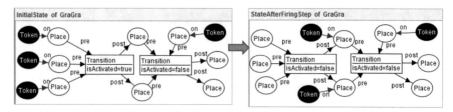

Fig. 12.6 Firing step resulting from applying a rule sequence with object flow

is neither simple nor very intuitive for modelling Petri net transition firing steps for transitions with an arbitrary number of pre- and post-domain places. A way closer to the inherent Petri net semantics (a transition removes the tokens from all of its pre-domain places and adds tokens to all of its post-domain places in *one atomic* step) would be to construct a *single* rule modelling the firing behaviour of a transition.

Construction of Concurrent Rules

A concurrent rule summarises a given rule sequence in one equivalent rule (see Def. 5.27). Recall, an E-concurrent rule $p *_E q = (L \leftarrow K \rightarrow R)$ is constructed from two rules, $p = (L_p \leftarrow K_p \rightarrow R_p)$ and $q = (L_q \leftarrow K_q \rightarrow R_q)$, that may be sequentially dependent via a graph E, where E is an overlapping of the right-hand side of the first rule and the left-hand side of the second rule (see Def. 2.27).

In AGG 2.0, we can construct a concurrent rule from a given rule sequence with the following options concerning the overlappings of one rule's RHS with the succeeding rule's LHS:

1. compute maximal overlappings according to rule dependencies;
2. compute all possible overlappings (this usually yields a large number of concurrent rules);
3. compute overlappings based on the previously defined object flow between the given rules;
4. compute the parallel rule (no overlappings), where rule graphs are disjointly unified, with NACs constructed according to [Lam10].

As an example, Fig. 12.7 shows the concurrent rule constructed from the rule sequence (ActivationCheck, RemovePre(3), AddPost(2), DeActivate).[2] Since an object flow is defined for this sequence, we choose option (3) *by object flow* for computing rule overlappings. Constraints on the type graph (e.g., "a Token node is connected to exactly one Place node") prevent the generation of unnecessary NACs.

---

[2] Note that a concurrent rule can be constructed only for a finite rule sequence.

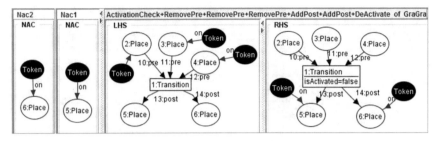

Fig. 12.7 Concurrent rule generated from a rule sequence in AGG

Construction of Amalgamated Rules

Still, the construction of concurrent rules from rule sequences is not flexible enough for defining Petri net firing, since, in general, the number of places in a transition's pre- and post-domain is not known a priori. What we need here is a concept of rule synthesis that allows us to apply rules to as many matches as we find in the current instance graph in a synchronised way. This kind of rule synthesis by so-called *amalgamated rules* is also supported in AGG 2.0.

If a set of rules $p_1, \ldots, p_n$ share a common subrule p_0, a set of multi-amalgamable[3] transformations $G \stackrel{(p_i,m_i)}{\Longrightarrow} G_i$ $(1 \le i \le n)$ leads to an amalgamated transformation $G \stackrel{\tilde{p},\tilde{m}}{\Longrightarrow} H$ via the amalgamated rule $\tilde{p} = p_1 +_{p_0} \ldots +_{p_0} p_n$, constructed as the gluing of $p_1, \ldots, p_n$ along p_0.

We call p_0 *kernel rule*, and $p_1, \ldots, p_n$ *multi rules* (see Def. 6.1). A kernel rule together with its embeddings in any number of multi rules is called *rule scheme (RS)* in AGG. An amalgamated transformation $G \stackrel{\tilde{p}}{\Longrightarrow} H$ is a transformation via the amalgamated rule $\tilde{p}$ (see Def. 6.11).

This concept is very useful for specifying $\forall$-quantified operations on recurring graph patterns (e.g., in model refactorings). The effect is that the kernel rule is applied only once while multi rules are applied as often as suitable matches are found.

Figure 12.8 depicts the rule scheme with a kernel rule and two multi rules specifying the firing of a Petri net transition with arbitrary many pre- and post-domain places. The kernel rule has the same application condition as the rule ActivationCheck (see Fig. 12.3). Note that we do not need the isActivated flag anymore because the check and the complete firing step are performed by a single application of the amalgamated rule. The amalgamated rule constructed, for example, along a match to an activated transition with three pre- and two post-domain places is similar to the rule in Fig. 12.7 but has no NACs because the match into the host graph is predefined by construction.

The definition of elementary Petri net firing by a rule scheme is the most intuitive and most flexible one: we simply state that if a transition fulfills a certain activation

---

[3] In multi-amalgamable transformations, the matches m_i agree on the p_0 and are independent outside (see Def. 6.13).

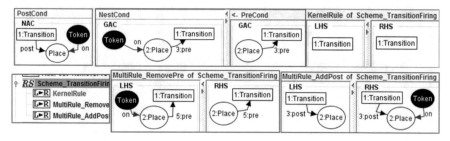

Fig. 12.8 Rule scheme defining a transition firing step in Petri nets

Fig. 12.9 Inverse rule of the rule AddPost

condition, from each pre-domain place (independently of their number) a token will be removed, and to each post-domain place (independently of their number) a token will be added.

For a larger case study using amalgamated rules to define the behavioural semantics of statecharts, see [GBEE11].

Construction of Inverse Rules

For a given rule, the inverse rule has LHS and RHS exchanged. Moreover, application conditions are shifted over the rule. Figure 12.9 displays the inverse rule of the rule AddPost (see Fig. 12.4), where an existing token is removed. The shifted NAC requires that there be exactly one token on the place for the rule to be applicable.

Construction of Minimal Rules

A new challenge from MDD comes from the field of *model versioning*, where the new notion of *graph modification* [TELW10], defined by a span $G \leftarrow D \rightarrow H$, has been established to formalise model differences for visual models. Based on graph modifications, so-called *minimal rules* may be extracted from a given span to exploit conflict detection techniques for rules. A minimal rule comprises the effects of a given rule in a minimal context. Via the context menu, AGG 2.0 supports the extraction of a minimal rule from a selected rule (interpreted as graph modification). For example, the minimal rule of the rule in Fig. 12.7 is depicted in Fig. 12.10. It does not contain the arcs connecting the places and transitions, since these arcs are not changed by the rule. It contains the place nodes (because edges connected to

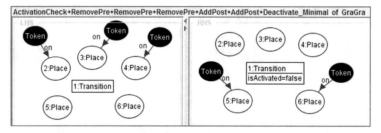

Fig. 12.10 Minimal rule of the concurrent rule in Fig. 12.7

them are deleted or generated), the token nodes (either deleted or generated) and the transition node (its attribute is possibly changed).

12.1.4 Analysis of Graph Transformation Systems

AGG supports several kinds of validations which comprise graph parsing, consistency checking of graphs (via graph constraints), applicability checking of rule sequences, and conflict and dependency detection by critical pair analysis of graph rules. We concentrate in this section on the critical pair analysis and the new usability features in AGG 2.0. Graph parsing and consistency checking is described already in [EEPT06]. The critical pair analysis (CPA) is the unique feature in AGG that is not supported by any other graph transformation tools. On the contrary, many tools (including our own developments described in Sects. 12.2 and 12.4) make use of the AGG API (application programming interface) to invoke the CPA of AGG and import the results into their development environments.

Critical Pair Analysis (CPA)

We recall from Sect. 5.2.4 that a sufficient condition for a graph transformation system to be confluent is to show local confluence for each pair of transformations $G \overset{p_1,m_1}{\Longrightarrow} H_1$ and $G \overset{p_2,m_2}{\Longrightarrow} H_2$, provided that the transformation system is terminating. The main idea behind showing local confluence for each possible transformation pair is to study critical pairs (see Def. 5.40). According to Theorem 5.44, a transformation system with ACs is locally confluent if all its critical pairs are strictly AC-confluent (Def. 5.43).

AGG 2.0 implements the critical pair analysis for rules with NACs and PACs, but not for rules with more complex general (nested) application conditions. Recall that, following from Def. 2.21, two direct transformations $H_1 \overset{p_1,m_1}{\Longrightarrow} G \overset{p_2,m_2}{\Longrightarrow} H_2$ are parallel dependent if we have one of the following kinds of conflict:

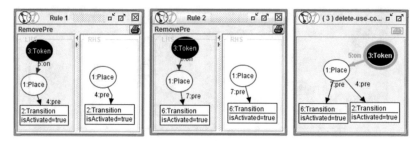

Fig. 12.11 CPA of rule pair (RemovePre, RemovePre), detecting a delete–use conflict

1. *Delete–use conflict*: The application of p_1 deletes a graph object that is used by p_2 (it is in the match of p_2).
2. *Produce–forbid conflict*: The application of p_1 generates graph objects in such a way that p_2 cannot be applied after p_1 because the NAC of p_2 is violated.
3. *Change–use-attribute conflict*: The application of p_1 changes attributes that are in the match of p_2.
4. *Change–forbid-attribute conflict*: The application of p_1 changes attributes to values that are forbidden by the NAC of p_2.

An example for a critical pair is illustrated in Fig. 12.11, where the classical *forward conflict* in Petri nets is detected when analyzing the rule pair (RemovePre, RemovePre). As indicated in the conflict view, we have a *delete–use conflict* since two transitions need to remove the same token from their common pre-domain place.

In analogy to conflict detection by critical pair analysis, AGG also supports the detection of sequential dependencies (again, rules with NACs and PACs are supported, but not with general application conditions). Recall that, following from Def. 2.22, two direct transformations $G \xrightarrow{p_1,m_1} H_1 \xrightarrow{p_2,m_2} G'$ are sequentially dependent (i.e., the second direct transformation is triggered by the first one) if we have one of the following kinds of dependency:

1. *Produce–use dependency*: The application of p_1 produces graph objects p_2 uses.
2. *Delete–forbid dependency*: The application of p_1 deletes graph objects, thereby initially validating the NAC of p_2.
3. *Change–use-attribute dependency*: The application of p_1 changes attributes that are in the match of p_2.
4. *Change–forbid-attribute dependency*: The application of p_1 changes attributes to values, thereby initially validating the NAC of p_2.

AGG exploits the following relationship between parallel and sequential dependency: $G \xrightarrow{p_1,m_1} H_1 \xrightarrow{p_2,m_2} G'$ are sequentially dependent iff $G \xleftarrow{p_1^{-1},m_1^*} H_1 \xrightarrow{p_2,m_2} G'$ are parallel dependent. According to this relationship, AGG constructs the inverse rule p_1^{-1} (shifting NACs and PACs if necessary) and applies critical pair analysis to compute potential conflicts for the rule pair (p_1^{-1}, p_2) in minimal context. These conflicts correspond to the potential sequential dependencies of the original rule pair (p_1, p_2) in the following way:

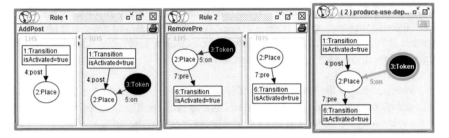

Fig. 12.12 CPA of rule pair (AddPost, RemovePre), detecting a produce–use dependency

1. A *delete–use conflict* of (p_1^{-1}, p_2) corresponds to a *produce–use dependency* of (p_1, p_2).
2. A *produce–forbid conflict* of (p_1^{-1}, p_2) corresponds to a *delete–forbid dependency* of (p_1, p_2).
3. A *change–use-attribute conflict* of (p_1^{-1}, p_2) corresponds to a *change–use-attribute dependency* of (p_1, p_2).
4. A *change–forbid-attribute conflict* of (p_1^{-1}, p_2) corresponds to a *change–forbid-attribute dependency* of (p_1, p_2).

An example of a sequential dependency detected by AGG for the Petri net firing rules AddPost and RemovePre is depicted in Fig. 12.12. As indicated in the dependency view, we have a *produce–use dependency* since the second transition needs to remove the same token that has been produced by the first transition. Obviously, this corresponds to a delete–use conflict of the inverse rule AddPost^{-1} and RemovePre.

New Usability Features for CPA in AGG 2.0

Selection of rules for CPA. So far, the CPA can be evoked on a graph grammar, yielding the critical pairs for each possible rule pair of the grammar. AGG 2.0 provides free selection of rule sets to be analysed. This feature has proved to be very convenient, e.g., for a case study, where self-healing systems are modelled by several rule sets (normal system behaviour rules, context-changing rules, repair rules). These sets are analysed for conflicts with each other (see Chap. 10).

Modularisation of a model into different sub-grammars. In AGG 2.0, rule sets may be imported into an existing grammar (provided the type graph of the imported grammar is a subgraph of the type graph of the importing grammar). This supports modularisation of a model without destroying the possibility to analyse the complete system. In a case study on modelling and analysing adaptive systems by graph transformation [BEE+13], we made heavy use of this modularisation feature by importing adaptation rule sets to the system grammar whenever the need for system adaptation arose.

Interrupt and resume running CPA. As a further usability feature, AGG 2.0 allows the user to interrupt a running critical pair analysis and to resume it later. This feature is very handy for complex computations which may take some time. A partial CPA result may be stored and reloaded.

Generation of Filter NACs. If during CPA, critical pairs are found that are inspected by the user and found not to be causing real conflicts, additional NACs for these rules may be generated automatically that contain the critical overlapping region. A new CPA of the rules together with these new *Filter NACs* does not show the previous critical pair anymore (see Def. 8.16). Possibly, new critical pairs are found due to the filter NACs. These may be extinguished by generating corresponding filter NACs for them, too. The procedure is repeated until no further critical pairs due to filter NACs are found. The construction always terminates if the structural part of each graph of a rule is finite, as shown in Fact 8.19.

Summarising, AGG 2.0 extends the existing features now coherently with support for application conditions and object flow, and for automatic construction of amalgamated, concurrent, inverse and minimal rules. Moreover, the critical pair analysis has become more usable due to experiences from several case studies. The critical pair analysis features offered by AGG are also used by our graph transformation-based tools ActiGra (a tool for checking consistency between control flow and functional behaviour; see Sect. 12.2), EMF Henshin (an EMF model transformation engine; see Sect. 12.3), as well as Henshin$_{TGG}$ (the extension of EMF Henshin to triple graph grammars; see Sect. 12.4).

12.2 ActiGra: Checking Consistency Between Control Flow and Functional Behaviour

In model-driven software engineering, models are key artifacts which serve as bases for automatic code generation. Moreover, they can be used for analyzing the system behaviour prior to implementing the system. In particular, it is interesting to know whether integrated parts of a model are consistent. For behavioural models, this means finding out whether the modelled system actions are executable in general or under certain conditions only. For example, an action in a model run might prevent one of the following actions occurring because the preconditions of this following action are not satisfied anymore. This situation is usually called a *conflict*. Correspondingly, it is interesting to know which actions do depend on other actions, i.e., an action may be performed only if another action has occurred before. We call such situations *causalities*. This section presents a plausibility checking approach regarding the consistency of the control flow and the functional behaviour given by actions modelled as graph transformation rules. Intuitively, consistency means that for a given initial state there is at least one model run that can be completed successfully.

We combine activity models defining the control flow and graph transformation rules in an *integrated behaviour model*, where a rule is assigned to each simple activity in the activity model. Given a system state typed over a given type graph, the behaviour of an integrated behaviour model can be executed by applying the specified actions in the predefined order. The plausibility check allows us to analyse an integrated behaviour model for *favourable* and *critical* signs concerning consistency. *Favourable* signs are, e.g., situations where rules are triggered by other rules that precede them in the control flow. On the other hand, *critical* signs are, e.g., situations where a rule causes a conflict with a second rule that should be applied after the first one along the control flow, or where a rule depends sequentially on the effects of a second rule which is scheduled by the control flow to be applied after the first one. An early feedback to the modeller indicating this kind of information in a natural way in the behavioural model is desirable to better understand the model.

For integrated behaviour models, sufficient consistency criteria have been developed already in [JLMT08]. The analysis consists of generating sets of rule sequences, describing potential model runs, and investigating them with respect to their applicability to initial states in a static way. The advantage thereof is that it is possible to declare consistency of a behavioural model without having to simulate each potential model run. However, especially for an infinite set of potential runs (in case of loops), this technique may lead to difficulties. Moreover, it is based on sufficient criteria leading to false negatives.

In this section, we follow a different approach, focusing on *plausibility* reasoning on integrated behaviour models and convenient visualisation of the static analysis results. This approach is complementary to [JLMT08], since we opt for back-annotating lightweight static analysis results. They do not only visualise the reasons for successful consistency analysis with more elaborate analysis techniques as presented in [JLMT08]; in addition, the visualisation of these lightweight results allows for plausibility reasoning on the integrated behaviour model also in the case of potential inconsistencies or false negatives or when consistency analysis results of the more elaborate techniques cannot be obtained. For plausibility reasoning, we determine conflicts and existing as well as nonexisting causalities (i.e., sequential dependencies) between rules depending on the control flow. On the one hand, they can lead to the detection of potential inconsistencies in the integrated behaviour model, and on the other hand they can lead to a better understanding of the reasons for model consistency. This lightweight technique seems to be very appropriate for allowing for early plausibility reasoning during development steps of integrated behaviour models. We visualise the results of our plausibility checks in an integrated development environment called ActiGra.[4] Potential inconsistencies and reasons for consistency are directly visualised within integrated behaviour models, e.g., as coloured arcs between activity nodes, and by detailed conflict and causality views.

The section is structured as follows: Section 12.2.1 presents our running example. In Sect. 12.2.2, we introduce our approach to integrated behaviour modelling and review the underlying formal concepts for static analysis based on graph trans-

---

[4] http://www.tfs.tu-berlin.de/actigra

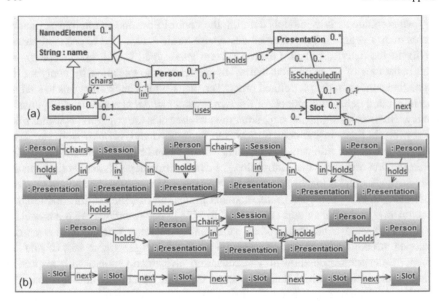

Fig. 12.13 Type graph (a) and instance graph (b) for the *Conference Scheduling System*, modelled with the ACTIGRA environment

formation as much as needed. Different forms of plausibility checking are presented in Sect. 12.2.3, where we validate our approach by checking a model of a conference scheduling system.[5]

12.2.1 Case Study: A Conference Scheduling System

This case study[6] models planning tasks for conferences. Its type graph is shown in Fig. 12.13 (a). A *Conference* contains *Persons*, *Presentations*, *Sessions* and *Slots*. A *Person* gives one or more *Presentations* and may chair arbitrary many *Sessions*. Note that a session chair may give one or more presentations in the session he or she chairs. A *Presentation* is in at most one *Session* and *scheduled* in at most one *Slot*. Slots are linked as a list by *next* arcs and *used* by *Sessions*.

Figure 12.13 (b) depicts a sample instance graph of an initial session plan before presentations are scheduled into time slots.[7] This instance graph conforms to the type graph. The obvious task is to find a valid assignment for situations like the one in Fig. 12.13 (b), assigning the presentations to available time slots such that the following conditions are satisfied: (1) there are no simultaneous presentations given by the same presenter, (2) no presenter is chairing another session running

---

[5] More details on the case study can be found in [EGLT11].

[6] Taken from the tool contest on *Graph-Based Tools 2008* [RV10].

[7] Due to space limitations, we do not show name attributes here.

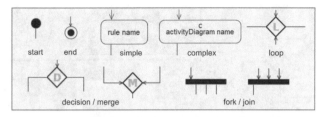

Fig. 12.14 Visual appearance of activity model building blocks

simultaneously, (3) nobody chairs two sessions simultaneously, (4) the presentations in one session are given not in parallel but in consecutive time slots, and (5) unused time slots are only at the begin or end of the conference. Moreover, it should be possible to generate arbitrary conference plans like the one in Fig. 12.13 (b). This is useful for testing the assignment procedure.

12.2.2 Integrating Activity Models with Graph Transformation

Our approach to behaviour modelling integrates activity models with graph transformation rules, i.e., the application order of rules is controlled by activity models. A rule describes the behaviour of a simple activity and is defined over a given type graph. The reader is supposed to be familiar with object-oriented modelling using the UML [UML15]. Therefore, we present our approach to integrated behaviour modelling from the perspective of its graph transformation-based semantics.

Integrated behaviour models

As in [JLM$^+$09], we define *well-structured activity models* as consisting of a start activity s, an activity block B, and an end activity e such that there is a transition between s and B and another one between B and e. Figure 12.14 illustrates the visual appearance of activity model building blocks.

An *activity block* can be a simple activity, a sequence of blocks, a fork–join structure, a decision–merge structure, or a loop. In addition, we allow complex activities which stand for nested well-structured activity models. In this hierarchy, we forbid nesting cycles. Activity blocks are connected by transitions (directed arcs). Decisions have an explicit *if*-guard and implicit *else*-guard which equals the negated *if*-guard, and loops have a *loop*-guard with corresponding implicit *else*-guard.

In our formalisation, an *integrated behaviour model* is a well-structured activity model A together with a type graph such that each *simple activity a* occurring in A is equipped with a *typed graph transformation rule* r_a and each *if* or *loop* guard is either *user-defined* or equipped with a typed *guard pattern*. We have *simple* and *application-checking* guard patterns: a simple guard pattern is a graph that has to

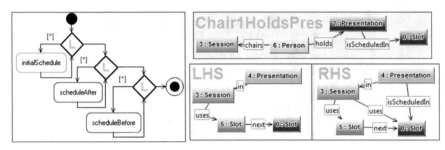

Fig. 12.15 Activity model *ScheduleControl* and rule *scheduleAfter*

Fig. 12.16 Graph rule *initialSchedule*

be found; an application-checking guard pattern is allowed for a transition entering a loop or decision followed by a simple activity in the loop-body or if-branch, respectively, and checks the applicability of this activity; it is formalised by a graph constraint (see Def. 5.10) and visualised by the symbol [*]. User-defined guards are evaluated by the user at run time to true or false. An *initial state* for an integrated behaviour model is given by a typed instance graph.

Example 12.1. Let us assume the system state depicted in Fig. 12.13 is the initial state of our integrated behaviour model. The activity diagram *ScheduleControl* is shown in the left part of Fig. 12.15. Its first step performs the initial scheduling of sessions and presentations into time slots by applying rule *initialSchedule* (see Fig. 12.16) as long as possible. The numerous conditions for this scheduling step are modelled by eight NACs. The sample NAC *twoPres* shown in Fig. 12.16 means that the rule must not be applied if the presenter holds already another presentation in the same slot.[8]

As second step, two loops are executed taking care to group the remaining presentations of a session into consecutive time slots, i.e., a presentation is scheduled in a free time slot either directly before or after a slot where there is already a scheduled presentation of the same session. The rule *scheduleAfter* is shown in the right part of Fig. 12.15. The rule *scheduleBefore* looks quite similar; only the direction of the *next* edge between the two slots is reversed. Both rules basically have the same NACs as rule *initialSchedule*, ensuring the required conditions for the schedule (see [BEL+10]). The NAC shown here ensures that the session chair does not hold a presentation in the time slot intended for the current scheduling.   △

---

[8] For the complete case study with all rules and NACs, see [BEL+10].

As in [JLM+09], we define a control flow relation on integrated behaviour models.[9] Intuitively, two activities or guards (a, b) are control flow-related whenever b is performed or checked after a. Moreover, we define an against-control flow relation which contains all pairs of activities or guards that are reverse to the control flow relation.

The *control flow relation* CFR_A of an activity model A contains all pairs (x, y) where x and y are activities or guards such that properties (1)–(4) hold:

(1) $(x, y) \in CFR_A$ if there is a transition from activity x to activity y.
(2) $(x, y) \in CFR_A$ if activity x has an outgoing transition with guard y.
(3) $(x, y) \in CFR_A$ if activity y has an incoming transition with guard x.
(4) If $(x, y) \in CFR_A$ and $(y, z) \in CFR_A$, then also $(x, z) \in CFR_A$.

The *against-control flow relation* $ACFR_A$ of an activity model A contains all pairs (x, y) such that (y, x) is in CFR_A.

The ACTIGRA tool

From the modeller's view, the main components of ACTIGRA are the following *views*, which are organised as a special ECLIPSE *perspective*, illustrated in Fig. 12.17.

1. The *Tree View* ① gives an overview of all elements of an ACTIGRA, as usual in ECLIPSE applications. This view offers support to add/delete elements such as object graphs, graph constraints or rules, and to edit element names. By double-clicking on an element, the visual view for this type of element is opened.
2. The *Type Graph View* ② visualises a type graph and supports freehand editing of node types, edge types, generalisation edges and multiplicities. The palette on the right-hand side contains the drawing tools.
3. The *Instance Graph View* ③ visualises an instance graph and supports free-hand editing of instance graphs. The palette on the right-hand side contains tools to draw nodes and edges of the corresponding types.
4. The *Activity Diagram View* ④ is a visual editor for well-structured activity diagrams. The panel contains all activity diagram elements. Simple activities contain the name of a rule which is applied when the corresponding activity is executed.
5. The *Rule View* ⑤ is a multi view editor consisting of editor panels for a rule's left- and right-hand sides (LHS, RHS) and (optionally) for one or more negative application conditions (NACs). Editing is supported as in the instance graph view, but in addition, mappings from the LHS to the RHS and to the NACs can be defined to identify common graph nodes. Mappings are visualised by equal node numbers and node colours. The mappings between edges can be inferred from the node mappings.
6. The *Graph Constraint View* (not shown in Fig. 12.17) is a visual editor for graph constraints which are used as guards for decision or loop activities. The execution of such an activity checks the guard on the current object graph and chooses the

---

[9] In contrast to [JLM+09], we include guards into the control flow relation.

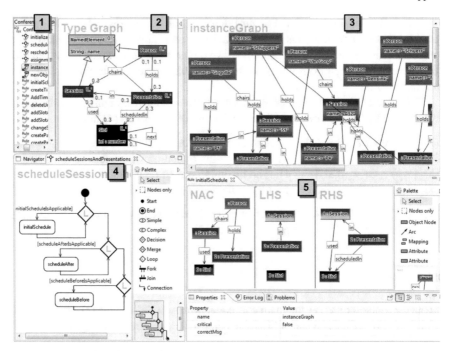

Fig. 12.17 The ACTIGRA perspective

corresponding branch if the graph constraint is fulfilled, or the alternative branch if it is violated.

Simulation of integrated behaviour models

The *semantics Sem(A)* of an integrated behaviour model A consisting of a start activity s, an activity block B, and an end activity e is the *set of sequences S_B*, where each sequence consists of *rules alternated with graph constraints* (stemming from guard patterns), generated by the main activity block B (for a formal definition of the semantics, see [JLM+09]).[10] For a block being a simple activity a inscribed by a rule r_a, $S_B = \{r_a\}$. For a sequence block $B = X \rightarrow Y$, we construct $S_B = S_X$ seq S_Y, i.e., the set of sequences that are concatenations of a sequence in S_X and a sequence in S_Y. For decision blocks, we construct the union of sequences of both branches (preceded by the if guard pattern and the negated guard pattern, respectively, in the case that the if guard is not user-defined); for loop blocks, we construct sequences containing the body of the loop i times ($0 \leq i \leq n$) (where each body sequence is preceded by the loop guard pattern and the repetition of body sequences is concluded

---

[10] Note that *Sem(A)* does not depend on the initial state of A. Moreover, we have a slightly more general semantics compared to [JLM+09], since we do not have only rules in the sequences of S_B, but also graph constraints.

with the negated guard pattern in the case that the loop guard is not user-defined). In contrast to [JLM+09], we restrict fork–join blocks to one simple activity in each branch and build a parallel rule from all branch rules [Lam10, EEPT06].[11] We plan to omit this restriction, however, when integrating object flow [JLM+09] into our approach, since then it would be possible to build unique concurrent rules for each fork–join branch. For B a complex activity inscribed by the name of the integrated behaviour model X, $S_B = Sem(X)$.

Consider a sequence $s \in Sem(A)$ of rules alternated with graph constraints and a start graph S, representing an initial state for A. We then say that each graph transformation sequence starting with S, applying each rule to the current instance graph and evaluating each graph constraint to true for the current instance graph in the order of occurrence in s, represents a *complete simulation run* of A. An integrated behaviour model A is *consistent* with respect to a start graph S, representing an initial state for A, if there is a sequence $s \in Sem(A)$ leading to a complete simulation run. In particular, if A contains user-defined guards, usually more than one complete simulation run should exist.

In ActiGra we can execute simulation runs on selected activity models. Chosen activities are highlighted and the completion of simulation runs is indicated. User-defined guards are evaluated interactively. If a simulation run cannot be completed, an error message tells the user which activity could not be executed.

12.2.3 Plausibility Checks for Integrated Behaviour Models

We now consider how to check plausibility regarding consistency of the control flow and the functional behaviour given by actions bundled in object rules. Thereby, we proceed as follows: We characterise *desired properties* for an integrated behaviour model and its initial state as being consistent. We determine the *favourable* as well as *critical signs*[12] for these properties to hold, show how the checks are supported by ActiGra and illustrate by our case study, which conclusions can be drawn by the modeller to validate our approach.

For the plausibility checks we wish to detect potential conflicts and causalities [EGLT11] between rules and guards occurring in the sequences of $Sem(A)$. Since in A simple activities, fork/joins as well as simple guard patterns correspond to rules,[13] we just call them rules for simplicity. Thereby, we disregard rules stemming from

---

[11] This fork–join semantics is slightly more severe than in [JLM+09], which allows all interleavings of rules from different branches no matter if they lead to the same result.

[12] In most cases, these favourable and critical signs merely describe *potential* reasons for the property to be fulfilled or not, respectively. For example, some critical pair describes which kind of rule overlap may be responsible for a critical conflict. By inspecting this overlap, the modeller may realise that the potential critical conflict may actually occur and adapt the model to avoid it. On the other hand, he may realise that it does not occur since the overlap corresponds to an invalid system state, intermediate rules deactivate the conflict, etc.

[13] For each *simple guard pattern* we can derive a *guard rule* (without side-effects) for the guarded branch and a negated guard rule for the alternative branch (as described in [JLM+09]). Application-

simple activities belonging to some fork/join block, since they do not occur as such in *Sem(A)*. Instead, the corresponding parallel rule for the fork/join is analysed. As an exception to this convention, the plausibility check in Sect. 12.2.3 inspects consistency of fork/joins and analyses also the enclosed simple activities.

Inspecting initialisation

If for some sequence in *Sem(A)* the first rule is applicable, then the corresponding sequence can lead to a complete simulation run. Otherwise, the corresponding sequence leads to an incomplete run. Given an integrated behaviour model *A* with initial state *S*, the first *plausibility check* computes automatically for which sequences in *Sem(A)* the first rule is applicable to *S*. The modeller then may inspect the simulation run(s) that should complete for correct initialisation (*desired property*). We identify the *favourable signs* as the set of possible initialisations: $FaI_A = \{r \mid r$ is first rule of sequence in $Sem(A)$ and r is applicable to $S\}$. We identify the *critical signs* as the set of impossible initialisations: $CrI_A = \{r \mid r$ is first rule of a sequence in $Sem(A)$ and r is not applicable to $S\}$.

ActiGra visualises the result of this plausibility check by highlighting the elements of FaI_A in green. Rules belonging to CrI_A are highlighted in *red*.[14]

Example 12.2. Let us assume the system state in Fig. 12.13 (b) is an initial state. The left part of Fig. 12.15 shows the initialisation check result for the activity model *ScheduleControl*. We have $FaI_{ScheduleControl} = \{initialSchedule\}$ and $CrI_{ScheduleControl} = \{scheduleAfter, scheduleBefore\}$. Thus, complete simulation runs on our initial state never start with *scheduleAfter* or *scheduleBefore*, but always with *initialSchedule*.   △

Inspecting trigger causalities along control flow direction

If a rule *a* may trigger a rule *b* and *b* is performed after *a*, then it may be advantageous for the completion of a corresponding simulation run. If for some rule *b* no rule *a* is performed before *b* that may trigger *b*, this may lead to an incomplete simulation run and the modeller may decide to add some triggering rule or adapt the post-condition of some previous rule in order to create a trigger for *b*. Alternatively, the initial state could be adapted such that *b* is applicable to the start graph. Given an integrated behaviour model *A* with initial state *S*, this *plausibility check* computes automatically, for each rule *a* in *A*, which predecessor rules may trigger *a*. The modeller may inspect each rule *a* for enough predecessor rules to trigger *a* then (*desired property*). We identify the *favourable signs* as the set of potential trigger causalities for some rule *a* along the control flow:

---

checking guard patterns are evaluated for simulation but disregarded by the plausibility checks, since they are not independent guards but check for the application of succeeding *rules* only.

[14] Concerning fork/join blocks in FaI_A or CrI_A, ActiGra colours the fork bar.

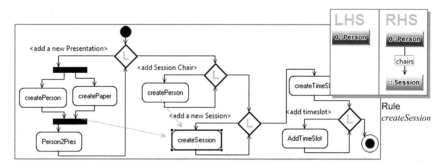

Fig. 12.18 Potential trigger causalities along the control flow in the activity model *GenConfPlans*

$FaTrAl_A(a) = \{(b, a) \mid (b, a) \in CFR_A$ such that b may trigger $a\}$. We say that $FaTrAl_A$ $= \{FaTrAl_A(a) \mid a$ is a rule in $A\}$ is the *set of potential trigger causalities in A along the control flow*. For this check, the dependency analysis based on critical pairs offered by AGG 2.0 (see Sect. 12.1.4) is used by ACTIGRA. We identify the *critical signs* as the set of nontriggered rules along the control flow that are not applicable to the initial state: $CrNonTrAl_A = \{a \mid a$ is rule in A such that $FaTrAl_A(a)$ $= \emptyset$ and a is not applicable to $S\}$.

ACTIGRA visualises the result of this plausibility check by displaying *dashed green arrows* from b to a selected rule a for each pair of rules (b, a) in $FaTrAl_A(a)$. If no rule is selected, then all pairs in $FaTrAl_A$ are displayed by dashed green arrows. Clicking on such an arrow from b to a opens a detail view, showing the reason(s) why b may trigger a as discovered by CPA. Conversely, ACTIGRA highlights each rule belonging to $CrNonTrAl_A$ in *red*. Moreover, when right-clicking on such a rule its applicability to the initial state can be checked.

Example 12.3. Consider the activity model *GenConfPlans* in Fig. 12.18 for generating conference plans, assuming an empty initial state. The set of potential trigger causalities along the control flow for *createSession* is given by

$$FaTrAl_{GenConfPlans}(createSession) = \{(createPerson + createPaper, createSession),$$
$$(createPerson, createSession)\}.$$

Here, we learn that we need at least one execution of a loop containing the rule *createPerson* (a rule with an empty left-hand side) to ensure a complete simulation run containing *createSession*. In AGG terms, we have a produce–use dependency of the rule pair (*createPerson, createSession*).

Consider the draft activity diagram B for generating conference plans shown in Fig. 12.19. Here, the set $FaTrAl_B(Person2Pres)$ of potential trigger causalities along the control flow is empty (in ACTIGRA, it is highlighted in red). Moreover, this rule is not applicable to the initial state of our model. Hence, we run into an incomplete simulation run if guard *<add a new Presentation>* is chosen by the user. One way to repair this situation is to insert one or more triggering rules before the rule *Person2Pres*(as, e.g., in Fig. 12.18). △

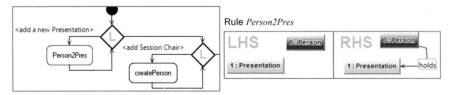

Fig. 12.19 Nontriggered rule *Person2Pres* along the control flow in activity model *B*

Inspecting conflicts along the control flow direction

If a rule *a* may disable a rule *b*, and *b* is performed after *a*, then this may lead to an incomplete simulation run. On the other hand, if for some rule *a* no rule *b* performed before *a* exists that may disable rule *a*, then the application of *a* is not impeded. Given an integrated behaviour model *A* with initial state *S*, this *plausibility check* computes automatically, for each rule *a* in *A*, which successor rules *b* in *A* may be disabled by *a*. The modeller then may inspect each rule *a* in *A* for the absence of rules performed before *a* disabling rule *a* (*desired property*). We identify the *critical signs* as the set of potential conflicts along the control flow caused by rule *a*: $CrDisAl_A(a) = \{(a, b) \mid a, b$ are rules in $A, (a, b) \in CFR_A$ and *a* may disable *b*}. We say that $CrDisAl_A = \{CrDisAl_A(a) \mid a$ is a rule in $A\}$ is the *set of potential conflicts along the control flow in A*. For this check, the conflict analysis based on critical pairs offered by AGG 2.0 (see Sect. 12.1.4) is used by ACTIGRA. We identify the *favourable signs* as the set of nondisabled rules along the control flow: $FaNonDisAl_A = \{a \mid a$ in A and $\nexists (b, a) \in CrDisAl_A \}$.

ACTIGRA visualises the result of this plausibility check by displaying faint red arrows from *a* to *b* for each pair of rules (a, b) in $CrDisAl_A$. If the rule *a* is selected, a bold red arrow from *a* to *b* for each pair of rules (a, b) in $CrDisAl_A(a)$ is shown. Clicking on such an arrow opens a detail view, showing the reason(s) why *a* may disable *b* as discovered by CPA. Each rule *a* in *A* belonging to $FaNonDisAl_A$ is highlighted in *green*.

Example 12.4. Consider the activity model *SchedulingControl* in Fig. 12.20 (a). Here, the set of potential conflicts along the control flow caused by the rule *initialSchedule* is given by

$$CrDisAl_{SchedulingControl}(initialSchedule) = \{(initialSchedule, initialSchedule),$$
$$(initialSchedule, scheduleAfter),$$
$$(initialSchedule, scheduleBefore)\}.^{[15]}$$

This gives the modeller a hint that in fact a scheduling might not terminate successfully in the case that rule *initialSchedule* creates a situation where not all remaining presentations can be scheduled in a way satisfying all conditions. The detail view of potential conflicts for the pair (*initialSchedule, scheduleAfter*) in Fig. 12.20 (b) shows a potential produce–forbid conflict where the rule *initialSchedule* (Fig. 12.16)

---

[15] Note that one pair in this set may indicate more than one conflict potentially occurring between the corresponding rules.

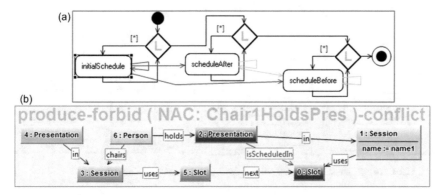

Fig. 12.20 (a) Potential conflicts along the control flow caused by rule *initialSchedule*; (b) Detail view of potential conflict of the rules *initialSchedule* and *scheduleAfter*

produces an edge from 2:Pres to 0:Slot, and the rule *scheduleAfter* then must not schedule 4:Pres to 0:Slot because of the NAC depicted in Fig. 12.15.   △

Inspecting trigger causalities against control flow direction

If a rule a may trigger a rule b and b is performed before a, then it might be the case that their order should be switched in order to obtain a complete simulation run. Given an integrated behaviour model A with initial state S, this *plausibility check* automatically computes for each rule a in A which successor rules of a may trigger a. The modeller then may inspect for each rule a in A that no rule performed after a exists that needs to be switched to a position before a in order to trigger its application (*desired property*). We identify the *critical signs* as the set of potential causalities against control flow triggered by a: $CrTrAg_A(a) = \{(a,b) \mid a, b \text{ rules in } A$ and $(a,b) \in ACFR_A$ such that a may trigger $b\}$. We say that $CrTrAg_A = \{CrTrAg_A(a) \mid a \text{ is a rule in } A\}$ is the *set of potential trigger causalities against control flow in A*. We identify the *favourable signs* as the set of rules not triggered against control flow: $FaNoTrAg_A = \{a \mid a \text{ is rule in } A \text{ and } \nexists(b,a) \in CrTrAg_A \}$.

ActiGra visualises the result of this plausibility check by displaying a *dashed red arrow* from a selected rule a to b for each pair of rules (a,b) in $CrTrAg_A(a)$. If no rule in particular is selected, then all pairs in $CrTrAg_A$ are displayed by dashed red arrows. Clicking on such an arrow from a to b opens a detail view, showing the reason(s) why a may trigger b as discovered by CPA. Conversely, each rule belonging to $FaNoTrAg_A$ is highlighted in *green*.

Example 12.5. In the activity diagram *GenConfPlans* in Fig. 12.21, we get the set of potential causalities against control flow $CrTrAg_{GenConfPlans} (createSession) = \{(createSession, Person2Pres)\}$. The causality *(createSession, Person2Pres)* indicates that the rule *Person2Pres* might be modelled too early in the control flow since the rule *createSession* is needed to trigger the rule *Person2Pres* completely.   △

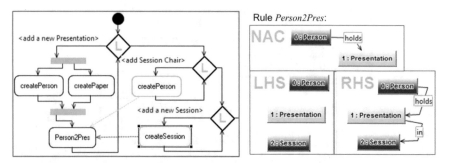

Fig. 12.21 Trigger causality against control flow *(createSession, Person2Pres)*

Inspecting causalities in fork/joins

We may not only consider the consistent sequential composition of rules as before, but consider also the parallel application of rules as specified by fork/join activities. Whenever a rule pair (a, b) belonging to the same fork/join may be causally dependent, it is not possible to change their application order in any situation without changing the result. However, the parallel application of the rules (a, b) implies that their application order should not matter.

Given an integrated behaviour model A with initial state S, this *plausibility check* determines automatically for each fork/join in A if potential causalities between the enclosed simple activities exist. The modeller may inspect each fork/join for its parallel execution not to be disturbed then *(desired property)*.

We need some more elaborate considerations for this case, since we wish to analyse simple activities within a fork/join block that are normally disregarded, as they only occur in the form of the corresponding parallel rule in $Sem(A)$. In particular, we define a *fork/join relation* FJR_A consisting of all rule pairs (a, b) belonging to the same fork/join block. We identify the *critical signs* as the set of potential causalities between different fork/join branches: $CrFJCa_A = \{(a, b)|(a, b) \in FJR_A$ and (a, b) causally dependent$\}$.[16] We identify the *favourable signs* as the set of fork/join structures with independent branches: $FaFJNoCa_A = \{fj|fj$ is fork/join in A and $(a, b) \notin CrFJCa_A$ for each (a, b) with a,b in different branches of $fj\}$.

ACTIGRA visualises the result of this plausibility check by displaying in each fork/join block a *dashed red arrow* from a to b for each $(a, b) \in CrFJCa_A$. The detail view shows the reason(s) why (a, b) are causally dependent, as discovered by CSA, and why this dependency might disturb parallel execution. On the other hand, each fork/join in $FaFJNoCa_A$ is highlighted by *green* fork and join bars.

Example 12.6. The set of potential causalities between different fork/join branches depicted in Fig. 12.22 is given by $\{(createPerson, Person2Pres)\}$. We may have a dependency (shown in the detail view) if the rule *createPerson* creates a *Person* node that is used by the rule *Person2Pres* to link it to a *Presentation* node.   △

---

[16] Here, we do not only regard trigger causalities between a and b, but also causalities making the application of the rule a irreversible, as described in [Lam10].

Fig. 12.22 Potential causality between different fork/join branches and its detail view

Summarising, ActiGra leverages the critical pair analysis of AGG to detect possibly unwanted interactions in integrated behavioural models. Please note that our approach to plausibility reasoning can easily be adapted to any other approach where modelling techniques describing the control flow of operations are integrated with operational rules, as, e.g., the integration of live sequence charts with object rules in [LMEP08]. The ActiGra approach and tool have recently been applied successfully to analyse aspect-oriented models [MHMT13]. In this case study, a crisis management system is modelled using ActiGra, and model consistency is analysed at the level of requirements modelling.

A further refinement step in activity-based behaviour modelling would be the specification of object flow between activities. Additionally specified object flow between two activities would further determine their interrelation. In this case, previously determined potential conflicts and causalities might not occur anymore. Thus, the plausibility checks would become more exact with additionally specified object flow. A first formalisation of integrated behaviour models with object flow based on graph transformation is presented in [JLM+09]. An extension of plausibility checks to this kind of activity models is left for future work.

To conclude, integrated behaviour models head towards a better integration of structural and behavioural modelling of (software) systems. Plausibility checks provide lightweight static analysis checks supporting the developer in constructing consistent models.

12.3 Controlled EMF Model Transformation with EMF Henshin

In graph transformation tools, the application of rules may be controlled implicitly as in AGG (Sect. 12.1), i.e., by a fixed strategy such as "apply rules in arbitrary order as long as possible" and by providing negative application conditions for rules. Alternatively, control strategies may be defined explicitly as in Fujaba [FNTZ00], where an activity diagram (story diagram) defines loops or conditions on rule applications. Explicit control structures raise the expressiveness of transformation systems since they provide means to regulate the transformation process without having to introduce helper structures into the rules.

In this section, we lift implicit and explicit control structures from graph transformation to EMF model transformation and introduce an extension of the tool EMF

Henshin[17] by visual editors for control structures. EMF Henshin is an Eclipse plug-in supporting visual modelling and execution of EMF model transformations, i.e., transformations of models conforming to a meta model given in the EMF Ecore format.[18] The transformation approach we use in our tool is based on graph transformation concepts which are lifted to EMF model transformation by also taking containment relations in meta models into account [ABJ+10, BESW10].

The Eclipse Modeling Framework EMF [EMF14] is a modelling and code generation facility for building tools and other applications based on a structured data model. Based on a meta model, EMF provides tools and runtime support to produce a set of Java classes for the meta model, a set of adapter classes that enable viewing and command-based editing of models conforming to the meta model, and a basic (tree-based) editor. EMF provides the foundation for interoperability with other EMF-based tools, e.g., OCL checkers.

The conceptual similarities of modelling based on typed attributed graphs and object-based modelling as performed by EMF are shown in Table 12.1, which is an extension of Table 3.1 by the EMF-specific concept of *containment*.

Table 12.1 Mapping EMF notions to typed attributed graph terminology

EMF notion	Typed attributed graph terminology
EMF model	Type graph with attribution, inheritance and multiplicities; edges can be marked as containments
Instance model	Typed attributed graph with containment edges
Class	Node in type graph
Object	Node in typed graph
Association	Edge in type graph (with possible multiplicities or containment mark)
Reference	Edge in typed graph that satisfies multiplicity and containment constraints

Containment relations, i.e., aggregations, define an ownership relation between objects. Thereby, they induce a tree structure in model instantiations. In MOF and EMF, this tree structure is further used to implement a mapping to XML, known as XMI (XML Meta data Interchange) [XMI08]. Containment implies a few constraints for model instantiations that must be ensured at runtime. As semantical constraints for containment edges, the MOF specification [Obj14a] states the following: *"An object may have at most one container"* and *"Cyclic containment is invalid"*. EMF provides full implementation of instance models that always ensures these constraints. In [BET12], containment constraints of EMF model transformations are translated to a special kind of well-formed graph transformation rule whose application leads to consistent transformation results only, i.e., they must not delete contained objects without deleting their containment relations as well, and they must

---

[17] http://www.eclipse.org/modeling/emft/henshin/, originating from EMF Tiger [EMT09, BEK+06, BEL+10]

[18] Note that we use the terms *meta model* and *model* in this section, which are called *EMF model* and *model instance* in the EMF documentation, respectively.

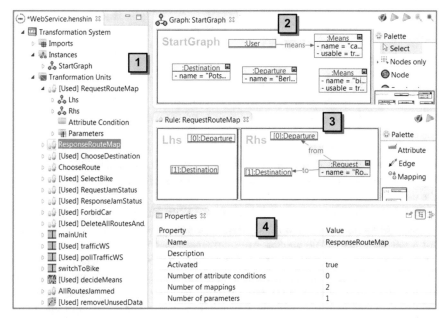

Fig. 12.23 EMF HENSHIN GUI with visual editors for graphs and rules

not generate objects without relating them to precisely one container. Moreover, containment cycles must not be produced by rule applications.

Applying EMF model transformation rules in EMF HENSHIN changes an EMF model *in-place*, i.e., the model is modified directly. Note that we speak of *EMF model transformation* in a general sense, comprising not only source-to-target model-to-model transformations but also model refactorings or simulation of the system's behaviour.[19] The EMF HENSHIN transformation engine provides classes that can freely be integrated into existing Java projects relying on EMF.

Figure 12.23 depicts the basic GUI of our EMF HENSHIN tool (without the extensions for control structure definition). The tree view ⓵ allows the modeller to import EMF EPackages containing the basic meta model(s) defining the domain of the transformation. The initial model is edited in a visual editor ⓶. In the rule editor ⓷, transformation rules can be created by editing a rule's left-hand side (LHS) and right-hand side (RHS), similarly to rule visualisation in AGG. The property view ⓸ shows additional information for selected objects. Note that all information edited using the editors in ⓶, ⓷ and ⓸ can also be obtained via the tree view ⓵.

The rule displayed in Fig. 12.23, ⓷, defines an operation adding a Request object and linking it to existing Departure and a Destination objects. This rule can now be applied to the current model (Fig. 12.23, ⓶), leading to the transformed graph shown in Fig. 12.24, where a Request object has now been created and linked to the

---

[19] As in our running example, where we handle the simulation of a personal mobility manager based on a web service.

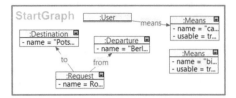

Fig. 12.24 Transformed graph after applying the rule *RequestRouteMap*

Departure object named "Berlin" and the Destination object named "Potsdam". The layout of newly added objects is computed automatically but may be adjusted by the user.

In this section, we describe the extension of EMF Henshin supporting the use of the control structures (called EMF Henshin *transformation units*), e.g., constructs for nondeterministic rule choices, rule sequences or conditional rule applications. Those constructs may be nested to define more complex control structures. Passing of model elements as parameters from one unit to another is also possible. Apart from control units defined over sets of rules, we now also support the graphical definition of application conditions for individual rules. These are application conditions in the sense of Def. 5.1, allowing for arbitrary nesting. Several application conditions can be combined by logical connectors.

The section on EMF Henshin is structured as follows: Section 12.3.1 presents our running example, the simulation of a personal mobility manager (PMM) based on a web service. Along the PMM example, we introduce the usage of transformation units and application conditions in Sects. 12.3.2 and 12.3.3, respectively.

12.3.1 Example: Personal Mobility Manager

As a running example, we specify and simulate the operational behaviour of a Personal Mobility Manager (PMM), a reactive service-based application designed to satisfy requirements related to individual user mobility [LMEP08]. The aim of the system is to help the user find an adequate route from a departure place to a destination and to propose an adequate means of transportation (either car or bike) by taking the current traffic intensity into account. We model the control flow of messages that are exchanged between the user, the PMM and the corresponding web service. To keep things simple, we do not model the actual web service here but simulate its responses by suitable variable assignments.

The modelling domain is specified by the meta model depicted in Fig. 12.25. We have model elements for a user, the user's departure and destination locations, the means of transport, and requests sent to the web service. A Route element contains a route given as response by the mobility web service, and a JamStatus element contains the response returned by the web service concerning the traffic on a given route. Edges with white triangles as arrowheads are inheritance relations; the re-

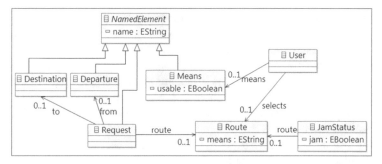

Fig. 12.25 Meta-model for the Personal Mobility Manager

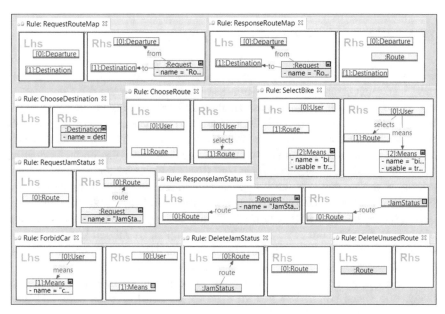

Fig. 12.26 EMF model transformation rules for the Personal Mobility Manager

maining edges denote associations with multiplicities, where multiplicity "0..1" at
the target of an association means that a source object of a correspondingly typed
reference is linked to at most one target object of the corresponding class. Note that
we do not use containment edges in this example. Utilising EMF HENSHIN to model
a case study with containment relations is described in [BET12].

Basic PMM actions are modelled by EMF model transformation rules, as shown
in Fig. 12.26.

The rule ChooseDestination creates a Destination object where the name of the destination is an input parameter; the rules RequestRouteMap and ResponseRouteMap
realise the creation of a route (modelled by a Route object) via a web service call.
Having called this web service more than once, one of the returned routes is chosen by the user using the rule ChooseRoute. For a given route, the web service is

used by the rules RequestJamStatus and ResponseJamStatus to get information about
the current traffic situation on this route. Depending on the information obtained by
the web service (and coded in the JamStatus node), the means of transport can be
changed from the default means "car" (as presented in the start graph in Fig. 12.23)
to the alternative means "bike". This is realised by applying the rules ForbidCar and
SelectBike. At last, the information about traffic (JamStatus node) and possible alternative routes which have not been chosen are deleted using the rules DeleteJamStatus
and DeleteUnusedRoute.

In the next subsection, we explain the use of EMF HENSHIN transformation units
to encapsulate and control the order of rule applications.

12.3.2 EMF HENSHIN *Transformation Units*

EMF HENSHIN transformation units may be arbitrarily nested inside each other. The
most basic unit is an EMF model transformation rule (like the rules depicted in
Fig. 12.26). An EMF HENSHIN transformation unit may be of type *IndependentUnit*
(of all applicable subunits, one is applied arbitrarily), *SequentialUnit* (all subunits
are applied sequentially[20] in a given order), *LoopUnit* (its subunit is applied as long
as possible), *ConditionalUnit* (its subunits are applied depending on the evaluation
of a given condition unit), and *PriorityUnit* (the applicable subunit with the highest
priority is applied next). A unit is applicable (and returns true) if it can be successfully executed. *PriorityUnits* and *IndependentUnits* are always applicable, while *Se-
quentialUnits* are applicable only if all subunits are applicable the defined number
of times in the given order; a *LoopUnit* is applicable if its subunit is applicable. A
ConditionalUnit is applicable if either the *then* subunit (in case the condition is true)
or the *else* subunit (in case the condition is false) is applicable.

EMF HENSHIN transformation units may be defined in the tree view or, alternatively, in a visual editor. The tree view shows all transformation units and their
nesting hierarchy. The visual editor for one unit shows the unit in a left view and
one selected subunit in a right view (see Fig. 12.27). The unit view shows the unit's
name as header, a set of parameters shown as boxes in the left column, and the
names and kinds of its subunits in the right column. Subunits of a *SequentialUnit*
have a counter indicating their number of applications. A unit and a subunit may
share parameters shown by the colouring of the parameter fields (see bottom part
of Fig. 12.27, where the unit trafficWS shares the parameter route with the subunit
RequestJamStatus). Arrows from (to) parameter boxes to (from) subunits indicate
which parameters are input (output) of which subunit. Parameter passing between
units serves as a match control mechanism; thus, the modeller may define, e.g., that
a unit must be applied to the very node that has been generated by the previously
applied unit.

---

[20] Optionally more than once by defining a *counter* for each subunit.

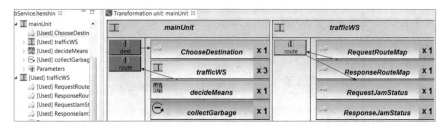

Fig. 12.27 EMF Henshin GUI with transformation unit editor

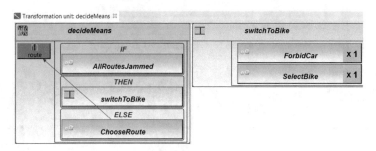

Fig. 12.28 EMF Henshin transformation unit decideMeans

The transformation unit mainUnit shown in Fig. 12.27 is the main control structure for the PMM example. It is a *SequentialUnit* (symbolised by a film strip as icon in the upper left corner) containing four subunits. This means that each subunit is applied as often as indicated by its counter, in the given order from top to bottom. The first subunit, ChooseDestination is a transformation rule, marked by gear wheels (see Fig. 12.26 for the rule definition). This rule has an input parameter, the destination dest, a user-defined parameter. The second subunit of the main unit is a *SequentialUnit* (to be applied three times). The unit trafficWS is shown with its contents in the view to the right: it contains in turn four rules realising the web service requests and processing the responses. The rule ResponseRouteMap produces an output parameter of type Route which serves again as input parameter for the rule RequestJamStatus.

The third subunit of mainUnit, decideMeans, is a *ConditionalUnit* (symbolised by an *if-then-else* icon). Clicking on its field, a detailed view of this unit is opened (see Fig. 12.28).

Here, a condition called AllRoutesJammed (which will be discussed in Sect. 12.3.3) is checked; it is given as application condition of the empty rule. If the condition is evaluated to true, the two rules ForbidCar and SelectBike in the sequential unit switchToBike are applied in this order. Otherwise, the rule ChooseRoute is applied and the parameter route is returned to the parent unit mainUnit.

Figure 12.29 illustrates the situation after applying the unit decideMeans, where all routes have been found to be jammed and the bicycle is selected as transport means.

The last child unit of mainUnit is the *LoopUnit* collectGarbage (with a loop as icon symbol). This unit applies its subunit as long as possible (see Fig. 12.30). It contains

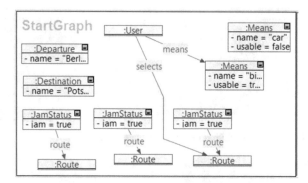

Fig. 12.29 Intermediate state where the bicycle is selected as transport means

Fig. 12.30 EMF HENSHIN transformation unit collectGarbage

Fig. 12.31 EMF HENSHIN interaction scheme DeleteAllRoutesAndJamsInOneStep

an *IndependentUnit* with two rules, DeleteJamStatus and DeleteUnusedRoute, which perform the garbage collection. One of the applicable subunits in an *IndependentUnit* is selected randomly (hence, it has a die as icon symbol).

Alternatively, an interaction scheme may be defined (see Def. 6.30), replacing the *LoopUnit* collectGarbage: the interaction scheme shown in Fig. 12.31 consists of an empty kernel rule DeleteAllRoutesAndJamInOneStep and one multi rule called DeleteRouteAndJamPair. Note that EMF HENSHIN implements maximal weakly disjoint matchings according to Def. 6.32. Thus, by specifying an empty kernel rule, we ensure that the connected Route-JamStatus node pairs, found by the matching process, do not overlap.

The application of the interaction scheme deletes all such pairs of a Route and a JamStatus node in one step, provided that they are not used anymore (i.e., there is no user connected to the corresponding Route node). The visualisation of rule schemes

is similar to the visualisation in AGG (see Fig. 12.8 in Sect. 12.1.3). But in addition
to AGG, EMF HENSHIN allows nesting of multi rules in arbitrary depth.

12.3.3 Application Conditions

In addition to (explicit) control structures, called transformation units, EMF HEN-
SHIN also implements application conditions for transformation rules according to
Def. 5.12. Recall that, like transformation units, application conditions can be
nested. Moreover, application conditions may be negated, and several application
conditions may be combined by using the logical connectors AND and OR.

Let us consider once more the *ConditionalUnit* decideMeans from our PMM example (see Fig. 12.28). Here, the condition AllRoutesJammed is expressed by an
empty rule[21] with a nested application condition, shown in Fig. 12.32. Since we
cannot use directly universal quantifiers " ∀ " in graph conditions, we reformulate
the condition *"All routes have a jam status and are jammed"* using existential quantifiers " ∃ " only. Hence, we get the logically equivalent expression *"There is no
route that has not been given a jam status or that is free"*.

In view $\boxed{1}$ of Fig. 12.32, the empty rule is shown together with the outermost
condition graph (a condition over *LHS*). In the tree view of $\boxed{1}$, it can be seen
that we require $\nexists$ *Route*, i.e., a morphism from the graph Route (consisting of a
single Route node) into the host graph must not exist for the rule to be applicable.
Since this application condition is nested, we require a further condition for the
Route graph, formulated as disjunction (OR-construct) over two more conditions:
($\nexists$ *HasNoJamStatus* ∨ ∃ *IsFree*). This formula can be seen in the tree view of
$\boxed{2}$, as well as in the corresponding visual hierarchical view where the formula is
depicted as an OR block with two compartments. Clicking on one of the two parts
of the disjunction in the visual view (or on one of the two OR branches in the
tree view) opens the next level, either for the formula $\nexists$HasNoJamStatus in $\boxed{3}$ or
for the formula ∃ *IsFree* in $\boxed{4}$. Here, we have arrived at the basic level of graph
morphisms. The complete nested application condition AllRoutesJammed means that
the empty rule is applicable (returns *true*) if there exists no route that has either no
JamStatus node or that has a JamStatus node with attribute jam=false. Recall that in
this case (all routes are jammed) the unit decideMeans (see Fig. 12.28) applies the
unit switchToBike; otherwise a route is chosen for the car as transport means.

Summarising, in this section, we have presented three extensions for supporting controlled EMF transformations in our EMF transformation environment EMF
HENSHIN. The first extension supports the visual definition of EMF HENSHIN transformation units, which may be hierarchically nested (the basic unit being a rule) and
which restrict the possible rule application sequences in a suitable way. The second
extension concerns the definition of (nested) interaction schemes to allow for amal-

---

[21] Note that we allow arbitrary transformation units as conditions in *ConditionalUnits*. While this
may lead to side effects if a unit different from the empty rule is used, the conceptual advantage is
that components of EMF HENSHIN transformation units always are transformation units in turn.

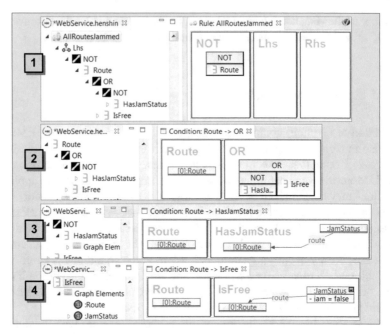

Fig. 12.32 Empty rule AllRoutesJammed with application condition

gamated EMF model transformations. This proved to be very useful for specifying operations on recurring model patterns. The third extension supports the definition of application conditions for transformation rules, where conditions may be nested and combined by logical connectors such as AND and OR.

For execution, EMF HENSHIN rules and transformation units can be used in other Java projects by instantiating the class RuleApplication or UnitApplication, respectively. The class RuleApplication requires a Rule instance from the EMF HENSHIN meta model. Once instantiated, the rule can be applied by calling the execute() method of RuleApplication. Transformation units can be executed in a similar way by using the class UnitApplication. The multi view EMF HENSHIN editor is available via the following GitHub repository: https://github.com/de-tu-berlin-tfs/Henshin-Editor.

Apart from the PMM example, EMF HENSHIN has been applied also for larger case studies, e.g., for model refactorings [AT13, ABJ+10] and model-to-model transformations such as the *Ecore2Genmodel* case study of the Transformation Tool Contest 2010 [BEJ10]. Recently, EMF HENSHIN has been used by the Luxembourg-based satellite operator SES to translate proprietary satellite control procedures into the open-source software SPELL (Satellite Procedure Execution Language and Library), a standardised satellite control language. For this purpose, the EMF HENSHIN editor has been extended to HENSHIN_TGG, supporting visual model transformation with triple graph grammars [HGN+14], which will be described in the next section.

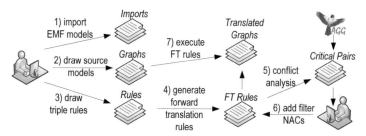

Fig. 12.33 Workflow overview of using HENSHIN_TGG for EMF model transformation

12.4 Bidirectional EMF Model Transformation with HENSHIN_TGG

In this section, we continue the idea of using graph transformation concepts to realise EMF model transformations. We consider a recent extension of the tool EMF HENSHIN to specify model transformations based on triple graph grammars (TGGs). We present the visual TGG modelling and analysis environment HENSHIN_TGG [EHGB12] building on the EMF model transformation engine EMF HENSHIN that was described in Sect. 12.3. In contrast to existing TGG implementations [GHL12, ALPS11, BGH⁺05, LAS⁺14b, HLG⁺13], HENSHIN_TGG does not only specify and perform EMF model transformations by TGGs but generates *forward translation rules* (synthesised from forward and source rules) according to Sect. 7.4, and offers a converter to translate forward translation rules to the graph transformation analyser AGG (see Sect. 12.1) in order to benefit from AGG's critical pair analysis for conflict detection. Figure 12.33 presents an overview of the overall workflow using the main tool features of HENSHIN_TGG.

The section on HENSHIN_TGG is structured as follows: we present our visual TGG editor in Sect. 12.4.1 and describe the generation of forward translation rules based on Sect. 7.4 in Sect. 12.4.2. An example for a conflict analysis of forward translation rules converted to AGG based on critical pairs is presented in Sect. 12.4.3, while Sect. 12.4.4 explains the automatic EMF model translation.

12.4.1 The Visual TGG Editor

Recall the main constructions and results of model transformations based on TGGs as given in Sect. 7.3. To demonstrate the features of our tool HENSHIN_TGG, we implement the well-known model transformation from class diagrams to database models (see Ex. 3.6). The triple type graph, underlying a model transformation, is defined by three different EMF models for the source, correspondence, and target components.

Example 12.7 (EMF models for CD2RDBM *model transformation).* Figure 12.34 depicts the three EMF models implementing the type graph *TG* of the triple graph grammar *TGG* for our *CD2RDBM* model transformation (see Ex. 3.6). The source

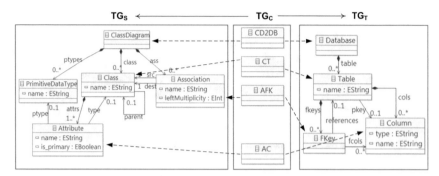

Fig. 12.34 EMF models comprising the triple type graph for the *CD2RDBM* model transformation

component TG_S defines the structure of class diagrams while in the target component the structure of relational database models is specified. Classes correspond to
tables, attributes to columns, and associations to foreign keys. Morphisms starting
at a correspondence part are indicated by dashed arrows and are modelled by intra-
model references in EMF.   △

The HENSHIN$_{\text{TGG}}$ editor uses EMF models as type graphs and EMF instance models conforming to the respective EMF models as typed (attributed) graphs.[22] The
three EMF models in Fig. 12.34 have been edited outside the visual TGG editor
using the graphical GMF editor for EMF, but any other EMF model editor or generator can be used as well. The morphisms are implemented as references between
the types of the three different EMF models. EMF models are imported into the visual TGG editor, which enables the use of previously produced EMF models. The
names of the three imported EMF models, *source, correspondence* and *target*, that
comprise the triple type graph are shown in the top compartment, *Imports*, of the
tree view ❶ in Fig. 12.35.

Once a triple type graph is available (i.e., the three EMF models have been imported), triple graphs typed over this type graph may be edited, e.g., for modifying
inputs and intermediate states when testing model transformations. The visual TGG
editor supports editing of triple graph nodes and edges by offering the available
types in the palette of the triple graph panel ❷. Only triple graphs conforming to
the triple type graph can be created. Moreover, only source triple graph elements
can be created and modified in the left-hand part of the editor, correspondence
graph elements in the center, and target graph elements in the right part. The separators between the different triple panels can be moved using the mouse. Morphisms
from correspondence to source and target elements are drawn as edges across the
separators. Figure 12.35 displays a sample triple graph, *OrderDetails*, containing a
complete source part (the class diagram) but incomplete corresponding target and
correspondence graphs.

---
[22] For more details on the formal correspondence of typed attributed graphs and EMF models, see
Table 12.1.

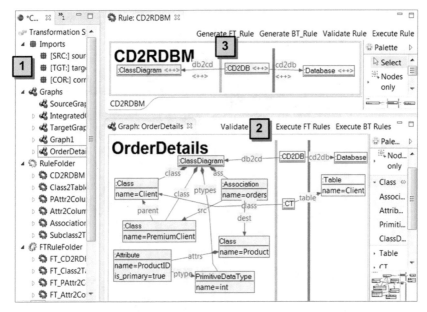

Fig. 12.35 Graphical user interface of the visual TGG editor

Recall that triple graphs can be generated by applying *triple rules* to a start triple graph *G* (see Def. 3.8). Triple rules synchronously build up the source, target and correspondence graph, i.e., they are nondeleting.

Example 12.8 (Triple rules). The triple rules shown in Fig. 12.36 are part of the rules of the grammar *TGG* for the model transformation *CD2RDBM* (see Ex. 3.9).

In Henshin_TGG, triple rules are drawn in short notation, i.e., left- and right-hand sides of a rule are depicted in one triple graph. Elements which are created by the rule are labeled by "++". The rule *CD2RDBM* (see ③ in Fig. 12.35) synchronously creates a class diagram together with the corresponding database. Analogously, the rule *Class2Table* creates a class with a name as input parameter together with the corresponding table in the relational database. A subclass is connected to the table of its superclass by the rule *Subclass2Table*. Attributes of a certain datatype are created together with their corresponding columns in the database component via the rule *Attr2Column*.   △

The visual Henshin_TGG editor for triple rules consists of three panel parts like the visual triple graph editor (see ② in Fig. 12.35). But in addition to the triple graph editor, the rule editor palette offers a "++" to mark elements as *created* (and to unmark marked elements if necessary). Note that Henshin_TGG checks triple rules for consistency at editing time, i.e., if a node is "++"-marked, all incident edges are marked automatically, as well.

Henshin_TGG supports negative application conditions for triple rules that forbid the presence of certain structures when applying a rule [EEHP09, GEH11]. A visual

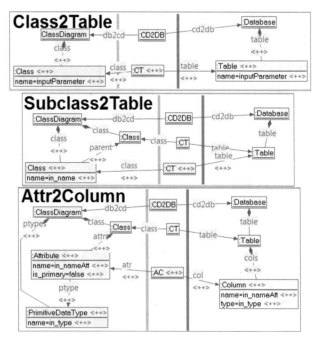

Fig. 12.36 Some rules for the model transformation *CD2RDBM* (HENSHIN_TGG screenshots)

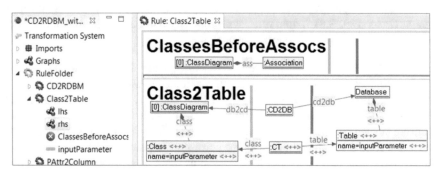

Fig. 12.37 Triple rule *C2T* with NAC *ClassesBeforeAssocs*

NAC editor can be opened via the tree view and consists of a three panel triple graph editor again. A rule may have several NACs; the one to be shown in the visual NAC editor has to be selected in the tree view. Figure 12.37 depicts the rule *Class2Table* with an additional NAC that forbids the synchronous creation of a class and a table if there are associations in the same class diagram.

The morphism from the rule to one of its NACs is indicated by equal numbers for mapped nodes (in Fig. 12.37, the *ClassDiagram* node is mapped to the NAC). Edges are mapped accordingly automatically. The rule palette entry *Mapping* supports the definition of a mapping from the triple rule to a NAC. Note that only unmarked elements (without "++") can be mapped to NAC elements, a consistency property which is also checked automatically by the editor.

A triple rule can be applied by clicking the button *Execute Rule* in the rule's tool bar (the upper right corner in Fig. 12.35), and selecting the graph the rule should be applied to. The result is shown in the view of the selected graph.

12.4.2 Generation of Forward Translation Rules

From each triple rule *tr*, so-called operational rules can be automatically derived [Sch94] for parsing a model of the source or target language (source and target rules) and for model transformations from source to target or backwards (forward and backward rules), as defined in Def. 7.12.

According to Sect. 7.4, the extension of forward rules to *forward translation rules* is based on additional Boolean attributes for all elements in the source component, called *translation attributes*, that control the translation process by keeping track of the elements which have been translated so far. This ensures that each element in the source graph is translated exactly once.

Let us recall the algorithm for constructing forward translation rules from triple rules from Def. 7.29: For each triple rule *tr*, initialise the forward translation rule $tr_{FT} = tr_F$ by the forward rule tr_F. Add an additional Boolean attribute *isTranslated* to each source element (node, edge or attribute) of tr_{FT}. In the left-hand side of tr_{FT}, for each source element, the value of the *isTranslated* attribute is set to **false** if the element is generated by the source rule tr_S of *tr*; otherwise it is set to **true**. In the right-hand side of tr_{FT}, the value of all *isTranslated* attributes is set to **true**. For all source elements in NACs, the attribute *isTranslated* is set to **true** as well.

Note that in contrast to forward translation rules, pure forward rules need additional control conditions to ensure correct executions, such as the source consistency condition defined in Def. 7.18.

In HENSHIN$_{\text{TGG}}$, forward translation rules are computed automatically. The translation attributes for nodes and edges and node attributes are kept separately as an external pointer structure in order to keep the source model unchanged. In the source graph editor panel of a forward translation rule, all elements that are still to be translated are marked by a "<tr>" tag.

Example 12.9 (Forward translation rule). Figure 12.38 shows the forward translation rule *FT_SubClass2Table* generated from the triple rule *SubClass2Table*. The *Class* node, its attribute and its incident edge are marked by a "<tr>" tag so as to be translated, since these model elements correspond to the model elements generated by the source rule of the triple rule *SubClass2Table*.   △

Forward translation rules can be edited in a restricted visual triple rule editor, which allows for a manual extension of additional NACs. All other rule editor operations are blocked because forward translation rules are generated automatically and should not be changed manually. Figure 12.39 shows the abstract syntax of the

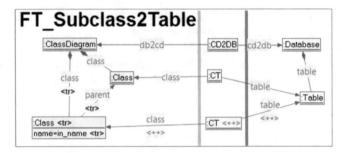

Fig. 12.38 Forward translation rule *FT_SubClass2Table*

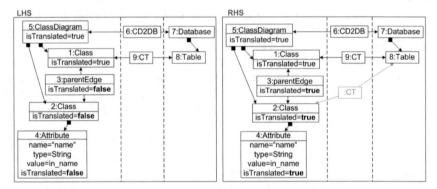

Fig. 12.39 Rule *FT_SubClass2Table* in abstract EMF Henshin syntax

forward translation rule *FT_SubClass2Table* from Fig. 12.38, as it is represented in EMF Henshin, where left-hand and right-hand sides of a rule are kept separately, with morphisms inbetween. We can see how the translation attributes of source elements are switched from **false** to **true**.

For matching, we internally keep two tables (hashmaps), "TranslatedNodes" and "TranslatedEdges", based on the IDs of the elements of an EMF instance model. These tables are constructed and updated dynamically during transformation execution. A match is valid if for each matched element we have one of the following cases:

- its translation attribute is **true** and its ID is present in the corresponding table of translated elements, or
- its translation attribute is **false** and its ID is not present in the corresponding table of translated elements.

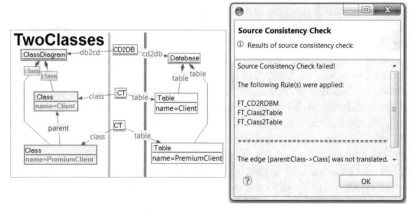

Fig. 12.40 Incomplete forward translation sequence: *parent* edge could not be translated

12.4.3 Conflict Analysis Based on AGG

According to Def. 7.33, a forward translation sequence $G_0 \overset{tr_{FT}^*}{\Longrightarrow} G_n$ is called *complete* if G_n is *completely translated*, i.e., all translation attributes of G_n are set to true. A model transformation based on forward translation rules with NACs (consisting of a source graph G_S, a target graph G_T, and a complete forward translation sequence $G_0 \overset{tr_{FT}^*}{\Longrightarrow} G_n$) is *terminating* if each forward translation rule changes at least one translation attribute from **false** to **true**; it is *correct* if each forward translation results in a triple graph that can be generated by triple rules, and it is *complete* if for each source graph there is a forward translation sequence that results in a triple graph that can be generated by triple rules.

However, not all terminating forward translation sequences are complete. A counterexample is a forward translation rule sequence applied to the triple graph *TwoClasses* consisting of a parent class named *Client* and a subclass named *PremiumClient* connected to class *Client* by a *parent* edge (see the source graph in Fig. 12.40).

The incomplete forward translation sequence is as follows: *FT_CD2DB* translates the ClassDiagram node to a Database node. In the next two steps, *FT_Class2Table* is applied to the class *PremiumClient* and to the class *Client* (in arbitrary order). The sequence is terminating (no forward translation rule can be applied anymore), but the result after applying this sequence is a triple graph where not all translation attributes are set to **true**, i.e., not all source model elements have been translated: the *parent* edge could not be translated (the result is a "misleading graph" in the sense of Def. 8.14). In HENSHIN_TGG, elements that could not be translated are highlighted in red and reported as error message in a separate window, showing the (partial) translation result (see Fig. 12.40). This allows the user to reason about possible conflicts between rule applications. The reason why the *parent* edge was not translated by the given forward translation sequence is a conflict between the rule *FT_Class2Table* (applied to the class *PremiumClient*) and *FT_SubClass2Table*, which could not be applied to the class *PremiumClient* after the application of the rule *FT_Class2Table*.

In order to ensure completeness in the general case, the execution of model transformations may require backtracking (not implemented in HENSHIN$_{TGG}$). However, as shown in Theorem 8.29, we get functional behaviour of the forward translation (i.e., the translation yields complete and unique results) without backtracking if the significant critical pairs between forward translation rules *extended by filter NACs* (see Def. 8.16) are strictly confluent and the system is terminating.

In order to make use of Theorem 8.29, we have to 1) perform a critical pair analysis on our set of forward translation rules to find rules in need of filter NACs, 2) extend these rules by adding filter NACs according the procedure in Fact 8.19, 3) repeat steps 1) and 2) on the extended rule set until no more filter NACs are necessary.

HENSHIN$_{TGG}$ implements a converter from triple rules in EMF HENSHIN to AGG, which provides a critical pair analysis engine (see Sect. 12.1.4). Recall that a critical pair (see Def. 5.40) is a conflict between two rules in minimal context, and it is significant if the overlapping graph can be embedded in a possible intermediate state of a model transformation sequence. In particular, it is not significant if a fragment in the source component cannot be embedded into a valid source model due to language constraints.

Figure 12.41 shows the (only) critical pair between the rules *FT_Class2Table* and *FT_SubClass2Table* as computed by the AGG critical pair analyser. In the view at the bottom, the critical overlapping graph of both rules' left-hand sides is shown, and it is indicated that we have a *change–use-attr* conflict, since both rules want to access and change the *isTranslated* attribute of the subclass. Note that this attribute is not plainly visible like other attributes since it is added internally when generating FT rules (see Fig. 12.38 and Fig. 12.39). Nevertheless, AGG treats *isTranslated* attributes as normal attributes and hence finds a change–use-attribute conflict for this attribute that both rules intend to change from *false* to *true*.

12.4.4 Performing Model Transformation in HENSHIN$_{TGG}$

The conflict shown in Fig. 12.41 is a *misleading graph* according to Def. 8.14; hence we add a filter NAC to the rule *FT_Class2Table* that forbids its application to classes which have a parent class. Figure 12.42 shows the rule *FT_Class2Table*, now extended by a filter NAC.

According to Fact 8.18, filter NACs can in principle be generated automatically. Note, however, that the generation of filter NACs is not yet supported by HENSHIN$_{TGG}$. A new run of the critical pair analysis does not find any more critical pairs. Hence, according to Theorem 8.29, our model transformation system has functional behaviour, since it is confluent and terminating (due to the "<tr>" tags, all elements are translated only once).

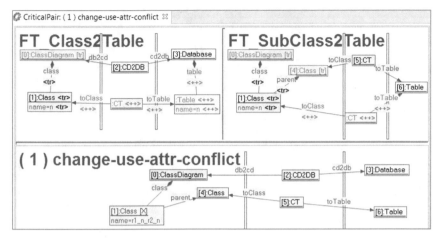

Fig. 12.41 Critical pair between rules *FT_Class2Table* and *FT_SubClass2Table* computed by AGG

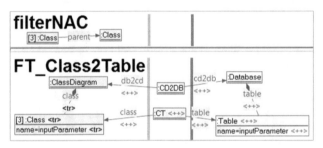

Fig. 12.42 Rule *FT_Class2Table* with filter NAC (top)

HENSHIN_TGG supports the automatic forward translation of a given source model by offering a button *Execute Forward Translation* in the toolbar of the EMF source model to be translated. On pressing the button the forward translation rules are executed in arbitrary order; confluence of the transformation system guarantees a unique result. The resulting target triple graph is shown in the same window as the source model since the translation is performed *in-place*. Figure 12.43 shows the target triple which is the result of translating the source model "TwoClasses". In addition, the sequence of applied forward translation rules is shown to the modeller in the message window. For debugging purposes, also single forward translation rule applications can be executed, analogously as for triple rules.

Summarising, HENSHIN_TGG extends the EMF HENSHIN engine by features based on triple graph grammars (TGGs) used for bidirectional model transformation, based on the formal definitions for TGGs from Chap. 7 (see also [Sch94, EEHP09, HEGO10]) and supports conflict analysis via the converter to AGG. The explicit marking of edges overcomes the restriction in [GHL12] that rules are required to create at least one node. Recently, HENSHIN_TGG has been extended to include also

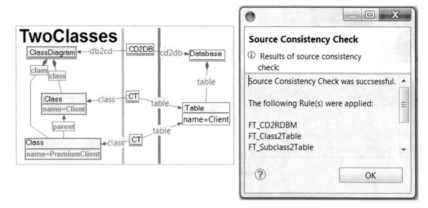

Fig. 12.43 Result of the forward translation of the source model *TwoClasses*

the generation of rules for backward translation (BT), for consistency creation (CC), and for model integration (IT). Furthermore, HENSHIN_TGG supports the execution and simulation of model synchronisation and integration operations on TGGs: forward model transformation (=FT=>), backward model transformation (=BT=>), model integration (=IT=), consistency checking (=CC=), state-based forward and backward propagation (=S-fPpg=> , <=S-bPpg=), and delta-based forward and backward propagation (=D-fPpg=>, <=D-bPpg=). HENSHIN_TGG allows the user to manually use the analysis and optimisations techniques presented in [HEGO10] in order to improve efficiency. The automated generation of filter NACs [HEGO10] can be implemented as a direct extension and is future work.

Recently, HENSHIN_TGG has been used by the Luxembourg-based satellite operator SES for developing the open-source software SPELL (Satellite Procedure Execution Language and Library), a standardised satellite control language [HGN+14]. The challenge was to convert existing control procedures from proprietary programming languages to SPELL. Using HENSHIN_TGG as model transformation engine to automate this process guaranteed a high-quality translation through automatic consistency testing.

HENSHIN_TGG is available at GitHub (see http://de-tu-berlin-tfs.github. io/Henshin-Editor/ and the WIKI manual of HENSHIN_TGG https://github. com/de-tu-berlin-tfs/Henshin-Editor/wiki for more information).

12.5 Related Tools

Tools Related to AGG 2.0

AGG is one of the standard graph transformation tools implementing the algebraic approach. Other graph transformation tools, such as Fujaba [Fuj07], Via-

Tra [VIA14], VMTS [VMT14], GrGen [GrG06], and Groove [Gro08], implement different kinds of graph transformation approaches. Some kinds of rule application control structures are offered by all of these tools, e.g., Fujaba uses story diagrams, a kind of activity diagram. Groove also supports nested application conditions as well as universal quantification using amalgamation.

Concerning the verification of graph transformation systems, VIATRA and Groove concentrate on some kind of model checking, while AGG is the only tool that consequently implements the theoretical results available for algebraic graph transformation. These results are mainly concerned with conflict and dependency detection of rules and static applicability checks for rule sequences.

Tools related to ACTIGRA

Our approach complements existing approaches that give a denotational semantics to activity diagrams by formal models. This semantics is used for validation purposes thereafter. For example, Eshuis [EW04] proposes a denotational semantics for a restricted class of activity models by means of labeled transition systems. Model checking is used to check properties. Störrle [Sto04] defines a denotational semantics for the control flow of UML 2.0 activity models including procedure calls by means of Petri nets. The standard Petri net theory provides an analysis of properties like reachability or deadlock-freeness. Both works stick to simple activities not further refined. In [EGSW07], business process models and web services are equipped with a combined graph transformation semantics and consistency can be validated by the model checker GROOVE. In contrast, we take integrated behaviour models and check for potential conflict and causality inconsistencies between activity-specifying rules directly. Thus, our technique is not a "pushbutton" technique which checks a temporal formula specifying a desired property, but offers additional views on activity models where users can conveniently investigate intended and unintended conflicts and causalities between activities. Conflicts and causalities are not just reported as such but reasons for consistencies and inconsistencies can also be investigated in depth.

Fujaba [FNTZ98], VMTS[23] and GReAT[24] are graph transformation tools for specifying and applying graph transformation rules along the control flow specified by activity models. However, controlled rule applications are not further validated with regard to conflict and causality inconsistencies within these tools. Conflicts and causalities of pairs of rule-specified activities have been considered in various application contexts such as use case integration [HHT02], feature modelling [JWEG07], model inconsistency detection [MSD06], and aspect-oriented modelling [MMT09]. Although sometimes embedded in explicit control flow, it has not been taken into account for inconsistency analysis. In the ACTIGRA approach, we analyse potential

---

[23] Visual Modeling and Transformation System: http://vmts.aut.bme.hu/

[24] Graph Rewriting and Transformation: http://www.isis.vanderbilt.edu/tools/great

conflict and causality inconsistencies between rule-specified activities w.r.t. the control flow to specify their execution order.

Tools related to EMF HENSHIN

There are a number of model transformation engines which can modify models in EMF format, such as ATL [JK05], EWL [KPPR07], Tefkat [LS05b], VIA-TRA2 [VB07], MOMENT [Bor07]. For ATL, a formal semantics based on Maude has been introduced recently [TV10]. Formal semantics defined in Maude for MOMENT and for ATL might be exploited for analyzing EMF model transformations. None of these tool environments supports visual editing of control structures.

Graph transformation tools like PROGRES [SWZ99], AGG [AGG14], Fujaba [FNTZ00] and MoTMoT [FOT10] feature visual editors which also support the definition of control structures, e.g., by story diagrams in Fujaba, which were extended by implicit control in [MV08]. The tool GrGen.NET [GK08] also supports the arbitrary nesting of application conditions but is based on a textual specification language. MoTMoT (Model-driven, Template-based, Model Transformer) is a compiler from visual model transformations to repository manipulation code. The compiler takes models conforming to a UML profile for Story Driven Modelling as input and outputs Java Metadata Interface (JMI) code. Control structures are expressed by activity diagrams. Since the MoTMoT code generator is built using AndroMDA, adding support for other repository platforms (like EMF) is possible in principle and consists of adding a new set of code templates.

To the best of our knowledge, none of the existing EMF model transformation approaches (whether based on graph transformation or not) support confluence and termination analysis of EMF model transformation rules yet. Here, the EMF HENSHIN approach and tool environment serves as a bridge to make well-established tool features and formal techniques for graph transformation available for model-driven development based on EMF.

Tools related to HENSHIN$_{TGG}$

General model transformation tools such as ATL [JABK08] and MOMENT2-MT [MOM12] are usually used to perform in-place model transformations and do not restrict the structure of transformation rules. Thus, they do not ensure TGG-specific properties like preservation of source models [Sch94] and syntactical correctness and completeness [EEHP09]. Moreover, the forward and backward transformations are manually specified and not generated from a single specification. While ATL and MOMENT2-MT use textual specification techniques, graph transformation tools like EMF HENSHIN (in-place) [ABJ+10] and Fujaba [Fuj07] offer the visual specification of transformation rules, i.e., a form of visual programming interface. The benchmarking framework in [ACG+14] provides means to compare tools for bidirectional model transformations on an abstract level.

In addition to HENSHIN$_{TGG}$, further TGG tools based on EMF are available [LAS$^+$14b, HLG$^+$13]. The TGG interpreter [GK10] provides a feature to define OCL expressions as rule conditions, while formal application conditions cannot be specified. However, the formal results concerning correctness and completeness [EEHP09] are not available for systems with OCL conditions. The TGG tools MOTE (model transformation engine) [GW09] and eMoflon [ALPS11] perform a compilation to the Fujaba tool suite [BGH$^+$05, Fuj07] for the execution of model transformations. While eMoflon supports the specification of TGGs with negative application conditions (NACs), this is not the case for MOTE. MOTE offers certain optimisation strategies concerning efficiency. Since correctness cannot be ensured for all optimisations, the tool executes dynamic runtime checks to validate that a model transformation sequence was executed correctly [GHL12]. Moreover, MOTE uses a relaxed notion of correspondences for triple graphs, where correspondence nodes may link an arbitrary number of source and target nodes [GHL12].

In order to improve efficiency of TGG tools, suitable static and dynamic conditions have been studied that allow us to completely avoid backtracking. Klar et al. [KLKS10] use a restricted class of TGGs for which they describe explicit dynamic conditions based on pre-checking contextual edges when translating a node. Lauder et al. [LAVS12] leverage these restrictions on TGGs and introduce the notion of precedence TGGs, where rules are required to form a partial order concerning the execution. However, these conditions are not checked statically. Giese et al. [GHL12] present efficiency conditions for a restricted class of TGGs, where each forward rule has to translate at least one source node and may not be in conflict with another rule via a critical pair. The first condition excludes examples where the translation of a single edge or attribute is handled separately by one rule [HEEO12], and the second condition excludes the well-studied case study on the object relational mapping [EEHP09]. The tool was extended by a prototypical export [GHL12] of so-called bookkeeping rules to AGG for conflict analysis, but it does not provide reimport and evaluation. An overview of further possible improvements concerning caching and reuse of existing structures for incremental model synchronisation is studied in [LAS14a].

Appendix A
Basic Notions of Category Theory

In this appendix, we give a short summary of the categorical terms used throughout this book based on [EEPT06]. We introduce categories, show how to construct them, and present some basic constructions such as pushouts and pullbacks. In addition, we give some specific categorical results which are needed for the main part of the book. For a more detailed introduction to category theory, see [EM85, EM90, AHS90, EMC$^+$01].

A.1 Categories

In general, a category is a mathematical structure that has objects and morphisms, with a composition operation on the morphisms and an identity morphism for each object.

Definition A.1 (Category). A *category* $\mathbf{C} = (Ob_C, Mor_C, \circ, id)$ is defined by

- a class Ob_C of *objects*;
- for each pair of objects $A, B \in Ob_C$, a set $Mor_C(A, B)$ of *morphisms*;
- for all objects $A, B, C \in Ob_C$, a *composition* operation $\circ_{(A,B,C)} : Mor_C(B, C) \times Mor_C(A, B) \to Mor_C(A, C)$; and
- for each object $A \in Ob_C$, an *identity* morphism $id_A \in Mor_C(A, A)$,

such that the following conditions hold:

1. *Associativity.* For all objects $A, B, C, D \in Ob_C$ and morphisms $f : A \to B$, $g : B \to C$ and $h : C \to D$, it holds that $(h \circ g) \circ f = h \circ (g \circ f)$.
2. *Identity.* For all objects $A, B \in Ob_C$ and morphisms $f : A \to B$, it holds that $f \circ id_A = f$ and $id_B \circ f = f$. $\triangle$

Remark A.2. Instead of $f \in Mor_C(A, B)$, we write $f : A \to B$ and leave out the index for the composition operation, since it is clear which one to use. For such a morphism f, A is called its domain and B its codomain. $\triangle$

Example A.3. 1. The basic example of a category is the category **Sets** , with the object class of all sets and with all functions $f : A \rightarrow B$ as morphisms. The composition is defined for $f : A \rightarrow B$ and $g : B \rightarrow C$ by $(g \circ f)(x) = g(f(x))$ for all $x \in A$, and the identity is the identical mapping $id_A : A \rightarrow A : x \mapsto x$.

2. The class of all graphs as objects and the class of all graph morphisms (as defined in Def. 2.1) form the category **Graphs**; the composition is given componentwise and the identities are the pairwise identities on nodes and edges.

3. Typed graphs and typed graph morphisms (see Def. 2.2) form the category **Graphs$_{TG}$**.   △

A.2 Construction of Categories, and Duality

There are various ways to construct new categories from given ones. The first way that we describe here is the Cartesian product of two categories, which is defined by the Cartesian products of the class of objects and the sets of morphisms with componentwise composition and identities.

Definition A.4 (Product category). Given two categories **C** and **D**, the *product category* **C** $\times$ **D** is defined by

- $Ob_{C \times D} = Ob_C \times Ob_D$;
- $Mor_{C \times D}((A, A'), (B, B')) = Mor_C(A, B) \times Mor_D(A', B')$;
- for morphisms $f : A \rightarrow B$, $g : B \rightarrow C \in Mor_C$ and $f' : A' \rightarrow B'$, $g' : B' \rightarrow C' \in Mor_D$, we define $(g, g') \circ (f, f') = (g \circ f, g' \circ f')$;
- $id_{(A,A')} = (id_A, id_{A'})$.   △

Another construction is that of a slice or a coslice category. Here the objects are morphisms of a category **C**, to or from a distinguished object X, respectively. The morphisms are morphisms in **C** that connect the object morphisms so as to lead to commutative diagrams.

Definition A.5 (Slice category). Given a category **C** and an object $X \in Ob_C$, the *slice category* **C** $\setminus X$ is defined as follows:

- $Ob_{C \setminus X} = \{f : A \rightarrow X | A \in Ob_C, f \in Mor_C(A, X)\}$;
- $Mor_{C \setminus X}(f : A \rightarrow X, g : B \rightarrow X) = \{m : A \rightarrow B| g \circ m = f\}$;
- for morphisms $m \in Mor_{C \setminus X}(f : A \rightarrow X, g : B \rightarrow X)$ and $n \in Mor_{C \setminus X}(g : B \rightarrow X, h : C \rightarrow X)$, we have $n \circ m$ as defined in **C** for $m : A \rightarrow B$ and $n : B \rightarrow C$;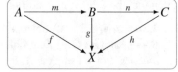
- $id_{f:A \rightarrow X} = id_A \in Mor_C$.   △

Example A.6. Given a type graph TG, the category **Graphs$_{TG}$** can be considered as the slice category **Graphs**$\setminus TG$. Each typed graph is represented in this slice category by its typing morphism, and the typed graph morphisms are exactly the morphisms in the slice category.   △

Definition A.7 (Coslice category). Given a category $\mathbf{C}$ and an object $X \in Ob_C$, then the *coslice category* $X \backslash \mathbf{C}$ is defined as follows:

- $Ob_{X \backslash C} = \{f : X \to A \mid A \in Ob_C, f \in Mor_C(X, A)\}$;
- $Mor_{X \backslash C}(f : X \to A, g : X \to B) = \{m : A \to B \mid g = m \circ f\}$;
- for morphisms $m \in Mor_{X \backslash C}(f : X \to A, g : X \to B)$ and $n \in Mor_{X \backslash C}(g : X \to B, h : X \to C)$, we have $n \circ m$ as defined in $\mathbf{C}$ for $m : A \to B$ and $n : B \to C$;
- $id_{f:X \to A} = id_A \in Mor_C$. △

As the last construction in this section, we introduce the dual category. For the dual category, we use the objects of a given category, but reverse all arrows, i.e., morphisms.

Definition A.8 (Dual category). Given a category $\mathbf{C}$, the *dual category* $\mathbf{C}^{op}$ is defined by

- $OB_{C^{op}} = Ob_C$;
- $Mor_{C^{op}}(A, B) = Mor_C(B, A)$;
- $f \circ^{C^{op}} g = g \circ^C f$ for all $f : A \to B, g : B \to C$;
- $id_A^{C^{op}} = id_A^C$ for all $A \in Ob_{C^{op}}$. △

The *duality principle* asserts that for each construction (statement) there is a dual construction. If a statement holds in all categories, then the dual statement holds in all categories, too. Some examples of dual constructions are monomorphisms and epimorphisms, pushouts and pullbacks, and initial and final objects, which will be described in the following sections.

A.3 Monomorphisms, Epimorphisms, and Isomorphisms

In this section, we consider a category $\mathbf{C}$ and analyse some important types of morphisms, namely monomorphisms, epimorphisms, and isomorphisms.

Intuitively speaking, two objects are isomorphic if they have the same structure. Morphisms that preserve this structure are called isomorphisms.

Definition A.9 (Isomorphism). A morphism $i : A \to B$ is called an *isomorphism* if there exists a morphism $i^{-1} : B \to A$ such that $i \circ i^{-1} = id_B$ and $i^{-1} \circ i = id_A$.

Two objects A and B are isomorphic, written $A \cong B$, if there is an isomorphism $i : A \to B$. △

Remark A.10. If i is an isomorphism, then i is both a monomorphism and an epimorphism. For every isomorphism i, the inverse morphism i^{-1} is unique. △

Example A.11. • In **Sets**, **Graphs**, and **Graphs$_{TG}$**, the isomorphisms are exactly those morphisms that are (componentwise) injective and surjective.

- In product, slice, and coslice categories, the isomorphisms are exactly those morphisms that are (componentwise) isomorphisms in the underlying category. △

Definition A.12 (Monomorphism and epimorphism). Given a category **C**, a morphism $m : B \to C$ is called a *monomorphism* if, for all morphisms $f, g : A \to B \in Mor_C$, it holds that $m \circ f = m \circ g \Rightarrow f = g$.

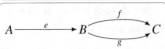

A morphism $e : A \to B \in Mor_C$ is called an *epimorphism* if, for all morphisms $f, g : B \to C \in Mor_C$, it holds that $f \circ e = g \circ e \Rightarrow f = g$. △

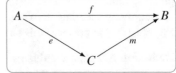

Remark A.13. Monomorphisms and epimorphisms are dual notions, i.e., a monomorphism in a category **C** is an epimorphism in the dual category $\mathbf{C^{op}}$, and vice versa. △

Fact A.14 (Monomorphisms and epimorphisms).

- *In **Sets**, the monomorphisms are all injective mappings, and the epimorphisms are all surjective mappings.*
- *In **Graphs** and **Graphs$_{TG}$**, the monomorphisms and epimorphisms are exactly those morphisms that are injective and surjective, respectively.*
- *In a slice category, the monomorphisms are exactly the monomorphisms of the underlying category. The epimorphisms of the underlying category are epimorphisms in the slice category, but not necessarily vice versa.*
- *In a coslice category, the epimorphisms are exactly the epimorphisms of the underlying category. The monomorphisms of the underlying category are monomorphisms in the slice category, but not necessarily vice versa.* △

In general, a factorisation of a morphism decomposes it into morphisms with special properties. In an epi–mono factorisation, these morphisms are an epimorphism and a monomorphism.

Definition A.15 (Epi–mono and (weak) $\mathcal{E}$–$\mathcal{M}$ factorisations). Given a category **C** and morphisms $f : A \to B$, $e : A \to C$, and $m : C \to B$ with $m \circ e = f$, if e is an epimorphism and m is a monomorphism then e and m are called an *epi–mono factorisation* of f:

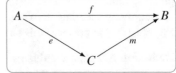

If for every morphism f we can find such morphisms e and m, with $f = m \circ e$, and this decomposition is unique up to isomorphism, then the category **C** is said to have an epi–mono factorisation.

C has an $\mathcal{E}$–$\mathcal{M}$ *factorisation* for given morphism classes $\mathcal{E}$ and $\mathcal{M}$ if for each f there is a decomposition, unique up to isomorphism, $f = m \circ e$ with $e \in \mathcal{E}$ and $m \in \mathcal{M}$. Usually $\mathcal{E}$ is a subclass of epimorphisms and $\mathcal{M}$ is a subclass of monomorphisms.

If we require only $f = m \circ e$ with $e \in \mathcal{E}$ and $m \in \mathcal{M}$, but not necessarily uniqueness up to isomorphism, we have a *weak $\mathcal{E}$–$\mathcal{M}$ factorisation*. △

The categories **Sets**, **Graphs**, and **Graphs**$_{TG}$, have epi–mono factorisations.

Definition A.16 (Jointly epimorphic). A morphism pair (e_1, e_2) with $e_i : A_i \rightarrow B$ $(i = 1, 2)$ is called *jointly epimorphic* if, for all $g, h : B \rightarrow C$ with $g \circ e_i = h \circ e_i$ for $i = 1, 2$, we have $g = h$. △

In the categories **Sets**, **Graphs**, and **Graphs**$_{TG}$, "jointly epimorphic" means "jointly surjective".

A.4 Pushouts and Pullbacks

Intuitively, a pushout is an object that emerges from gluing two objects along a common subobject. In addition, we introduce the dual concept of a pullback and the construction of both in specific categories.

Definition A.17 (Pushout). Given morphisms $f : A \rightarrow B$ and $g : A \rightarrow C \in Mor_C$, a *pushout* (D, f', g') over f and g is defined by

- a pushout object D and
- morphisms $f' : C \rightarrow D$ and $g' : B \rightarrow D$ with $f' \circ g = g' \circ f$,

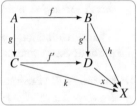

such that the following universal property is fulfilled: for all objects X with morphisms $h : B \rightarrow X$ and $k : C \rightarrow X$ with $k \circ g = h \circ f$, there is a unique morphism $x : D \rightarrow X$ such that $x \circ g' = h$ and $x \circ f' = k$: △

Remark A.18. The pushout object D is unique up to isomorphism. This means that if (X, k, h) is also a pushout over f and g, then $x : D \xrightarrow{\sim} X$ is an isomorphism with $x \circ g' = h$ and $x \circ f' = k$. Vice versa, if (D, f', g') is a pushout over f and g and $x : D \xrightarrow{\sim} X$ is an isomorphism, then (X, k, h) is also a pushout over f and g, where $k = x \circ f'$ and $h = x \circ g'$. Uniqueness up to isomorphism follows directly from the corresponding universal properties (see Lem. A.21). △

Fact A.19 (Pushout constructions).

1. *In **Sets**, a pushout over morphisms $f : A \rightarrow B$ and $g : A \rightarrow C$ can be constructed as follows. Let*

$$\sim_{f,g} = t(\{(a_1, a_2) \in A \times A |\ f(a_1) = f(a_2) \vee g(a_1) = g(a_2)\})$$

be the transitive closure of $Kern(f)$ and $Kern(g)$; $\sim_{f,g}$ is an equivalence relation. We define the object D and the morphisms as:

- $D = A|_{\sim_{f,g}} \dot\cup B\backslash f(A) \dot\cup C\backslash g(A)$,
- $f' : C \rightarrow D : x \mapsto \begin{cases} [a] & : & \exists\, a \in A : g(a) = x \\ x & : & otherwise \end{cases}$,

- $g' : B \to D : x \mapsto \begin{cases} [a] & : & \exists\, a \in A : f(a) = x \\ x & : & otherwise \end{cases}$.

2. *In* **Graphs** *and* **Graphs**$_{TG}$*, pushouts can be constructed componentwise in* **Sets***.*
3. *If the categories* **C** *and* **D** *have pushouts, the pushouts in the product category can be constructed componentwise.*
4. *If the category* **C** *has pushouts, the pushouts in the slice category* **C** \X *can be constructed over the pushouts in* **C***. Given objects* $f : A \to X$, $g : B \to X$, *and* $h : C \to X$, *and morphisms* m *and* n *in* **C** \X *as in* (1)*, it holds that* $g \circ m = f = h \circ n$ *by the definition of morphisms in* **C** \X*.*

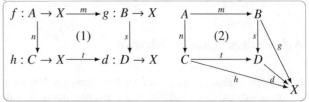

We construct the pushout (2) *in* **C** *over* $C \xleftarrow{n} A \xrightarrow{m} B$*. From* (2)*, we obtain the induced morphism* $d : D \to X$ *as the pushout object, and morphisms* s *and* t *with* $d \circ s = g$ *and* $d \circ t = h$*, leading to the pushout* (1) *in* **C** \X*.*
This construction works analogously for the coslice category X**C***.   △*

In various situations, we need a reverse construction of a pushout. This is called the pushout complement.

Definition A.20 (Pushout complement). Given morphisms $f : A \to B$ and $n : B \to D$, $A \xrightarrow{g} C \xrightarrow{m} D$ is the *pushout complement* of f and n if (D, m, n) is the pushout over f and g.   △

Pushout squares can be decomposed if the first square is a pushout, and can be composed, preserving their pushout properties.

Lemma A.21 (Pushout composition and decomposition). *Given the following commutative diagram, the following hold:*

- Pushout composition.
  If (1) *and* (2) *are pushouts,*
  then (1) + (2) *is also a pushout.*
- Pushout decomposition. *If* (1) *and* (1) + (2) *are pushouts, then* (2) *is also a pushout.*   △

The dual construction of a pushout is a pullback. Pullbacks can be seen as a generalised intersection of objects over a common object.

Definition A.22 (Pullback). Given morphisms $f : C \to D$ and $g : B \to D$, a *pullback* (A, f', g') over f and g is defined by

- a pullback object A and
- morphisms $f' : A \rightarrow B$ and $g' : A \rightarrow C$
  with $g \circ f' = f \circ g'$,

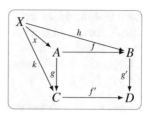

such that the following universal property is fulfilled: for all objects X with morphisms $h : X \rightarrow B$ and $k : X \rightarrow C$, with $f \circ k = g \circ h$, there is a unique morphism $x : X \rightarrow A$ such that $f' \circ x = h$ and $g' \circ x = k$:   △

Fact A.23 (Pullback constructions).

1. *In* **Sets***, the pullback $C \xleftarrow{\pi_g} A \xrightarrow{\pi_f} B$ over morphisms $f : C \rightarrow D$ and $g : B \rightarrow D$ is constructed by $A = \bigcup_{d \in D} f^{-1}(d) \times g^{-1}(d)$ with morphisms $f' : A \rightarrow B :$ $(x, y) \mapsto y$ and $g' : A \rightarrow C : (x, y) \mapsto x$.*
2. *In* **Graphs** *and* **Graphs$_{TG}$***, pullbacks can be constructed componentwise in* **Sets***.*
3. *The category* **PTNets** *has pullbacks, but they cannot be constructed componentwise (see [EEPT06]).*
4. *In a product, slice, or coslice category, the construction of pullbacks is dual to the construction of pushouts if the underlying categories have pullbacks.*   △

Pullback squares can be decomposed if the last square is a pushout, and can be composed, preserving their pullback properties.

Lemma A.24 (Pullback composition and decomposition). *Given the following commutative diagram, the following hold:*

- Pullback composition.
  If (1) and (2) are pullbacks, then (1) + (2) is also a pullback.

$$
\begin{array}{ccccc}
A & \longrightarrow & B & \longrightarrow & E \\
\downarrow & (1) & \downarrow & (2) & \downarrow \\
C & \longrightarrow & D & \longrightarrow & F
\end{array}
$$

- Pullback decomposition. *If (2) and (1) + (2) are pullbacks, then (1) is also a pullback.*   △

A.5 Binary Coproducts and Initial Objects

Binary coproducts can be seen as a generalisation of the disjoint union of sets and graphs in a categorical framework. Analogously, initial objects are the categorical representation of the empty set and the empty graph. Note, however, that the construction of binary coproducts and initial objects of algebras is much more difficult.

Definition A.25 (Binary coproduct). Given two objects $A, B \in Ob_C$, the *binary coproduct* $(A + B, i_A, i_B)$ is given by

- a coproduct object $A + B$ and
- morphisms $i_A : A \rightarrow A + B$ and $i_B : B \rightarrow A + B$,

such that the following universal property is fulfilled: for all objects X with morphisms f :
$A \to X$ and $g : B \to X$, there is a morphism
$[f, g] : A + B \to X$ such that $[f, g] \circ i_A = f$ and
$[f, g] \circ i_B = g$:   △

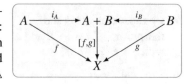

Remark A.26. Given two morphisms $f : A \to A'$ and $g : B \to B'$, there is a unique
coproduct morphism $f + g : A + B \to A' + B'$,
induced by the binary coproduct $A + B$ and the
morphisms $i_{A'} \circ f$ and $i_{B'} \circ g$:   △

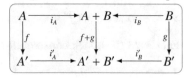

Example A.27. • In **Sets**, the coproduct object $A + B$ is the disjoint union $A \cup B$ of
  A and B, and i_A and i_B are inclusions. For $A \cap B = \varnothing$, we use the representation
  $A \cup B = A \cup B$, and for $A \cap B \neq \varnothing$, we use $A \cup B = A \times \{1\} \cup B \times \{2\}$.
• In **Graphs** and **Graphs$_{TG}$**, the coproduct can be constructed componentwise in
  Sets.
• In a product or slice category, coproducts can be constructed componentwise if
  the underlying categories have coproducts.
• In a coslice category, the coproduct of objects $f : X \to A$ and $g : X \to B$ is
  constructed as the pushout of f and g in the underlying category.   △

Definition A.28 (Initial object). In a category **C**, an object I is called *initial* if, for
each object A, there exists a unique morphism $i_A : I \to A$.   △

Example A.29. • In **Sets**, the initial object is the empty set.
• In **Graphs** and **Graphs$_{TG}$**, the initial object is the empty graph.
• In a product category $A \times B$, the initial object is the tuple (I_1, I_2), where I_1, I_2 are
  the initial objects in A and B (if they exist).
• If **C** has an initial object I, the initial object in a slice category $\mathbf{C} \backslash X$ is the unique
  morphism $I \to X$.
• In general, a coslice category has no initial object.   △

Remark A.30. The dual concept of an initial object is that of a final object, i.e., an
object Z such that there exists a unique morphism $z_A : A \to Z$ for each object A.
Each set Z with $card(Z) = 1$ is final in **Sets**.
   Initial objects are unique up to isomorphism.   △

A.6 Functors, Functor Categories, and Comma Categories

Functors are mappings between different categories which are compatible with composition and the identities. Together with natural transformations, this leads to the
concept of functor categories. Another interesting construction for building new categories is that of comma categories.

Definition A.31 (Functor). Given two categories **C** and **D**, a *functor* $F : \mathbf{C} \to \mathbf{D}$ is given by $F = (F_{Ob}, F_{Mor})$, with

- a mapping $F_{Ob} : Ob_C \to Ob_D$ and
- a mapping $F_{Mor(A,B)} : Mor_C(A, B) \to Mor_D(F_{Ob}(A), F_{Ob}(B))$ of the morphisms for each pair of objects $A, B \in Ob_C$,

such that the following apply:

1. For all morphisms $f : A \to B$ and $g : B \to C \in Mor_C$, it holds that $F(g \circ f) = F(g) \circ F(f)$.
2. For all objects $A \in Ob_C$, it holds that $F(id_A) = id_{F(A)}$.   △

Remark A.32. For simplicity, we have left out the indices and have written $F(A)$ and $F(f)$ for both objects and morphisms.   △

To compare functors, natural transformations are used. Functors and natural transformations form a category, called a functor category.

Definition A.33 (Natural transformation). Given two categories **C** and **D** and functors $F, G : \mathbf{C} \to \mathbf{D}$, a *natural transformation* α : $F \Rightarrow G$ is a family of morphisms $\alpha = (\alpha_A)_{A \in Ob_C}$ with $\alpha_A : F(A) \to G(A) \in Mor_D$, such that, for all morphisms $f : A \to B \in Mor_C$, it holds that $\alpha_B \circ F(f) = G(f) \circ \alpha_A$:   △

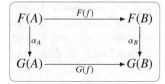

Definition A.34 (Functor category). Given two categories **C** and **D**, the *functor category* $[\mathbf{C}, \mathbf{D}]$ is defined by the class of all functors $F : \mathbf{C} \to \mathbf{D}$ as the objects, and by natural transformations as the morphisms. The composition of the natural transformations $\alpha : F \Rightarrow G$ and $\beta : G \Rightarrow H$ is the componentwise composition in **D**, which means that $\beta \circ \alpha = (\beta_A \circ \alpha_A)_{A \in Ob_C}$, and the identities are given by the identical natural transformations defined componentwise over the identities $id_{F(A)} \in$ **D**.   △

Fact A.35 (Constructions in functor categories).

- *In a functor category $[\mathbf{C}, \mathbf{D}]$, natural transformations are monomorphisms, epimorphisms, and isomorphisms if they are componentwise monomorphisms, epimorphisms, and isomorphisms, respectively, in* **D**.
- *If the category* **D** *has pushouts, then pushouts can be constructed "pointwise" in a functor category $[\mathbf{C}, \mathbf{D}]$.*
- *The construction of pullbacks is dual to the construction of pushouts if the underlying category* **D** *has pullbacks.*   △

In the following, we define comma categories and show under what conditions pushouts and pullbacks can be constructed.

Definition A.36 (Comma category). Given two functors $F : \mathbf{A} \to \mathbf{C}$ and $G :$ $\mathbf{B} \to \mathbf{C}$ and an index set $\mathcal{I}$, the comma category **ComCat(F, G; $\mathcal{I}$)** is defined by the class of all triples (A, B, op), with $A \in Ob_A$, $B \in Ob_B$, and $op = [op_i]_{i \in I}$, where $op_i \in Mor_C(F(A), G(B))$, as objects; a morphism $f : (A, B, op) \to (A', B', op')$ in **ComCat**$(F, G; I)$ is a pair $f = (f_A : A \to A', f_B : B \to B')$ of morphisms in **A** and **B** such that $G(f_B) \circ op_i = op'_i \circ F(f_A)$ for all $i \in I$.

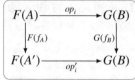

The composition of morphisms in **ComCat**$(F, G; I)$ is defined componentwise, and identities are pairs of identities in the component categories **A** and **B**. △

Remark A.37. The short notation (F, G) for **ComCat**$(F, G; I)$, where $|I| = 1$, explains the name "comma category".

Note that we have **ComCat**$(F, G; \varnothing) = A \times B$. △

Fact A.38 (Constructions in comma categories).

1. *In a comma category* **ComCat**$(F, G; I)$ *with* $F : A \to C$, $G : B \to C$, *and an index set* I, *morphisms are monomorphisms, epimorphisms and isomorphisms if they are componentwise monomorphisms, epimorphisms and isomorphisms, respectively, in* **A** *and* **B**.
2. *If the categories* **A** *and* **B** *have pushouts and* F *preserves pushouts, then* **ComCat**$(F, G; I)$ *has pushouts, which can be constructed componentwise.*
3. *If the categories* **A** *and* **B** *have pullbacks and* G *preserves pullbacks, then* **ComCat**$(F, G; I)$ *has pullbacks, which can be constructed componentwise.* △

A.7 Isomorphism and Equivalence of Categories

In the following, we define the isomorphism and equivalence of categories.

Definition A.39 (Isomorphism of categories). Two categories **C** and **D** are called *isomorphic*, written $C \cong D$, if there are functors $F : C \to D$ and $G : D \to C$ such that $G \circ F = ID_C$ and $F \circ G = ID_D$, where ID_C and ID_D are the identity functors on **C** and **D**, respectively. △

Remark A.40. Isomorphisms of categories can be considered as isomorphisms in the "category of all categories" **Cat**, where the objects are all categories and the morphisms are all functors. Note, however, that the collection of all categories is, in general, no longer a "proper" class in the sense of axiomatic set theory. For this reason, **Cat** is not a "proper" category. △

Fact A.41 (Isomorphic categories). *The category* **Graphs** *of graphs is isomorphic to the functor category* $[S, \textbf{Sets}]$, *where the "schema category"* S *is given by the schema* $S : \cdot \rightrightarrows \cdot$. △

Definition A.42 (Equivalence of categories). Two categories **C** and **D** are called *equivalent*, written $C \equiv D$, if there are functors $F : C \to D$ and $G : D \to C$ and natural transformations $\alpha : G \circ F \Rightarrow ID_C$ and $\beta : F \circ G \Rightarrow ID_D$ that are componentwise isomorphisms, i.e., $\alpha_A : G(F(A)) \overset{\sim}{\longrightarrow} A$ and $\beta_B : F(G(B)) \overset{\sim}{\longrightarrow} B$ are isomorphisms for all $A \in \mathbf{C}$ and $B \in \mathbf{D}$, respectively. $\triangle$

Remark A.43. If $\mathbf{C}$ and $\mathbf{D}$ are isomorphic or equivalent then all "categorical" properties of $\mathbf{C}$ are shared by $\mathbf{D}$, and vice versa. If $\mathbf{C}$ and $\mathbf{D}$ are isomorphic, then we have a bijection between objects and between morphisms of $\mathbf{C}$ and $\mathbf{D}$. If they are only equivalent, then there is only a bijection of the corresponding isomorphism classes of objects and morphisms of $\mathbf{C}$ and $\mathbf{D}$. However, the cardinalities of corresponding isomorphism classes may be different; for example all sets M with cardinality $|M| = n$ are represented by the set $M_n = \{0, \ldots, n-1\}$. Taking the sets M_n ($n \in \mathbb{N}$) as objects and all functions between these sets as morphisms, we obtain a category $\mathbf{N}$, which is equivalent—but not isomorphic—to the category $\mathbf{FinSets}$ of all finite sets and functions between finite sets. $\triangle$

Appendix B

Proofs and Additional Properties for Parts II and III

In this chapter, we present different properties as well as some more technical proofs for the results in Parts II and III.

B.1 Derived Properties of Limits and Colimits

In the following, we formulate and prove different properties of diagrams concerning pullbacks, pushouts, pushout complements, and colimits in M-adhesive categories where the additional HLR properties (see Def. 4.23) hold.

Lemma B.1. *If* (1) *is a pushout,* (2) *is a pullback, and* $n' \in M$ *then there exists a unique morphism* $c : C' \to C$ *such that* $c \circ f' = f$, $n \circ c = n'$, *and* $c \in M$. △

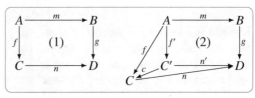

Proof. Since (2) is a pullback, $n' \in M$ implies that $m \in M$, and then also $n \in M$ because (1) is a pushout. Construct the pullback (3) with $v, v' \in M$, and since $n' \circ f = g \circ m = n \circ f$ there exists a unique morphism $f^* : A \to C''$ with $v \circ f^* = f'$ and $v' \circ f^* = f$. Now consider the following cube (4), where the bottom face is pushout (1), the back left face is a pullback because $m \in M$, the front left face is

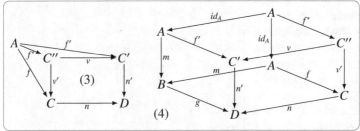

pullback (2), and the front right face is pullback (3). By pullback composition and decomposition also the back right face is a pullback, and then the M-van Kampen

property implies that the top face is a pushout. Since (5)
is a pushout and pushout objects are unique up to isomorphism this implies that v is an isomorphism and $C'' \cong C'$.
Now define $c := v' \circ v^{-1} : C \to C'$ and we have that
$c \circ f' = v' \circ v^{-1} \circ f' = v' \circ f^* = f$, $n \circ c = n \circ v' \circ v^{-1} = n'$,
and $c \in M$ by decomposition of M-morphisms. $\square$

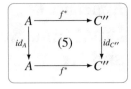

Lemma B.2. *If* (1) + (2) *is a pullback*, (1) *is a
pushout*, (2) *commutes, and* $o \in M$ *then also*
(2) *is a pullback.* $\triangle$

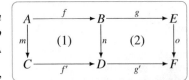

Proof. With $o \in M$, (1) + (2) a pullback,
and (1) a pushout, we have that $m, n \in$
M. Construct the pullback (3) of o and
g'; it follows that $\bar{n} \in M$ and we get an
induced morphism $b : B \to \bar{B}$ with $\bar{g} \circ$
$b = g, \bar{n} \circ b = n$, and by decomposition
of M-morphisms $b \in M$.

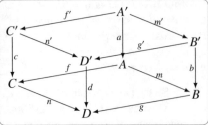

   By pullback decomposition, also (4) is a pullback
and we can apply Lem. B.1 with pushout (1) and $\bar{n} \in$
M to obtain a unique morphism $\bar{b} \in M$ with $n \circ \bar{b} = \bar{n}$
and $\bar{b} \circ b \circ f = f$. Now $n \in M$ and $n \circ \bar{b} \circ b = \bar{n} \circ b = n$
implies that $\bar{b} \circ b = id_B$, and similarly $\bar{n} \in M$ and
$\bar{n} \circ b \circ \bar{b} = n \circ \bar{b} = \bar{n}$ implies that $b \circ \bar{b} = id_{\bar{B}}$, which
means that B and $\bar{B}$ are isomorphic such that also (2) is a pullback. $\square$

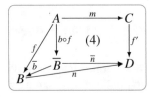

Lemma B.3. *Given the following commutative cube with the bottom face as
a pushout, the front right face has a
pushout complement over* $g \circ b$ *if the back
left face has a pushout complement over*
$f \circ a$. $\triangle$

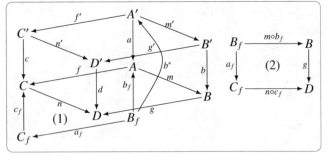

Proof. Construct the initial pushout (1)
over f. Since the back left face has a pushout complement there is a morphism
$b^* : B_f \to A'$ such
that $a \circ b^* = b_f$. The
bottom face being a
pushout implies that
(2) as the composi-
tion is the initial
pushout over g. Now
$b \circ m' \circ b^* = m \circ$
$a \circ b^* = m \circ b_f$, and
the pushout comple-
ment of $g \circ b$ exists. $\square$

Lemma B.4. *Given pullbacks* (1) *and* (2) *with pushout complements over* $f' \circ m$ *and* $g' \circ n$, *respectively, also* (1) + (2) *has a pushout complement over* $(g' \circ f') \circ m$. △

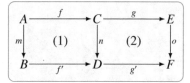

Proof. Let C' and E' be the pushout complements of (1) and (2), respectively. By Lem. B.1 there are morphisms c and e such that $c \circ f = f^*$, $n^* \circ c = n$, $e \circ g = g*$, and $o^* \circ e = o$. Now (2') can be decomposed into pushouts (3) and (4), and (1') + (4) is also a pushout and the pushout complement of $(g' \circ f') \circ m$. □

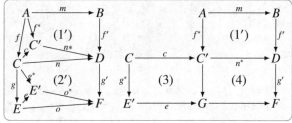

Lemma B.5. *Given the following pushouts* (1_i) *and* (3_i) *with* $b_i \in \mathcal{M}$ *for* $i = 1, \ldots, n$, *morphisms* $f_{ij} : B_i \to C_j$ *with* $c_j \circ f_{ij} = d_i$ *for all* $i \neq j$, *and the limit* (2) *of* $(c_j)_{j=1,\ldots,n}$ *such that* g_i *is the induced morphism into* E *using* $c_j \circ f_{ij} \circ b_i = d_i \circ b_i = c_i \circ a_i$, (4) *is the colimit of* $(h_i)_{i=1,\ldots,n}$, *where* l_i *is the induced morphism from pushout* (3_i) *compared with* $\bar{e} \circ g_i = c_i \circ e_i \circ g_i = c_i \circ a_i = d_i \circ b_i$.

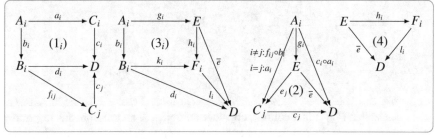

△

Proof. We prove this by induction over n.

I.B. $n = 1$: For $n = 1$, we have that C_1 is the limit of c_1, i.e., $E = C_1$; it follows that $F_1 = C_1$ for the pushout $(3_1) = (1_1)$, and obviously (4_1) is a colimit.

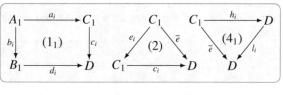

I.S. $n \to n + 1$: Consider the pushouts (1_i) with $b_i \in \mathcal{M}$ for $i = 1, \ldots, n+1$, morphisms $f_{ij} : B_i \to C_j$ with $c_j \circ f_{ij} = d_i$ for all $i \neq j$, the limits (2_n) and (2_{n+1}) of $(c_i)_{i=1,\ldots,n}$ and $(c_i)_{i=1,\ldots,n+1}$, respectively, leading to pullback (5_{n+1}) by construction of limits. Moreover, g_{in} and g_{in+1} are the induced morphisms into E_n and E_{n+1}, respectively, leading to pushouts (3_{in}) and (3_{in+1}). By induction hypothesis, (4_n) is the colimit of $(h_{in})_{i=1,\ldots,n}$, and we have to show that (4_{n+1}) is the colimit of $(h_{in+1})_{i=1,\ldots,n+1}$.

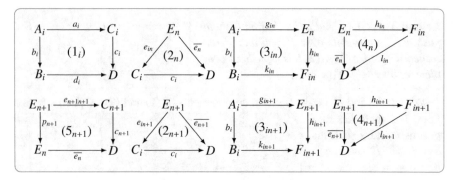

Since (2_n) is a limit and $c_i \circ f_{n+1i} = d_{n+1}$ for all $i = 1,\ldots,n$, we obtain a unique morphism m_{n+1} with $e_{in} \circ m_{n+1} = f_{n+1i}$ and $\overline{e_n} \circ m_{n+1} = d_{n+1}$. Since $(1_{n+1}) = (6_{n+1}) + (5_{n+1})$ is a pushout and (5_{n+1}) is a pullback, by $\mathcal{M}$-pushout–pullback decomposition (see Def. 4.21) also (5_{n+1}) and (6_{n+1}) are pushouts, and it follows that $F_{n+1n+1} = E_n$. From pushout (3_{in+1}) and $h_{in} \circ p_{n+1} \circ g_{in+1} = h_{in} \circ g_{in} = k_{in} \circ b_i$ we get an induced morphism q_{in+1} with $q_{in+1} \circ h_{in+1} = h_{in} \circ p_{n+1}$ and $q_{in+1} \circ k_{in+1} = k_{in}$, and from pushout decomposition with $(3_{in+1}) + (7_{in+1}) = (3_{in})$ also (7_{in+1}) is a pushout.

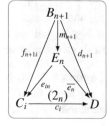

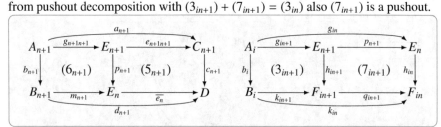

To show that (4_{n+1}) is a colimit, consider an object X and morphisms (x_i) and y with $x_i \circ h_{in+1} = y$ for $i = 1,\ldots,n$ and $x_{n+1} \circ p_{n+1} = y$. From pushout (7_{in+1}) we obtain a unique morphism z_i with $z_i \circ q_{in+1} = x_i$ and $z_i \circ h_{in} = x_{n+1}$. Now colimit (4_n) induces a unique morphism z with $z \circ \overline{e_n} = x_{n+1}$ and $z \circ l_{in} = z_i$. It follows directly that $z \circ l_{in+1} = z \circ l_{in} \circ q_{in+1} = z_i \circ q_{in+1} = x_i$ and $z \circ \overline{e_{n+1}} = z \circ \overline{e_n} \circ p_{n+1} = x_{n+1} \circ p_{n+1} = y$. The uniqueness of z follows directly from the construction; thus (4_{n+1}) is the required colimit.

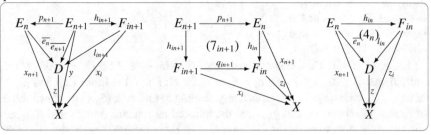

□

Lemma B.6. *Given the diagrams* (1_i) *for* $i = 1,\dots,n$, (2), *and* (3), *with* $b = +b_i$, *and* a *and* e *induced by the coproducts* $+A_i$ *and* $+B_i$, *respectively, we have:*

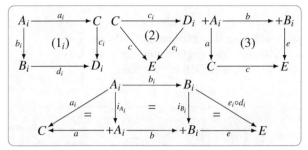

1. *If* (1_i) *are pushouts and* (2) *a colimit then also* (3) *is a pushout.*
2. *If* (3) *is a pushout then we find a decomposition into pushouts* (1_i) *and colimit* (2) *with* $e_i \circ d_i = e \circ i_{B_i}$. △

Proof. 1. Given an object X and morphisms y, z with $y \circ a = z \circ b$. From pushout (1_i) we obtain with $z \circ i_{B_i} \circ b_i = z \circ b \circ i_{A_i} = y \circ a \circ i_{A_i} = y \circ a_i$ a unique morphism x_i with $x_i \circ c_i = y$ and $x_i \circ d_i = z \circ i_{B_i}$. Now colimit (2) implies a unique morphism x with $x \circ c = y$ and $x \circ e_i = x_i$. It follows that $x \circ e \circ i_{B_i} = x \circ e_i \circ d_i = x_i \circ d_i = z \circ i_{B_i}$, and since z is unique w. r. t. $z \circ i_{B_i}$ it follows that $z = x \circ e$. Uniqueness of x follows from the uniqueness of x and x_i, and hence (3) is a pushout.

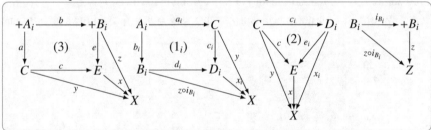

2. Define $a_i := a \circ i_{A_i}$. Now construct pushouts (1_i). With $e \circ i_{B_i} \circ b_i = e \circ b \circ i_{A_i} = c \circ a_i$, each pushout (1_i) induces a unique morphism e_i with $e_i \circ d_i = e \circ i_{B_i}$ and $e_i \circ c_i = c$. Given an object X and morphisms y, y_i with $y_i \circ c_i = y$, we obtain a morphism z with $z \circ i_{B_i} = y_i \circ d_i$ from coproduct $+B_i$. Then we have that $y \circ a \circ i_{A_i} = y_i \circ c_i \circ a_i = y_i \circ d_i \circ b_i = z \circ i_{B_i} \circ b_i = z \circ b \circ i_{A_i}$, and from coproduct $+A_i$ it follows that $y \circ a = z \circ b$. Now pushout (3) implies a unique morphism x with $x \circ c = y$ and $x \circ e = z$. From pushout (1_i), using $x \circ e_i \circ d_i = x \circ e \circ i_{B_i} = z \circ i_{B_i} = y_i \circ d_i$ and $x \circ e_i \circ c_i = x \circ c = y = y_i \circ c_i$, it follows that $x \circ e_i = y_i$; thus (2) is a colimit.

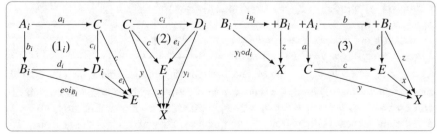

□

Lemma B.7. *Given   colimits   (1)–(4) such that (5_i) is a pushout for all $i = 1,\ldots,n$ and (6_k)–(9_k) commute for all $k = 1,\ldots,m$, also (10) is a pushout.*

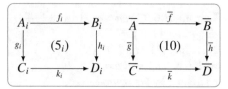

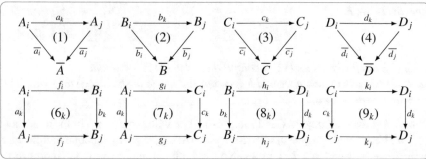

△

Proof. The morphisms $\overline{f}, \overline{g}, \overline{h}$, and $\overline{k}$ are uniquely induced by the colimits. We show this exemplarily for the morphism $\overline{f}$: From colimit (1), with $\overline{b}_j \circ f_j \circ a_k = \overline{b}_j \circ b_k \circ f_i = \overline{b}_i \circ f_i$ we obtain a unique morphism $\overline{f}$ with $\overline{f} \circ \overline{a}_i = \overline{b}_i \circ f_i$. It follows directly that $\overline{k} \circ \overline{g} = \overline{h} \circ \overline{f}$.

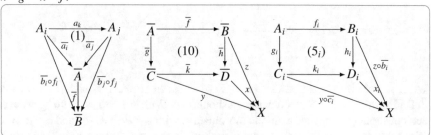

Now consider an object X and morphisms y, z with $y \circ \overline{g} = z \circ \overline{f}$. From each pushout (5_i) with $y \circ \overline{c}_i \circ g_i = y \circ \overline{g} \circ \overline{a}_i = z \circ \overline{f} \circ \overline{a}_i = z \circ \overline{b}_i \circ f_i$ we obtain a unique morphism x_i with $x_i \circ k_i = y \circ \overline{c}_i$ and $x_i \circ h_i = z \circ \overline{b}_i$.

For all $k = 1,\ldots,m$, $x_j \circ d_k \circ k_i = x_j \circ k_j \circ c_k = y \circ \overline{c}_j \circ c_k = y \circ \overline{c}_i$ and $x_j \circ d_k \circ h_i = x_j \circ h_j \circ b_k = z \circ \overline{b}_j \circ b_k = z \circ \overline{b}_i$, and pushout (5_i) implies that $x_i = x_j \circ d_k$. This

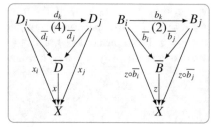

means that colimit (4) implies a unique x with $x \circ \overline{d}_i = x_i$. Now consider colimit (2), and $x \circ \overline{h} \circ \overline{b}_i = x \circ \overline{d}_i \circ h_i = x_i \circ h_i = z \circ \overline{b}_i$ implies that $x \circ \overline{h} = z$. Similarly, $x \circ \overline{k} = y$, and the uniqueness follows from the uniqueness of x with respect to (4). Thus, (10) is indeed a pushout.   □

Lemma B.8. *Consider colimits* (1) *and* (2) *such that* (3_i) *commutes for all* $i = 1,\dots,n$, f *is an epimorphism, and* (4) *is a pushout with* $\overline{f}$ *induced by colimit* (1). *Then also* (5) *is a pushout, where* c *and* d *are induced from the coproducts.*

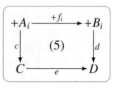

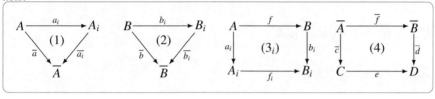

$\triangle$

Proof. Since (1) is a colimit and $\overline{b}_i \circ \overline{f}_i \circ a_i = \overline{b}_i \circ b_i \circ f = \overline{b} \circ f$, we actually get an induced $\overline{f}$ with $\overline{f} \circ \overline{a}_i = \overline{b}_i \circ f_i$ and $\overline{f} \circ \overline{a} = \overline{b} \circ f$. From the coproducts, we obtain induced morphisms c with $c \circ i_{A_i} = \overline{c} \circ \overline{a}_i$ and d with $d \circ i_{B_i} = \overline{d} \circ \overline{b}_i$. Moreover, for all $i = 1,\dots,n$ we have that $d \circ (+f_i) \circ i_{A_i} = d \circ i_{B_i} \circ f_i = \overline{d} \circ \overline{b}_i \circ f_i = \overline{d} \circ \overline{f} \circ \overline{a}_i = e \circ \overline{c} \circ \overline{a}_i = e \circ c \circ i_{A_i}$. Uniqueness of the induced coproduct morphisms leads to $d \circ (+f_i) = e \circ c$, i.e., (5) commutes.

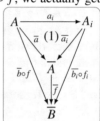

We have to show that (5) is a pushout. Given morphisms x, y with $x \circ c = y \circ (+f_i)$, we have that $y \circ i_{B_i} \circ b_i \circ f = y \circ i_{B_i} \circ f_i \circ a_i = y \circ (+f_i) \circ i_{A_i} \circ a_i = x \circ c \circ i_{A_i} \circ a_i = x \circ \overline{c} \circ \overline{a}_i \circ a_i = x \circ \overline{c} \circ \overline{a}$ for all $i = 1,\dots,n$. Since f is an epimorphism we have that $y \circ i_{B_i} \circ b_i = y \circ i_{B_j} \circ b_j$ for all i, j. Now define $y' := y \circ i_{B_i} \circ b_i$, and from colimit (2) we obtain a unique morphism $\overline{y}$ with $\overline{y} \circ \overline{b}_i = y \circ i_{B_i}$ and $\overline{y} \circ \overline{b} = y'$.

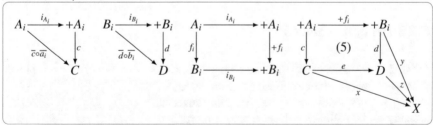

Now $x \circ \overline{c} \circ \overline{a}_i = x \circ c \circ i_{A_i} = y \circ (+f_i) \circ i_{A_i} = y \circ i_{B_i} \circ f_i = \overline{y} \circ \overline{b}_i \circ f_i = \overline{y} \circ \overline{f} \circ \overline{a}_i$ and $x \circ \overline{c} \circ \overline{a} = x \circ \overline{c} \circ \overline{a}_i \circ a_i = \overline{y} \circ \overline{f} \circ \overline{i} \circ a_i = \overline{y} \circ \overline{f} \circ \overline{a}$, and the uniqueness of the induced colimit morphism implies that $\overline{y} \circ \overline{f} = x \circ \overline{c}$.

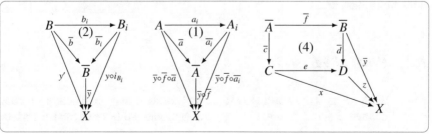

This means that X can be compared to pushout (4), and we obtain a unique morphism z with $z \circ \overline{d} = \overline{y}$ and $z \circ e = x$. Now $z \circ d \circ i_{B_i} = z \circ \overline{d} \circ \overline{b}_i = \overline{y} \circ \overline{b}_i = y \circ i_{B_i}$, and it follows that $z \circ d = y$. Similarly, the uniqueness of z w. r. t. to the pushout property of (5) follows; thus (5) is a pushout.   □

B.2 Proofs for Sect. 4.4

B.2.1 Proof of Fact 4.36

Proof. The construction is always well defined, since there is at least the trivial $\mathcal{M}$-subobject $m_i = \mathrm{id}_B$ with $e_i = f$ and at most finitely many $\mathcal{M}$-subobjects. It follows that $\overline{m_i} \in \mathcal{M}$ and therefore also $m \in \mathcal{M}$, because $\mathcal{M}$ is closed under composition.

It remains to show that $e \in \mathcal{E}$. Let $e = m' \circ e'$ be a factorisation of e with $m' \in \mathcal{M}$. Then we have that $m \circ m'$ is an $\mathcal{M}$-subobject of B and $m \circ m' \circ e' = f$, and, hence, $B' = B_i$, $m \circ m' = m_i$, and $e' = e_i$ for some $i \in \mathcal{I}$. This implies that there exists $\overline{m_i} : \overline{B} \to B_i = B'$ with $m_i \circ \overline{m_i} = m$. Now,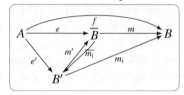

$m_i \circ \overline{m_i} \circ m' = m \circ m' = m_i$ and since $m_i \in \mathcal{M}$ we have that m_i is a monomorphism which implies that $\overline{m_i} \circ m' = \mathrm{id}_{B'}$. Similarly, $m \circ m' \circ \overline{m_i} = m_i \circ \overline{m_i} = m$ and since $m \in \mathcal{M}$ we have that m is a monomorphism which implies that $m' \circ \overline{m_i} = \mathrm{id}_{\overline{B}}$. Hence, m' and $\overline{m_i}$ are mutually inverse isomorphisms and $e \in \mathcal{E}$.   □

B.2.2 Proof of Fact 4.38

Proof. For two extremal $\mathcal{E}$–$\mathcal{M}$ factorisations $m_1 \circ e_1 = m_2 \circ e_2 = f$ of a morphism $f : A \to B$, we construct a pullback over $m_1, m_2 \in \mathcal{M}$, leading to morphisms $n_1, n_2 \in \mathcal{M}$. The universal property of the pullback induces a unique morphism $e : A \to \overline{A}$ which, together with n_1, n_2, factors $e_1, e_2 \in \mathcal{E}$. Since these are extremal, n_1 and n_2 are isomorphisms and the two extremal $\mathcal{E}$–$\mathcal{M}$ factorisations are isomorphic.   □

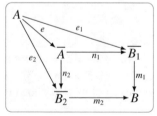

B.2.3 Proof of Fact 4.40

Proof. We have to show that (1) is the initial pushout over m. The construction is well-defined, since $\mathcal{I}$ is finite by construction, and we have at least the trivial

pushout (4) over m. Since finite $\mathcal{M}$-intersections can be constructed by iterated pullbacks, we show by induction that (Q_i)
and hence also $(1) = (Q_i) + (P_i)$ are pushouts.

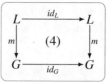

I.B. For $\mathcal{I} = \{1\}$, we have that $(Q_1) = (P_1) = (1) = (4)$ by
construction.

I.S. Let $\mathcal{I} = \{1, \ldots, n+1\}$. Now consider the $\mathcal{M}$-intersections B_{1n} and C_{1n} for
pushouts $(P_i)_{i=1,\ldots,n}$, leading to the pushout in the front left face of the commutative
cube by the induction hypothesis. The top and bottom faces are pullbacks using the
$\mathcal{M}$-intersection construction. The right
front face is the pushout (P_{n+1}). Since
all horizontal morphisms are in $\mathcal{M}$, using the cube pushout–pullback prop-
erty (see Theorem 4.22) it follows that
the back faces are pushouts. This
means, in particular, that (Q_{n+1}) is a
pushout, and by pushout composition
so is (1).

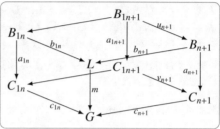

(1) is initial, because every other pushout
$(1')$ over m with $b' \in \mathcal{M}$ is equal to (P_{i_0}) for
some $i_0 \in \mathcal{I}$. Hence, the initiality property
is given by the pushout (Q_{i_0}) as constructed
above.   □

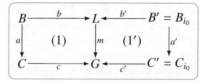

B.3 Construction of $\mathcal{M}$-Adhesive Categories

To enhance flexibility, we use an extension of comma categories [Pra07], where we
relax the restrictions on the domain of the functors compared to standard comma
categories, which allows us to adjust the category to describe different operations
on the objects.

Definition B.9 (General comma category). Given index sets $\mathcal{I}$ and $\mathcal{J}$, cate-
gories $\mathbf{C}_j$ for $j \in \mathcal{J}$ and $\mathbf{X}_i$ for $i \in \mathcal{I}$, and for each $i \in \mathcal{I}$ two functors
$F_i : \mathbf{C}_{k_i} \to \mathbf{X}_i$, $G_i : \mathbf{C}_{\ell_i} \to \mathbf{X}_i$ with $k_i, \ell_i \in \mathcal{J}$, the *general comma category*
$GComCat((\mathbf{C}_j)_{j \in \mathcal{J}}, (F_i, G_i)_{i \in \mathcal{I}}; \mathcal{I}, \mathcal{J})$ is defined by

- objects $((A_j \in \mathbf{C}_j)_{j \in \mathcal{J}}, (op_i)_{i \in \mathcal{I}})$, where $op_i : F_i(A_{k_i})$
  $\to G_i(A_{\ell_i})$ is a morphism in $\mathbf{X}_i$,
- morphisms $h : ((A_j), (op_i)) \to ((A'_j), (op'_i))$ as tu-
  ples $h = ((h_j : A_j \to A'_j)_{j \in \mathcal{J}})$ such that for all $i \in \mathcal{I}$
  we have that $op'_i \circ F_i(h_{k_i}) = G_i(h_{\ell_i}) \circ op_i$.   △

$$
\begin{array}{ccc}
F_i(A_{k_i}) & \xrightarrow{\;op_i\;} & G_i(A_{\ell_i}) \\
\downarrow{\scriptstyle F_i(h_{k_i})} & & \downarrow{\scriptstyle G_i(h_{\ell_i})} \\
F_i(A'_{k_i}) & \xrightarrow{\;op'_i\;} & G_i(A'_{\ell_i})
\end{array}
$$

A standard comma category is an instantiation of a general comma category.

Lemma B.10. *A comma category* $\mathbf{A} = ComCat(F : \mathbf{C} \to \mathbf{X}, G : \mathbf{D} \to \mathbf{X}, \mathcal{I})$ *is a
special case of a general comma category.*   △

Proof. With I as given, $\mathcal{J} = \{1, 2\}$, $\mathbf{C}_1 = \mathbf{C}$, $\mathbf{C}_2 = \mathbf{D}$, $X_i = X$, $F_i = F$ and $G_i = G$ for all $i \in I$; the resulting general comma category is obviously isomorphic to $\mathbf{A}$.

$\square$

Product, slice and coslice categories are special cases of comma categories.

Lemma B.11. *For product, slice and coslice categories, we have the following isomorphic comma categories:*

1. $\mathbf{C} \times \mathbf{D} \cong ComCat(!_{\mathbf{C}} : \mathbf{C} \to \mathbf{1}, !_{\mathbf{D}} : \mathbf{D} \to \mathbf{1}, \varnothing)$,
2. $\mathbf{C} \backslash X \cong ComCat(id_{\mathbf{C}} : \mathbf{C} \to \mathbf{C}, X : \mathbf{1} \to \mathbf{C}, \{1\})$ *and*
3. $X \backslash \mathbf{C} \cong ComCat(X : \mathbf{1} \to \mathbf{C}, id_{\mathbf{C}} : \mathbf{C} \to \mathbf{C}, \{1\})$,

where $\mathbf{1}$ is the final category, $!_{\mathbf{C}} : \mathbf{C} \to \mathbf{1}$ is the final morphism from $\mathbf{C}$, and $X : \mathbf{1} \to \mathbf{C}$ maps $1 \in \mathbf{1}$ to $X \in \mathbf{C}$.

$\triangle$

Proof. This is obvious.

$\square$

In a general comma category, pushouts can be constructed componentwise in the underlying categories if the domain functors of the operations preserve pushouts. This is a generalisation of the corresponding result in [PEL08] for comma categories.

Lemma B.12. *Consider a general comma category* $\mathbf{G} = GComCat((\mathbf{C}_j)_{j \in \mathcal{J}},$ $(F_i, G_i)_{i \in I}; I, \mathcal{J})$ *based on $\mathcal{M}$-adhesive categories $(\mathbf{C}_j, \mathcal{M}_j)$, where F_i preserves pushouts along $\mathcal{M}_{k_i}$-morphisms.*

For objects $A = ((A_j), (op_i^A))$, $B = ((B_j), (op_i^B))$, *and* $C = ((C_j), (op_i^C)) \in \mathbf{G}$ *and morphisms* $f = (f_j) : A \to B$, $g = (g_j) : A \to C$ *with* $f \in \times_{j \in \mathcal{J}} \mathcal{M}_j$, *we have: The diagram (1) is a pushout in $\mathbf{G}$ iff for all $j \in \mathcal{J}$ (1)$_j$ is a pushout in $\mathbf{C}_j$, with $D = ((D_j), (op_j^D))$, $f' = (f_j')$, and $g' = (g_j')$.*

$$
\begin{array}{ccc}
A_j & \xrightarrow{\ f_j\ } & B_j \\
{\scriptstyle g_j} \downarrow & (1)_j & \downarrow {\scriptstyle g_j'} \\
C_j & \xrightarrow{\ f_j'\ } & D_j
\end{array}
\qquad
\begin{array}{ccc}
A & \xrightarrow{\ f\ } & B \\
{\scriptstyle g} \downarrow & (1) & \downarrow {\scriptstyle g'} \\
C & \xrightarrow{\ f'\ } & D
\end{array}
$$

$\triangle$

Proof. "$\Leftarrow$" Given the morphisms f and g in (1), and the pushouts (1)$_j$ in $\mathbf{C}_j$ for $j \in \mathcal{J}$. We have to show that (1) is a pushout in $\mathbf{G}$.

Since F_i preserves pushouts along $\mathcal{M}_{k_i}$-morphisms, with $f_{k_i} \in \mathcal{M}_{k_i}$ the diagram (2)$_i$ is a pushout for all $i \in I$. Then $D = ((D_j), (op_j^D))$ is an object in $\mathbf{G}$, where, for $i \in I$, op_i^D is induced by pushout (2)$_i$ and $G_i(f_{\ell_i}') \circ op_i^C \circ F_i(g_{k_i}) = G_i(f_{\ell_i}') \circ G_i(g_{\ell_i}) \circ op_i^A = G_i(g_{\ell_i}') \circ G_i(f_{\ell_i}) \circ op_i^A = G_i(g_{\ell_i}') \circ op_i^B \circ F_i(f_{k_i})$. It holds that $op_i^D \circ F_i(f_{k_i}') = G_i(f_{\ell_i}') \circ op_i^C$ and $op_i^D \circ F_i(g_{k_i}') = G_i(g_{\ell_i}') \circ op_i^B$. Therefore $f' = (f_j')$ and $g' = (g_j')$ are morphisms in $\mathbf{G}$ such that (1) commutes.

It remains to show that (1) is a pushout. Given an object $X = ((X_j), (op_i^X))$ and morphisms $h = (h_j) : B \to X$ and $k = (k_j) : C \to X$ in $\mathbf{G}$ such that $h \circ f = k \circ g$, from pushouts (1)$_j$ we obtain unique morphisms $x_j : D_j \to X_j$ such that $x_j \circ g_j' = h_j$ and $x_j \circ f_j' = k_j$ for all $j \in \mathcal{J}$.

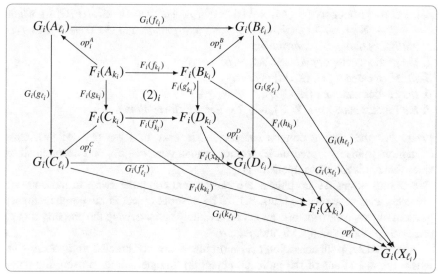

Since $(2)_i$ is a pushout, from $G_i(x_{\ell_i}) \circ op_i^D \circ F_i(g_i') = G_i(x_{\ell_i}) \circ G_i(g_{\ell_i}') \circ op_i^B = G_i(h_{\ell_i}) \circ op_i^B = op_i^X \circ F(h_{k_i}) = op_i^X \circ F_i(x_{k_i}) \circ F_i(g_{k_i}')$ and $G_i(x_{\ell_i}) \circ op_i^D \circ F_i(f_{k_i}') = G_i(x_{\ell_i}) \circ G_i(f_{\ell_i}') \circ op_i^C = G_i(k_{\ell_i}) \circ op_i^C = op_i^X \circ F_i(k_{\ell_i}) = op_i^X \circ F_i(x_{k_i}) \circ F_i(f_{k_i}')$ it follows that $G_i(x_{\ell_i}) \circ op_i^D = op_i^X \circ F_i(x_{k_i})$. Therefore $x = (x_j) \in \mathbf{G}$, and x is unique with respect to $x \circ g' = h$ and $x \circ f' = k$.

"$\Rightarrow$" Given the pushout (1) in $\mathbf{G}$ we have to show that $(1)_j$ are pushouts in $\mathbf{C}_j$ for all $j \in \mathcal{J}$. Since $(\mathbf{C}_j, \mathcal{M}_j)$ is an $\mathcal{M}$-adhesive category there exists a pushout $(1')_j$ over $f_j \in \mathcal{M}_j$ and g_j in $\mathbf{C}_j$.

Therefore (using "$\Leftarrow$") there is a corresponding pushout $(1')$ in $\mathbf{G}$ over f
and g with $E = ((E_j), (op_i^E))$, $f^* = (f_j^*)$ and $g^* = (g_j^*)$. Since pushouts are
unique up to isomorphism it follows
that $E \cong D$, which means $E_j \cong D_j$ and
therefore $(1)_j$ is a pushout in $\mathbf{C}_j$ for all $j \in \mathcal{J}$.

$$
\begin{array}{ccc}
A_j & \xrightarrow{f_j} & B_j \\
{\scriptstyle g_j}\downarrow & (1')_j & \downarrow{\scriptstyle g_j^*} \\
C_j & \xrightarrow{f_j^*} & E_j
\end{array}
\qquad
\begin{array}{ccc}
A & \xrightarrow{f} & B \\
{\scriptstyle g}\downarrow & (1') & \downarrow{\scriptstyle g^*} \\
C & \xrightarrow{f^*} & E
\end{array}
$$

$\square$

We extend the Construction Theorem in [EEPT06] to general comma categories and full subcategories. Basically, it holds that, under some consistency properties, if the underlying categories are $\mathcal{M}$-adhesive categories so are the constructed ones.

Theorem B.13 (Construction Theorem). *If $(\mathbf{C}, \mathcal{M}_1)$, $(\mathbf{D}, \mathcal{M}_2)$, and $(\mathbf{C}_j, \mathcal{M}_j)$ for $j \in \mathcal{J}$ are $\mathcal{M}$-adhesive categories, then also the following categories are $\mathcal{M}$-adhesive categories:*

1. *the* general comma category $(\mathbf{G}, (\times_{j \in \mathcal{J}} \mathcal{M}_j) \cap Mor_\mathbf{G})$ with $\mathbf{G} = GComCat((\mathbf{C}_j)_{j \in \mathcal{J}}, (F_i, G_i)_{i \in I}; I, \mathcal{J})$, where, for all $i \in I$, $F_i : \mathbf{C}_{k_i} \to \mathbf{X}_i$ preserves pushouts along $\mathcal{M}_{k_i}$-morphisms and $G_i : \mathbf{C}_{\ell_i} \to \mathbf{X}_i$ preserves pullbacks along $\mathcal{M}_{\ell_i}$-morphisms,
2. *any full subcategory $(\mathbf{C}', \mathcal{M}_1|_{\mathbf{C}'})$ of $\mathbf{C}$, where pushouts and pullbacks along $\mathcal{M}_1$ are created and reflected by the inclusion functor,*

3. *the* comma category $(\mathbf{F}, (\mathcal{M}_1 \times \mathcal{M}_2) \cap Mor_{\mathbf{F}})$, *with* $\mathbf{F} = ComCat(F, G; \mathcal{I})$, *where* $F : \mathbf{C} \to \mathbf{X}$ *preserves pushouts along* $\mathcal{M}_1$-*morphisms and* $G : \mathbf{D} \to \mathbf{X}$ *preserves pullbacks along* $\mathcal{M}_2$-*morphisms,*
4. *the* product category $(\mathbf{C} \times \mathbf{D}, \mathcal{M}_1 \times \mathcal{M}_2)$,
5. *the* slice category $(\mathbf{C} \backslash X, \mathcal{M}_1 \cap Mor_{\mathbf{C} \backslash X})$,
6. *the* coslice category $(X \backslash \mathbf{C}, \mathcal{M}_1 \cap Mor_{X \backslash \mathbf{C}})$,
7. *the* functor category $([\mathbf{X}, \mathbf{C}], \mathcal{M}_1$-*functor transformations).*   △

Proof. For the general comma category, it is easy to show that $\mathcal{M}$ is a class of monomorphisms closed under isomorphisms, composition, and decomposition since this holds for all components $\mathcal{M}_j$.

Pushouts along $\mathcal{M}$-morphisms are constructed componentwise in the underlying categories as shown in Lem. B.12. The pushout object is the componentwise pushout object, where the operations are uniquely defined using the property that F_i preserves pushouts along $\mathcal{M}_{k_i}$ morphisms.

Analogously, pullbacks along $\mathcal{M}$-morphisms are constructed componentwise, where the operations of the pullback object are uniquely defined using the property that G_i preserves pullbacks along $\mathcal{M}_{\ell_i}$-morphisms.

The $\mathcal{M}$-van Kampen property follows, since in a proper cube, all pushouts and pullbacks can be decomposed, leading to proper cubes in the underlying categories, where the $\mathcal{M}$-van Kampen property holds. The subsequent recomposition yields the $\mathcal{M}$-van Kampen property for the general comma category.

For a full subcategory $\mathbf{C}'$ of $\mathbf{C}$ define $\mathcal{M}' = \mathcal{M}_1|_{\mathbf{C}'}$. By reflection, pushouts and pullbacks along $\mathcal{M}'$-morphisms in $\mathbf{C}'$ exist. Obviously, $\mathcal{M}'$ is a class of monomorphisms with the required properties. Since we only restrict the objects and morphisms, the $\mathcal{M}$-van Kampen property is inherited from $\mathbf{C}$.

As shown in Lemmas B.10 and B.11, product, slice, coslice, and comma categories are instantiations of general comma categories. Obviously, the final category $\mathbf{1}$ is an $\mathcal{M}$-adhesive category and the functors $!_{\mathbf{C}}$, $!_{\mathbf{D}}$, $id_{\mathbf{C}}$, and X preserve pushouts and pullbacks. Thus, the proposition follows directly from the general comma category for these constructions.

The proof for the functor category is explicitly given in [EEPT06].   □

B.4 Proofs for Sect. 4.5

B.4.1 Proof of Fact 4.51

Proof. 1. If $\mathbf{C}_j$ has binary coproducts for all $j \in \mathcal{J}$ and F_i preserves binary coproducts for all $i \in \mathcal{I}$, then the coproduct of two objects $A = ((A_j), (op_i^A))$ and $B = ((B_j), (op_i^B))$ in $\mathbf{G}$ is the object $A + B = ((A_j + B_j), (op_i^{A+B}))$, where op_i^{A+B} is the unique morphism induced by $G_i(i_{A_{\ell_i}}) \circ op_i^A$ and $G_i(i_{B_{\ell_i}}) \circ op_i^B$. If also G_i preserves coproducts then $op_i^{A+B} = op_i^A + op_i^B$.

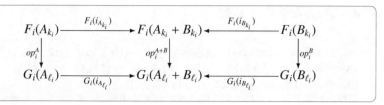

2. If the inclusion functor reflects binary coproducts this is obvious. Otherwise, if we have an initial object I, given $A, B \in \mathbf{C'}$ we can construct the pushout over $i_A : I \to A$, $i_B : I \to B$, which exists because $i_A, i_B \in \mathcal{M}$ or due to general pushouts. In this case, the pushout object is also the coproduct of A and B, because for any object in compari-

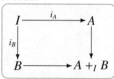

son to the coproduct the morphisms agree via i_A and i_B on I, and the constructed pushout induces also the coproduct morphism.

3. This follows directly from Item 1, since the comma category is an instantiation of general comma categories. The coproduct of objects $(A_1, A_2, (op_i^A))$ and $(B_1, B_2, (op_i^B))$ of the comma category is the object $A + B = (A_1 + B_1, A_2 + B_2, op_i^{A+B})$.

4. Since $\mathbf{C} \times \mathbf{D} \cong ComCat(!_{\mathbf{C}} : \mathbf{C} \to \mathbf{1}, !_{\mathbf{D}} : \mathbf{D} \to \mathbf{1}, \varnothing)$ (see Lem. B.11) and $!_{\mathbf{C}}$ preserves coproducts this follows from Item 3. The coproduct of objects (A_1, A_2) and (B_1, B_2) of the product category is the componentwise coproduct $(A_1 + B_1, A_2 + B_2)$ in $\mathbf{C}$ and $\mathbf{D}$, respectively.

5. Since $\mathbf{C} \backslash X \cong ComCat(id_{\mathbf{C}} : \mathbf{C} \to \mathbf{C}, X : \mathbf{1} \to \mathbf{C}, \{1\})$ (see Lem. B.11) and $id_{\mathbf{C}}$ preserves coproducts this follows from Item 3. In the slice category, the coproduct of (A, a') and (B, b') is the object $(A + B, [a', b'])$ which consists of

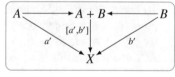

the coproduct $A + B$ in $\mathbf{C}$ together with the morphism $[a', b'] : A + B \to X$ induced by a' and b'.

6. If $\mathbf{C}$ has general pushouts, given two objects (A, a') and (B, b') in $X \backslash \mathbf{C}$ we construct the pushout over a' and b' in $\mathbf{C}$. The coproduct of (A, a') and (B, b') is the pushout object $A +_X B$ together with the coslice morphism $b \circ a' = a \circ b'$. For any object (C, c') in comparison to the

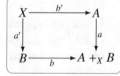

coproduct, the coslice morphism c' ensures that the morphisms agree via a' and b' in X such that the pushout also induces the coproduct morphism.

7. If $\mathbf{C}$ has binary coproducts, the coproduct of two functors $A, B : \mathbf{X} \to \mathbf{C}$ in $[\mathbf{X}, \mathbf{C}]$ is the componentwise coproduct functor $A + B$ with $A + B(x) = A(x) + B(x)$ for an object $x \in \mathbf{X}$ and $A + B(h) = A(h) + B(h)$ for a morphism $h \in \mathbf{X}$. $\square$

B.4.2 Proof of Fact 4.54

Proof. 1. Given objects $A = ((A_j), (op_i^A))$, $B = ((B_j), (op_i^B))$, $C = ((C_j), (op_i^C))$, and morphisms $f = (f_j) : A \to C$, $g = (g_j) : B \to C$ in **G**, we have an $\mathcal{E}_j'$–$\mathcal{M}_j'$ pair factorisation $((e_j, e_j'), m_j)$ of f_j, g_j in $\mathbf{C}_j$.

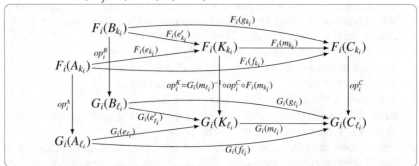

If $G_i(m_{\ell_i})$ is an isomorphism, we have an object $K = ((K_j), (op_i^K = G_i(m_{\ell_i})^{-1} \circ op_i^C \circ F_i(m_{k_i})))$ in **G**. By definition, $m = (m_j) : K \to C$ is a morphism in **G**. For $e = (e_j)$ we have $op_i^K \circ F_i(e_{k_i}) = G_i(m_{\ell_i})^{-1} \circ op_i^C \circ F_i(m_{k_i}) \circ F_i(e_{k_i}) = G_i(m_{\ell_i})^{-1} \circ op_i^C \circ F_i(f_{k_i}) = G_i(m_{\ell_i})^{-1} \circ G_i(f_{\ell_i}) \circ op_i^A = G_i(e_{\ell_i}) \circ op_i^A$, and an analogous result for $e' = (e_j')$; therefore e and e' are morphisms in **G**. This means that $((e, e'), m)$ is an $\mathcal{E}'$–$\mathcal{M}'$ pair factorisation in **G**.

To show the $\mathcal{E}'$–$\mathcal{M}'$ diagonal property, we consider $(e, e') = ((e_j), (e_j')) \in \mathcal{E}'$, $m = (m_j) \in \mathcal{M}'$, and morphisms $a = (a_j)$, $b = (b_j), n = (n_j)$ in **G**. Since $(e_j, e_j') \in \mathcal{E}_j'$ and $m_j \in \mathcal{M}_j'$, we get a unique morphism d_j :

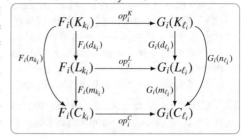

$K_j \to L_j$ in $\mathbf{C}_j$ with $m_j \circ d_j = n_j$, $d_j \circ e_j = a_j$, and $d_j \circ e_j' = b_j$.

It remains to show that $d = (d_j) \in$ **G**, i.e., the compatibility with the operations. For all $i \in \mathcal{I}$ we have that $G_i(m_{\ell_i}) \circ op_i^L \circ F_i(d_{k_i}) = op_i^C \circ F_i(m_{k_i}) \circ F_i(d_{k_i}) = op_i^C \circ F_i(n_{k_i}) = G_i(n_{\ell_i}) \circ op_i^K = G_i(m_{\ell_i}) \circ G_i(d_{\ell_i}) \circ op_i^K$, and since $G_i(m_{\ell_i})$ is an isomorphism it follows that $op_i^L \circ F_i(d_{k_i}) = G_i(d_{\ell_i}) \circ op_i^K$, i.e., $d \in$ **G**.

2. This is obvious.

3. This follows directly from Item 1, since any comma category is an instantiation of a general comma category. For morphisms $f = (f_1, f_2)$ and $g = (g_1, g_2)$ in **F** we construct the componentwise pair factorisations $((e_1, e_1'), m_1)$ of f_1, g_1 with $(e_1, e_1') \in \mathcal{E}_1'$ and $m_1 \in \mathcal{M}_1'$, and $((e_2, e_2'), m_2)$ of f_2, g_2 with $(e_2, e_2') \in \mathcal{E}_2'$ and $m_2 \in \mathcal{M}_2'$. This leads to morphisms $e = (e_1, e_2)$, $e' = (e_1', e_2')$, and $m = (m_1, m_2)$ in

$\mathbf{F}$, and an $\mathcal{E}'$–$\mathcal{M}'$ pair factorisation with $(e, e') \in \mathcal{E}'$ and $m \in \mathcal{M}'$. If the $\mathcal{E}'_1$–$\mathcal{M}'_1$ and the $\mathcal{E}'_2$–$\mathcal{M}'_2$ pair factorisations are strong then also $\mathcal{E}'$–$\mathcal{M}'$ is a strong pair factorisation.

4. Since $\mathbf{C} \times \mathbf{D} \cong ComCat(!_{\mathbf{C}} : \mathbf{C} \to \mathbf{1}, !_{\mathbf{D}} : \mathbf{D} \to \mathbf{1}, \varnothing)$ (see Lem. B.11) and $!_{\mathbf{D}}(\mathcal{M}'_2) \subseteq \{id_1\} = Isos$ this follows from Item 3. For morphisms $f = (f_1, f_2)$ and $g = (g_1, g_2)$ in $\mathbf{C} \times \mathbf{D}$ we construct the componentwise pair factorisations $((e_1, e'_1), m_1)$ of f_1, g_1 with $(e_1, e'_1) \in \mathcal{E}'_1$ and $m_1 \in \mathcal{M}'_1$, and $((e_2, e'_2), m_2)$ of f_2, g_2 with $(e_2, e'_2) \in \mathcal{E}'_2$ and $m_2 \in \mathcal{M}'_2$. This leads to morphisms $e = (e_1, e_2)$, $e' = (e'_1, e'_2)$, and $m = (m_1, m_2)$ in $\mathbf{C} \times \mathbf{D}$, and an $\mathcal{E}'$–$\mathcal{M}'$ pair factorisation with $(e, e') \in \mathcal{E}'$ and $m \in \mathcal{M}'$. If the $\mathcal{E}'_1$–$\mathcal{M}'_1$ and the $\mathcal{E}'_2$–$\mathcal{M}'_2$ pair factorisations are strong then also $\mathcal{E}'$–$\mathcal{M}'$ is a strong pair factorisation.

5. Since $\mathbf{C} \backslash X \cong ComCat(id_{\mathbf{C}} : \mathbf{C} \to \mathbf{C}, X : \mathbf{1} \to \mathbf{C}, \{1\})$ (see Lem. B.11) and $X(\mathcal{M}'_2) \subseteq X(\{id_1\}) = \{id_X\} \subseteq Isos$ this follows from Item 3. Given morphisms f and g in $\mathbf{C} \backslash X$, an $\mathcal{E}'_1$–$\mathcal{M}'_1$ pair factorisation of f and g in $\mathbf{C}$ is also an $\mathcal{E}'_1$–$\mathcal{M}'_1$ of f and g in $\mathbf{C} \backslash X$. If the $\mathcal{E}'_1$–$\mathcal{M}'_1$ pair factorisation is strong in $\mathbf{C}$ this is also true for $\mathbf{C} \backslash X$.

6. Given morphisms $f : (A, a') \to (C, c')$ and $g : (B, b') \to (C, c')$ in $X \backslash \mathbf{C}$, we have an $\mathcal{E}'_1$–$\mathcal{M}'_1$ pair factorisation $((e, e'), m)$ of f and g in $\mathbf{C}$. This is a pair factorisation in $X \backslash \mathbf{C}$ if $e \circ a' = e' \circ b'$, because then $(K, e \circ a')$ and $(K, e' \circ b')$ is the same object in $X \backslash \mathbf{C}$. If m is a monomorphism, this follows from $m \circ e \circ a' = f \circ a' = c' = g \circ b' = m \circ e' \circ b'$.

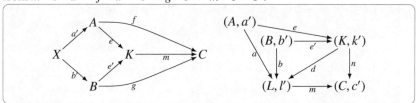

To prove that strongness is preserved we have to show the $\mathcal{E}'_1$–$\mathcal{M}'_1$ diagonal property in $X \backslash \mathbf{C}$. Since it holds in $\mathbf{C}$, given $(e, e') \in \mathcal{E}'$, $m \in \mathcal{M}'$, and morphisms a, b, n in $X \backslash \mathbf{C}$ with $n \circ e = m \circ a$ and $n \circ e' = m \circ b$ we get an induced unique $d : K \to L$ with $d \circ e = a$, $d \circ e' = b$, and $m \circ d = n$ from the diagonal property in $\mathbf{C}$. It remains to show that d is a valid morphism in $X \backslash \mathbf{C}$. Since $m \circ d \circ k' = n \circ k' = c' = m \circ l'$ and m is a monomorphism it follows that $d \circ k' = l'$ and thus $d \in X \backslash \mathbf{C}$.

7. Given morphisms $f = (f(x))_{x \in \mathbf{X}}$ and $g = (g(x))_{x \in \mathbf{X}}$ in $[\mathbf{X}, \mathbf{C}]$, we have an $\mathcal{E}'_1$–$\mathcal{M}'_1$ pair factorisation $((e_x, e'_x), m_x)$ with $m_x : K_x \to C(x)$ of $f(x), g(x)$ in $\mathbf{C}$ for all $x \in \mathbf{X}$. We have to show that $K(x) = K_x$ can be extended to a functor and that $e = (e_x)_{x \in \mathbf{X}}$, $e' = (e'_x)_{x \in \mathbf{X}}$, and $m = (m_x)_{x \in \mathbf{X}}$ are functor transformations. For a morphism $h : x \to y$ in $\mathbf{X}$ we use the $\mathcal{E}'_1$–$\mathcal{M}'_1$ diagonal property in $\mathbf{C}$ with $(e_x, e'_x) \in \mathcal{E}'_1$, $m_y \in \mathcal{M}'_1$ to define $K_h : K_x \to K_y$ as the unique induced morphism with $m_y \circ K_h = C(h) \circ m_x$, $K_h \circ e_x = e_y \circ A(h)$, and $K_h \circ e'_x = e'_y \circ B(h)$.

Using the uniqueness property of the strong pair factorisation in **C**, we can show that K with $K(x) = K_x$, $K(h) = K_h$ is a functor and by construction e, e', and m are functor transformations. This means that $(e, e') \in \mathcal{E}'$ and $m \in \mathcal{M}'$, i.e., this is an $\mathcal{E}'$–$\mathcal{M}'$ pair factorisation of f and g.

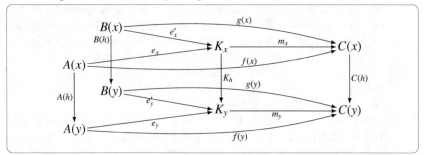

The $\mathcal{E}'$–$\mathcal{M}'$ diagonal property can be shown as follows. Given $(e, e') \in \mathcal{E}'$, $m \in \mathcal{M}'$, and morphisms a, b, n in $[\mathbf{X}, \mathbf{C}]$ from the $\mathcal{E}'_1$–$\mathcal{M}'_1$ diagonal property in **C**, we obtain a unique morphism $d_x : K(x) \rightarrow L(x)$ for $x \in \mathbf{X}$. It remains to show that $d = (d_x)_{x \in \mathbf{X}}$ is a functor transformation, i.e., we have to show for all $h : x \rightarrow y \in \mathbf{X}$ that $L(h) \circ d_x = d_y \circ K(h)$.

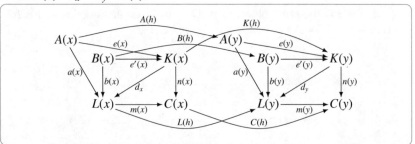

Because $(e(x), e'(x)) \in \mathcal{E}'_1$ and $m(y) \in \mathcal{M}'_1$, the $\mathcal{E}'_1$–$\mathcal{M}'_1$ diagonal property can be applied. This means that there is a unique $k : K(x) \rightarrow L(y)$ with $k \circ e(x) = L(h) \circ a(x)$, $k \circ e'(x) = L(h) \circ b(x)$, and $m(y) \circ k = n(y) \circ K(h)$.

For $L(h) \circ d_x$ we have that $L(h) \circ d_x \circ e(x) = L(h) \circ a(x)$, $L(h) \circ d_x \circ e'(x) = L(h) \circ b(x)$ and $m(y) \circ L(h) \circ d_x = C(h) \circ m(x) \circ d_x = C(h) \circ n(x) = n(y) \circ K(h)$. Similarly, for $d_y \circ K(h)$ we have that $d_y \circ K(h) \circ e(x) = d_y \circ e(y) \circ A(h) = a(y) \circ A(h) = L(h) \circ a(x)$, $d_y \circ K(h) \circ e'(x) = d_y \circ e'(y) \circ B(h) = b(y) \circ B(h) =$ 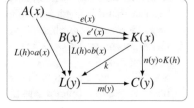 $L(h) \circ b(x)$, and $m(y) \circ d_y \circ K(h) = n(y) \circ K(h)$. Thus, from the uniqueness of k it follows that $k = L(h) \circ d_x = d_y \circ K(h)$ and d is a functor transformation. $\square$

B.4.3 Proof of Fact 4.55

Proof. 1. Given $f = (f_j) : A \to D \in \mathcal{M}'$, we have initial pushouts $(1)_j$ over $f_j \in \mathcal{M}'_j$ in $\mathbf{C}_j$ with $b_j, c_j \in \mathcal{M}_j$. Since $G_i(\mathcal{M}_{\ell_i}) \subseteq Isos$, $G_i(b_{\ell_i})^{-1}$ and $G_i(c_{\ell_i})^{-1}$ exist. Define objects $B = ((B_j), (op_i^B = G_i(b_{\ell_i})^{-1} \circ op_i^A \circ F_i(b_{k_i})))$ and $C = ((C_j), (op_i^C = G_i(c_{\ell_i})^{-1} \circ op_i^D \circ F_i(c_{k_i}))$ in $\mathbf{G}$. Then we have that

$$\begin{array}{ccc} B_j & \xrightarrow{b_j} & A_j \\ a_j\downarrow & (1)_j & \downarrow f_j \\ C_j & \xrightarrow{c_j} & D_j \end{array}$$

- $G_i(b_{\ell_i}) \circ op_i^B = G_i(b_{\ell_i}) \circ G_i(b_{\ell_i})^{-1} \circ op_i^A \circ F_i(b_{k_i}) = op_i^A \circ F_i(b_{k_i})$,
- $G_i(c_{\ell_i}) \circ op_i^C = G_i(c_{\ell_i}) \circ G_i(c_{\ell_i})^{-1} \circ op_i^D \circ F_i(c_{k_i}) = op_i^D \circ F_i(c_{k_i})$,
- $G_i(c_{\ell_i}) \circ G_i(a_{\ell_i}) \circ op_i^B = G_i(f_{\ell_i}) \circ G_i(b_{\ell_i}) \circ op_i^B = G_i(f_{\ell_i}) \circ op_i^A \circ F_i(b_{k_i}) = op_i^D \circ F_i(f_{k_i}) \circ F_i(b_{k_i}) = op_i^D \circ F_i(c_{k_i}) \circ F_i(a_{k_i}) = G_i(c_{\ell_i}) \circ op_i^C \circ F_i(a_{k_i})$ and since $G_i(c_{\ell_i})$ is an isomorphism this implies that $G_i(a_{\ell_i}) \circ op_i^B = op_i^C \circ F_i(a_{k_i})$, which

means that $a = (a_j)$, $b = (b_j)$, and $c = (c_j)$ are morphisms in $\mathbf{G}$ with $b, c \in \mathcal{M}'$, (1) is a valid square in $\mathbf{G}$, and by Lem. B.12 also a pushout.

$$\begin{array}{ccccc} ((B_j),(op_i^B)) & \xrightarrow{b} & ((A_j),(op_i^A)) & & ((A_j),(op_i^A)) & \xleftarrow{d} & ((E_j),(op_i^E)) \\ a\downarrow & (1) & \downarrow f & & f\downarrow & (2) & \downarrow g \\ ((C_j),(op_i^C)) & \xrightarrow{c} & ((D_j),(op_i^D)) & & ((D_j),(op_i^D)) & \xleftarrow{e} & ((F_j),(op_i^F)) \end{array}$$

It remains to show the initiality. For any pushout (2) in $\mathbf{G}$ with $d = (d_j), e = (e_j) \in \mathcal{M}$, Lem. B.12 implies that the components $(2)_j$ are pushouts in $\mathbf{C}_j$. The initiality of pushout $(1)_j$ implies that there are unique morphisms $b_j^* : B_j \to E_j$ and $c_j^* : C_j \to F_j$ with $d_j \circ b_j^* = b_j$, $e_j \circ c_j^* = c_j$, and $b_j^*, c_j^* \in \mathcal{M}_j$ such that $(3)_j$ is a pushout.

$$\begin{array}{ccccccc} A_j \xleftarrow{d_j} E_j & & B_j \xrightarrow{b_j^*} E_j & & ((B_j),(op_i^B)) \xrightarrow{b^*} ((A_j),(op_i^A)) \\ f_j\downarrow (2)_j \downarrow g_j & & a_j\downarrow (3)_j \downarrow g_j & & a\downarrow (3) \downarrow g \\ D_j \xleftarrow{e_j} F_j & & C_j \xrightarrow{c_j^*} F_j & & ((E_j),(op_i^E)) \xrightarrow{c^*} ((F_j),(op_i^F)) \end{array}$$

With $G_i(d_{\ell_i}) \circ G_i(b_{\ell_i}^*) \circ op_i^B = G_i(b_{\ell_i}) \circ op_i^B = op_i^A \circ F_i(b_{k_i}) = op_i^A \circ F_i(d_{k_i}) \circ F_i(b_{k_i}^*) = G_i(d_{\ell_i}) \circ op_i^E \circ F_i(b_{k_i}^*)$ and $G_i(d_{\ell_i})$ being an isomorphism it follows that $G_i(b_{\ell_i}^*) \circ op_i^B = op_i^E \circ F_i(b_{k_i}^*)$ and therefore $b^* = (b_j^*) \in \mathbf{G}$, and analogously $c^* = (c_j^*) \in \mathbf{G}$. This means that we have unique morphisms $b^*, c^* \in \mathcal{M}'$ with $d \circ b^* = b$ and $e \circ c^* = c$, and by Lem. B.12 (3) composed of $(3)_j$ is a pushout. Therefore (1) is the initial pushout over f in $\mathbf{G}$.

2. This is obvious.

3. Since comma categories are an instantiation of general comma categories, this follows directly from Item 1. The initial pushout of $f = (f_1, f_2) : (A_1, A_2, (op_i^A)) \to (D_1, D_2, (op_i^D)) \in \mathcal{M}'_1 \times \mathcal{M}'_2$ is the componentwise initial pushout in $\mathbf{C}$ and $\mathbf{D}$, with $B = (B_1, B_2, (op_i^B = G(b_2)^{-1} \circ op_i^A \circ F(b_1)))$ and $C = (C_1, C_2, (op_i^C = G(c_1)^{-1} \circ op_i^D \circ F(c_1)))$.

4. Since $\mathbf{C} \times \mathbf{D} \cong ComCat(!_{\mathbf{C}} : \mathbf{C} \to \mathbf{1}, !_{\mathbf{D}} : \mathbf{D} \to \mathbf{1}, \varnothing)$ (see Lem. B.11), $!_{\mathbf{C}}$ preserves pushouts, and $!_{\mathbf{D}}(\mathcal{M}_2) \subseteq \{id_1\} = Isos$, this follows from Item 3. The initial pushout (3) over a morphism $(f_1, f_2) : (A_1, A_2) \to (D_1, D_2) \in \mathcal{M}'_1 \times \mathcal{M}'_2$ is the componentwise product of the initial pushouts over f_1 in $\mathbf{C}$ and f_2 in $\mathbf{D}$.

5. Since $\mathbf{C}\backslash X \cong ComCat(id_{\mathbf{C}} : \mathbf{C} \to \mathbf{C}, X : \mathbf{1} \to \mathbf{C}, \{1\})$, $id_{\mathbf{C}}$ preserves pushouts, and $X(\mathcal{M}_2) = X(\{id_1\}) = \{id_X\} \subseteq Isos$, this follows from Item 3. The initial pushout over $f : (A, a') \to (D, d') \in \mathcal{M}'_1$ in $\mathbf{C}\backslash X$ is given by the initial pushout over f in $\mathbf{C}$, with objects (B, b'), (C, c'), $b' = a' \circ b$, and $c' = d' \circ c$.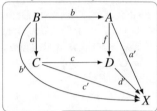

6. Given objects (A, a') and (D, d') and a morphism $f : A \to D$ in $X\backslash\mathbf{C}$ with $f \in \mathcal{M}'_1$, the initial pushout (1) over f in $\mathbf{C}$ exists by assumption. For any pushout (2) in $X\backslash\mathbf{C}$ with $d, e \in \mathcal{M}_1$, the corresponding diagram (3) is a pushout in $\mathbf{C}$. Since (1) is an initial pushout in $\mathbf{C}$ there exist unique morphisms $b^* : B \to E$ and $c^* : C \to F$ such that $d \circ b^* = b$, $e \circ c^* = c$, $b^*, c^* \in \mathcal{M}_1$, and (4) is a pushout in $\mathbf{C}$.

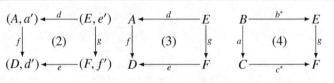

(i) If diagram (1) has a valid extension via morphisms $b' : X \to B$, $c' : X \to C$ in $X\backslash\mathbf{C}$, then this is also a pushout in $X\backslash\mathbf{C}$. With $d \circ b^* \circ b' = b \circ b' = a' = d \circ e'$ and d being a monomorphism it follows that $b^* \circ b = e'$ and thus $b^* \in X\backslash\mathbf{C}$, and analogously $c^* \in X\backslash\mathbf{C}$. This means that (4) is also a pushout in $X\backslash\mathbf{C}$.

(ii) If $a' : X \to A \in \mathcal{M}_1$ and the pushout complement of $f \circ a'$ in $\mathbf{C}$ exists, we can construct the unique pushout complement (5) in $\mathbf{C}$, and the corresponding diagram (6) is a pushout in $X\backslash\mathbf{C}$.

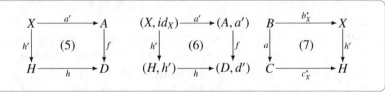

It remains to show the initiality of (6). For any pushout (2), $e' : X \to E$ is unique with respect to $d \circ e' = a'$ because d is a monomorphism. Since (1) is an initial pushout in $\mathbf{C}$ and (5) is a pushout, there are morphisms $b^*_X : B \to X$ and $c^*_X : C \to H$ such that $b^*_X, c^*_X \in \mathcal{M}_1$, $a' \circ b^*_X = b$, $h \circ c^*_X = c$, and (7) is a pushout in $\mathbf{C}$. With $e \circ c^* \circ a = c \circ a = h \circ c^*_X \circ a = h \circ h' \circ b^*_X = f \circ a' \circ b^*_X = f \circ d \circ e' \circ b^*_X = e \circ g \circ e' \circ b^*_X$ and e being a monomorphism it follows that $c^* \circ a = g \circ e' \circ b^*_X$.

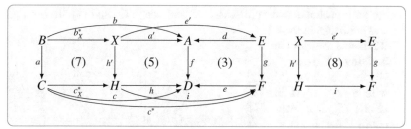

Pushout (7) implies that there is a unique $i : H \to F$ with $c^* = i \circ c_X^*$ and $i \circ h' = g \circ e'$. It further follows that $e \circ i = h$ using the pushout properties of H. By pushout decomposition, (8) is a pushout in $\mathbf{C}$ and the corresponding square in $X\backslash\mathbf{C}$ is also a pushout. Therefore, (6) is an initial pushout over f in $X\backslash\mathbf{C}$.

7. If $\mathbf{C}$ has intersections of $\mathcal{M}_1$-subobjects this means that given $c_i : C_i \to D \in \mathcal{M}_1$ with $i \in \mathcal{I}$ for some index set $\mathcal{I}$ the corresponding diagram has a limit $(C, (c_i' : C \to C_i)_{i \in \mathcal{I}}, c : C \to D)$ in $\mathbf{C}$ with $c_i \circ c_i' = c$ and $c, c_i' \in \mathcal{M}_1$ for all $i \in \mathcal{I}$.
   Let M denote the class of all $\mathcal{M}_1$-functor transformations. Given $f : A \to D \in \mathcal{M}'$, by assumption we can construct componentwise the initial pushout (1_x) over $f(x)$ in $\mathbf{C}$ for all $x \in \mathbf{X}$, with $b_0(x), c_0(x) \in \mathcal{M}_1$.

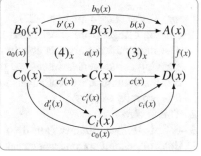

Define $(C, (c_i' : C \to C_i)_{i \in \mathcal{I}}, c : C \to D)$ as the limit in $[\mathbf{X}, \mathbf{C}]$ of all those $c_i : C_i \to D \in M$ such that for all $x \in \mathbf{X}$ there exists a $d_i'(x) : C_0(x) \to C_i(X) \in \mathcal{M}_1$ with $c_i(x) \circ d_i'(x) = c_0(x)$ (2), which defines the index set $\mathcal{I}$. Limits in $[\mathbf{X}, \mathbf{C}]$ are constructed componentwise in $\mathbf{C}$, and if $\mathbf{C}$ has intersections of $\mathcal{M}_1$-subobjects it follows that also $[\mathbf{X}, \mathbf{C}]$ has intersections of M-subobjects. Hence $c, c_i' \in M$ and $C(x)$ is the limit of $c_i(x)$ in $\mathbf{C}$. Now we construct the pullback (3) over $c \in M$ and f in $[\mathbf{X}, \mathbf{C}]$, and since M-morphisms are closed under pullbacks also $b \in M$.

For $x \in X$, $C(x)$ being the limit of $c_i(x)$, the family $(d_i'(x))_{i \in \mathcal{I}}$ with (2) implies that there is a unique morphism $c'(x) : C_0(x) \to C(x)$ with $c_i'(x) \circ c'(x) = d_i'(x)$ and $c(x) \circ c'(x) = c_0(x)$. Then $(3)_x$ is a pullback and $c(x) \circ c'(x) \circ a_0(x) = c_0(x) \circ a_0(x) = f(x) \circ b_0(x)$ implies the existence of a unique $b'(x) : B_0(x) \to B(x)$ with $b(x) \circ b'(x) = b_0(x)$ and $a(x) \circ b'(x) = c'(x) \circ a_0(x)$. $\mathcal{M}_1$ is closed under decomposition, $b_0(x) \in \mathcal{M}_1$, and $b(x) \in \mathcal{M}_1$ implies that $b'(x) \in \mathcal{M}_1$. Since $(1)_x$ is a pushout, $(3)_x$ is a pullback, the whole diagram commutes, and $c(x), b'(x) \in \mathcal{M}_1$, the $\mathcal{M}_1$ pushout–pullback property implies that $(3)_x$ and $(4)_x$

are both pushouts and pullbacks in $\mathbf{C}$ and hence (3) and (4) are both pushouts
and pullbacks in $[\mathbf{X}, \mathbf{C}]$.

It remains to show the
initiality of (3) over f.
Given a pushout (5)
with $b_1, c_1 \in \mathcal{M}$ in
$[\mathbf{X}, \mathbf{C}]$, (5_x) is a pushout in $\mathbf{C}$ for all $x \in \mathbf{X}$.
Since (1_x) is an initial
pushout in $\mathbf{C}$, there ex-

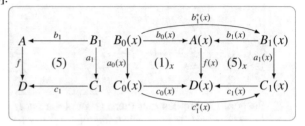

ist morphisms $b_1^*(x) : B_0(x) \to B_1(x)$, $c_1^* : C_0(x) \to C_1(x)$ with $b_1^*(x), c_1^*(x) \in$
$\mathcal{M}_1$, $b_1(x) \circ b_1^*(x) = b_0(x)$, and $c_1(x) \circ c_1^*(x) = c_0(x)$. Hence $c_1(x)$ satisfies (2) for
$i = 1$ and $d_1'(x) = c_1^*(x)$. This means that c_1 is one of the morphisms the limit C
was built of and there is a morphism $c_1' : C \to C_1$ with $c_1 \circ c_1' = c$ by construction
of the limit C.

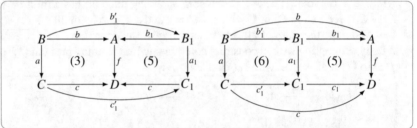

Since (5) is a pushout along $\mathcal{M}$-morphisms, it is also a pullback, and $f \circ b = c \circ a =$
$c_1 \circ c_1' \circ a$ implies that there exists a unique $b_1' : B \to B_1$ with $b_1 \circ b_1' = b$ and
$a_1 \circ b_1' = c_1' \circ a$. By $\mathcal{M}$-decomposition, also $b_1' \in \mathcal{M}$. Now using also $c_1 \in \mathcal{M}$, the
$\mathcal{M}$ pushout–pullback decomposition property implies that also (6) is a pushout,
which shows the initiality of (3). □

B.4.4 Proof of Fact 4.56

Proof. 1. As shown in Lem. B.12, pushouts over $\mathcal{M}$-morphisms in the general
    comma category are constructed componentwise in the underlying categories.
    The induced morphism is constructed from the induced morphisms in the un-
    derlying components. Since also pullbacks over $\mathcal{M}$-morphisms are constructed
    componentwise, the effective pushout property of the categories $(\mathbf{C}_j, \mathcal{M}_j)$ implies
    this property in $(\mathbf{G}, \mathcal{M})$.
2. This is obvious.
3.–6. This follows directly from Item 1, because all these categories are instantia-
    tions of general comma categories.
7. Pushouts and pullbacks over $\mathcal{M}$-morphisms as well as the induced morphisms are
    constructed pointwise in the functor category; thus the effective pushout property
    is directly induced. □

B.5 Proofs for Sect. 6.2

B.5.1 Proof of Fact 6.10

Proof. Consider the colimits $(\tilde{L}_s, (t_{i,L})_{i=0,...,n})$ of $(s_{i,L})_{i=1,...,n}$, $(\tilde{K}_s, (t_{i,K})_{i=0,...,n})$ of $(s_{i,K})_{i=1,...,n}$, and $(\tilde{R}_s, (t_{i,R})_{i=0,...,n})$ of $(s_{i,R})_{i=1,...,n}$, with $t_{0,*} = t_{i,*} \circ s_{i,*}$ for $* \in \{L, K, R\}$. Since $t_{i,L} \circ l_i \circ s_{i,K} = t_{i,L} \circ s_{i,L} \circ l_0 = t_{0,L} \circ l_0$, we get an induced morphism $\tilde{l}_s : \tilde{K}_s \to \tilde{L}_s$ with $\tilde{l}_s \circ t_{i,K} = t_{i,L} \circ l_i$ for $i = 0,\ldots,n$. Similarly, we obtain $\tilde{r}_s : \tilde{K}_s \to \tilde{R}_s$ with $\tilde{r}_s \circ t_{i,K} = t_{i,R} \circ r_i$ for $i = 0,\ldots,n$.

The colimit of a bundle of n morphisms can be constructed by iterated pushout constructions, which means that we only have to require pushouts over $\mathcal{M}$-morphisms. Since pushouts are closed under $\mathcal{M}$-morphisms, the iterated pushout construction leads to $t \in \mathcal{M}$.

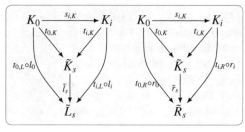

It remains to show that (14_i) (and $(14_i) + (1_i)$) and (15_i) (and $(15_i) + (2_i)$) are pullbacks, and (14_i) (and $(14_i) + (1_i)$) has a pushout complement for $t_{i,L} \circ l_i$. We prove this by induction over j for (14_i) (and $(14_i) + (1_i)$), the pullback property of (15_i) follows analogously. We prove: Let $\tilde{L}_j$ and $\tilde{K}_j$ be the colimits of $(s_{i,L})_{i=1,...,j}$ and $(s_{i,K})_{i=1,...,j}$, respectively. Then (16_{ij}) is

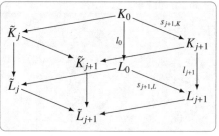

a pullback with pushout complement property for all $i = 0,\ldots,j$.

Basis $j = 1$: The colimits of $s_{1,L}$ and $s_{1,K}$ are L_1 and K_1, respectively, which means that $(16_{01}) = (1) + (16_{11})$ and (16_{11}) are both pushouts and pullbacks.

Induction step $j \to j + 1$: Construct $\tilde{L}_{j+1} = \tilde{L}_j +_{L_0} L_{j+1}$ and $\tilde{K}_{j+1} = \tilde{K}_j +_{K_0} K_{j+1}$ as pushouts, and we have the right cube with the top and bottom faces as pushouts, the back faces as pullbacks, and by the $\mathcal{M}$-van Kampen property also the front faces are pullbacks. Moreover, by Lem. B.3 the front faces have the pushout complement property, and by Lem. B.4 this also holds for (16_{0j}) and (16_{ij}) as compositions. Thus, for a given n, (16_{in}) is the required pullback (14_i) (and $(14_i) + (1_i)$) with pushout complement property, using $\tilde{K}_n = \tilde{K}_s$ and $\tilde{L}_n = \tilde{L}_s$. Obviously, $\tilde{ac}_s = \bigwedge_{i=1,...,n} \text{Shift}(t_{i,L}, ac_i) \Rightarrow \text{Shift}(t_{i,L}, ac_i)$ for all $i = 1,\ldots,n$, which completes the first part of the proof.

If ac_0 and ac_i are complement-compatible we have that $ac_i \cong \mathrm{Shift}(s_{i,L}, ac_0) \wedge L(p_i^*, \mathrm{Shift}(v_i, ac_i'))$. Consider the pullback (17_i), which is a pushout by M-push-out–pullback decomposition and the uniqueness of pushout complements, and the pushout (18_i). For ac_i', it holds that $\mathrm{Shift}(t_{i,L}, L(p_i^*, \mathrm{Shift}(v_i, ac_i'))) \cong L(\tilde{p}_s^*, \mathrm{Shift}(\tilde{k}_i \circ v_i, ac_i')) \cong L(\tilde{p}_s^*, \mathrm{Shift}(\tilde{v}, \mathrm{Shift}(\tilde{l}_i, ac_i')))$. Define $ac_i^* := \mathrm{Shift}(\tilde{l}_i, ac_i')$ as an application condition on $\tilde{L}_0$. It follows that $\tilde{ac}_s = \bigwedge_{i=1,\ldots,n} \mathrm{Shift}(t_{i,L}, ac_i) \cong \bigwedge_{i=1,\ldots,n}(\mathrm{Shift}(t_{i,L} \circ s_{i,L}, ac_0) \wedge \mathrm{Shift}(t_{i,L}, L(p_i^*, \mathrm{Shift}(v_i, ac_i')))) \cong \mathrm{Shift}(t_{0,L}, ac_0) \wedge \bigwedge_{i=1,\ldots,n} L(\tilde{p}_s^*, \mathrm{Shift}(\tilde{v}, ac_i^*))$.

For $i = 0$ define $ac'_{s0} = \bigwedge_{j=1,\ldots,n} ac_j^*$, and hence $\tilde{ac}_s \cong \mathrm{Shift}(t_{0,L}, ac_0) \wedge L(\tilde{p}_s^*, \mathrm{Shift}(\tilde{v}, ac'_{s0}))$ implies the complement-compatibility of ac_0 and $\tilde{ac}_s$. For $i > 0$, we have that $\mathrm{Shift}(t_{0,L}, ac_0) \wedge L(\tilde{p}_s^*, \mathrm{Shift}(\tilde{v}, ac_i^*)) \cong \mathrm{Shift}(t_{i,L}, ac_i)$. Define $ac'_{si} = \bigwedge_{j=1,\ldots,n \setminus i} ac_j^*$, and hence $\tilde{ac}_s \cong \mathrm{Shift}(t_{i,L}, ac_i) \wedge L(\tilde{p}_s^*, \mathrm{Shift}(\tilde{v}, ac'_{si}))$ implies the complement-compatibility of ac_i and $\tilde{ac}_s$. $\qquad\square$

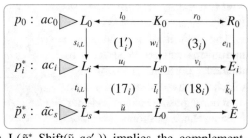

B.5.2 Proof of Fact 6.16

Proof. From Fact 6.8 it follows that each single direct transformation $G \xrightarrow{p_i, m_i} G_i$ can be decomposed into a transformation $G \xrightarrow{p_0, m_0^i} G_0^i \xrightarrow{\bar{p}_i, \bar{m}_i} G_i$ with $m_0^i = m_i \circ s_{i,L}$ and, since the bundle is s-amalgamable, $m_0 = m_i \circ s_{i,L} = m_0^i$ and $G_0 := G_0^i$ for all $i = 1, \ldots, n$.

It remains to show the pairwise parallel independence. From the constructions of the complement rule and the Concurrency Theorem we obtain the following diagram for all $i = 1, \ldots, n$.

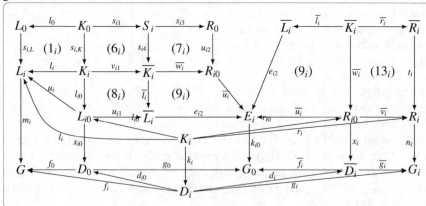

For $i \neq j$, from weakly independent matches it follows that we have a morphism $p_{ij} : L_{i0} \to D_j$ with $f_j \circ p_{ij} = m_i \circ u_i$. It follows that $f_j \circ p_{ij} \circ w_i = m_i \circ u_i \circ w_i = m_i \circ s_{i,L} \circ l_0 = m_0 \circ l_0 = m_j \circ s_{j,L} \circ l_0 = m_j \circ u_j \circ w_j = m_j \circ u_j \circ l_{j0} \circ s_{j,K} = m_j \circ l_j \circ s_{j,K} = f_j \circ k_j \circ s_{j,K}$ and with $f_j \in \mathcal{M}$ we have that $p_{ij} \circ w_i = k_j \circ s_{jk}$ (∗).

Now consider the pushout $(19_i) = (6_i) + (8_i)$ in comparison with object $\overline{D}_j$ and morphisms $d_j \circ p_{ij}$ and $x_j \circ u_{j2} \circ s_{i3}$. We have that $d_j \circ p_{ij} \circ l_{i0} \circ s_{i,K} = d_j \circ p_{ij} \circ w_i \overset{(*)}{=} d_j \circ k_j \circ s_{j,K} = x_j \circ r_{j0} \circ s_{j,K} = x_j \circ \overline{w}_j \circ v_{j1} \circ s_{j,K} = x_j \circ u_{j2} \circ s_{j3} \circ s_{j1} = x_j \circ u_{j2} \circ r_0 = x_j \circ u_{j2} \circ s_{i3} \circ s_{i1}$. Now pushout (19_i) induces a unique morphism q_{ij} with $q_{ij} \circ u_{i1} = d_j \circ p_{ij}$ and $q_{ij} \circ \overline{l}_i \circ s_{i4} = x_j \circ u_{j2} \circ s_{i3}$.

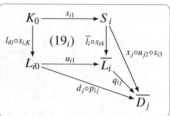

For the parallel independence of $G_0 \overset{\overline{p_i},\overline{m_i}}{\Longrightarrow} G_i$, $G_0 \overset{\overline{p_j},\overline{m_j}}{\Longrightarrow} G_j$, we have to show that $q_{ij} : \overline{L}_i \to \overline{D}_j$ satisfies $\overline{f}_j \circ q_{ij} = k_{i0} \circ e_{i2} =: \overline{m}_i$.

With $f_0 \in \mathcal{M}$ and $f_0 \circ d_{j0} \circ p_{ij} = f_j \circ p_{ij} = m_i \circ u_i = f_0 \circ c_{i0}$ it follows that $d_{j0} \circ p_{ij} = x_{i0}$ (∗∗). This means that $\overline{f}_j \circ q_{ij} \circ u_{i1} = \overline{f}_j \circ d_j \circ p_{ij} = g_0 \circ d_0 \circ p_{ij} \overset{(**)}{=} g_0 \circ x_{i0} = k_{i0} \circ e_{i2} \circ u_{i1}$. In addition, we have that $\overline{f}_j \circ q_{ij} \circ \overline{l}_i \circ s_{i4} = \overline{f}_j \circ x_j \circ u_{j2} \circ s_{i3} = k_{j0} \circ \overline{u}_j \circ u_{j2} \circ s_{i3} = k_{i0} \circ \overline{u}_i \circ u_{i2} \circ s_{i3} = k_{i0} \circ e_{i2} \circ \overline{l}_i \circ s_{i4}$. Since (19_i) is a pushout we have that u_{i1} and $\overline{l}_i \circ s_{i4}$ are jointly epimorphic and it follows that $\overline{f}_j \circ q_{ij} \circ e_{i2} = k_{i0} \circ e_{i2}$.

If ac_0 and ac_i are not complement-compatible then $\overline{ac}_i =$ true and trivially $\overline{g}_j \circ q_{ij} \models \overline{ac}_i$ for all $j \neq i$. Otherwise, we have that $g_j \circ p_{ij} \models ac'_i$, and with $g_j \circ p_{ij} = \overline{g}_j \circ d_j \circ p_{ij} = \overline{g}_j \circ q_{ij} \circ u_{i1}$ it follows that $\overline{g}_j \circ q_{ij} \circ u_{i1} \models ac'_i$, which is equivalent to $\overline{g}_j \circ q_{ij} \models \text{Shift}(u_{i1}, ac'_1) = \overline{ac}_i$.   □

B.5.3 Proof of Theorem 6.17

Proof. 1. We have to show that $\tilde{p}_s$ is applicable to G leading to an amalgamated transformation $G \overset{\tilde{p}_s,\tilde{m}}{\Longrightarrow} H$ with $m_i = \tilde{m} \circ t_{i,L}$, where $t_i : p_i \to \tilde{p}_i$ are the kernel morphisms constructed in Fact 6.10. Then we can apply Fact 6.8, which implies the decomposition of $G \overset{\tilde{p}_s,\tilde{m}}{\Longrightarrow} H$ into $G \overset{p_i,m_i}{\Longrightarrow} G_i \overset{q_i}{\Longrightarrow} H$, where q_i is the (weak) complement rule of the kernel morphism t_i.

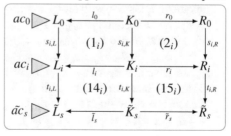

Given the kernel morphisms, the amalgamated rule, and the bundle of direct transformations, we have pullbacks (1_i), (2_i), (14_i), (15_i) and pushouts (20_i), (21_i).

Using Fact 6.16, we know that we can apply p_0 via m_0, leading to a direct transformation $G \overset{p_0,m_0}{\Longrightarrow} G_0$ given by pushouts (20_0) and (21_0). Moreover, we find decompositions of pushouts (20_0) and (20_i) into pushouts $(1'_i) + (22_i)$ and $(22_i) + (23_i)$ by $\mathcal{M}$-pushout–pullback decomposition and uniqueness of pushout complements.

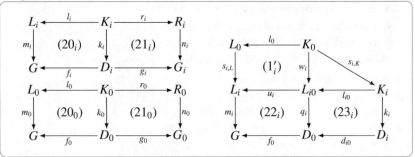

Since we have consistent matches, $m_i \circ s_{i,L} = m_0$ for all $i = 1, \ldots, n$. Then the colimit $\tilde{L}_s$ implies that there is a unique morphism $\tilde{m} : \tilde{L}_s \to G$ with $\tilde{m} \circ t_{i,L} = m_i$ and $\tilde{m} \circ t_{0,L} = m_0$ (a). Moreover, $m_i \models ac_i \Rightarrow \tilde{m} \circ t_{i,L} \models ac_i \Rightarrow \tilde{m} \models$ Shift$(t_{i,L}, ac_i)$ for all $i = 1, \ldots, n$, and thus $\tilde{m} \models \tilde{ac}_s = \bigwedge_{i=1,\ldots,n}$ Shift$(t_{i,L}, ac_i)$.

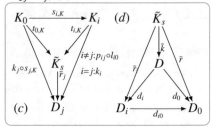

Weakly independent matches means that there exist morphisms p_{ij} with $f_j \circ p_{ij} = m_i \circ u_i$ for $i \neq j$. Construct D as the limit of $(d_{i0})_{i=1,\ldots,n}$ with morphisms d_i. Now f_0 is a monomorphism with $f_0 \circ d_{i0} \circ p_{ji} = f_i \circ p_{ji} = m_i \circ u_j = f_0 \circ q_j$, which implies that $d_{i0} \circ p_{ji} = q_j$. It follows that $d_{i0} \circ p_{ji} \circ l_{j0} = q_j \circ l_{j0}$ and, together with $d_{i0} \circ k_i = q_i \circ l_{i0}$, limit D implies that there exists a unique morphism r_j with $d_i \circ r_j = p_{ji} \circ l_{ji}$, $d_i \circ r_i = k_i$, and $d_0 \circ r_j = q_j \circ l_{j0}$ (b).

Similarly, f_j is a monomorphism with $f_j \circ p_{ij} \circ l_{i0} \circ s_{i,K} = m_i \circ u_i \circ w_i = m_i \circ s_{i,L} \circ l_0 = m_0 \circ l_0 = m_j \circ s_{j,L} \circ l_0 = m_j \circ l_j \circ s_{j,K} = f_j \circ k_j \circ s_{j,K}$, which implies that $p_{ij} \circ l_{i0} \circ s_{i,K} = k_j \circ s_{j,K}$. Now colimit $\tilde{K}_s$ implies that there is a unique morphism $\tilde{r}_j$ with $\tilde{r}_j \circ t_{i,K} = p_{ij} \circ l_{i0}$, $\tilde{r}_j \circ t_{j,K} = k_j$, and $\tilde{r}_j \circ t_{0,K} = k_j \circ s_{j,K}$

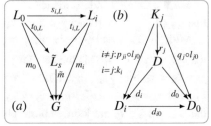

(c). Since $d_{i0} \circ \tilde{r}_i \circ t_{i,K} = d_{i0} \circ k_i = q_i \circ l_{i0} = d_{j0} \circ p_{ij} \circ l_{i0} = d_{j0} \circ \tilde{r}_j \circ t_{i,K}$ and $d_{i0} \circ \tilde{r}_i \circ t_{0,K} = d_{i0} \circ k_i \circ s_{i,K} = k_0 = d_{j0} \circ \tilde{r}_j \circ t_{0,K}$, colimit $\tilde{K}_s$ implies that for all i, j we have that $d_{i0} \circ \tilde{r}_i = d_{j0} \circ \tilde{r}_j =: \tilde{r}$. From limit D it now follows that there exists a unique morphism $\tilde{k}$ with $d_i \circ \tilde{k} = \tilde{r}_i$ and $d_0 \circ \tilde{k} = \tilde{r}$ (d).

We have to show that (20_s) with $f = f_0 \circ d_0$ is a pushout. With $f \circ \tilde{k} \circ t_{i,K} = f_0 \circ d_0 \circ \tilde{k} \circ t_{i,K} = f_0 \circ \tilde{r} \circ t_{i,K} = f_0 \circ d_{i0} \circ \tilde{r}_i \circ t_{i,K} = f_0 \circ d_{i0} \circ k_i = f_i \circ k_i = m_i \circ l_i = \tilde{m} \circ t_{i,L} \circ l_i = \tilde{m} \circ \tilde{l}_s \circ t_{i,K}$, $f \circ \tilde{k} \circ t_{0,K} = f_0 \circ d_0 \circ \tilde{k} \circ t_{0,K} = f_0 \tilde{r} \circ t_{0,K} = f_0 \circ d_{i0} \circ \tilde{r}_i \circ t_{0,K} = f_0 \circ d_{i0} \circ k_i \circ s_{i,K} = f_0 \circ k_0 = m_0 \circ l_0 = \tilde{m} \circ t_{0,L} \circ l_0 = \tilde{m} \circ \tilde{l}_s \circ t_{0,K}$, and $\tilde{K}_s$ as colimit, it follows that $f \circ \tilde{k} = \tilde{m} \circ \tilde{l}_s$, thus the square commutes.

$$
\begin{array}{ccccccccc}
L_s & \xleftarrow{\;\bar{l}_s\;} & K_s & K_i & \xrightarrow{\;r_i\;} & D & \xrightarrow{\;d_i\;} & D_i & +K_i & \xrightarrow{\;+l_{i0}\;} & +L_{i0} \\
\downarrow{\scriptstyle\bar{m}} & (20_s) & \downarrow{\scriptstyle\bar{k}} & \downarrow{\scriptstyle l_{i0}} & (24_i) & \downarrow{\scriptstyle x_i} & (25_i) & \downarrow{\scriptstyle d_{i0}} & \downarrow{\scriptstyle r} & (25) & \downarrow{\scriptstyle\bar{d}} \\
G & \xleftarrow{\;f\;} & D & L_{i0} & \xrightarrow{\;x_{i0}\;} & P_i & \xrightarrow{\;y_{i0}\;} & D_0 & D & \xrightarrow{\;d_0\;} & D_0
\end{array}
$$

Pushout (23_i) can be decomposed into pushouts (24_i) and (25_i). Using Lem. B.5 it follows that D_0 is the colimit of $(x_i)_{i=1,\dots,n}$, because (23_i) is a pushout, D is the limit of $(d_{i0})_{i=1,\dots,n}$, and we have morphisms p_{ij} with $d_{j0} \circ p_{ij} = q_i$. Then Lem. B.6 implies that also (25) is a pushout.

Now consider the coequalisers $\tilde{K}_s$ of $(i_{K_i} \circ s_{i,K} : K_0 \to +K_i)_{i=1,\dots,n}$ (which is actually $\tilde{K}_s$ by construction of colimits), $\tilde{L}_0$ of $(i_{L_{i0}} \circ w_i : K_0 \to +L_{i0})_{i=1,\dots,n}$ (as already constructed in Fact 6.10), D of $(\bar{k} \circ t_{0,K} : K_0 \to D)_{i=1,\dots,n}$, and D_0 of $(k_0 : K_0 \to D_0)_{i=1,\dots,n}$.

In the right cube, the top square with identical morphisms is a pushout, the top cube commutes, and the middle square is pushout (25). Using Lem. B.7 it follows that also the bottom face (26) constructed of the four coequalisers is a pushout.

In the cube below, the top and middle squares are pushouts and the two top cubes commute. Using again Lem. B.7 it follows that (20_s) in the bottom face is

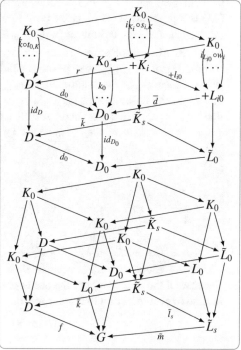

actually a pushout, where $(27) = (1'_i) + (17_i)$ is a pushout by composition. Now we can construct pushout (21_s), which completes the direct transformation $G \overset{\bar{p}_s,\bar{m}}{\Longrightarrow} H$.

$$
\begin{array}{ccccccccccccc}
\tilde{K}_s & \longrightarrow & \tilde{L}_0 & K_0 & \longrightarrow & \tilde{L}_0 & \tilde{L}_s & \xleftarrow{\;\bar{l}_s\;} & \tilde{K}_s & \xrightarrow{\;\bar{r}_s\;} & \tilde{R}_s \\
\downarrow{\scriptstyle\bar{k}} & (26) & \downarrow & \downarrow{\scriptstyle l_0} & (27) & \downarrow & \downarrow{\scriptstyle\bar{m}} & (20_s) & \downarrow{\scriptstyle\bar{k}} & (21_s) & \downarrow{\scriptstyle\bar{n}} \\
D & \xrightarrow{\;d_0\;} & D_0 & L_0 & \xrightarrow{\;t_{0,K}\;} & \tilde{L}_s & G & \xleftarrow{\;f\;} & D & \xrightarrow{\;g\;} & H
\end{array}
$$

2. Using the kernel morphisms t_i we obtain transformations $G \overset{p_i,m_i}{\Longrightarrow} G_i \overset{q_i}{\Longrightarrow} H$ from Fact 6.8 with $m_i = \bar{m} \circ t_{i,L}$. We have to show that this bundle of transformations is s-amalgamable. Applying again Fact 6.8 we obtain transformations $G \overset{p_0,m_0^i}{\Longrightarrow}$

$G_0^i \xrightarrow{\bar{p}_i} G_i$ with $m_0^i = m_i \circ s_{i,L}$. It follows that $m_0^i = m_i \circ s_{i,L} = \tilde{m} \circ t_{i,L} \circ s_{i,L} = \tilde{m} \circ t_{0,L} = \tilde{m} \circ t_{j,L} \circ s_{j,L} = m_j \circ s_{j,L}$ and thus we have consistent matches with $m_0 := m_0^i$ well defined and $G_0 = G_0^i$.

It remains to show the weakly independent matches. Given the above transformations we have pushouts (20_0), (20_i), (20_s) as above. Then we can find decompositions of (20_0) and (20_s) into pushouts $(27) + (28)$ and $(26) + (28)$, respectively. Using pushout (26) and Lem. B.8 it follows that (25) is a pushout, since $\tilde{K}_s$ is the colimit of

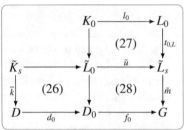

$(s_{i,L})_{i=1,\dots,n}$, $\tilde{L}_0$ is the colimit of $(w_i)_{i=1,\dots,n}$, and id_{K_0} is obviously an epimorphism. Now Lem. B.6 implies that there is a decomposition into pushouts (24_i) with colimit D_0 of $(x_i)_{i=1,\dots,n}$ and pushout (25_i) by M-pushout–pullback decomposition. Since D_0 is the colimit of $(x_i)_{i=1,\dots,n}$ and (25_j) is a pushout it follows that D_j is the colimit of $(x_i)_{i=1,\dots,j-1,j+1,\dots,n}$ with morphisms $q_{ij} : P_i \to D_j$ and $d_{j0} \circ q_{ij} = y_{i0}$. Thus we obtain for all $i \neq j$ a morphism $p_{ij} = q_{ij} \circ x_{i0}$ and $f_j \circ p_{ij} = f_0 \circ d_{j0} \circ q_{ij} \circ x_{i0} = f_0 \circ y_{i0} \circ x_{i0} = m_i \circ u_i$.

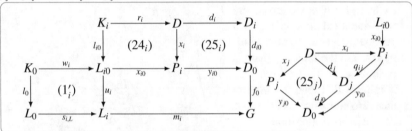

3. Because of the uniqueness of the used constructions, the above constructions are inverse to each other up to isomorphism.   □

B.5.4 Proof of Theorem 6.24

Proof. "if": If $G \xrightarrow{\bar{p}_s, \bar{m}} H$ and $G \xrightarrow{\bar{p}_{s'}, \bar{m}'} H'$ are parallel amalgamation independent define $r_{ij} = d'_j \circ \tilde{r}_s \circ t_{i,L}$ and $r'_{ji} = d_i \circ \tilde{r}_{s'} \circ t'_{j,L}$. It follows that $f_i \circ r'_{ji} = f_i \circ d_i \circ \tilde{r}_{s'} \circ t'_{j,L} = f \circ \tilde{r}_{s'} \circ t'_{j,L} = \tilde{m}' \circ t'_{j,L} = m'_j$, $f'_j \circ r_{ij} = f'_j \circ d'_j \circ \tilde{r}_s \circ t_{i,L} = f' \circ \tilde{r}_s \circ t_{i,L} = \tilde{m} \circ t_{i,L} = m_i$, and by precondition we have that $g_i \circ d_i \circ \tilde{r}_{s'} \models \text{Shift}(t_{j,L}, ac'_j)$, which means that $g_i \circ d_i \circ \tilde{r}_{s'} \circ t'_{j,L} = g_i \circ r'_{ji} \models ac'_j$. Similarly, $g'_j \circ d'_j \circ \tilde{r}_s \models \text{Shift}(t_{i,L}, ac_i)$ implies that $g'_j \circ d'_j \circ \tilde{r}_s \circ t_{i,L} = g'_j \circ r_{ij} \models ac_i$. This means that $G \xrightarrow{p_i, m_i} G_i$ and $G \xrightarrow{p'_j, m'_j} G'_j$ are pairwise parallel independent for all i, j.

The induced morphisms $\tilde{r}_s : \tilde{L}_s \to D'$ and $\tilde{r}_{s'} : \tilde{L}_{s'} \to D$ are exactly the morphisms $\tilde{r}_s$ and $\tilde{r}_{s'}$ given by parallel independence with $g' \circ \tilde{r}_s \models \tilde{ac}_s$ and $g \circ \tilde{r}_{s'} \models \tilde{ac}_{s'}$.

This means that $(G \xrightarrow{p_i,m_i} G_i)_{i=1,\dots,n}$ and $(G \xrightarrow{p'_j,m'_j} G'_j)_{j=1,\dots,n'}$ are parallel bundle independent.

"only if": Suppose $(G \xrightarrow{p_i,m_i} G_i)_{i=1,\dots,n}$ and $(G \xrightarrow{p'_j,m'_j} G'_j)_{j=1,\dots,n'}$ are parallel bundle independent. We have to show that the morphisms $\tilde{r}_s$ and $\tilde{r}_{s'}$ actually exist. D is the limit of $(d_{i0})_{i=1,\dots,n}$ as already constructed in the proof of Theorem 6.17. f_0 is an M-morphism, and $f_0 \circ d_{i0} \circ r'_{ji} = f_i \circ r'_{ji} = m_i = \tilde{m} \circ t_{i,L} = m_k = f_k \circ r'_{jk} = f_0 \circ d_{k0} \circ r'_{jk}$ implies that $d_{i0} \circ r'_{ji} = d_{k0} \circ r'_{jk} =: r'_{j0}$ for all i, k. Now the limit D implies that there exists a unique morphism r'_{js} such that $d_i \circ r'_{js} = r'_{ji}$ and $d_0 \circ r'_{js} = r'_{j0}$ (a).

Similarly, M-morphism f_i and $f_i \circ r'_{ji} \circ s_{j,L} = m'_j \circ s_{j,L} = m'_0 = m'_k \circ s'_{k,L} = f_i \circ r'_{ki} \circ s'_{k,L}$ imply that $r'_{ji} \circ s'_{j,L} = r'_{ki} \circ s'_{k,L}$ for all i, k. It follows that $d_i \circ r'_{js} \circ s_{j,L} = r'_{ji} \circ s'_{j,L} = r'_{ki} \circ s'_{k,L} = d_i \circ r'_{ks} \circ s'_{k,L}$, and with M-morphism d_i we have that $r'_{js} \circ s_{j,L} = r'_{ks} \circ s_{k,L} =: r'_{0s}$. From colimit $\tilde{L}_{s'}$ we obtain a morphism $\tilde{r}_{s'}$ with $\tilde{r}_{s'} \circ t'_{j,L} = r'_{j,s}$ and $\tilde{r}_{s'} \circ t'_{0,L} = r'_{0,s}$ (b).

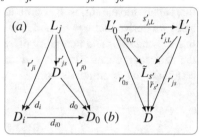

It follows that $f \circ \tilde{r}_{s'} \circ t'_{0,L} = f_i \circ d_i \circ r'_{0s} = f_i \circ d_i \circ r'_{js} \circ s'_{j,L} = f_i \circ r'_{ji} \circ s'_{j,L} = m'_j \circ s_{j,L} = m'_0 = \tilde{m}' \circ t_{0,L}$ and $f \circ \tilde{r}_{s'} \circ t_{j,L} = f_i \circ d_i \circ r'_{js} = f_i \circ r'_{ji} = m'_j = \tilde{m}' \circ t'_{j,L}$. The colimit property of $\tilde{L}_{s'}$ implies now that $f \circ \tilde{r}_{s'} = \tilde{m}'$. Similarly, we obtain the required morphism $\tilde{r}_s$ with $f' \circ \tilde{r}_s = \tilde{m}$.

Since we have already required that $g \circ \tilde{r}_{s'} \models \tilde{ac}_{s'}$ and $g' \circ \tilde{r}_s \models \tilde{ac}_s$, this means that $G \xrightarrow{\tilde{p}_s,\tilde{m}} H$ and $G \xrightarrow{\tilde{p}_{s'},\tilde{m}'} H'$ are parallel independent. Moreover, from the pairwise independence we know that $g_i \circ r'_{ji} = g_i \circ d_i \circ r'_{js} = g_i \circ d_i \circ \tilde{r}_{s'} \circ t'_{j,L} \models ac'_j$, which implies that $g_i \circ d_i \circ \tilde{r}_{s'} \models \text{Shift}(t'_{j,L} ac'_j)$. Similarly, $g'_j \circ r_{ij} \models ac_i$ implies that $g'_j \circ d'_j \circ \tilde{r}_s \models \text{Shift}(t_{i,L}, ac_i)$, which leads to parallel amalgamation independence of the amalgamated transformations.   □

B.6 Proofs for Chap. 7

In this section, we provide proofs for Lem. 7.9 and Facts 7.36 and 7.52 in Chap. 7.

B.6.1 Proof for Lem. 7.9

Proof. We show the result in two steps, starting with the data component (data nodes V_D and algebra D) and then handling the graph component (nodes V, edges E, attribution edges E_{NA}, E_{EA}). Let us consider an application condition $ac = \exists (L \rightarrow P, \vee_{i \in I} \exists (P \rightarrow C_i, \text{true}))$. The result for application conditions with more nesting structures and Boolean operators follows by induction while simpler conditions can be equivalently represented using identities as morphisms.

Data component: Let us consider one element of the disjunction of the shift construction. We derive the morphisms depicted in the commuting diagrams $(3), (4)$ below. Note that the vertical morphism $V_D^{L_S} \to V_D^L$ is an identity by the definition of *toS* (see Fig. 7.1). The further vertical morphisms are $\mathcal{M}$-morphisms by the shift construction, and thus isomorphisms on the data component. Note that the mapping for the data nodes is induced by the mapping for the algebra $(f_{V_D}(x) = f_D(x))$. Without loss of generality, we assume that the isomorphisms are identities as depicted in diagrams $(3D), (4D)$ for the data component. Moreover, we can create diagrams $(1), (2)$ for the original application condition, where Fig. 7.1 ensures that the vertical morphisms are identities. By commutativity of $(1D)$ and $(2D)$ and neutrality of identities we derive that $a = a_S$ and $a_i = a_{i,S}$ on the data component. By commutativity of $(3D)$ and $(4D)$ and neutrality of identities we derive that $a_S = a'$ and $a_{i,S} = a_i'$ on the data component. Thus, $a = a'$ and $a_i = a_i'$ on the data component and we can conclude that ac coincides with ac' on the data component.

Graph component: We derive diagrams $(1), (2)$ below by the definition of *toS* (see Fig. 7.1) and diagrams $(3), (4)$ by the shift construction using the general assumption that the application condition of the triple rule contains almost injective morphisms only. We now consider each triple component and derive diagrams $(1S)$–$(4S)$ for the source component, $(1C)$–$(4C)$ for the correspondence component and $(1T)$–$(4T)$ for the target component.

Source component: Note that p'^S is injective and the morphism pair (p'^S, a'^S) in diagram $(3S)$ is jointly surjective due to the shift construction and $j^S = id$ is an isomorphism. This implies that p'^S is an isomorphism. For diagram $(4S)$ we can apply the same arguments and derive that $c_i'^S$ is an isomorphism. Therefore, we derive that $ac \cong ac'$ on the source component for the graph component.

Correspondence component: We inspect the derived diagrams $(1C)$–$(4C)$. Recall that ac is an S-application condition, such that the C- and T-components consist of identities (id_L^C). The shift construction yields a disjunction with several solutions for diagrams $(3C)$ and $(4C)$ (jointly surjective). One solution for diagram $(3C)$ is given by $(1C)$ and one solution for diagram $(4C)$ is given by $(2C)$. Let $m\colon L \to G$ be a match satisfying ac, i.e., we have M-morphims $q^C\colon P^C = L^C \to G^C$ and $q_i^C\colon C_i^C = L^C \to G^C$. Since $(1C)$ and $(2C)$ are solutions for $(3C)$ and $(4C)$ as explained above, we know that morphisms q^C and q_i^C are compatible with m, such that $m \models ac'$. Vice versa, let $m\colon L \to G$ be a match satisfying ac', i.e., we have one sequence $(L \to P' \to C_i')$ of the disjunction by the shift construction and M-morphisms $q^C\colon P'^C \to G^C, q_i'^C\colon C_i'^C \to G^C$ compatible with m and (a', a_i'). If a'^C would not be injective, we could conclude the contradiction that $q'^C \circ a'^C \neq m^C$. Therefore, q'^C is injective and since a'^C is surjective $((\varnothing, a'^C)$ is jointly surjective), we can conclude that a'^C is an isomorphism. Therefore, $P'^C \cong P^C$ and we can obtain an isomorphism $isop\colon P^C \to P'^C$, such that $q^C = q'^C \circ isop \in M$ is compatible with m^C. Since $m \models ac'$ we have that $q'^C \circ a'^C = m^C$ and $q_i'^C \circ a_i'^C = q'^C$. If $a_i'^C$ would not be injective, we could conclude the contradiction $q_i'^C \circ a_i'^C \neq q'^C$. Therefore, $q_i'^C$ is injective and since $a_i'^C$ is surjective $((\varnothing, a_i'^C)$ is jointly surjective), we can conclude that $a_i'^C$ is an isomorphism. Therefore, $C_i'^C \cong C_i^C$ and we can obtain an isomorphism $isoc_{,i}\colon C_i^C \to C_i'^C$, such that $q_i^C = q_i'^C \circ isoc_{,i} \in M$ is compatible with q^C. Therefore, $ac \models m$. Combining both cases we have that $ac \equiv ac$ on the correspondence component for the graph component.

For the target component, we can perform the same steps as for the correpondence component and derive that $ac \equiv ac'$ on the target component for the graph component. Thus, $ac \equiv ac'$ for the graph component. Together with the proof for the data component above, we derive that $ac \equiv ac'$. □

In order to prove Fact 7.36, we use Def. B.14 and Lem. B.15 below concerning the equivalence of single transformation steps using the on-the-fly construction of model transformations based on forward rules presented in [EEHP09]. In this context, forward sequences are constructed with an on-the-fly check for partial source consistency. Partial source consistency requires that the constructed forward sequence $G_0 \overset{tr_F^*}{\Longrightarrow} G_k$ be partially match consistent, meaning that for each intermediate forward step $G_{k-1} \overset{tr_{k,F}}{\Longrightarrow} G_k$ the compatibility with the corresponding source step $G_{k-1,0} \overset{tr_{k,S}}{\Longrightarrow} G_{k,0}$ of the simultaneously created source sequence $G_{00} \overset{tr_S^*}{\Longrightarrow} G_{k,0}$ be checked. Compatibility requires that the forward match $m_{k,F}$ be forward consistent, which means that the co-match $n_{k,S}$ of the source step and the match $m_{k,F}$ of the forward step coincide on the source component with respect to the inclusion $G_{k-1,0} \hookrightarrow G_0 \hookrightarrow G_{k-1}$. The formal condition of a forward consistent match is given in Def. B.14 by a pullback diagram where both matches satisfy the corresponding NACs, and, intuitively, it specifies that the effective elements of the forward rule are matched for the first time in the forward sequence.

Definition B.14 (Forward-consistent match). Given a partially match-consistent

sequence $\varnothing = G_{00} \overset{tr_S^*}{\Longrightarrow} G_{n-1,0} \overset{g_n}{\hookrightarrow} G_0 \overset{tr_F^*}{\Longrightarrow} G_{n-1}$,
a match $m_{n,F} : L_{n,F} \to G_{n-1}$ for $tr_{n,F} : L_{n,F} \to$
$R_{n,F}$ is called *forward-consistent* if there is a source
match $m_{n,S}$ such that diagram (1) is a pullback and
the matches $m_{n,F}$ and $m_{n,S}$ satisfy the corresponding
target and source *NACs*, respectively. $\triangle$

$$
\begin{array}{ccc}
L_{n,S} \hookrightarrow R_{n,S} \hookrightarrow L_{n,F} \\
m_{n,S} \downarrow \quad (1) \qquad \downarrow m_{n,F} \\
G_{n-1,0} \underset{g_{n-1}}{\hookrightarrow} G_0 \hookrightarrow G_{n-1}
\end{array}
$$

Lemma B.15 (Forward translation step without ACs). *Let TR be a set of triple
rules without ACs with $tr_i \in TR$, and let TR_F, TR_{FT} be the derived sets of forward
and forward translation rules, repectively. Given a partially match consistent forward sequence $\varnothing = G_{00} \overset{tr_S^*}{\Longrightarrow} G_{i-1,0} \overset{g_{i-1}}{\hookrightarrow} G_0 \overset{tr_F^*}{\Longrightarrow} G_{i-1}$ and a corresponding forward translation sequence $G_0' \overset{tr_{FT}^*}{\Longrightarrow} G_{i-1}'$, both with almost injective matches, such
that $G_{i-1}' = G_{i-1} \oplus Att_{G_0 \setminus G_{i-1,0}}^F \oplus Att_{G_{i-1,0}}^T$, the following are equivalent:*

1. *There is a TGT step $G_{i-1} \overset{tr_{i,F},m_{i,F}}{\Longrightarrow} G_i$ with forward consistent match $m_{i,F}$*
2. *There is a forward translation TGT step $G_{i-1}' \overset{tr_{i,FT},m_{i,FT}}{\Longrightarrow} G_i'$*

and we have $G_i' = G_i \oplus Att_{G_0 \setminus G_{i,0}}^F \oplus Att_{G_{i,0}}^T$. $\triangle$

Proof. For simpler notation we assume w.l.o.g. that rule morphisms are inclusions
and matches are inclusions except for the data value component.
Constructions:
1. TGT step $G_{i-1} \overset{tr_F}{\Longrightarrow} G_i$ with forward consistent match is given by

$$
\begin{array}{ccccccc}
L_{i,S} & \overset{tr_{i,S}}{\hookrightarrow} & R_{i,S} & \hookrightarrow & L_{i,F} & \overset{tr_{i,F}}{\hookrightarrow} & R_{i,F} \\
m_{i,S} \downarrow & (1) & \downarrow n_{i,S} & (2) & m_{i,F} \downarrow & (3) & \downarrow n_{i,F} \\
G_{i-1,0} & \underset{t_{i,S}}{\hookrightarrow} & G_{i,0} & \underset{g_i}{\hookrightarrow} & G_0 \hookrightarrow G_{i-1} & \underset{t_{i,F}}{\hookrightarrow} & G_i
\end{array}
$$

where (1) and (3) are pushouts and pullbacks, (2) commutes, and since $m_{i,F}$ is forward consistent we have by Def. B.14 that (2) and therefore also (1+2) is a pullback.
    (1 + 2) is a pullback
$\Leftrightarrow m_{i,F}(L_{i,F}) \cap G_{i-1,0} = m_{i,F}(L_{i,S})$
$\Rightarrow m_{i,F}(L_{i,F} \setminus L_{i,S}) \cap G_{i-1,0} = \varnothing.$
2. Translation TGT step $G_{i-1}' \overset{tr_{i,FT},m_{i,FT}}{\Longrightarrow} G_i'$ is given by (PO_1), (PO_2)

$$
\begin{array}{ccccc}
L_{i,FT} & \longleftarrow & K_{i,FT} & \longrightarrow & R_{i,FT} \\
\downarrow & (PO_1) & \downarrow & (PO_2) & \downarrow \\
G_{i-1}' & \longleftarrow & D_{i-1}' & \longrightarrow & G_i'
\end{array}
$$

$L_{i,FT} = L_{i,F} \oplus Att_{L_{i,S}}^T \oplus Att_{R_{i,S} \setminus L_{i,S}}^F$ $K_{i,FT} = L_{i,F} \oplus Att_{L_{i,S}}^T$
$R_{i,FT} = R_{i,F} \oplus Att_{L_{i,S}}^T \oplus Att_{R_{i,S} \setminus L_{i,S}}^T = R_{i,F} \oplus Att_{R_{i,S}}^T$

Direction $1. \Rightarrow 2.$: We construct $(PO_1), (PO_2)$ as follows from diagrams (1)–(3):

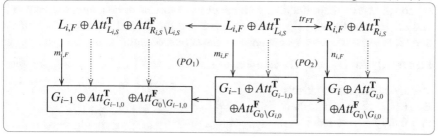

The match $m_{i,FT}$ is constructed as follows:

$$m_{i,FT}(x) = \begin{cases} m_{i,F}(x), & x \in L_{i,F} \\ \mathtt{tr\_}m_{i,F}(y), & x = \mathtt{tr\_}y, src_{L_{FT}}(x) = y \\ \mathtt{tr\_}m_{i,F}(y)\_a, & x = \mathtt{tr\_}y\_a, src_{L_{FT}}(x) = y \end{cases}$$

The match $m_{i,F}$ is injective except for the data value nodes. For this reason, the match $m_{i,FT}$ is an almost injective match, i.e., possibly noninjective on the data values.

Pushouts $(PO_1), (PO_2)$ are equivalent to pushouts $(0), (3)$ without translation attributes. Thus, the additional translation attributes are not involved in these pushouts.

$$\begin{array}{ccccc} L_{i,F} & \xleftarrow{\;id\;} & L_{i,F} & \longrightarrow & R_{i,F} \\ \downarrow & {\scriptstyle(0)} & \downarrow & {\scriptstyle(3)} & \downarrow \\ G_{i-1} & \xleftarrow{\;id\;} & G_{i-1} & \longrightarrow & G_i \end{array}$$

We now consider the translation attributes. Let $E_{i,0} = (G_0 \setminus G_{i-1,0}) \setminus (n_{i,S}(R_{i,S} \setminus L_{i,S}))$, constructed componentwise on the sets of nodes and edges. This implies that $E_{i,0}$ is a family of sets and not necessarily a graph, because some edges could be dangling. However, we only need to show the pushout properties for these sets, because the boundary nodes and context are handled properly in pushouts $(0), (3)$ earlier and the translation attribute edges for the items in $E_{i,0}$ are derived uniquely according to Def. 7.26. Thus, we have the following pushouts for the translation attributes:

$$\begin{array}{cc} L_{i,S} \longleftarrow L_{i,S} \\ \downarrow \; {\scriptstyle PO_1^T} \; \downarrow \\ G_{i-1,0} \longleftarrow G_{i-1,0} \end{array} \qquad \begin{array}{cc} (R_{i,S} \setminus L_{i,S}) \longleftarrow \emptyset \\ \downarrow \; {\scriptstyle PO_1^F} \; \downarrow \\ (G_0 \setminus G_{i-1,0}) \longleftarrow E_{i,0} \end{array} \qquad \begin{array}{cc} L_{i,S} \longrightarrow R_{i,S} \\ \downarrow \; {\scriptstyle PO_2^T} \; \downarrow \\ G_{i-1,0} \longrightarrow G_{i,0} \end{array} \qquad \begin{array}{cc} \emptyset \longrightarrow \emptyset \\ \downarrow \; {\scriptstyle PO_2^F} \; \downarrow \\ E_{i,0} \longrightarrow E_{i,0} \end{array}$$

Pushout (PO_1^T) is a trivial pushout, (PO_1^F) is pushout by the definition of $E_{i,0}$, (PO_2^T) is a pushout by (1) and (PO_2^F) is a trivial pushout. Using pushout (1) for the source step we have $G_{i,0} = G_{i-1,0} \cup (n_{i,S}(R_{i,S} \setminus L_{i,S}))$, and thus $E_{i,0} = (G_0 \setminus G_{i-1,0}) \setminus (n_{i,S}(R_{i,S} \setminus L_{i,S})) = G_0 \setminus (G_{i-1,0} \cup n_{i,S}(R_{i,S} \setminus L_{i,S})) = (G_0 \setminus G_{i,0})$. This implies $G_i' = G_i \oplus Att_{G_{i,0}}^\mathsf{T} \oplus Att_{G_0 \setminus G_{i,0}}^\mathsf{F}$.

Direction $2. \Rightarrow 1.$:

We construct diagrams (1)–(3) from pushouts $(PO_1), (PO_2)$. The pushouts (PO_1) and (PO_2) without translation attributes are equivalent to the pushouts $(0), (3)$ and $(PO_1^T), (PO_1^F), (PO_2^T), (PO_2^F)$ for families of sets. They do not overlap, because the have different types according to the construction of the type graph with attributes by Def. 7.26. The match is a forward translation match, and thus it is injective on all components except the data value nodes. It remains to construct diagrams (1) and (2) for graphs with (1) as a pushout. Since the C- and T-components of (1) and (2) are trivial it remains to construct the corresponding S-components, denoted here by $L_{i,S}^S$ for $L_{i,S}$, etc. The morphisms $L_{i,S}^S \xrightarrow{tr_{i,S}^S} R_{i,S}^S \xrightarrow{id} L_{i,F}^S \xrightarrow{m_{i,F}^S} G_{i-1}^S$ are given

already as graph morphisms. By (PO_2^T) we have a pushout in a family of sets and $G_{i-1,0}^S \subseteq G_0^S = G_{i-1}^S$ by assumption leads to a unique $G_{i-1,0}^S \hookrightarrow$ $G_{i-1}^S = G_0^S$, such that (4) and (5) below commute for fami-

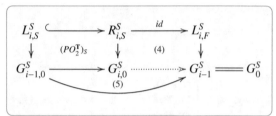

lies of sets, using the fact that $(PO_2^T)_S$ is a pullback, and hence also $(PO_2^T)_S + (4)$ is a pullback for families of sets.

Since $L_{i,S}^S \xrightarrow{tr_{i,S}^S} R_{i,S}^S \xrightarrow{id} L_{i,F}^S \xrightarrow{m_{i,F}^S} G_{i-1}^S = G_0^S$ and $G_{i-1,0}^S \hookrightarrow G_0^S$ are graph morphisms by assumption and $G_{i-1,0}^S \hookrightarrow G_0^S$ is injective, we also have that $L_{i,S}^S \to G_{i-1}^S$ is a graph morphism such that $(PO_2^T)_S$ becomes a pushout in **Graphs** with unique source and target maps for $G_{i,0}^S$. Finally, this implies that $G_{i,0}^S \to G_{i-1}^S = G_0^S$ is an injective graph morphism and w.l.o.g. an inclusion. Hence, we obtain the diagrams (1) and (2) for triple graphs from $(PO_2^T)_S$ and (4) for graphs, where (4) is a pushout and a pullback and (1)+(2) is a pullback by pullback (1) and injective $G_{i,0} \hookrightarrow G_0 \hookrightarrow G_{i-1}$.

Using pushout (1) for families of sets given by $(PO_2^T)_S$ we have $G_{i,0} = G_{i-1,0} \cup n_{i,S}(R_{i,S} \setminus L_{i,S})$, and thus $E_{i,0} = (G_0 \setminus G_{i,0})$, implying $G_i' = G_i \oplus Att_{G_{i,0}}^T \oplus Att_{G_0 \setminus G_{i,0}}^F$.

□

B.6.2 Proof for Fact 7.36

Proof. We first show the equivalence of the sequences disregarding the NACs. Item 1 is equivalent to the existence of the sequence $G_0 \xRightarrow{tr_{1,F},m_{1,F}} G_1 \xRightarrow{tr_{2,F},m_{2,F}} G_2 \ldots \xRightarrow{tr_{n,F},m_{n,F}} G_n$ with $G_n^S = G^S$, where each match is forward consistent according to Def. B.14. Item 2 is equivalent to the existence of the complete forward translation sequence $G_0' \xRightarrow{tr_{1,FT},m_{1,FT}} G_1' \xRightarrow{tr_{2,FT},m_{2,FT}} G_2' \ldots \xRightarrow{tr_{n,FT},m_{n,FT}} G_n'$ via TR_{FT}.

Disregarding the NACs, it remains to show that $G_0'^S = Att^F(G^S)$ and $G_n'^S = Att^T(G^S)$. We apply Lem. B.15 for $i = 0$ with $G_{0,0} = \varnothing$ up to $i = n$ with $G_{n,0} = G_0$, and using $G_0^S = G^S$ we derive:

$$G_0'^S = G_0^S \oplus Att_{G_{0,0}}^T \oplus Att_{G_0^S \setminus G_{0,0}}^F = G_0^S \oplus Att_{G_0^S}^F = G^S \oplus Att_{G^S}^F = Att^F(G^S).$$

$$G_n'^S = G_n^S \oplus Att_{G_{n,0}}^T \oplus Att_{G_0^S \setminus G_{n,0}}^F = G_n^S \oplus Att_{G_0^S}^T = G^S \oplus Att_{G^S}^T = Att^T(G^S).$$

Now, we show that the single steps are also NAC consistent. For each step, we have transformations $G_{i-1,0} \xrightarrow{tr_{i,S}, m_{i,S}} G_{i,0}$, $G_{i-1} \xrightarrow{tr_{i,F}, m_{i,F}} G_i$, $G'_{i-1} \xrightarrow{tr_{i,FT}, m_{i,FT}} G'_i$ with $G'_{i-1} = G_{i-1} \oplus Att^{\mathbf{F}}_{G_0 \setminus G_{i-1,0}} \oplus Att^{\mathbf{T}}_{G_{i-1,0}}$, $G'_i = G_i \oplus Att^{\mathbf{F}}_{G_0 \setminus G_{i,0}} \oplus Att^{\mathbf{T}}_{G_{i,0}}$, and $m_{i,FT}|_{L_{i,F}} = m_{i,F}$.

For a target NAC $n : L_i \to N$, we have to show that $m_{i,F} \models n$ iff $m_{i,FT} \models n_{FT}$, where n_{FT} is the corresponding forward translation NAC of n. If $m_{i,FT} \not\models n_{F_T}$, we find a monomorphism q' with $q' \circ n_{FT} = m_{i,FT}$. Since $n = n_{FT}|_N$, define $q = q'|_N$, and it follows that $q \circ n = m_{i,F}$, i.e., $m_{i,F} \not\models n$. Vice versa, if $m_{i,F} \not\models n$, we find a monomorphism q with $q \circ n = m_{i,F}$. Since $N^S = L_i^S$, we do not have any additional translation attributes in N_{FT}. Thus $m_{i,FT}$ can be extended by q to $q' : N_{FT} \to G'_{i-1}$ such that $m_{i,FT} \not\models n_{FT}$.

Similarly, we have to show that for a source NAC $n : L \to N$, $m_{i,S} \models n$ iff $m_{i,FT} \models n_{FT}$. As for target NACs, if $m_{i,FT} \not\models n_{FT}$, we find a monomorphism q' with $q' \circ n_{FT} = m_{i,FT}$ and for the restriction to L_i^S and N^S it follows that $q^S \circ n^S = m_{i,FT}^S$, i.e., $m_{i,S} \not\models n$. Vice versa, if $m_{i,S} \not\models n$, we find a monomorphism q with $q \circ n = m_{i,S}$. Now define q' with $q'(x) = m_{i,FT}(x)$ for $x \in L_{FT}$ and $q'(x) = q(x)$ for $x \in N \setminus L_i$, and for each $x \in N^S \setminus L_i^S$ we have that $q(x) \in G_{i-1,0}$. From the above characterisation of G'_{i-1} it follows that the corresponding translation attributes $\mathtt{tr}\_x$ and $\mathtt{tr}\_x\_a$ are set to $\mathbf{T}$ in G'_{i-1}. Thus, q' is well defined and $q' \circ n_{FT} = m_{i,FT}$, i.e., $m_{i,FT} \not\models n_{FT}$.

The equality of the model transformation relations follows by the equality of the pairs (G^S, G^T) in the model transformation sequences in both cases. $\square$

B.6.3 Proof for Fact 7.52

Proof. **1.** It is straightforward to show that $\mathcal{F}$ is a well-defined functor $\mathcal{F}$:

ATrGraphs $\to$ **AGraphs**. Given pushout (1) in **ATrGraphs** we have to show pushout (2) in **AGraphs**.

Pushout (1) in **ATrGraphs** is equivalent to commutativity of double cube (3) in **AGraphs** with three vertical pushouts, because pushouts in **ATrGraphs** are constructed componentwise as pushouts in **AGraphs**.

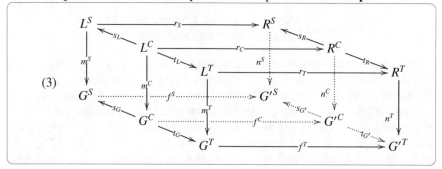

Diagram (2) is a pushout in **Graphs** iff the V- and E-components are pushouts in **Sets**. In the following we show this for the E-component, while the proof for the V-component is similar and even simpler, because we have no $Links$- and $Link_T$-parts. The E-component of (2) is shown in diagram (4) in **Sets**.

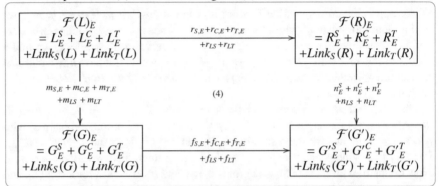

Diagram (4) is the disjoint union of five diagrams, where the S-, C- and T-part are the E-components of corresponding pushouts in (3) in **Graphs**, and hence pushouts in **Sets**. We will show that the $Links$-part (and similarly the $Link_T$-part) is pushout in **Sets**. Since coproducts of pushouts are again pushouts (in any category) (4) is a pushout in **Sets**. First of all, (5) commutes, because $\mathcal{F}$ is a functor and (5) is the LS-part of (2). In order to show the universal properties we assume we have $h_1 : Links(R) \rightarrow X$ and $h_2 : Links(G) \rightarrow X$ with $h_1 \circ r_{LS} = h_2 \circ m_{LS}$ and we have to construct a unique $h : Links(G') \rightarrow X$ such that (6) and (7) commute.

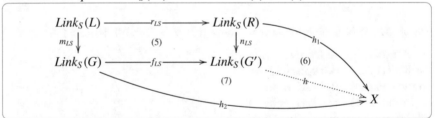

Given $(x, y) \in Links(G')$, we have $x \in G'^C_V$ and $y \in G'^S_V$ with $s_{G'}(x) = y$. Since G'^C in (3) is a pushout object we have either $x_1 \in R^C$ with $n^C(x_1) = x$ or $x_2 \in G^C$ with $f^C(x_2) = x$, leading to $h(x, y)$ defined by

$$h(x, y) = \begin{cases} h_1(x_1, y_1), & \text{for } n^C(x_1) = x \text{ and } y_1 = s_R(x_1) \\ h_2(x_2, y_2), & \text{for } f^C(x_2) = x \text{ and } y_2 = s_G(x_2). \end{cases}$$

$h_1(x_1, y_1)$ is well defined and (6) commutes, because $(x_1, y_1) \in Links(R)$ and $n^S(y_1) = n^S \circ s_R(x_1) = s_{G'}(x) = y$ implies $h \circ n_{LS}(x_1, y_1) = h(n^C(x_1), n^S(y_1)) = h(x, y)$. Similarly, $h_2(x_2, y_2)$ is well-defined and (7) commutes. It remains to show that h is well defined. For this purpose it is sufficient to show that for $x = n^C(x_1) = f^C(x_2)$ with $y_1 = s_R(x_1)$ and $y_2 = s_G(x_2)$ we can show $h_1(x_1, y_1) = h_2(x_2, y_2)$. The pushout of the C-component and $n^C(x_1) = f^C(x_2)$ imply existence of $x_0 \in L^C$ with $r^C(x_0) = x_1$ and $m^C(x_0) = x_2$, if r^C or n^C are injective. If both are not injective we have a chain $x_{01}, \ldots, x_{0n}$ connecting x_1 and x_2 and the proof is similar. In the injective case we have $y_0 = s_L(x_0)$ with $(x_0, y_0) \in Links(L)$ and $r_{LS}(x_0, y_0) = (r^C(x_0), r^S(y_0)) = (x_1, y_1)$, and similarly $m_{LS}(x_0, y_0) = (x_2, y_2)$. Using $h_1 \circ r_{LS} = h_2 \circ m_{LS}$ this implies $h_1(x_1, y_1) = h_1 \circ r_{LS}(x_0, y_0) = h_2 \circ m_{LS}(x_0, y_0) = h_2(x_2, y_2)$.

2. $\mathcal{F}: \textbf{ATrGraphs} \to \textbf{AGraphs}$ defines $\mathcal{F}_{TG}: \textbf{ATrGraphs}_{TG} \to \textbf{AGraphs}_{\mathcal{F}(TG)}$ because for each $(G, t: G \to TG)$ in $\textbf{ATrGraphs}_{TG}$ we have $(\mathcal{F}(G), \mathcal{F}(t)): \mathcal{F}(G) \to \mathcal{F}(TG))$ in $\textbf{AGraphs}_{\mathcal{F}(TG)}$ and for each morphism $f : (G, t) \to (G', t')$ with $f : G \to G'$ and $t' \circ f = t$ we have $\mathcal{F}(f): (\mathcal{F}(G), \mathcal{F}(t)) \to (\mathcal{F}(G'), \mathcal{F}(t'))$ with $\mathcal{F}(f): \mathcal{F}(G) \to \mathcal{F}(G')$ and $\mathcal{F}(t') \circ \mathcal{F}(f) = \mathcal{F}(t)$.

3. First, we show that $\mathcal{F}_{TG}$ is injective on objects. By construction of $\mathcal{F}(TG)$ we have $\mathcal{F}(TG)_V = TG_V^S + TG_V^C + TG_V^T$ and $\mathcal{F}(TG)_E = TG_E^S + TG_E^C + TG_E^T + Links(TG) + Link_T(TG)$ with $s_{\mathcal{F}(TG)}(x, y) = x$, $t_{\mathcal{F}(TG)}(x, y) = y$ and $TG_E^C = \emptyset$, and corresponding coproduct embeddings in $\textbf{AGraphs}$ and $\textbf{Sets}$. For (G, t) in $\textbf{TrGraphs}_{TG}$ with $t : G \to TG$ we have the following pullbacks (1)–(3) in $\textbf{Graphs}$ and (4)–(5) in $\textbf{Sets}$, because $\mathcal{F}(t) = t_S + t_C + t_T + t_{LS} + t_{LT}$.

These distinguished pullback constructions with inclusions in the upper and lower rows determine completely (G, t) with $G = (G^S \xleftarrow{s_G} G^C \xrightarrow{t_G} G^T)$ and $G_E^C = \emptyset$, where for each $e \in Links(G) \subseteq \mathcal{F}(G)_E$ with $s_{\mathcal{F}(G)}(e) = x$ and $t_{\mathcal{F}(G)}(e) = y$ we have $s_{G,V}(x) = y$. Vice versa, for each $x \in G_V^C, y \in G_V^S$ with $s_{G,V}(x) = y$ we have $e = (x, y) \in Links(G)$. Similarly, $Link_T(G)$ completely determines $t_{G,V}$, while $s_{G,E}$ and $t_{G,E}$ are empty. This implies for $(G, t), (G', t') \in \textbf{ATrGraphs}_{TG}$ with $(\mathcal{F}(G), \mathcal{F}(t)) = (\mathcal{F}(G'), \mathcal{F}(t'))$ that $(G, t) = (G', t')$, and hence the injectivity of $\mathcal{F}_{TG}$ on objects.

Next, we show that $\mathcal{F}_{TG}$ is injective on morphisms and creates morphisms. Using $\mathcal{F}(L) = L^S + L^C + L^T + Links(L) + Link_T(L)$ and $\mathcal{F}(G) = G^S + G^C + G^T + Links(G) + Link_T(G)$ we obtain unique $m^S : L^S \to G^S, m^C : L^C \to G^C, m^T : L^T \to G^T, m_{LS} : Links(L) \to Links(G), m_{LT} : Link_T(L) \to Link_T(G)$, which are type-compatible. For example, concerning m^S, graphs G^S and L^S are given by pullbacks $G^S = \mathcal{F}(type_G)^{-1}(TG^S) \subseteq \mathcal{F}(G)$ and $L^S = \mathcal{F}(type_L)^{-1}(TG^S) \subseteq \mathcal{F}(L)$, and m^S is the unique induced morphism leading to type compatibility.

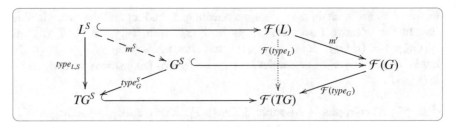

The triple graph morphism $m = (m^S,$
$m^C, m^T)$ is given by the right diagram,
where commutativity of (1) is shown below for the V-component. It is trivial for
the E-component, because $L_E^C = G_E^C = \emptyset$.

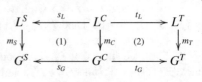

This construction implies $\mathcal{F}(m) =$
$m^S + m^C + m^T + m_{LS} + m_{LT} = m'$, because $m' : \mathcal{F}(L) \to \mathcal{F}(G)$ is uniquely determined
by the S-, C-, T-, $Links$- and $Link_T$-components. This also implies uniqueness of
m with $\mathcal{F}(m) = m'$. For commutativity of (1) (and similarly for (2)) we use the assumption that $m' : \mathcal{F}(L) \to \mathcal{F}(G)$ is a graph morphism in $\mathbf{AGraphs}_{\mathcal{F}(TG)}$ and hence
also in $\mathbf{AGraphs}$. This implies commutativity of (3).

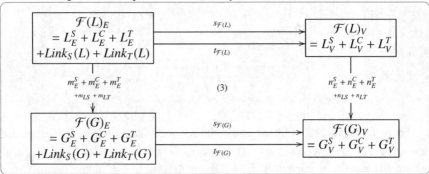

For all $(x, y) \in Links(L)$ we have $m_V \circ s_{\mathcal{F}(L)}(x, y) = s_{\mathcal{F}(G)} \circ m_E(x, y)$. This implies $s_{\mathcal{F}(G)}(m_{LS}(x, y)) = m_V^C(x)$, because we have $m_V \circ s_{\mathcal{F}(L)}(x, y) = m_V(x) = m_V^C(x)$
and $s_{\mathcal{F}(G)} \circ m_E(x, y) = s_{\mathcal{F}(G)}(m_{LS}(x, y))$. Similarly, $t_{\mathcal{F}(G)}(m_{LS}(x, y)) = m_V^S(y)$, which
implies $m_{LS}(x, y) = (m_V^C(x), m_V^S(y))$.

Now, we show that the V-component of (1) commutes. Given $x \in L_V^C$, we have
$s_{L,V}(x) = y$ and $(x, y) \in Links(L)$. For $(x, y) \in Links(L)$ we have $m_{LS}(x, y) =$
$(m_V^C(x), m_V^S(y)) \in Links(G')$. This implies $s_{G,V} \circ m_V^C(x) = m_V^S(y) = m_V^S \circ s_{L,V}(x)$.
By $\mathcal{F}_{TG}(f) = \mathcal{F}_{TG}(g)$ we can conlude that $f = g$ by the uniqueness of the creation
property, and hence the injectivity of $\mathcal{F}_{TG}$.

$A \cong B$ implies $\mathcal{F}_{TG}(A) \cong \mathcal{F}_{TG}(B)$, because $\mathcal{F}_{TG}$ is a functor. Vice versa, $\mathcal{F}_{TG}(A) \cong$
$\mathcal{F}_{TG}(B)$ implies isomorphisms $m_1' : \mathcal{F}_{TG}(A) \xrightarrow{\sim} \mathcal{F}_{TG}(B)$ and $m_2' : \mathcal{F}_{TG}(B) \xrightarrow{\sim} \mathcal{F}_{TG}(A)$
leading to unique morphisms $m_1 : A \to B$ and $m_2 : B \to A$ with $\mathcal{F}_{TG}(m_1) = m_1'$ and
$\mathcal{F}_{TG}(m_2) = m_2'$, because $\mathcal{F}_{TG}$ creates morphisms. Finally, $m_2 \circ m_1 = id_A$ (and similarly $m_1 \circ m_2 = id_B$), and hence $A \cong B$, because $\mathcal{F}_{TG}(m_2 \circ m_1) = \mathcal{F}_{TG}(m_2) \circ \mathcal{F}_{TG}(m_1) =$

$m'_2 \circ m'_1 = id_{\mathcal{F}_{TG}(A)} = \mathcal{F}_{TG}(id_A).$

4. $\mathcal{F}_{TG}$: **ATrGraphs**$_{TG}$ → **AGraphs**$_{\mathcal{F}(TG)}$ preserves pushout (1), because $\mathcal{F}$: **ATrGraphs** → **AGraphs** preserves pushouts by part 1 and pushouts in **ATrGraphs**$_{TG}$ and **AGraphs**$_{\mathcal{F}(TG)}$ are based on those in **ATrGraphs** and **AGraphs**, respectively. Vice versa, if (2) is a pushout in **ATrGraphs**$_{TG}$ and we let (1') be a pushout in **ATrGraphs**$_{TG}$, then also (2') is a pushout in **AGraphs**$_{\mathcal{F}(TG)}$.

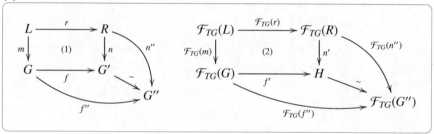

Uniqueness of pushouts implies $\mathcal{F}_{TG}(G') \cong \mathcal{F}_{TG}(G'')$ and hence $G' \cong G''$ by part 3, where $G' \cong G''$ is compatible with n, n' and f, f', respectively, showing that also (1) is pushout. Finally, we show that $\mathcal{F}_{TG}$ creates pushouts, given $r : L \to R$ and $m : L \to G$ in **TrGraphs**$_{TG}$, and that H is the pushout object of $\mathcal{F}_{TG}(r), \mathcal{F}_{TG}(m)$ in (2).

Let G'' with n'', f'' be pushout of r and m. Then $\mathcal{F}_{TG}(G'')$ is also pushout, and hence $H \cong \mathcal{F}_{TG}(G'')$. According to the construction in part 3 we can construct G' in **TrGraphs**$_{TG}$ with $\mathcal{F}_{TG}(G') \cong H$. Note that in this construction $s_{G',V}$ and $t_{G',V}$ are functions defined by $Link_S(G') = type_H^{-1}(Link_S(TG))$ and $Link_T(G') = type_H^{-1}(Link_T(TG))$, respectively, because $H \cong \mathcal{F}_{TG}(G'')$ and this functional property holds for $\mathcal{F}_{TG}$ by construction. Hence, we have $n' : \mathcal{F}_{TG}(R) \to \mathcal{F}_{TG}(G')$ and $f' : \mathcal{F}_{TG}(G) \to \mathcal{F}_{TG}(G')$ and by part 3 we have unique $n : R \to G'$ and $f : G \to G'$ with $\mathcal{F}_{TG}(n) = n'$ and $\mathcal{F}_{TG}(f) = f'$. Now reflection of pushouts implies that (1) is a pushout in **TrGraphs**$_{TG}$ with $G' \cong G''$.

5. Similarly to Item 1 we can show that $\mathcal{F}$ preserves pullbacks, because the $Link_S$ and $Link_T$ diagrams can be shown to be pullbacks and the disjoint union of pullbacks in **Sets** is again a pullback. This allows us to show preservation, reflection and creation of pullbacks similar to those of pushouts in Item 4. □

References

ABJ⁺10.   T. Arendt, E. Biermann, S. Jurack, C. Krause, and G. Taentzer. Henshin: Advanced concepts and tools for in-place EMF model transformations. In D. Petriu, N. Rouquette, and O. Haugen, editors, *Proc. of the ACM/IEEE 13th Intern. Conf. on Model Driven Engineering Languages and Systems (MoDELS'10)*, volume 6394 of *LNCS*, pages 121–135, 2010.

ABV07.   Víctor Anaya, Giuseppe Berio, and Maria Jose Verdecho. Evaluating Quality of Enterprise Modelling Languages: The UEML solution. In Ricardo Jardim-Gonçalves, Jörg P. Müller, Kai Mertins, and Martin Zelm, editors, *Enterprise Interoperability II - New Challenges and Industrial Approaches, Proc. Int. Conf. on Interoperability for Enterprise Software and Applications (IESA 2007)*, pages 237–240. Springer, 2007.

ACG⁺14.   Anthony Anjorin, Alcino Cunha, Holger Giese, Frank Hermann, Arend Rensink, and Andy Schürr. Benchmarx. In *Proceedings of the Workshops of the EDBT/ICDT 2014 Joint Conference (EDBT/ICDT 2014), Athens, Greece, March 28, 2014.*, volume 1133 of *CEUR Workshop Proceedings*, pages 82–86. CEUR-WS.org, 2014.

AGG14.   TFS-Group, TU Berlin. *AGG*, 2014. http://tfs.tu-berlin.de/agg.

AHRT14.   Thorsten Arendt, Annegret Habel, Hendrik Radke, and Gabriele Taentzer. From Core OCL invariants to nested graph constraints. In Holger Giese and Barbara König, editors, *Proc. Int. Conf. on Graph Transformation, ICGT 2014*, volume 8571 of *Lecture Notes in Computer Science*, pages 97–112. Springer, 2014.

AHS90.   J. Adámek, H. Herrlich, and G. Strecker. *Abstract and Concrete Categories*. Wiley, 1990.

ALMW09.   Jesper Andersson, Rogério Lemos, Sam Malek, and Danny Weyns. Modeling dimensions of self-adaptive software systems. In *Software Engineering for Self-Adaptive Systems*, pages 27–47. Springer, 2009.

ALPS11.   A. Anjorin, M. Lauder, S. Patzina, and A. Schürr. eMoflon: Leveraging EMF and Professional CASE Tools. In *INFORMATIK 2011*, volume 192 of *Lecture Notes in Informatics*, page 281. Gesellschaft für Informatik, 2011. Extended abstract.

AR02.   Farhad Arbab and Jan J. M. M. Rutten. A coinductive calculus of component connectors. In Martin Wirsing, Dirk Pattinson, and Rolf Hennicker, editors, *WADT*, volume 2755 of *LNCS*, pages 34–55. Springer, 2002.

Arb04.   Farhad Arbab. Reo: A Channel-based Coordination Model for Component Composition. *Mathematical Structures in Computer Science*, 14(3):329–366, June 2004. Preprint available at http://homepages.cwi.nl/~farhad/MSCS03Reo.pdf.

Arb05.   Farhad Arbab. Abstract Behavior Types: A Foundation Model for Components and Their Composition. *Science of Computer Programming*, 55:3–52, March 2005.

ASLS14.   Anthony Anjorin, Karsten Saller, Malte Lochau, and Andy Schürr. Modularizing triple graph grammars using rule refinement. In Stefania Gnesi and Arend Rensink, editors, *Fundamental Approaches to Software Engineering - 17th International Conference, FASE 2014, Held as Part of the European Joint Conferences on Theory and Practice of Software, ETAPS 2014, Grenoble, France, April 5-13, 2014, Proceedings*, Lecture Notes in Computer Science, pages 340–354. Springer, 2014.

ASW09.   Kerstin Altmanninger, Martina Seidl, and Manuel Wimmer. A survey on model versioning approaches. *IJWIS*, 5(3):271–304, 2009.

AT13.   Thorsten Arendt and Gabriele Taentzer. A tool environment for quality assurance based on the Eclipse Modeling Framework. *Automated Software Engineering*, 20(2):141–184, 2013.

Bal00.   Paolo Baldan. *Modelling Concurrent Computations: from Contextual Petri Nets to Graph Grammars*. PhD thesis, Computer Science Department - University of Pisa, 2000.

BB09.   Nelly Bencomo and Gordon S. Blair. Using architecture models to support the generation and operation of component-based adaptive systems. In *Software Engineering for Self-Adaptive Systems*, pages 183–200, 2009.

BBCP04.   Paolo Baldan, Nadia Busi, Andrea Corradini, and G. Michele Pinna. Domain and event structure semantics for Petri nets with read and inhibitor arcs. *Theor. Comput. Sci.*, 323(1-3):129–189, 2004.

BCDW04.   Jeremy S. Bradbury, James R. Cordy, Juergen Dingel, and Michel Wermelinger. A survey of self-management in dynamic software architecture specifications. In *Proceedings of the 1st ACM SIGSOFT workshop on Self-managed systems (WOSS '04)*, pages 28–33. ACM, 2004.

BCF⁺10.   Davi M. J. Barbosa, Julien Cretin, Nate Foster, Michael Greenberg, and Benjamin C. Pierce. Matching lenses: alignment and view update. In Paul Hudak and Stephanie Weirich, editors, *Proceeding of the 15th ACM SIGPLAN international conference on Functional programming, ICFP 2010, Baltimore, Maryland, USA, September 27-29, 2010*, ICFP '10, pages 193–204, New York, NY, USA, 2010. ACM.

BCG⁺12.   Roberto Bruni, Andrea Corradini, Fabio Gadducci, Alberto Lluch-Lafuente, and Andrea Vandin. A conceptual framework for adaptation. In *Fundamental Approaches to Software Engineering - 15th International Conference, FASE 2012*, pages 240–254, Tallinn, Estonia, March 24 - April 1, 2012.

BCH⁺06.   Paolo Baldan, Andrea Corradini, Tobias Heindel, Barbara König, and Pawel Sobocinski. Processes for adhesive rewriting systems. In Luca Aceto and Anna Ingólfsdóttir, editors, *FoSSaCS*, volume 3921 of *Lecture Notes in Computer Science*, pages 202–216. Springer, 2006.

BEdLT04.   R. Bardohl, H. Ehrig, J. de Lara, and G. Taentzer. Integrating Meta Modelling with Graph Transformation for Efficient Visual Language Definition and Model Manipulation. In M. Wermelinger and T. Margaria-Steffen, editors, *Proc. Fundamental Aspects of Software Engineering 2004*, volume 2984 of *LNCS*. Springer, 2004.

BEE⁺13.   Antonio Bucchiarone, Hartmut Ehrig, Claudia Ermel, Patrizio Pelliccione, and Olga Runge. Modeling and analysis of self-adaptive systems based on graph transformation. Technical Report 2013/03, TU Berlin, 2013.

BEE⁺15.   Antonio Bucchiarone, Hartmut Ehrig, Claudia Ermel, Patrizio Pelliccione, and Olga Runge. Rule-based modeling and static analysis of self-adaptive systems by graph transformation. In Rocco De Nicola and Rolf Hennicker, editors, *Software, Services, and Systems*, volume 8950 of *Lecture Notes in Computer Science*, pages 582–601. Springer International Publishing, 2015.

BEGG10.   B. Braatz, H. Ehrig, K. Gabriel, and U. Golas. Finitary M-Adhesive Categories. In H. Ehrig, A. Rensink, G. Rozenberg, and A. Schürr, editors, *Proceedings of Intern. Conf. on Graph Transformation (ICGT' 10)*, volume 6372 of *LNCS*, pages 234–249. Springer, 2010.

BEJ10.   E. Biermann, C. Ermel, and S. Jurack. Modeling the "Ecore to GenModel" transformation with EMF Henshin. In *Proc. Transformation Tool Contest 2010 (TTC'10)*, 2010. `http://planet-research20.org/ttc2010/index.php?option=com_content&view=article&id=110&Itemid=152`.

BEK+06.   E. Biermann, K. Ehrig, C. Köhler, G. Kuhns, G. Taentzer, and E. Weiss. Graphical
            definition of in-place transformations in the Eclipse Modeling Framework. In O. Nier-
            strasz, J. Whittle, D. Harel, and G. Reggio, editors, *Proc. of the International Confer-
            ence on Model Driven Engineering Languages and Systems (MoDELS'06)*, volume
            4199 of *LNCS*, pages 425–439. Springer, Berlin, 2006.

BEL+10.   E. Biermann, C. Ermel, L. Lambers, U. Prange, and G. Taentzer. Introduction to
            AGG and EMF Tiger by modeling a conference scheduling system. *Int. Journal on
            Software Tools for Technology Transfer*, 12(3-4):245–261, July 2010.

BESW10.   E. Biermann, C. Ermel, J. Schmidt, and A. Warning. Visual modeling of controlled
            EMF model transformation using Henshin. *ECEASST*, 32:1–14, 2010.

BET12.   E. Biermann, C. Ermel, and G. Taentzer. Formal foundation of consistent EMF model
            transformations by algebraic graph transformation. *Software and Systems Modeling
            (SoSyM)*, 11(2):227–250, 2012.

Béz05.   Jean Bézivin. On the unification power of models. *Software and System Modeling*,
            4(2):171–188, 2005.

BFH87.   P. Böhm, H.-R. Fonio, and A. Habel. Amalgamation of graph transformations: a
            synchronization mechanism. *Computer and System Sciences (JCSS)*, 34:377–408,
            1987.

BFS00.   Peter Buneman, Mary Fernandez, and Dan Suciu. UnQL: a query language and alge-
            bra for semistructured data based on structural recursion. *The VLDB Journal*, 9(1):76–
            110, 2000.

BG08.   Antonio Bucchiarone and Juan Pablo Galeotti. Dynamic software architectures veri-
            fication using DynAlloy. In *GT-VMT 2008*, 2008.

BGH+05.   S. Burmester, H. Giese, M. Hirsch, D. Schilling, and M. Tichy. The Fujaba real-time
            tool suite: Model-driven development of safety-critical, real-time systems. In *Proc.
            27th Intern. Conf. on Software Engineering (ICSE), St. Louis, Missouri, USA*, May
            2005.

BGP07.   Luciano Baresi, Sam Guinea, and Liliana Pasquale. Self-healing BPEL processes
            with Dynamo and the JBoss rule engine. In *ESSPE'07*, pages 11–20. ACM, 2007.

BH10.   Christoph Brandt and Frank Hermann. How Far Can Enterprise Modeling for Banking
            Be Supported by Graph Transformation? In Hartmut Ehrig, Arend Rensink, Grzegorz
            Rozenberg, and Andy Schürr, editors, *Int. Conf. on Graph Transformation (ICGT
            2010)*, volume 6372 of *Lecture Notes in Computer Science*, pages 3–26. Springer,
            2010.

BHE09a.   Denes Bisztray, Reiko Heckel, and Hartmut Ehrig. Verification of architectural refac-
            torings: Rule extraction and tool support. *Electronic Communications of the EASST*,
            16, 2009.

BHE09b.   Christoph Brandt, Frank Hermann, and Thomas Engel. Modeling and Reconfigura-
            tion of critical Business Processes for the purpose of a Business Continuity Manage-
            ment respecting Security, Risk and Compliance requirements at Credit Suisse using
            Algebraic Graph Transformation. In *Enterprise Distributed Object Computing Con-
            ference Workshops, 2009. EDOCW 2009. 13th, Proc. International Workshop on Dy-
            namic and Declarative Business Processes (DDBP 2009)*, pages 64–71. IEEE Xplore
            Digital Library, 2009.

BHE09c.   Christoph Brandt, Frank Hermann, and Thomas Engel. Security and Consistency of
            IT and Business Models at Credit Suisse realized by Graph Constraints, Transforma-
            tion and Integration using Algebraic Graph Theory. In *Proc. Int. Conf. on Exploring
            Modeling Methods in Systems Analysis and Design 2009 (EMMSAD'09)*, volume 29
            of *LNBIP*, pages 339–352, Heidelberg, 2009. Springer.

BHEE10.   Christoph Brandt, Frank Hermann, Hartmut Ehrig, and Thomas Engel. Enterprise
            Modelling using Algebraic Graph Transformation - Extended Version. Technical Re-
            port 2010/06, TU Berlin, Fak. IV, 2010.

BHG11.   Christoph Brandt, Frank Hermann, and Jan Friso Groote. Generation and Evalua-
            tion of Business Continuity Processes; Using Algebraic Graph Transformation and

the mCRL2 Process Algebra. *Journal of Research and Practice in Information Technology*, pages 65–86, 2011.

BHTV05.   Luciano Baresi, Reiko Heckel, Sebastian Thöne, and Dániel Varró. Style-based modeling and refinement of service-oriented architectures. *Journal of Software and Systems Modeling (SOSYM)*, 5(2):187–207, 2005.

BKM⁺12.   Antonio Bucchiarone, Nawaz Khurshid, Annapaola Marconi, Marco Pistore, and Heorhi Raik. A car logistics scenario for context-aware adaptive service-based systems. In *ICSE Workshop on Principles of Engineering Service Oriented Systems*, pages 65–66, 2012.

BKPPT00.   P. Bottoni, M. Koch, F. Parisi-Presicce, and G. Taentzer. Consistency Checking and Visualization of OCL Constraints. In *UML 2000 - The Unified Modeling Language*, volume 1939 of *LNCS*. Springer, 2000.

BMSG⁺09.   Yuriy Brun, Giovanna Marzo Serugendo, Cristina Gacek, Holger Giese, Holger Kienle, Marin Litoiu, Hausi Müller, Mauro Pezzè, and Mary Shaw. Engineering self-adaptive systems through feedback loops. In Betty H.C. Cheng, Rogério de Lemos, Holger Giese, Paola Inverardi, and Jeff Magee, editors, *Software Engineering for Self-Adaptive Systems*, volume 5525 of *Lecture Notes in Computer Science*, pages 48–70. Springer, 2009.

BN89.   D. F. C. Brewer and M. J. Nash. The Chinese Wall Security Policy. In *IEEE Symposium on Security and Privacy*, pages 206–214, 1989.

BN96.   Peter Bernus and Laszlo Nemes. A framework to define a generic enterprise reference architecture and methodology. *Computer Integrated Manufacturing Systems*, 9(3):179 – 191, 1996.

BNS⁺05.   András Balogh, Attila Németh, András Schmidt, István Rath, Dávid Vágó, Dániel Varró, and András Pataricza. The VIATRA2 model transformation framework. In *Proc. European Conference on Model Driven Architecture (ECMDA'05)*, 2005.

BNW08.   S. Becker, M. Nagl, and B. Westfechtel. Incremental and interactive integrator tools for design product consistency. In Manfred Nagl and Wolfgang Marquardt, editors, *Collaborative and Distributed Chemical Engineering. From Understanding to Substantial Design Process Support*, volume 4970 of *LNCS*, pages 224–267. Springer, 2008.

Bor07.   Artur Boronat. *MOMENT: A Formal Framework for Model Management*. PhD thesis, Universitat Politècnica de València, 2007.

BP99.   Nadia Busi and G. Michele Pinna. Process semantics for Place/Transition nets with inhibitor and read arcs. *Fundamenta Informaticae*, 40(2-3):165–197, 1999.

BPSR09.   Felix Böse, Jakub Piotrowski, and Bernd Scholz-Reiter. Autonomously controlled storage management in vehicle logistics - applications of RFID and mobile computing systems. *International Journal of RT Technologies: Research an Application*, 1(1):57–76, 2009.

BPVR09.   Antonio Bucchiarone, Patrizio Pelliccione, Charlie Vattani, and Olga Runge. Self-repairing systems modeling and verification using AGG. In *Joint Working IEEE/IFIP Conference on Software Architecture 2009 & European Conference on Software Architecture (WICSA'09)*, 2009.

Bra13.   Christoph Brandt. *An Enterprise Modeling Framework for Banks using Algebraic Graph Transformation*. PhD thesis, TU Berlin, 2013.

CFH⁺09.   Krzysztof Czarnecki, J. Foster, Zhenjiang Hu, Ralf Lämmel, Andy Schürr, and James Terwilliger. Bidirectional Transformations: A Cross-Discipline Perspective. In *Proc. ICMT'09*, volume 5563 of *LNCS*, pages 260–283. Springer, 2009.

CH06.   Krzysztof Czarnecki and Simon Helsen. Feature-based survey of model transformation approaches. *IBM Systems Journal*, 45(3):621–645, 2006.

CHK⁺01.   Ned Chapin, Joanne E. Hale, Khaled Md. Kham, Juan F. Ramil, and Wui-Gee Tan. Types of software evolution and software maintenance. *Journal of Software Maintenance*, 13:3–30, January 2001.

CHS08.   Andrea Corradini, Frank Hermann, and Pawel Sobociński. Subobject Transformation Systems. *Applied Categorical Structures*, 16(3):389–419, February 2008.

CIO99.   The Chief Information Officers Council. *Federal Enterprise Architecture Framework Version 1.1*, September 1999. http://www.enterprise-architecture.info/Images/Documents/Federal%20EA%20Framework.pdf.

CMR96.   Andrea Corradini, Ugo Montanari, and Francesca Rossi. Graph processes. *Fundamenta Informaticae*, 26(3/4):241–265, 1996.

CNM06.   Massimiliano Colombo, Elisabetta Di Nitto, and Marco Mauri. Scene: A service composition execution environment supporting dynamic changes disciplined through rules. In *ICSOC*, pages 191–202, 2006.

CPEV05.   Gerardo Canfora, Massimiliano Di Penta, Raffaele Esposito, and Maria Luisa Villani. An approach for QoS-aware service composition based on genetic algorithms. In *Proceedings of the 2005 Conference on Genetic and Evolutionary Computation (GECCO '05)*, pages 1069–1075, 2005.

CSW08.   Tony Clark, Paul Sammut, and James Willans. *Applied metamodelling: a foundation for language driven development*, volume 2005. Ceteva, 2008.

Dij65.   E.W. Dijkstra. Solution of a problem in concurrent programming control. *Communications of the ACM*, 8(9):569, 1965.

Dis08.   Zinovy Diskin. Algebraic Models for Bidirectional Model Synchronization. In Krzysztof Czarnecki, Ileana Ober, Jean-Michel Bruel, Axel Uhl, and Markus Völter, editors, *Model Driven Engineering Languages and Systems*, volume 5301 of *Lecture Notes in Computer Science*, pages 21–36. Springer, 2008. 10.1007/978-3-540-87875-9_2.

Dis11.   Zinovy Diskin. Model Synchronization: Mappings, Tiles, and Categories. In *Generative and Transformational Techniques in Software Engineering III*, volume 6491 of *LNCS*, pages 92–165. Springer, 2011.

dLVA04.   J. de Lara, H. Vangheluwe, and M. Alfonseca. Meta-Modelling and Graph Grammars for Multi-Paradigm Modelling in AToM^3. *Software and System Modeling: Special Section on Graph Transformations and Visual Modeling Techniques*, 3(3):194–209, 2004.

DXC11a.   Zinovy Diskin, Yingfei Xiong, and Krzysztof Czarnecki. From state- to delta-based bidirectional model transformations: the asymmetric case. *Journal of Object Technology*, 10:6: 1–25, 2011.

DXC+11b.   Zinovy Diskin, Yingfei Xiong, Krzysztof Czarnecki, Hartmut Ehrig, Frank Hermann, and Fernando Orejas. From state- to delta-based bidirectional model transformations: The symmetric case. In Jon Whittle, Tony Clark, and Thomas Kühne, editors, *Model Driven Engineering Languages and Systems, 14th International Conference, MODELS 2011, Wellington, New Zealand, October 16-21, 2011. Proceedings*, volume 6981 of *Lecture Notes in Computer Science*, pages 304–318. Springer, 2011.

EDG+11.   Alexander Egyed, Andreas Demuth, Achraf Ghabi, Roberto Erick Lopez-Herrejon, Patrick Mäder, Alexander Nöhrer, and Alexander Reder. Fine-tuning model transformation: Change propagation in context of consistency, completeness, and human guidance. In *ICMT'11*, volume 6707 of *LNCS*, pages 1–14. Springer, 2011.

EE05.   H. Ehrig and K. Ehrig. Overview of Formal Concepts for Model Transformations based on Typed Attributed Graph Transformation. In *Proc. Int. Workshop on Graph and Model Transformation (GraMoT'05)*, volume 152 of *ENTCS*. Elsevier, September 2005.

EE08.   H. Ehrig and C. Ermel. Semantical Correctness and Completeness of Model Transformations using Graph and Rule Transformation. In *Proc. International Conference on Graph Transformation (ICGT'08)*, volume 5214 of *LNCS*, pages 194–210. Springer, 2008.

EE10.   C. Ermel and K. Ehrig. Graph modelling and transformation: Theory meets practice. *ECEASST*, 30:1–22, 2010.

EEE+07.   Hartmut Ehrig, Karsten Ehrig, Claudia Ermel, Frank Hermann, and Gabriele Taentzer. Information preserving bidirectional model transformations. In Matthew B. Dwyer and Antónia Lopes, editors, *Fundamental Approaches to Software Engineering*, volume 4422 of *LNCS*, pages 72–86. Springer, 2007.

EEH08a.   H. Ehrig, K. Ehrig, and F. Hermann. From Model Transformation to Model Integration based on the Algebraic Approach to Triple Graph Grammars. *ECEASST*, 10, 2008.

EEH08b.   H. Ehrig, K. Ehrig, and F. Hermann. From Model Transformation to Model Integration based on the Algebraic Approach to Triple Graph Grammars (Long Version). Technical Report 2008/03, Technische Universität Berlin, Fakultät IV, 2008.

EEH08c.   H. Ehrig, C. Ermel, and F. Hermann. On the Relationship of Model Transformations Based on Triple and Plain Graph Grammars. In G. Karsai and G. Taentzer, editors, *Proc. Third International Workshop on Graph and Model Transformation (GRaMoT'08)*, GRaMoT '08, pages 9–16, New York, NY, USA, 2008. ACM.

EEH08d.   H. Ehrig, C. Ermel, and F. Hermann. On the Relationship of Model Transformations Based on Triple and Plain Graph Grammars (Long Version). Technical Report 2008/05, Technische Universität Berlin, Fakultät IV, 2008.

EEHP09.   H. Ehrig, C. Ermel, F. Hermann, and U. Prange. On-the-Fly Construction, Correctness and Completeness of Model Transformations based on Triple Graph Grammars. In A. Schürr and B. Selic, editors, *ACM/IEEE 12th Int. Conf. on Model Driven Engineering Languages and Systems (MODELS'09)*, volume 5795 of *LNCS*, pages 241–255. Springer, 2009.

EEKR99.   H. Ehrig, G. Engels, H.-J. Kreowski, and G. Rozenberg, editors. *Handbook of Graph Grammars and Computing by Graph Transformation, Volume 2: Applications, Languages and Tools*. World Scientific, 1999.

EEPT05.   H. Ehrig, K. Ehrig, U. Prange, and G. Taentzer. Formal Integration of Inheritance with Typed Attributed Graph Transformation for Efficient VL Definition and Model Manipulation. In *Proc. IEEE Symposium on Visual Languages and Human-Centric Computing (VL/HCC'05)*, IEEE Computer Society, Dallas, Texas, USA, September 2005.

EEPT06.   H. Ehrig, K. Ehrig, U. Prange, and G. Taentzer. *Fundamentals of Algebraic Graph Transformation*. EATCS Monographs in Theor. Comp. Science. Springer, 2006.

EER⁺10.   H. Ehrig, C. Ermel, O. Runge, A. Bucchiarone, and P. Pelliccione. Formal analysis and verification of self-healing systems. In D. Rosenblum and G. Taentzer, editors, *Proc. Intern. Conf. on Fundamental Aspects of Software Engineering (FASE'10)*, volume 6013 of *LNCS*, pages 139–153. Springer, 2010.

EET11.   Hartmut Ehrig, Claudia Ermel, and Gabriele Taentzer. A formal resolution strategy for operation-based conflicts in model versioning using graph modifications. In Dimitra Giannakopoulou and Fernando Orejas, editors, *Int. Conf. on Fundamental Approaches to Software Engineering (FASE'11)*, volume 6603 of *Lecture Notes in Computer Science*, pages 202–216. Springer, 2011.

EGH10.   Hartmut Ehrig, Ulrike Golas, and Frank Hermann. Categorical Frameworks for Graph Transformation and HLR Systems based on the DPO Approach. *Bulletin of the EATCS*, 102:111–121, 2010.

EGH⁺12.   Hartmut Ehrig, Ulrike Golas, Annegret Habel, Leen Lambers, and Fernando Orejas. M-Adhesive Transformation Systems with Nested Application Conditions. Part 2: Embedding, Critical Pairs and Local Confluence. *Fundam. Inform.*, 118(1-2):35–63, 2012.

EGH⁺14.   Hartmut Ehrig, Ulrike Golas, Annegret Habel, Leen Lambers, and Fernando Orejas. M-adhesive transformation systems with nested application conditions. part 1: parallelism, concurrency and amalgamation. *Mathematical Structures in Computer Science*, 24(4):1–48, 2014.

EGLT11.   Claudia Ermel, Jürgen Gall, Leen Lambers, and Gabriele Taentzer. Modeling with plausibility checking: Inspecting favorable and critical signs for consistency between control flow and functional behavior. Technical Report 2011/2, TU Berlin, 2011.

EGSW07.   Gregor Engels, Baris Güldali, Christian Soltenborn, and Heike Wehrheim. Assuring consistency of business process models and web services using visual contracts.

In *Applications of Graph Transformations with Industrial Relevance, Third International Symposium, AGTIVE 2007, Kassel, Germany, October 10-12, 2007, Revised Selected and Invited Papers*, volume 5088 of *LNCS*, pages 17–31. Springer, 2007.

EHGB12. Claudia Ermel, Frank Hermann, Jürgen Gall, and Daniel Binanzer. Visual modeling and analysis of EMF model transformations based on triple graph grammars. *ECE-ASST*, 54:1–14, 2012.

EHK+96. H. Ehrig, R. Heckel, M. Korff, M. Löwe, L. Ribeiro, A. Wagner, and A. Corradini. Algebraic approaches to graph transformation II: Single pushout approach and comparison with double pushout approach. In G. Rozenberg, editor, *The Handbook of Graph Grammars and Computing by Graph Transformations, Volume 1: Foundations*, pages 247–312. World Scientific, 1996.

EHKP91a. H. Ehrig, A. Habel, H.-J. Kreowski, and F. Parisi-Presicce. From graph grammars to high level replacement systems. In *4th Int. Workshop on Graph Grammars and their Application to Computer Science*, volume 532 of *LNCS*, pages 269–291. Springer, 1991.

EHKP91b. H. Ehrig, A. Habel, H.-J. Kreowski, and F. Parisi-Presicce. Parallelism and concurrency in high-level replacement systems. *Math. Struct. in Comp. Science*, 1:361–404, 1991.

EHL10. Hartmut Ehrig, Annegret Habel, and Leen Lambers. Parallelism and Concurrency Theorems for Rules with Nested Application Conditions. *Electr. Communications of the EASST*, 26:1–24, 2010.

EHPP04. H. Ehrig, A. Habel, J. Padberg, and U. Prange. Adhesive high-level replacement categories and systems. In F. Parisi-Presicce, P. Bottoni, and G. Engels, editors, *Proc. 2nd Int. Conference on Graph Transformation (ICGT'04)*, volume 3256 of *LNCS*, pages 144–160, Rome, Italy, October 2004. Springer.

Ehr79. H. Ehrig. Introduction to the Algebraic Theory of Graph Grammars (A Survey). In *Graph Grammars and their Application to Computer Science and Biology*, volume 73 of *LNCS*, pages 1–69. Springer, 1979.

EHS09. Hartmut Ehrig, Frank Hermann, and Christoph Sartorius. Completeness and Correctness of Model Transformations based on Triple Graph Grammars with Negative Application Conditions. *ECEASST*, 18, 2009.

EHSB11. H. Ehrig, F. Hermann, H. Schölzel, and C. Brandt. Propagation of Constraints along Model Transformations Based on Triple Graph Grammars. *ECEASST*, 41, 2011.

EHSB13. Hartmut Ehrig, Frank Hermann, Hanna Schölzel, and Christoph Brandt. Propagation of constraints along model transformations using triple graph grammars and borrowed context. *Visual Languages and Computing*, 24(5):365–388, 2013.

EKMR99. H. Ehrig, H.-J. Kreowski, U. Montanari, and G. Rozenberg, editors. *Handbook of Graph Grammars and Computing by Graph Transformation. Vol 3: Concurrency, Parallelism and Distribution*. World Scientific, 1999.

EM85. H. Ehrig and B. Mahr. *Fundamentals of Algebraic Specification 1: Equations and Initial Semantics*, volume 6 of *EATCS Monographs on Theoretical Computer Science*. Springer, Berlin, 1985.

EM90. H. Ehrig and B. Mahr. *Fundamentals of Algebraic Specification 2: Module Specifications and Constraints*, volume 21 of *EATCS Monographs on Theoretical Computer Science*. Springer, Berlin, 1990.

EMC+01. H. Ehrig, B. Mahr, F. Cornelius, M. Grosse-Rhode, P. Zeitz, G. Schröter, and K. Robering. *Mathematisch Strukturelle Grundlagen der Informatik, 2. überarbeitete Auflage*. Springer, 2001.

EMF14. Eclipse Consortium. *Eclipse Modeling Framework (EMF) – Version 2.9.2*, 2014. http://www.eclipse.org/modeling/emf/.

EMT09. TFS-Group, TU Berlin. *EMF Tiger*, 2009. http://tfs.cs.tu-berlin.de/emftrans.

EPS73. H. Ehrig, M. Pfender, and H.J. Schneider. Graph grammars: an algebraic approach. In *14th Annual IEEE Symposium on Switching and Automata Theory*, pages 167–180. IEEE, 1973.

EPT04.   H. Ehrig, U. Prange, and G. Taentzer. Fundamental theory for typed attributed graph transformation. In F. Parisi-Presicce, P. Bottoni, and G. Engels, editors, *Proc. 2nd Int. Conference on Graph Transformation (ICGT'04), Rome, Italy*, volume 3256 of *LNCS*. Springer, 2004.

ER76.   H. Ehrig and B.K. Rosen. Commutativity of independent transformations on complex objects. Research Report RC 6251, IBM T. J. Watson Research Center, Yorktown Heights, 1976.

Erm06.   C. Ermel. *Simulation and Animation of Visual Languages based on Typed Algebraic Graph Transformation*. PhD thesis, Technische Universität Berlin, Fak. IV, Books on Demand, Norderstedt, 2006.

Erm09.   Claudia Ermel. Visual modelling and analysis of model transformations based on graph transformation. *Bulletin of the EATCS*, 99:135 – 152, 2009.

EW04.   R. Eshuis and R. Wieringa. Tool support for verifying uml activity diagrams. *IEEE Trans. on Software Eng.*, 7(30), 2004.

FG98.   Mark S. Fox and Michael Grüninger. Enterprise Modeling. *AI Magazine*, 19(3):109–121, 1998.

FGM+07.   J. Nathan Foster, Michael B. Greenwald, Jonathan T. Moore, Benjamin C. Pierce, and Alan Schmitt. Combinators for bidirectional tree transformations: A linguistic approach to the view-update problem. *ACM Trans. Program. Lang. Syst.*, 29(3), May 2007.

FNTZ98.   T. Fischer, Jörg Niere, L. Torunski, and Albert Zündorf. Story Diagrams: A new Graph Rewrite Language based on the Unified Modeling Language. In G. Engels and G. Rozenberg, editors, *Proc. of the 6th Int. Workshop on Theory and Application of Graph Transformation*, LNCS 1764, pages 296–309. Springer, November 1998.

FNTZ00.   T. Fischer, Jörg Niere, L. Torunski, and Albert Zündorf. Story diagrams: A new graph rewrite language based on the Unified Modeling Language. In G. Engels and G. Rozenberg, editors, *Proc. of the 6th International Workshop on Theory and Application of Graph Transformation (TAGT)*, volume 1764 of *LNCS*, pages 296–309. Springer, Berlin, 2000.

FOT10.   FOTS-Group, University of Antwerp. *MoTMoT: Model driven, Template based, Model Transformer*, 2010. http://www.fots.ua.ac.be/motmot/index.php.

Fra02.   Ulrich Frank. Multi-perspective Enterprise Modeling (MEMO) - Conceptual Framework and Modeling Languages. In *Proc. Hawaii Int. Conf. on System Sciences (HICSS 2002)*, page 72, 2002.

Fuj07.   Software Engineering Group, University of Paderborn. *Fujaba Tool Suite*, 2007.

GBEE11.   Ulrike Golas, Enrico Biermann, Hartmut Ehrig, and Claudia Ermel. A Visual Interpreter Semantics for Statecharts Based on Amalgamated Graph Transformation. *ECEASST*, 39, 2011.

GBEG14.   Karsten Gabriel, Benjamin Braatz, Hartmut Ehrig, and Ulrike Golas. Finitary M-Adhesive Categories. *Mathematical Structures in Computer Science*, 24(4):1–40, 2014.

GCH+04.   David Garlan, Shang-Wen Cheng, An-Cheng Huang, Bradley Schmerl, and Peter Steenkiste. Rainbow: Architecture-based self-adaptation with reusable infrastructure. *Computer*, 37(10):46–54, 2004.

GEH11.   Ulrike Golas, Hartmut Ehrig, and Frank Hermann. Formal Specification of Model Transformations by Triple Graph Grammars with Application Conditions. *ECEASST*, 39, 2011.

GH09.   H. Giese and S. Hildebrandt. Efficient Model Synchronization of Large-Scale Models. Technical Report 28, Hasso Plattner Institute at the University of Potsdam, 2009.

GHE14.   Ulrike Golas, Annegret Habel, and Hartmut Ehrig. Multi-amalgamation of rules with application conditions in M-adhesive categories. *Mathematical Structures in Computer Science*, 24(4):1–68, 2014.

GHL10.   Holger Giese, Stephan Hildebrandt, and Leen Lambers. Toward bridging the gap between formal semantics and implementation of triple graph grammars. Technical Report 37, Hasso Plattner Institute at the University of Potsdam, 0 2010.

GHL12.   Holger Giese, Stephan Hildebrandt, and Leen Lambers. Bridging the gap be-
            tween formal semantics and implementation of triple graph grammars. *Soft-
            ware and Systems Modeling*, pages 1–27, 2012. http://dx.doi.org/10.1007/
            s10270-012-0247-y.
GHN+13.   Susann Gottmann, Frank Hermann, Nico Nachtigall, Braatz Benjamin, Claudia Er-
            mel, Hartmut Ehrig, and Thomas Engel. Correctness and Completeness of Gener-
            alised Concurrent Model Synchronisation Based on Triple Graph Grammars. In *Proc.
            Int. Workshop on Analysis of Model Transformations 2013 (AMT'13)*. CEUR-WS.
            org, 2013.
GK08.   Rubino Geiß and Moritz Kroll. GrGen.net: A fast, expressive, and general purpose
            graph rewrite tool. In A. Schürr, M. Nagl, and A. Zündorf, editors, *Proc. 3rd Intl.
            Workshop on Applications of Graph Transformation with Industrial Relevance (AG-
            TIVE'07)*, volume 5088 of *LNCS*. Springer, 2008.
GK10.   J. Greenyer and E. Kindler. Comparing relational model transformation technologies:
            implementing query/view/transformation with triple graph grammars. *Software and
            Systems Modeling (SoSyM)*, 9(1):21–46, 2010.
Gol10.   U. Golas. Multi-amalgamation in M-adhesive categories: Long version. Technical
            Report 2010/05, Technical University of Berlin, 2010.
Gol11.   Ulrike Golas. *Analysis and Correctness of Algebraic Graph and Model Transforma-
            tions*. PhD thesis, Technische Universität Berlin, 2011.
GPR11.   J. Greenyer, S. Pook, and J. Rieke. Preventing information loss in incremental model
            synchronization by reusing elements. In *Proc. ECMFA 2011*, volume 6698 of *LNCS*,
            pages 144–159. Springer, 2011.
GrG06.   Universität Karlsruhe. *Graph Rewrite GENerator (GrGen)*, 2006. http://www.
            info.uni-karlsruhe.de/software.php/id=7&lang=en.
Gro08.   *GRaphs for Object-Oriented VErification (GROOVE)*, 2008. http://groove.
            sourceforge.net/groove-index.html.
GW09.   Holger Giese and Robert Wagner. From model transformation to incremental bidirec-
            tional model synchronization. *Software and Systems Modeling*, 8:21–43, 2009.
HCE14.   Frank Hermann, Andrea Corradini, and Hartmut Ehrig. Analysis of Permutation
            Equivalence in M-adhesive Transformation Systems with Negative Application Con-
            ditions. *Mathematical Structures in Computer Science*, 24(4):1–47, 2014.
HCEK10.   Frank Hermann, Andrea Corradini, Hartmut Ehrig, and Barbara König. Efficient
            Analysis of Permutation Equivalence of Graph Derivations Based on Petri Nets. *ECE-
            ASST*, 29:1–15, 2010.
HEEO11.   Frank Hermann, Hartmut Ehrig, Claudia Ermel, and Fernando Orejas. Concurrent
            model synchronization with conflict resolution based on triple graph grammars - ex-
            tended version. Technical Report 2011/14, TU Berlin, Fak. IV, 2011.
HEEO12.   Frank Hermann, Hartmut Ehrig, Claudia Ermel, and Fernando Orejas. Concurrent
            model synchronization with conflict resolution based on triple graph grammars. In
            Juan de Lara and Andrea Zisman, editors, *Int. Conf. on Fundamental Approaches to
            Software Engineering (FASE'12)*, volume 7212 of *Lecture Notes in Computer Sci-
            ence*, pages 178–193. Springer, 2012.
HEGO10.   Frank Hermann, Hartmut Ehrig, Ulrike Golas, and Fernando Orejas. Efficient Anal-
            ysis and Execution of Correct and Complete Model Transformations Based on Triple
            Graph Grammars. In J. Bézivin, R.M. Soley, and A. Vallecillo, editors, *Proc. Int.
            Workshop on Model Driven Interoperability (MDI'10)*, MDI '10, pages 22–31, New
            York, NY, USA, 2010. ACM.
HEGO14.   Frank Hermann, Hartmut Ehrig, Ulrike Golas, and Fernando Orejas. Formal analysis
            of model transformations based on triple graph grammars. *Mathematical Structures
            in Computer Science*, 24(4):1–57, 2014.
Hei09.   Tobias Heindel. *A Category Theoretical Approach to the Concurrent Semantics
            of Rewriting: Adhesive Categories and Related Concepts*. PhD thesis, Universität
            Duisburg-Essen, 2009.

Hei10.   T. Heindel. Hereditary Pushouts Reconsidered. In *Proceedings of ICGT 2010*, volume
                6372 of *LNCS*, pages 250–265. Springer, 2010.
HEO⁺11a.   Frank Hermann, Hartmut Ehrig, Fernando Orejas, Krzysztof Czarnecki, Zinovy
                Diskin, and Yingfei Xiong. Correctness of model synchronization based on triple
                graph grammars. In Jon Whittle, Tony Clark, and Thomas Kühne, editors, *ACM/IEEE
                14th Int. Conf. on Model Driven Engineering Languages and Systems (MoDELS'11)*,
                volume 6981 of *LNCS*, pages 668–682. Springer, 2011.
HEO⁺11b.   Frank Hermann, Hartmut Ehrig, Fernando Orejas, Krzysztof Czarnecki, Zinovy
                Diskin, and Yingfei Xiong. Correctness of model synchronization based on triple
                graph grammars - extended version. Technical Report 2011/07, TU Berlin, 2011.
HEO⁺13.   Frank Hermann, Hartmut Ehrig, Fernando Orejas, Krzysztof Czarnecki, Zinovy
                Diskin, Yingfei Xiong, Susann Gottmann, and Thomas Engel. Model synchronization
                based on triple graph grammars: correctness, completeness and invertibility. *Software
                & Systems Modeling*, pages 1–29, 2013.
HEOG10a.   F. Hermann, H. Ehrig, F. Orejas, and U. Golas. Formal analysis of functional be-
                haviour for model transformations based on triple graph grammars - extended version.
                Technical Report 2010/08, Technical University of Berlin, 2010.
HEOG10b.   Frank Hermann, Hartmut Ehrig, Fernando Orejas, and Ulrike Golas. Formal Analysis
                of Functional Behaviour of Model Transformations Based on Triple Graph Gram-
                mars. In H. Ehrig, A. Rensink, G. Rozenberg, and A. Schürr, editors, *Proceedings
                of Intern. Conf. on Graph Transformation (ICGT' 10)*, volume 6372 of *LNCS*, pages
                155–170. Springer, 2010.
Her09.   Frank Hermann. Permutation Equivalence of DPO Derivations with Negative Appli-
                cation Conditions based on Subobject Transformation Systems. *Electronic Commu-
                nications of the EASST*, 16, 2009.
Her11.   Frank Hermann. *Analysis and Optimization of Visual Enterprise Models Based on
                Graph and Model Transformation*. PhD thesis, TU Berlin, 2011.
HGN⁺14.   Frank Hermann, Susann Gottmann, Nico Nachtigall, Hartmut Ehrig, Benjamin
                Braatz, Gianluigi Morelli, Alain Pierre, Thomas Engel, and Claudia Ermel. Triple
                graph grammars in the large for translating satellite procedures. In D. Di Ruscio
                and D. Varro, editors, *Proc. Int. Conf. on Model Transformations (ICMT 2014)*, num-
                ber 8568 in Lecture Notes of Computer Science, pages 122–137, Switzerland, 2014.
                Springer International Publishing.
HHI⁺10.   Soichiro Hidaka, Zhenjiang Hu, Kazuhiro Inaba, Hiroyuki Kato, Kazutaka Matsuda,
                and Keisuke Nakano. Bidirectionalizing graph transformations. In *Proceedings of
                the 15th ACM SIGPLAN international conference on Functional programming*, ICFP
                2010, pages 205–216. ACM, 2010.
HHK10.   Frank Hermann, Mathias Hülsbusch, and Barbara König. Specification and verifica-
                tion of model transformations. *ECEASST*, 30:1–21, 2010.
HHT96.   A. Habel, R. Heckel, and G. Taentzer. Graph Grammars with Negative Application
                Conditions. *Special issue of Fundamenta Informaticae*, 26(3,4):287–313, 1996.
HHT02.   J.H. Hausmann, R. Heckel, and G. Taentzer. Detection of Conflicting Functional Re-
                quirements in a Use Case-Driven Approach. In *Proc. of Int. Conference on Software
                Engineering 2002*, pages 105 – 115, Orlando, USA, 2002.
HIM00.   Dan Hirsch, Paola Inverardi, and Ugo Montanari. Reconfiguration of software ar-
                chitecture styles with name mobility. In *Proceedings of the 4th International Con-
                ference on Coordination Languages and Models*, COORDINATION '00, pages 148–
                163, London, UK, 2000. Springer.
HKT02.   R. Heckel, J. Küster, and G. Taentzer. Towards Automatic Translation of UML Mod-
                els into Semantic Domains . In H.-J. Kreowski, editor, *Proc. of APPLIGRAPH Work-
                shop on Applied Graph Transformation (AGT 2002)*, pages 11 – 22, 2002.
HLG⁺13.   Stephan Hildebrandt, Leen Lambers, Holger Giese, Jan Rieke, Joel Greenyer, Wil-
                helm Schäfer, Marius Lauder, Anthony Anjorin, and Andy Schürr. A survey of triple
                graph grammar tools. *ECEASST*, 57, 2013.

HMT08.   Zhenjiang Hu, Shin-Cheng Mu, and Masato Takeichi. A programmable editor for developing structured documents based on bidirectional transformations. *Higher-Order and Symbolic Computation*, 21(1-2):89–118, 2008.

HP05.   A. Habel and K.-H. Pennemann. Nested constraints and application conditions for high-level structures. In H.-J. Kreowski, U. Montanari, F. Orejas, G. Rozenberg, and G. Taentzer, editors, *Formal Methods in Software and Systems Modeling*, volume 3393 of *Lecture Notes in Computer Science*, pages 294–308. Springer, 2005.

HP09.   Annegret Habel and Karl-Heinz Pennemann. Correctness of high-level transformation systems relative to nested conditions. *Mathematical Structures in Computer Science*, 19:1–52, 2009.

HPW11.   Martin Hofmann, Benjamin C. Pierce, and Daniel Wagner. Symmetric lenses. In Thomas Ball and Mooly Sagiv, editors, *Proceedings of the 38th ACM SIGPLAN-SIGACT Symposium on Principles of Programming Languages, POPL 2011, Austin, TX, USA, January 26-28, 2011*, pages 371–384. ACM, 2011.

HPW12.   Martin Hofmann, Benjamin Pierce, and Daniel Wagner. Edit lenses. *SIGPLAN Not.*, 47(1):495–508, January 2012.

ISO04.   ISO/IEC. *ISO/IEC 15009-1:2004, Software and system engineering - High-level Petri nets - Part 1: Concepts, definitions and graphical notation*. ISO/IEC, 2004.

ISO06.   International Organization for Standardization (ISO). *ISO Standard 19439:2006: Enterprise integration – Framework for enterprise modelling*, 2006.

JABK08.   Frédéric Jouault, Freddy Allilaire, Jean Bézivin, and Ivan Kurtev. ATL: A model transformation tool. *Science of Computer Programming*, 72(1-2):31 – 39, 2008.

JK95.   Ryszard Janicki and Maciej Koutny. Semantics of inhibitor nets. *Inf. Comput.*, 123(1):1–16, 1995.

JK05.   F. Jouault and I. Kurtev. Transforming models with ATL. In *MoDELS Satellite Events*, volume 3844 of *LNCS*, pages 128–138. Springer, Berlin, 2005.

JLM+09.   S. Jurack, L. Lambers, K. Mehner, G. Taentzer, and G. Wierse. Object Flow Definition for Refined Activity Diagrams. In M. Chechik and M. Wirsing, editors, *Proc. Fundamental Approaches to Software Engineering (FASE'09)*, volume 5503 of *LNCS*, pages 49–63. Springer, 2009.

JLMT08.   S. Jurack, L. Lambers, K. Mehner, and G. Taentzer. Sufficient Criteria for Consistent Behavior Modeling with Refined Activity Diagrams. In K. Czarnecki, editor, *Proc. ACM/IEEE 11th International Conference on Model Driven Engineering Languages and Systems (MoDELS'08)*, volume 5301 of *LNCS*, pages 341–355. Springer, 2008.

JLS07.   P.T. Johnstone, S. Lack, and P. Sobociński. Quasitoposes, Quasiadhesive Categories and Artin Glueing. In T. Mossakowski, U. Montanari, and M. Haveraaen, editors, *Algebra and Coalgebra in Computer Science. Proceedings of CALCO 2007*, volume 4626 of *LNCS*, pages 312–326. Springer, 2007.

JWEG07.   Praveen K. Jayaraman, Jon Whittle, Ahmed M. Elkhodary, and Hassan Gomaa. Model composition in product lines and feature interaction detection using critical pair analysis. In *Model Driven Engineering Languages and Systems, 10th International Conference, MoDELS 2007, Nashville, USA, September 30 - October 5, 2007, Proceedings*, volume 4735 of *LNCS*, pages 151–165. Springer, 2007.

KB06.   Sascha Klüppelholz and Christel Baier. Symbolic model checking for channel-based component connectors. In *Proc. Int. Workshop on the Foundations of Coordination Languages and Software Architectures (FOCLASA'06)*, 2006.

KC03.   Jeffrey O. Kephart and David M. Chess. The vision of autonomic computing. *Computer*, 36(1):41–50, 2003.

Ken91.   R. Kennaway. Graph Rewriting in Some Categories of Partial Morphisms. In H. Ehrig, H.-J. Kreowski, and G. Rozenberg, editors, *Proceeding of Int. Workshop Graph-Grammars and Their Application to Computer Science 1990*, volume 532 of *LNCS*, pages 490–504. Springer, 1991.

KHM06.   Harmen Kastenberg, Frank Hermann, and Tony Modica. Towards Translating Graph Transformation Systems by Model Transformation. In *Proc. International Workshop* on Graph and Model Transformation (GraMoT'06), Satellite Event of the IEEE Symposium on Visual Languages and Human-Centric Computing, volume 4, Brighton, UK, September 2006. Electronic Communications of the EASST.

KK04. H. C. M. Kleijn and M. Koutny. Process semantics of general inhibitor nets. *Information and Computation*, 190(1):18–69, 2004.

KKvT10. Hans-Jörg Kreowski, Sabine Kuske, and Caroline von Totth. Stepping from graph transformation units to model transformation units. *ECEASST*, 30, 2010.

KLKS10. F. Klar, M. Lauder, A. Königs, and A. Schürr. Extended Triple Graph Grammars with Efficient and Compatible Graph Translators. In *Graph Transformations and Model Driven Enginering - Essays Dedicated to Manfred Nagl on the Occasion of his 65th Birthday*, volume 5765 of *LNCS*, pages 144–177. Springer, 2010.

KM07. Jeff Kramer and Jeff Magee. Self-managed systems: an architectural challenge. In *2007 Future of Software Engineering*, FOSE '07, pages 259–268, Washington, DC, USA, 2007. IEEE Computer Society.

KPPR07. D. Kolovos, R. Paige, F Polack, and L. Rose. Update transformations in the small with Epsilon wizard language. *Journal of Object Technology*, 6(9):53–69, 2007.

KS06. A. Königs and A. Schürr. Tool Integration with Triple Graph Grammars - A Survey. In *Proc. SegraVis School on Foundations of Visual Modelling Techniques*, volume 148, pages 113–150, Amsterdam, 2006. Electronic Notes in Theoretical Computer Science, Elsevier.

KW07. E. Kindler and R. Wagner. Triple graph grammars: Concepts, extensions, implementations, and application scenarios. Technical Report TR-ri-07-284, Department of Computer Science, University of Paderborn, Germany, 2007.

Lam10. Leen Lambers. *Certifying Rule-Based Models using Graph Transformation* . PhD thesis, Technische Universität Berlin, 2010. Also as book available: Südwestdeutscher Verlag für Hochschulschriften, ISBN: 978-3-8381-1650-1.

LAS14a. Erhan Leblebici, Anthony Anjorin, and Andy Schürr. A catalogue of optimization techniques for triple graph grammars. In Hans-Georg Fill, Dimitris Karagiannis, and Ulrich Reimer, editors, *Modellierung 2014, 19.-21. March 2014, Vienna, Austria*, volume 225 of *LNI*, pages 225–240. GI, 2014.

LAS+14b. Erhan Leblebici, Anthony Anjorin, Andy Schürr, Stephan Hildebrandt, Jan Rieke, and Joel Greenyer. A comparison of incremental triple graph grammar tools. *ECEASST*, 67, 2014.

LAVS12. Marius Lauder, Anthony Anjorin, Gergely Varró, and Andy Schürr. Bidirectional Model Transformation with Precedence Triple Graph Grammars. In Antonio Vallecillo, Juha-Pekka Tolvanen, Ekkart Kindler, Harald Störrle, and Dimitris Kolovos, editors, *Proc. of the 8th European Conf. on Modelling Foundations and Applications*, volume 7349 of *LNCS*, pages 287–302. Springer, 2012.

LB93. M. Löwe and M. Beyer. AGG — an implementation of algebraic graph rewriting. In *Proc. Fifth Int. Conf. Rewriting Techniques and Applications*, volume 690 of *LNCS*, pages 451–456. Springer, 1993.

LBM10. Ivan Lanese, Antonio Bucchiarone, and Fabrizio Montesi. A framework for rule-based dynamic adaptation. In *Trustworthly Global Computing - 5th International Symposium, TGC 2010, Munich, Germany, February 24-26, 2010, Revised Selected Papers*, volume 6084 of *Lecture Notes in Computer Science*, pages 284–300. Springer, 2010.

LEOP08. L. Lambers, H. Ehrig, F. Orejas, and U. Prange. Parallelism and Concurrency in Adhesive High-Level Replacement Systems with Negative Application Conditions. In H. Ehrig, J. Pfalzgraf, and U. Prange, editors, *Proceedings of the ACCAT workshop at ETAPS 2007*, volume 203 / 6 of *ENTCS*, pages 43–66. Elsevier, 2008.

LMEP08. L. Lambers, L. Mariani, H. Ehrig, and M. Pezze. A Formal Framework for Developing Adaptable Service-Based Applications. In J.L. Fiadeiro and P. Inverardi, editors, *Proc. Fundamental Approaches to Software Engineering (FASE'08)*, volume 4961 of *LNCS*, pages 392–406. Springer, 2008.

LS04.   S. Lack and P. Sobociński. Adhesive Categories. In *Proc. FOSSACS 2004*, volume
            2987 of *LNCS*, pages 273–288. Springer, 2004.
LS05a.   S. Lack and P. Sobociński. Adhesive and quasiadhesive categories. *Theoretical Infor-
            matics and Applications*, 39(2):511–546, 2005.
LS05b.   M. Lawley and J. Steel. Practical declarative model transformation with Tefkat. In
            MoDELS Satellite Events, volume 3844 of *LNCS*, pages 139–150. Springer, Berlin,
            2005.
Mar00.   Chris Marshall. *Enterprise modeling with UML: designing successful software
            through business analysis.* Addison-Wesley Longman, 2000.
MDA15.   Object Management Group. *MDA Specifications*, 2015. `http://www.omg.org/`
            `mda/specs.htm`.
MEE12.   Maria Maximova, Hartmut Ehrig, and Claudia Ermel. Transfer of Local Confluence
            and Termination between Petri Net and Graph Transformation Systems Based on *M*-
            Functors. *ECEASST*, 51:1–12, 2012.
MEE13.   Maria Maximova, Hartmut Ehrig, and Claudia Ermel. Analysis of Hypergraph Trans-
            formation Systems in AGG based on M-Functors. *ECEASST*, 58, 2013.
MG06.   Tom Mens and Pieter Van Gorp. A taxonomy of model transformation. *Electr. Notes
            Theor. Comput. Sci.*, 152:125–142, 2006.
MHMT13.   Katharina Mehner-Heindl, Mattia Monga, and Gabriele Taentzer. Analysis of aspect-
            oriented models using graph transformation systems. In Ana Moreira, Ruzanna
            Chitchyan, João Araújo, and Awais Rashid, editors, *Aspect-Oriented Requirements
            Engineering*, pages 243–270. Springer, 2013.
MM90.   J. Meseguer and U. Montanari. Petri Nets are Monoids. *Information and Computa-
            tion*, 88(2):105–155, 1990.
MMT09.   Katharina Mehner, Mattia Monga, and Gabriele Taentzer. Analysis of aspect-oriented
            model weaving. In *T. Aspect-Oriented Software Development V*, volume 5490 of
            LNCS, pages 235–263. Springer, 2009.
MOF15.   Object Management Group. *Meta-Object Facility (MOF), Version 2.5*, 2015. `http:`
            `//www.omg.org/spec/MOF/2.5/`.
MOM12.   University of Leicester. *MOMENT2-MT*, 2012. `http://www.cs.le.ac.uk/`
            `people/aboronat/tools/moment2-gt/`.
MSD06.   Tom Mens, Ragnhild Van Der Straeten, and Maja D'Hondt. Detecting and resolving
            model inconsistencies using transformation dependency analysis. In *Model Driven
            Engineering Languages and Systems, 9th International Conference, MoDELS 2006,
            Genova, Italy, October 1-6, 2006, Proceedings*, volume 4199 of *LNCS*, pages 200–
            214. Springer, 2006.
MV08.   B. Meyers and P. Van Gorp. Towards a hybrid transformation language: Implicit and
            explicit rule scheduling in story diagrams. In *Proceedings of the 6th International
            Fujaba Days*, 2008.
MVDJ05.   T. Mens, N. Van Eetvelde, S. Demeyer, and D. Janssens. Formalizing refactorings
            with graph transformations. *Software Tools for Technology Transfer*, 17:247–276,
            2005.
MVVK05.   T. Mens, P. Van Gorp, D. Varró, and G. Karsai. Applying a Model Transformation
            Taxonomy to Graph Transformation Technology. In *Proc. International Workshop
            on Graph and Model Transformation (GraMoT'05)*, volume 152 of *ENTCS*, pages
            143–159. Elsevier, 2005.
Nag79.   Manfred Nagl. *Graph-Grammatiken: Theorie, Anwendungen, Implementierung.*
            Vieweg, 1979.
New42.   Maxwell Herman Alexander Newman. On theories with a combinatorial definition of
            "equivalence". *Annals of Mathematics*, 43(2):223–243, 1942.
Obj14a.   Object Management Group. Meta Object Facility (MOF) Core Specification Version
            2.4.2. `http://www.omg.org/spec/MOF/`, 2014.
Obj14b.   Object Management Group. *Object Constraint Language, Version 2.4*, 2014.

OEP08.   F. Orejas, H. Ehrig, and U. Prange. A Logic of Graph Constraints. In J.L. Fiadeiro
            and P. Inverardi, editors, *Proc. Fundamental Approaches to Software Engineering
            (FASE'08)*, volume 4961 of *LNCS*, pages 179–198. Springer, 2008.
OMG14.   Object Management Group. *Business Process Model and Notation (BPMN), Version
            2.0.2*, 2014. formal/2013-12-09, http://www.omg.org/spec/BPMN/2.0.2/.
PE07.   U. Prange and H. Ehrig. From Algebraic Graph Transformation to Adhesive HLR
            Categories and Systems. In S. Bozapalidis and G. Rahonis, editors, *Algebraic Infor-
            matics. Proceedings of CAI 2007*, volume 4728 of *LNCS*, pages 122–146. Springer,
            2007.
PEL08.   U. Prange, H. Ehrig, and L. Lambers. Construction and Properties of Adhesive and
            Weak Adhesive High-Level Replacement Categories. *Applied Categorical Structures*,
            16(3):365–388, 2008.
Pen08.   K.H. Pennemann. An Algorithm for Approximating the Satisfiability Problem of
            High-level Conditions. In *Proc. International Workshop on Graph Transformation
            for Verification and Concurrency (GT-VC'07)*, volume 213 of *ENTCS*, pages 75–94.
            Elsevier Science, 2008.
PLS08.   Heiko Pfeffer, David Linner, and Stephan Steglich. Dynamic adaptation of workflow
            based service compositions. In *Proceedings of the 4th international conference on
            Intelligent Computing: Advanced Intelligent Computing Theories and Applications -
            with Aspects of Theoretical and Methodological Issues (ICIC '08)*, pages 763–774.
            Springer, 2008.
Plu93.   Detlef Plump. Hypergraph Rewriting: Critical Pairs and Undecidability of Conflu-
            ence. In *Term Graph Rewriting: Theory and Practice*, pages 201–213. John Wiley,
            1993.
Plu95.   D. Plump. On Termination of Graph Rewriting. In M. Nagl, editor, *Graph-Theoretic
            Concepts in Computer Science. Proceedings of WG 1995*, volume 1017 of *LNCS*,
            pages 88–100. Springer, 1995.
Plu05.   Detlef Plump. Confluence of Graph Transformation Revisited. In *Processes, Terms
            and Cycles: Steps on the Road to Infinity*, volume 3838 of *LNCS*, pages 280–308.
            Springer, 2005.
PM07.   Oscar Pastor and Juan Carlos Molina. *Model-Driven Architecture in Practice: A Soft-
            ware Production Environment Based on Conceptual Modeling*. Springer, 2007.
PR69.   J.L. Pfaltz and A. Rosenfeld. Web grammars. In *Proceedings of Int. Joint Conf. on
            Artificial Intelligence 1969*, pages 609–620, 1969.
Pra71.   T. W. Pratt. Pair Grammars, Graph Languages and String-to-Graph Translations.
            Journal of Computer and System Sciences, 5:560–595, 1971.
Pra07.   U. Prange. Algebraic High-Level Nets as Weak Adhesive HLR Categories. *Electronic
            Communications of the EASST*, 2:1–13, 2007.
QVT15.   Object Management Group. *Meta Object Facility (MOF) 2.0 Query/View/Transfor-
            mation Specification. Version 1.2*, 2015. http://www.omg.org/spec/QVT/1.2/.
RE96.   G. Rozenberg and J. Engelfriet. Elementary Net Systems. In W. Reisig and G. Rozen-
            berg, editors, *Lectures on Petri Nets I: Basic Models*, volume 1491 of *LNCS*, pages
            12–121. Springer, 1996.
Rei85.   Wolfgang Reisig. *Petri Nets: An Introduction*, volume 4 of *EATCS Monographs on
            Theoretical Computer Science*. Springer, 1985.
RET12.   O. Runge, C. Ermel, and G Taentzer. AGG 2.0 – new features for specifying
            and analyzing algebraic graph transformations. In Andy Schürr, Daniel Varro, and
            Gergely Varro, editors, *Applications of Graph Transformation with Industrial Rel-
            evance, 4th International Symposium, (AGTIVE'11), Proceedings*, volume 7233 of
            LNCS. Springer, 2012.
RGSS09.   Gabi Dreo Rodosek, Kurt Geihs, Hartmut Schmeck, and Burkhard Stiller. Self-
            healing systems: Foundations and challenges. In *Self-Healing and Self-Adaptive Sys-
            tems*, number 09201 in Dagstuhl Seminar Proceedings. Germany, 2009.
Roz97.   Grzegorz Rozenberg. *Handbook of Graph Grammars and Computing by Graph
            Transformations, Volume 1: Foundations*. World Scientific, 1997.

RRLW09.   Adrian Rutle, Alessandro Rossini, Yngve Lamo, and Uwe Wolter. A Category-Theoretical Approach to the Formalisation of Version Control in MDE. In *Proc. of Int. Conf. on Fundamental Approaches to Software Engineering (FASE'09)*, volume 5503 of *LNCS*, pages 64–78. Springer, 2009.

RV10.   A. Rensink and P. Van Gorp, editors. *International Journal on Software Tools for Technology Transfer (STTT), Special Section on Graph Transformation Tool Contest 2008*, volume 12(3-4). Springer, 2010.

SAB98.   Monique Snoeck, Rakesh Agarwal, and Chiranjit Basu. Enterprise Modelling. In Serge Demeyer and Jan Bosch, editors, *Proc. Workshops at the Europ. Conf. on Object-Oriented Programming (ECOOP 1998)*, volume 1543 of *Lecture Notes in Computer Science*, pages 222–227. Springer, 1998.

SAL⁺03.   J. Sprinkle, A. Agrawal, T. Levendovszky, F. Shi, and G. Karsai. Domain model translation using graph transformations. In *Int. Conf. on Engineering of Computer-Based Systems*, pages 159–168, 2003.

Sch94.   A. Schürr. Specification of Graph Translators with Triple Graph Grammars. In G. Tinhofer, editor, *WG94 20th Int. Workshop on Graph-Theoretic Concepts in Computer Science*, volume 903 of *Lecture Notes in Computer Science*, pages 151–163, Heidelberg, 1994. Springer.

Sch06.   D. C. Schmidt. Model-driven engineering. *IEEE Computer*, 39(2):25–31, 2006.

Sel08.   Bran Selic. MDA manifestations. *UPGRADE: The European Journal for Informatics Professional*, IX(2):12–16, 2008.

SK08.   Andy Schürr and Felix Klar. 15 years of triple graph grammars. In *Intern. Conf. on Graph Transformation (ICGT 2008)*, pages 411–425, 2008.

Ste10.   Perdita Stevens. Bidirectional Model Transformations in QVT: Semantic Issues and Open Questions. *Software and Systems Modeling*, 9:7–20, 2010.

Sto04.   H. Stoerrle. Semantics of UML 2.0 activity diagrams. In *International Conference on Visual Languages and Human Centric Computing VLHCC*. IEEE, 2004.

SWZ99.   A. Schürr, A. Winter, and A. Zündorf. The PROGRES-approach: Language and environment. In H. Ehrig, G. Engels, H.-J. Kreowski, and G. Rozenberg, editors, *Handbook of Graph Grammars and Computing by Graph Transformation, Volume 2: Applications, Languages and Tools*, pages 487 – 550. World Scientific, 1999.

Sys14.   SysML Open Source Specification Project. *SysML.org*, 2014.

SZK05.   George Spanoudakis, Andrea Zisman, and Alexander Kozlenkov. A service discovery framework for service centric systems. In *Proceedings of the 2005 IEEE International Conference on Services Computing (SCC '05)*, pages 251–259, 2005.

Tae04.   G. Taentzer. AGG: A graph transformation environment for modeling and validation of software. In J. Pfaltz, M. Nagl, and B. Boehlen, editors, *Application of Graph Transformations with Industrial Relevance (AGTIVE'03)*, volume 3062 of *LNCS*, pages 446 – 456. Springer, Berlin, 2004.

Tae10.   Gabriele Taentzer. What algebraic graph transformations can do for model transformations. *ECEASST*, 30, 2010.

TELW10.   G. Taentzer, C. Ermel, P. Langer, and M. Wimmer. Conflict detection for model versioning based on graph modifications. In H. Ehrig, A. Rensink, G. Rozenberg, and A. Schürr, editors, *Proc. of Int. Conf. on Graph Transformations (ICGT'10)*, volume 6372 of *LNCS*, pages 171–186. Springer, 2010.

TER99.   G. Taentzer, C. Ermel, and M. Rudolf. The AGG-Approach: Language and Tool Environment. In H. Ehrig, G. Engels, H.-J. Kreowski, and G. Rozenberg, editors, *Handbook of Graph Grammars and Computing by Graph Transformation, volume 2: Applications, Languages and Tools*, pages 551–603. World Scientific, 1999.

TV10.   J. Troya and A. Vallecillo. Towards a rewriting logic semantics for ATL. In L. Tratt and M. Gogolla, editors, *Proc. of the Intern. Conf. on Model Transformation (ICMT'10)*, volume 6142 of *LNCS*, pages 230–244. Springer, 2010.

UML15.   Object Management Group. *Unified Modeling Language (UML)—Version 2.5*, 2015. http://www.omg.org/spec/UML/.

Var02.   Dániel Varró. A formal semantics of UML Statecharts by model transition systems. In Andrea Corradini, Hartmut Ehrig, Hans-Jörg Kreowski, and Grzegorz Rozenberg, editors, *Proc. ICGT 2002: 1st Int. Conf. on Graph Transformation*, volume 2505 of *LNCS*, pages 378–392. Springer, 2002.

VB07.   Dániel Varró and András Balogh. The model transformation language of the VIATRA2 framework. *Science of Computer Programming*, 68(3):214–234, 2007.

Ver02.   F. Vernadat. UEML: Towards a Unified Enterprise Modelling Language. *International Journal of Production Research*, 40(17):4309 – 4321, 2002.

VGS+05.   Kunal Verma, Karthik Gomadam, Amit P. Sheth, John A. Miller, and Zixin Wu. The METEOR-S approach for configuring and executing dynamic web processes. Technical report, University of Georgia, Athens, 2005.

VIA14.   VIATRA2 developers. VIATRA2 (VIsual Automated model TRAnsformations) framework http://www.eclipse.org/viatra2/viatra2, 2014.

VMT14.   Budapest University of Technology and Economics, HUN. *Visual Modeling and Transformation System (VMTS)*, 2014. https://www.aut.bme.hu/Pages/Research/VMTS/Introduction.

VSB+06.   Markus Völter, Thomas Stahl, Jorn Bettin, Arno Haase, and Simon Helsen. *Model-Driven Software Development: Technology, Engineering, Management*. John Wiley, 2006.

WHW+06.   Steve R. White, James E. Hanson, Ian Whalley, David M. Chess, Alla Segal, and Jeffrey O. Kephart. Autonomic computing: Architectural approach and prototype. *Integr. Comput.-Aided Eng.*, 13(2):173–188, 2006.

WTEK08.   J. Winkelmann, G. Taentzer, K. Ehrig, and J. Küster. Translation of Restricted OCL Constraints into Graph Constraints for Generating Meta Model Instances by Graph Grammars. *ENTCS*, 211:159–170, 2008.

WWW04.   WWW Consortium (W3C). *XML Schema Part 1: Structures (Second Edition)*, 2004.

XMI08.   Object Management Group. *MOF 2.0 / XMI Mapping Specification*, 2008. http://www.omg.org/technology/documents/formal/xmi.htm.

XSHT11.   Yingfei Xiong, Hui Song, Zhenjiang Hu, and Masato Takeichi. Synchronizing concurrent model updates based on bidirectional transformation. *Software and Systems Modeling*, pages 1–16, 2011.

Index

Printed in the United States
By Bookmasters